Lecture Notes in Computer Sci

T0237906

Commenced Publication in 1973
Founding and Former Series Editors:
Gerhard Goos, Juris Hartmanis, and Jan van Leeuwen

Editorial Board

Advanced Research in Computing and Software Science
Subline of Lectures Notes in Computer Science

Subline Series Editors

Subline Advisory Board

Ning Chen   Edith Elkind   Elias Koutsoupias
(Eds.)

Internet
and Network
Economics

7th International Workshop, WINE 2011
Singapore, December 11-14, 2011
Proceedings

 Springer

Volume Editors

Ning Chen
Nanyang Technological University
School of Physical and Mathematical Sciences
SPMS-MAS-03-01
21 Nanyang Link, Singapore 637371
E-mail: ningc@ntu.edu.sg

Edith Elkind
Nanyang Technological University
School of Physical and Mathematical Sciences
SPMS-MAS-03-01
21 Nanyang Link, Singapore 637371
E-mail: eelkind@ntu.edu.sg

Elias Koutsoupias
University of Athens
Panepistimiopolis, Ilissia
Athens 15784, Greece
E-mail: elias@di.uoa.gr

ISSN 0302-9743   e-ISSN 1611-3349
ISBN 978-3-642-25509-0   e-ISBN 978-3-642-25510-6
DOI 10.1007/978-3-642-25510-6
Springer Heidelberg Dordrecht London New York

Library of Congress Control Number: 2011941124

CR Subject Classification (1998): C.2, F.2, D.2, H.4, F.1, H.3

LNCS Sublibrary: SL 3- Information Systems and Application, incl. Internet/Web
and HCI

Typesetting: Camera-ready by author, data conversion by Scientific Publishing Services, Chennai, India

Printed on acid-free paper

Springer is part of Springer Science+Business Media (www.springer.com)

Preface

This volume contains the papers presented at WINE 2011: 7th Workshop on Internet and Network Economics held during December 11–14, 2011 in Singapore.

Over the past decade, there has been a growing interaction between researchers in theoretical computer science, networking and security, economics, mathematics, sociology, and management sciences devoted to the analysis of problems arising from the Internet and the World Wide Web. The Workshop on Internet and Network Economics (WINE) is an interdisciplinary forum for the exchange of ideas and results arising from these various fields.

There were 100 submissions to this edition of the workshop, including both long (12 pages) and short (8 pages) papers. All papers were rigorously reviewed by the Program Committee members and/or external referees; almost all papers received at least three detailed reviews. The papers were evaluated on the basis of their significance, novelty, soundness and relevance to the workshop.

A new feature of this year's WINE was that the authors were allowed to designate their papers as working papers: such papers had to be submitted in the usual format (12 or 8 pages) and reviewed in the same way as regular submissions, but only a 1–2-page abstract appears in the proceedings. This allows the authors to subsequently publish the full versions of their papers in journals that do not permit prior publication of the same material in conference proceedings. The Program Committee accepted 3 such papers, in addition to 31 full papers and 5 short papers.

Besides the regular talks, the program also included three invited talks by Cynthia Dwork (MSR Silicon Valley, USA), Preston McAfee (Yahoo! Research, USA) and Herve Moulin (Rice University, USA). The program of the workshop also included tutorials by Jason Hartline (Northwestern University) and Nicole Immorlica (Northwestern University).

We are very grateful to Google Research, Yahoo! Research and Microsoft Research for their generous financial contribution to the conference. We would also like to thank the School of Physical and Mathematical Sciences of Nanyang Technological University (Singapore) for hosting the tutorials and providing organizational support.

We also acknowledge EasyChair, a powerful and flexible system for managing all stages of the paper-handling process, from the submission stage to the preparation of the final version of the proceedings.

September 2011

Ning Chen
Edith Elkind
Elias Koutsoupias

Organization

Program Committee

Suzanne Albers	Humboldt University of Berlin, Germany
Niv Buchbinder	Open University, Israel
Ioannis Caragiannis	University of Patras, Greece
Ning Chen	Nanyang Technological University, Singapore
Giorgos Christodoulou	University of Liverpool, UK
Jose Correa	University of Chile, Chile
Shahar Dobzinski	Cornell University, USA
Edith Elkind	Nanyang Technological University, Singapore
Angelo Fanelli	Nanyang Technological University, Singapore
Arpita Ghosh	Yahoo! Research, USA
Paul Goldberg	University of Liverpool, UK
Laurent Gourves	LAMSADE, France
Gianluigi Greco	University of Calabria, Italy
Tobias Harks	TU Berlin, Germany
Martin Hoefer	RWTH Aachen University, Germany
Anna Karlin	University of Washington, USA
Elias Koutsoupias	University of Athens, Greece
Ron Lavi	Technion, Israel
Pinyan Lu	Microsoft Research Asia, China
Yishay Mansour	Tel Aviv University, Israel
Preston Mcafee	Yahoo! Research, USA
Vahab Mirrokni	Google Research, USA
Herve Moulin	Rice University, USA
David Pennock	Yahoo! Research, USA
Troels Bjerre Sorensen	University of Warwick, UK
Vijay Vazirani	Georgia Tech, USA
Adrian Vetta	McGill University, Canada

Additional Reviewers

Alaei, Saeed	Barbay, Jérémy	Dean, Brian
Anari, Nima	Bei, Xiaohui	Engel, Yagil
Arava, Radhika	Bilo, Davide	Escoffier, Bruno
Ashlagi, Itai	Bilo, Vittorio	Faliszewski, Piotr
Augustine, John	Celis, Elisa	Feldman, Jon
Badanidiyuru, Ashwinkumar	Chen, Xue	Feldman, Moran
	Chen, Yiling	Ferraioli, Diodato
Balseiro, Santiago	Dams, Johannes	Frongillo, Rafael

Fu, Hu
Gairing, Martin
Gallotti, Vasco
Goel, Gagan
Gonen, Rica
Gravin, Nikolay
Guo, Mingyu
Haghpanah, Nima
Hua, Xia
Jain, Kamal
Kanellopoulos,
    Panagiotis
Karande, Chinmay
Koch, Ronald
Kovacs, Annamaria
Kyropoulou, Maria
Lai, John
Le Truc, Viet
Lenzner, Pascal
Lokshtanov, Daniel

Markakis, Evangelos
Mehta, Aranyak
Mertzios, George
Mihalak, Matus
Monaco, Gianpiero
Monnot, Jerome
Moscardelli, Luca
Mu'Alem, Ahuva
Nguyen, Kim Thang
Nick, Bobo
Nikzad, Afshin
Olver, Neil
Paes Leme, Renato
Palopoli, Luigi
Pascual, Fanny
Pasquale, Francesco
Pierrakos, George
Procaccia, Ariel
Pyrga, Evangelia
Reichman, Daniel

Savani, Rahul
Scarcello, Francesco
Schapira, Michael
Schwartz, Roy
Sivan, Balasubramanian
Skopalik, Alexander
Stier-Moses, Nicolas
Stiller, Sebastian
Suri, Siddharth
Tang, Bo
Toft, Tomas
Tripathi, Pushkar
Wang, Lei
Wang, Yajun
Yan, Qiqi
Zhang, Hongyang
Zhang, Jialin
Zhang, Shengyu
Zhu, Jiamin
Zhu, Zeyuan Allen

Table of Contents

Full Papers

Short Papers

Working Papers

The Snowball Effect of Uncertainty in Potential Games

Maria-Florina Balcan, Florin Constantin, and Steven Ehrlich

College of Computing, Georgia Institute of Technology
{ninamf,florin,sehrlich}@cc.gatech.edu

Abstract. Uncertainty is present in different guises in many settings, in particular in environments with strategic interactions. However, most game-theoretic models assume that players can accurately observe interactions and their own costs. In this paper we quantify the effect on social costs of two different types of uncertainty: adversarial perturbations of small magnitude to costs (effect called the Price of Uncertainty (PoU) [3]) and the presence of several players with Byzantine, i.e. arbitrary, behavior (effect we call the Price of Byzantine behavior (PoB)). We provide lower and upper bounds on PoU and PoB in two well-studied classes of potential games: consensus games and set-covering games.

1 Introduction

Uncertainty, in many manifestations and to different degrees, arises naturally in applications modeled by games. In such settings, players can rarely observe accurately and assign a precise cost or value to a given action in a specific state. For example a player who shares costs for a service (e.g. usage of a supercomputer or of a lab facility) with others may not know the exact cost of this service. Furthermore, this cost may fluctuate over time due to unplanned expenses or auxiliary periodic costs associated with the service. In a large environment (e.g. the Senate or a social network), a player may only have an estimate of the behaviors of other players who are relevant to its own interests. Another type of uncertainty arises when some players are misbehaving, i.e., they are Byzantine.

The main contribution of this paper is to assess the long-term effect of small local uncertainty on cost-minimization *potential games* [7]. We show that uncertainty can have a strong *snowball effect*, analogous to the increase in size and destructive force of a snowball rolling down a snowy slope. Namely, we show that small perturbations of costs on a per-player basis or a handful of players with Byzantine (i.e. adversarial) behavior can cause a population of players to go from a good state (even a good equilibrium state) to a state of much higher cost. We complement these results highlighting the lack of robustness under uncertainty with guarantees of resilience to uncertainty. We assess the effects of uncertainty in two important classes of potential games using the framework introduced by [3]. The first class we analyze is that of *consensus games* [2,6] for which relatively little was previously known on the effect of uncertainty. The second class we analyze is that of *set-covering games* [5], for which we improve

N. Chen, E. Elkind, and E. Koutsoupias (Eds.): WINE 2011, LNCS 7090, pp. 1–12, 2011.

on the previously known bounds of Balcan et al. [3]. We review in detail the uncertainty models and these classes of games, as well as our results below.

We consider both *improved-response* (IR) dynamics in which at each time step exactly one player may update strategy in order to lower his (apparent) cost and *best-response* (BR) dynamics in which the updating player chooses what appears to be the least costly strategy. Any state is assigned a *social cost*, which for most of our paper is defined as the sum of all players' costs in that state. We measure the effect of uncertainty as the maximum multiplicative increase in social cost when following these dynamics. We instantiate this measure to each type of uncertainty.

For the first uncertainty type, we assume adversarial perturbations of costs of magnitude at most $1 + \epsilon$ for $\epsilon > 0$ (a small quantity that may depend on game parameters). That is, a true cost of c may be perceived as any value within $[\frac{1}{1+\epsilon}c, (1 + \epsilon)c]$. Consider a game G and an initial state S_0 in G. We call a state S (ϵ, IR)-*reachable* from S_0 if there exists a valid ordering of updates in IR dynamics and corresponding perturbations (of magnitude at most ϵ) leading from S_0 to S. The *Price of Uncertainty* [3] (for IR dynamics) given ϵ of game G is defined as the ratio of the highest social cost of an (ϵ, IR)-reachable state S to the social cost of starting state S_0.

$$PoU_{IR}(\epsilon, G) = \max \left\{ \frac{cost(S)}{cost(S_0)} \; : \; S_0; \; S \; (\epsilon, IR)\text{-reachable from } S_0 \right\}$$

For a class $\mathcal{G}$ of games and $\epsilon > 0$ we define $PoU_{IR}(\epsilon, \mathcal{G}) = \sup_{G \in \mathcal{G}} PoU_{IR}(\epsilon, G)$ as the highest PoU of any game G in $\mathcal{G}$ for ϵ. PoU_{BR} is defined analogously.

For the second uncertainty type, we assume B additional players with arbitrary, or Byzantine [8] behavior. We define the *Price of Byzantine behavior* $(PoB(B))$ as the effect of the B Byzantine players on social cost, namely the maximum ratio of the cost of a state reachable in the presence of B Byzantine agents to that of the starting state.

$$PoB(B, G) = \max \left\{ \frac{cost(S)}{cost(S_0)} \; : \; S_0; \; S \; B\text{-Byz-reachable from } S_0 \right\}$$

where state S of G is B-*Byz-reachable* from S_0 if some valid ordering of updates by players (including the B Byzantine ones) goes from S_0 to S. $PoB(B, \mathcal{G}) = \sup_{G \in \mathcal{G}} PoB(B, G)$ for class $\mathcal{G}$. PoB, like PoU, may depend on the dynamics[1].

A low PoU or PoB shows resilience of a system to small errors by players in estimating costs or behavior of others. In the games we study, social costs cannot increase much without uncertainty (namely $PoU(0) = PoB(0)$ are small), yet modest instances of uncertainty (in terms of ϵ or B) can lead to significant increases in costs (i.e. large $PoU(\epsilon)$ and $PoB(B)$). We introduce in the following the classes of games we study and summarize our results.

Consensus games [6] model a basic strategic interaction: choosing one side or the other (e.g. in a debate) and incurring a positive cost for interacting with

---

[1] We omit parameters from *PoU* and *PoB* when they are clear from context.

each agent that chose the other side. More formally, there are two colors (or strategies), white (W) and red (R), which each player may choose; hence IR and BR dynamics coincide. Each player occupies a different vertex in an undirected graph G with vertices $\{1, \ldots, n\}$ and edges $E(G)$ (without self-loops). A player's cost is defined as the number of neighbors with the other color. We establish $PoU(\epsilon) = \Omega(n^2 \epsilon^3)$ for $\epsilon = \Omega(n^{-1/3})$ and $PoU(\epsilon) = O(n^2 \epsilon)$ for any ϵ. These bounds are asymptotically tight for constant ϵ. We exactly quantify $PoB(B)$ as $\Theta(n\sqrt{nB})$ by exhibiting an instance with $\Theta(n\sqrt{nB})$ edges that is flippable (i.e. it can flip from one monochromatic state to the other given B Byzantine players) and then reducing any other consensus game to this instance.

Set-covering games [5] model many applications where all users of a resource share fairly its base cost. These natural games fall in the widely studied class of fair-cost sharing games [1]. In a set-covering game, there are m sets, each set j with its own fixed weight (i.e. base cost) w_j. Each of the n players must choose exactly one set j (different players may have access to different sets) and share its weight equally with its other users, i.e incur cost $w_j / n_j(S)$ where $n_j(S)$ denotes the number of users of set j in state S. We prove $PoU_{IR}(\epsilon) = (1 + \epsilon)^{O(m^2)} O(\log m)$ for $\epsilon = O(\frac{1}{m})$ — this it is small for a small number of resources even if there are many players. This improves over the previous bounds of [3], which had an additional dependence on the number of players n. We also improve the existing lower bound for these games (due to [3]) to $PoU_{IR}(\epsilon) = \Omega(\log^p m)$ for $\epsilon = \Theta(\frac{1}{m})$ and any constant $p > 0$. Our new lower bound is a subtle construction that uses an intricate "pump" gadget with finely tuned parameters. A pump replaces, in a non-trivial recursive manner with identical initial and final pump states, one chip of small cost with one chip of high cost. Finally, we show a lower bound of $PoU_{BR}(\epsilon) = \Omega(\epsilon n^{1/3} / \log n)$ for $\epsilon = \Omega(n^{-1/3})$ and $m = \Omega(n)$ which is valid even if an arbitrary ordering of player updates is specified a priori, unlike the existing lower bound of [3].

We note that our lower bounds use judiciously tuned gadgets that create the desired snowball effect of uncertainty. Most of them hold even if players must update in a specified order, e.g. round-robin (i.e. cyclically) or the player to update is the one with the largest reduction in (perceived) cost. Our upper bounds on PoU hold no matter which player updates at any given step.

Due to the lack of space we only provide sketches for most proofs in this paper. Full proofs appear in the long version of the paper [4].

2   Consensus Games

In this section, we provide lower and upper bounds regarding the effect of uncertainty on consensus games. Throughout the section, we call an edge *good* if its endpoints have the same color and *bad* if they have different colors. The social cost is the number of bad edges (i.e. half the sum of all players' costs) plus one, which coincides with the game's potential. Thus $PoU(0) = PoB(0) = 1$. Since the social cost is in $[1, n^2]$, $PoU(\epsilon) = O(n^2), \forall \epsilon$ and $PoB(B) = O(n^2), \forall B$.

2.1   Lower Bound and Upper Bound for Perturbation Model

Perturbation model. The natural uncertainty here is in the number of neighbors of each color that a vertex perceives. We assume that if a vertex i has n' neighbors of some color, then a perturbation may cause i to perceive instead an arbitrary integer in $[\frac{1}{1+\epsilon}n', (1+\epsilon)n']$. Since each action's cost is the number of neighbors of the other color, this is a cost perturbation model. In this model, only an $\epsilon = \Omega(\frac{1}{n})$ effectively introduces uncertainty.[2] We also assume $\epsilon \leq 1$ in this section.

Theorem 1. $PoU(\epsilon, consensus) = \Omega(n^2 \epsilon^3)$ *for* $\epsilon = \Omega(n^{-1/3})$ *even for arbitrary orderings of player updates.*

Proof sketch. We sketch below the three components of our construction assuming that the adversary can choose which player updates at any step. We can show that the adversary can reduce an arbitrary ordering (that he cannot control) to his desired schedule by compelling any player other than the next one in his schedule not to move.

- The *output* component has $k_{out} = \Theta(\frac{1}{\epsilon}(\log n - 2\log\frac{1}{\epsilon}))$ levels, where any level $i \geq 0$ has $\frac{1}{\epsilon}(1+\epsilon)^i$ nodes. Each node on level $i \geq 1$ is connected to all the nodes on levels $i - 1$ and $i + 1$.
- The *input* component consists of two cliques, K_{red} and K_{white}, each of size $1/\epsilon^2$, and each of these nodes is connected to the first output level. The dynamics are "seeded" from the input component.
- There is an *initializer* gadget[3] with $k_{init} = \frac{1}{\epsilon}\log\frac{2}{\epsilon}$ levels, where level i has $\frac{1}{\epsilon}(1+\epsilon)^i$ nodes. Again, each node on level $i \geq 1$ is connected to all the nodes on levels $i - 1$ and $i + 1$. Every node of the final level of the initializer is connected to all nodes in the clique K_{red}.

We require $\epsilon \geq n^{-1/3}$ so that the number $\Theta(\frac{1}{\epsilon^3})$ of initializer nodes is at most a constant fraction of the total n nodes. Then the number $\Theta(\frac{1}{\epsilon^2}(1+\epsilon)^{k_{out}})$ of output nodes will be a large fraction of the n nodes.

The initial coloring is: all output nodes white, K_{red} and K_{white} white, and all initializer nodes white except for the first two levels which are red. We thus initially have $\frac{(1+\epsilon)^3}{\epsilon^2}$ bad edges, all in the initializer. Throughout the dynamics, for both the initializer and output components, the nodes in each level have the same color, except for the level that is currently being updated. The schedule consists of two epochs. The first epoch is for all the initializer nodes and the clique K_{red}

---

[2] If $\epsilon < 1/3n$, consider a node with r red neighbors, w white neighbors, and $r > w$. Since r and w are both integers, $r \geq w + 1$. For a cost increasing move to occur, we must have $(1+\epsilon)^2 w \geq r \geq w + 1$. This implies that $n \geq w \geq 1/(2\epsilon + \epsilon^2) \geq 1/3\epsilon > n$.

[3] Without this initializer, we can get a worse lower bound $PoU(\epsilon, consensus) = \Omega(n^2 \epsilon^4)$, for a wider range of $\epsilon = \Omega(n^{-1/2})$, again for an arbitrary ordering. The main difference is that K_{red} is initially red and the initial state has $\Theta(\frac{1}{\epsilon^3})$ bad edges. In the long version of our paper, we additionally show that when the adversary can control the ordering of updates to match its schedule, we can improve both this lower bound and that of Theorem 1 to $\Omega(n^2\epsilon^3)$ for the range $\epsilon = \Omega(n^{-1/2})$.

to change color. Thereafter they are left alone. At this point, the adversary can prevent the clique nodes from changing their color, and he can change the color of nodes in the first output level at will. Indeed, these nodes have $\frac{1}{\epsilon^2}$ neighbors of either color in each clique and $\Theta(\frac{1}{\epsilon})$ neighbors in the second output level, difference small enough to be overcome by a $(1 + \epsilon)$-factor perturbation. The second epoch has a phase for each two consecutive output levels i and $i + 1$ in which these levels obtain their final color and then are never considered again. This is achieved by changing all prior levels to the intended color of i and $i + 1$.

In the final state we have the first two output levels colored red, the next two colored white, then the next two red, and so on. The final number of bad edges is $\Omega(n^2\epsilon)$. Since we started with only $\Theta(\frac{1}{\epsilon})$ bad edges the number of bad edges has increased by a factor of $\Omega(n^2\epsilon^3)$. Thus $PoU(\epsilon, consensus) = \Omega(n^2\epsilon^3)$.   □

We note that the previously known lower bound of $\Omega(1 + n\epsilon)$ due to Balcan et. al [3] was based on a much simpler construction. Our new bound is better by a factor of at least $n^{1/3}$ for $\epsilon = \Omega(1/n^3)$. We also note that since $PoU(\epsilon) = O(n^2)$ for any ϵ, it implies a tight PoU bound of $\Theta(n^2)$ for *any* constant ϵ. We also provide a PoU *upper bound* for consensus games that depends on ϵ. It implies that the existing $\Theta(n^2)$ lower bound cannot be replicated for any $\epsilon = o(1)$. The proof is based on comparing the numbers of good and bad edges at the first move that increases the social cost.

Theorem 2. $PoU(\epsilon, consensus) = O(n^2\epsilon)$.

2.2 Tight Bound for Byzantine Players

As described earlier, Byzantine players can choose their color ignoring their neighbors' colors (and therefore their own cost). Note however the Byzantine players cannot alter the graph[4]. In this section we show a tight bound on the effect of B Byzantine players, for any B: the effect of one Byzantine player is very high, of order $n\sqrt{n}$ and that the subsequent effect of $B \leq n$ Byzantine players is proportional to the square root of B. As was the case for PoU, the effect of uncertainty is decomposed multiplicatively into a power of n and a power of the extent of uncertainty (ϵ for PoU, B for PoB).

Theorem 3. $PoB(B, consensus) = \Theta(n\sqrt{n \cdot B})$.

The proof of the $O(n\sqrt{n \cdot B})$ upper bound follows from Lemmas 1, 2 and 3 below. The key to this bound is the notion of a flippable graph. For any consensus game, let S_{red} be the configuration where all nodes are red, and similarly let S_{white} be the configuration where all nodes are white.

---

[4] For the lower bound, we assume that a player will break ties in our favor when he chooses between two actions of equal cost. With one more Byzantine player the same bound holds even if players break ties in the worst possible way for us. For the upper bound, we assume worst possible players' moves from the social cost point of view.

Definition 1 (*B*-Flippable graph). *Consider graph G on n vertices of which B are designated special nodes and the other $n - B$ nodes are called normal. We say G is B-flippable (or just flippable when B is clear from context) if in the consensus game defined on G where the special nodes are the Byzantine agents, the state S_{white} is B-Byz-reachable from S_{red}.*

We now describe the concept of a *conversion dynamics* in a consensus game which we use in several of our proofs. In such a dynamics, we start in a state where all vertices are red and have Byzantine players change their color to white. Then all normal nodes are allowed in a repeated round-robin fashion to update, so long as they are currently red. This ends when either every vertex is white or no vertex will update its color.

We note that in a flippable graph the conversion dynamics induces an ordering of the normal vertices: nodes are indexed by how many other white nodes are present in total at the state when they change color to white. We note that there may be more than one valid ordering. In the following with each B-flippable graph, we shall arbitrarily fix a canonical ordering (by running the conversion dynamics). Where there is sufficient context, we shall use v a vertex interchangeably with its index in this ordering. Using this ordering we induce a canonical orientation by orienting edge uv from u to v if and only if $u < v$. We also orient all edges away from the B special nodes. To simplify notation, we shall write $v_{in} = |\delta^-(v)|$ and $v_{out} = |\delta^+(v)|$ for a vertex v's in-degree and out-degree respectively. We note that by construction, for a flippable graph we have $v_{in} \geq v_{out}$. One can easily show the following:

Claim 1. *A graph is B-flippable if and only if there exists an ordering on the $n - B$ normal vertices of the graph such that, in the canonical orientation of the edges, every normal vertex v has $v_{in} \geq v_{out}$. A graph is B-flippable if and only if for every pair of states S, S', S is B-Byz-reachable from S'.*

Lemma 1. *Fix a game G on n vertices, B of which are Byzantine, and a pair of configurations S_0 and S_T such that S_T is B-Byz-reachable from S_0. If $cost(S_0) \leq n$, then there exists a B-flippable graph F with at most $3n$ nodes and at least $cost(S_T)$ edges (in total).*

Proof Sketch. The proof has two stages. In the first stage, we construct a consensus game G' and configuration S'_T such that $cost(S'_T) \geq k$, S'_T is B-Byz-reachable from S_{red}, but $V(G') \leq 3n$. In the second stage, we delete some edges of G' to create a graph G'', showing that S_{white} is B-Byz-reachable from S_{red} in G'' (thus G'' is flippable) while ensuring that $E(G'') \geq k$.

We first describe the construction of G'. We separate the nodes of G into two sets I_r and I_w based on their color in the initial configuration S_0. For each edge that is bad in S_0 we introduce a 'mirror' gadget. A mirror consists of a single node whose color the adversary can easily control and a helper node to change the color. The nodes of G' are all the nodes of G and at most $2n$ nodes from mirror gadgets. The edges of G' are all edges that are good in state S_0 of G, and at most $5n$ edges introduced by the mirrors.

In G', the nodes of I_r and I_w interact with each other only indirectly, via mirrors that are controlled by the adversary. Using this fact, the adversary can simulate the dynamics from S_0 to S_T on I_r. For any state S, let $\bar{S}$ be the state in which every node has the opposite color from in S. The adversary can also simulate the dynamics over states in which the color of every node has been reversed. Thus the adversary can simulate dynamics from $\bar{S}_0$ to $\bar{S}_T$ in I_w.

At the end of this process, every edge that was bad in S_T is also bad in this final state S'_T. Note that the initial state is one in which all nodes in I_r are red, and all nodes in I_w are red (since they are red in $\bar{S}_0$). Thus the dynamics lead to S'_T from S_{red}. Thus we have created G' and a state S_T where $cost(S'_T) \geq k$ and G' is flippable, but $|V(G')| \leq 3n$.

In the second stage, we identify a set of edges in G', and delete them to form G''. This set of edges are precisely those whose endpoints both remain red in the conversion dynamics. We show that these edges are not bad in any state, hence none of these edges are bad in S'_T, and secondly in G'', S_{white} is B-Byz-reachable from S_{red}. The first statement implies that $cost(S'_T) \leq |E(G'')|$, and the second statement implies that G'' is flippable, which is what we wanted. $\qquad\square$

Definition 2 ($F_{seq}(n, B)$). *Let $F_{seq}(n, B)$ be the B-flippable graph with $n - B$ normal nodes with labels $\{1, 2, \ldots, n - B\}$. There is an edge from each special node to each normal node. Every normal node v satisfies $v_{out} = \min(v_{in}, (n - B) - v)$, and v is connected to the nodes of $\{v + 1, \ldots, v + v_{out}\}$. This is called the* no-gap *property. In general, if $k = \min(v_{in}, n - v)$ then v has out-arc set $\{v + 1, \ldots, v + k\}$.*

By claim 1 we immediately get that F_{seq} is B-flippable. Our upper bound follows by showing $|E(F)| \leq |E(F_{seq}(n, B))|$ for any flippable graph F on n vertices. For this, we take a generic flippable graph and transform it into F_{seq} without reducing the number of edges. We say there is a *gap*(a, b, c) for $a < b < c$ if vertex a does not have an edge to b but does have an edge to c. Note that this is defined in terms of an ordering on the vertices; we use the conversion ordering for each graph.

Lemma 2. *A flippable graph on n vertices has at most as many edges as $F_{seq}(n, B)$.*

Proof sketch. We prove this by inducting on the lexicographically minimal gap of flippable graphs. If a gap is present, then we can either add or move edges to create a lexicographically greater gap. Eventually this eliminates all gaps without reducing the number of edges. Since a graph with no gaps is a subgraph of $F_{seq}(n, B)$, we have bounded the number of edges in any flippable graph. $\qquad\square$

Our last lemma tightly counts the number of edges in $F_{seq}(n, B)$ via an inductive argument and thus, by Lemma 2, it also upper bounds the number of edges in any flippable graph.

Lemma 3. *If $B \leq \frac{n}{2}$, the flippable graph $F_{seq}(n, B)$ has $\Theta(n\sqrt{nB})$ edges.*

Proof sketch. We count the edges by counting the number of in-edges to a given node. By induction, we show that the first node to have jB in-edges has index $\binom{j+1}{2}B + 1$. This implies that any node $k \in [n]$ has $\Theta(\sqrt{kB})$ in-edges. Summing over all n nodes, we find there are $\Theta(n\sqrt{nB})$ edges in the graph in total. □

Proof of Theorem 3. We first argue that the $PoB(B, consensus) = O(n\sqrt{nB})$. Consider a consensus graph G on n nodes, and a pair of configurations S_0 and S_T B-Byz-reachable from S_0. If $B \geq n/2$, then the statement is trivial, so we may assume that $B < n/2$. We assume $cost(S_0) < n$: if $cost(S_0) \geq n$, since G has fewer than n^2 edges, we get $PoB(B, G) \leq n^2/n = n$. Denote by $k := cost(S_T) - 1$ the number of bad edges in S_T. By Lemma 1, we demonstrate a flippable graph F on fewer than $3n$ nodes, with at least k edges. By Lemma 2, F has at most as many edges as $F_{seq}(3n, B)$, which has only $O(n\sqrt{nB})$ edges by Lemma 3. We get $PoB(B) = O(n\sqrt{nB})$.

It will now be enough to prove that $PoB(B, F_{seq}(n, B)) = \Omega(n\sqrt{nB})$. We claim now that if G is a flippable graph with m edges, then $PoB(B, G) \geq \frac{m}{2}$. We get this via the following probabilistic argument using the fact that the adversary can color G arbitrarily (by claim 1). Consider a random coloring of the graph, where each node is colored white independently with probability $1/2$. The probability an edge is bad is $1/2$, so in expectation, there are $m/2$ bad edges. Thus some state has at least $m/2$ bad edges and it is reachable via dynamics from any other state (claim 1) since G is a flippable graph. This establishes $PoB(B, G) \geq \frac{m}{2}$. Since $F_{seq}(n, B)$ is flippable and it has $m = \Theta(n\sqrt{nB})$ edges, we get $PoB(B, F_{seq}(n, B)) = \Omega(n\sqrt{nB})$. □

In contrast to the existing bound $PoB(1) = \Omega(n)$, our bound is parametrized by B, sharper (by a $\Theta(\sqrt{n})$ factor for $B = 1$) and asymptotically tight.

3   Set-Covering Games and Extensions

Set-covering games (SCG) are a basic model for fair division of costs, and have wide applicability, ranging e.g. from a rental car to advanced military equipment shared by allied combatants. A set-covering game admits the potential function $\Phi(S) = \sum_{j=1}^{m} \sum_{i=1}^{n_j(S)} \frac{w_j}{i} = \sum_{j=1}^{m} \Phi^j(S)$ where $\Phi^j(S) = \sum_{i=1}^{n_j(S)} \frac{w_j}{i}$. $\Phi^j(S)$ has an intuitive representation as a stack of $n_j(S)$ *chips*, where the i-th chip from the bottom has a cost of w_j/i. When a player i moves from set j to j' one can simply move the topmost chip for set j to the top of stack j'. This tracks the change in i's costs, which equals by definition the change in potential Φ. We will only retain the global state (number of players using each set) and discard player identities. This representation has been introduced for an existing PoU_{IR} upper bound of [3]; we refine it for our improved upper bound.

SCGs have quite a small gap between potential and cost [1]: $cost(S) \leq \Phi(S) \leq cost(S)\Theta(\log n), \forall S$. Hence without uncertainty, the social cost can only increase by a logarithmic factor: $PoU(0) = PoB(0) = \Theta(\log n)$.

3.1 Upper Bound for Improved-Response

We start with an upper bound on PoU_{IR} in set-covering games that only depends on the number m of sets.

Theorem 4. $PoU_{IR}(\epsilon, \text{set-covering}) = (1 + \epsilon)^{O(m^2)} O(\log m)$ *for* $\epsilon = O(\frac{1}{m})$.

In particular for $\epsilon = O(\frac{1}{m^2})$ we obtain a logarithmic $PoU_{IR}(\epsilon)$.

Proof of Theorem 4. We let J_0 denote the sets initially occupied and $W_0 = cost(S_0) = \sum_{j \in J_0} w_j$ be their total weight. We discard any set not used during the dynamics.

With each possible location of a chip at some height i (from bottom) in some stack j, we assign a *position* of value[5] w_j/i. Thus a chip's cost equals the value of its position in the current state. We will bound the cost of the m most expensive chips by bounding costs of expensive positions and moves among them.

It is easy to see that any set has weight at most $W_0(1 + \epsilon)^{2(m-1)}$ (clearly the case for sets in J_0). Indeed, whenever a player moves to a previously unoccupied set j' from a set j, the weight of j' is at most $(1 + \epsilon)^2$ times the weight of j; one can trace back each set to an initial set using at most $m - 1$ steps (there are m sets in all). We also claim that at most $mi(1+\epsilon)^{2m}$ positions have value at least $\frac{W_0}{i}, \forall i$: indeed positions of height $i(1 + \epsilon)^{2m}$ or more on any set have value less than $\frac{W_0}{i}$ since any set has weight at most $W_0(1 + \epsilon)^{2(m-1)}$.

Fix a constant $C > (1 + \epsilon)^{2m}$ (recall $\epsilon = O(\frac{1}{m})$). Note that any chip on a position of value less than $\frac{W_0}{m}$ in S_0 never achieves a cost greater than $\frac{W_0}{m}(1 + \epsilon)^{2Cm^2}$. Indeed, by the reasoning above for $i = m$, there are at most $m \cdot m \cdot (1 + \epsilon)^{2m} \le Cm^2$ positions of greater value. Thus the chip's cost never exceeds $\frac{W_0}{m}(1 + \epsilon)^{2Cm^2}$ as it can increase at most Cm^2 times (by an $(1 + \epsilon)^2$ factor).

We upper bound the total cost of the *final* m most expensive chips, as it is no less than the final social cost: for a set, its weight equals the cost of its most expensive chip. We reason based on chips' initial costs. Namely, we claim $h(i) \le \frac{W_0}{i-1} \cdot (1 + \epsilon)^{2Cm^2}, \forall i \in [m]$, where $h(i)$ denotes the cost of ith most expensive chip in the final configuration. If this chip's initial cost is less than $\frac{W_0}{m}$ then the bound follows from the claim above. Now consider all chips with an initial cost at least $\frac{W_0}{m}$. As argued above, at most Cm^2 positions have value $\frac{W_0}{m}$ or more, and any of these chips increased in cost by at most $(1 + \epsilon)^{2Cm^2}$. A simple counting argument[6] shows that for any i, there are at most i chips of initial cost at least $\frac{W_0}{i}$ and thus $h(i) \le \frac{W_0}{i-1} \cdot (1 + \epsilon)^{2Cm^2}, \forall i$.

---

[5] We refer to the weight of a set, the cost of a chip and the value of a position.

[6] We claim that for any k, there are at most k chips of initial cost at least $\frac{W_0}{k}$. Let J_0, be the set of initially used resources. For each $j \in J_0$, let r_j be set j's fraction of the initial weight (i.e. $w_j = r_j W_0$), and let p_j be the number of initial positions with value greater than $\frac{W_0}{k}$ in set j. We have $\frac{w_j}{p_j} = \frac{r_j W_0}{p_j} \ge \frac{W_0}{k}$, implying $p_j \le kr_j$. Counting the number of initial positions with sufficient value yields $\sum_j p_j \le \sum_j kr_j = k \sum_j r_j = k$ since $\sum_j r_j = 1$.

As the ith most expensive chip has cost at most $\frac{W_0}{i-1}(1+\epsilon)^{2Cm^2}$ (at most $i-1$ chips have higher final cost),

$$\sum_{i=1}^{m} h(i) = h(1) + \sum_{i=2}^{m} h(i) \leq h(1) + \sum_{i=2}^{m} \frac{W_0}{i-1}(1+\epsilon)^{2Cm^2}$$

$$= O(W_0(1+\epsilon)^{2m} + W_0(1+\epsilon)^{2Cm^2} \log m) = W_0 \cdot (1+\epsilon)^{O(m^2)} O(\log m)$$

As desired, $PoU_{IR}(\epsilon, set\text{-}covering) = (1+\epsilon)^{O(m^2)} O(\log m)$ as the final social cost is at most $\sum_{i=1}^{m} h(i)$. $\qquad\square$

The existing bound [3] is $PoU_{IR}(\epsilon) = O((1+\epsilon)^{2mn} \log n)$. Unlike our bound, it depends on n (exponentially) and it does not guarantee a small $PoU_{IR}(\epsilon)$ for $\epsilon = \Theta(\frac{1}{m^2})$ and $m = o(n)$. This bound uses chips in a less sophisticated way, noting that any chip can increase its cost (by $(1+\epsilon)^2$) at most mn times.

Our technique also yields a bound of $PoU_{BR}(\epsilon) = (1+\epsilon)^{O(m^2)} O(\log m)$ for $\epsilon = O(\frac{1}{m})$ in matroid congestion games [3] – see the full version for details [4]. These games are important in that they precisely characterize congestion games for which arbitrary BR dynamics (without uncertainty) converge to a Nash equilibrium in polynomial time.

3.2   Lower Bound for Improved-Response

Our upper bound showed that $PoU_{IR}(\epsilon)$ is logarithmic for $\epsilon = O(\frac{1}{m^2})$. A basic example (one player hopping along sets of cost $1, (1+\epsilon)^2, \ldots, (1+\epsilon)^{2(m-1)}$), applicable to many classes of games, yields the lower bound $(1+\epsilon)^{2(m-1)} \leq PoU_{IR}(\epsilon, set\text{-}covering)$. In fact, this immediate lower bound was the best known on $PoU_{IR}(\epsilon)$. For $\epsilon = \omega(\frac{1}{m})$, we get that $PoU_{IR}(\epsilon)$ is large. An intriguing question is what happens in the range $[\omega(\frac{1}{m^2}), \Theta(\frac{1}{m})]$, in particular for natural uncertainty magnitudes such as $\epsilon = \Theta(\frac{1}{m})$ or $\epsilon = \Theta(\frac{1}{n})$.

In this section we show that for $\epsilon = \Theta(\frac{1}{\min(m,n)})$, PoU can be as high as polylogarithmic. We provide a construction that repeatedly uses the snowball effect to locally increase one chip's cost, without other changes to the state. Our main gadget is a *pump*, which is used as a black box in the proof. A pump increases a chip's cost by $\alpha = \log n'$, where $n' = \min(m, n)$. We use p pumps to increase each chip's cost by a $\Omega(\log^p n')$ factor. As pumps are "small", the total cost increase is $\Omega(\log^p n')$.

Theorem 5. $PoU_{IR}(\epsilon, set\text{-}covering) = \Omega(\log^p \min(m, n))$, for $\epsilon = \Theta(\frac{1}{\min(m,n)})$ and constant $p > 0$.

Before providing a sketch of Theorem 5, we provide the formal definition of a pump. An (α, W)-pump uses $O(\frac{1}{\epsilon})$ sets and $O(2^\alpha)$ players to increase, one by one, an arbitrary number of chip costs by an α factor from W/α to W. For ease of exposition, we assume $m = \Theta(n)$ and we only treat $p = 2$, i.e. how to achieve $PoU_{IR}(\frac{1}{n}) = \Omega(\log^2 n)$. For general p, we use p pump gadgets instead of two.

Definition 3 (Pump). *An (α, W)-pump P is an instance of a set-covering game specified as follows:*

- The number m_P of sets used is $O(\frac{1}{\epsilon})$. For our choice of ϵ, $m_P = O(n)$. The total weight W_P of all sets in P that are initially used is in $(2^\alpha W, e2^\alpha W)$. The number of players used is $n_P = 2^{\alpha+1} - 2$.
- Within $O(n^3)$ moves of IR dynamics contained within the pump, and with a final state identical to its initial state, a pump can consume any chip of cost at least W/α to produce a chip of cost W.

Proof sketch of Theorem 5. Let $N := \alpha^2 2^\alpha$. The number of players will be $n := N + n_{P_1} + n_{P_2}$. Thus $\alpha = \Theta(\log n)$. Note that each player can use any set.

We use two pumps, an $(\alpha, 1/\alpha)$ pump P_1, and a $(\alpha, 1)$ pump P_2. Aside from the pumps, we have Type-I, Type-II and Type-III sets, each with a weight of $1/\alpha^2$, $1/\alpha$ and 1 respectively. At any state of the dynamics, each such set will be used by no player or exactly one player. In the latter case, we call the set *occupied*. We have N Type-I sets, 1 Type-II set, and N Type-III sets, i.e. $m := 2N + 1 + m_{P_1} + m_{P_2} = \Theta(n)$ sets in all.

Let $\mathrm{cfg}(i, j, k)$ refer to the configuration with i Type-I sets occupied, j Type-II sets occupied, and k Type-III sets occupied. We shall use $2N + 1$ intermediate states, denoted $state_i$. Our initial state is $state_0 = \mathrm{cfg}(N, 0, 0)$, and our final configuration will be $state_{2N} = \mathrm{cfg}(0, 0, N)$. In general, $state_{2i} = \mathrm{cfg}(N - i, 0, i)$ and $state_{2i+1} = \mathrm{cfg}(N - i - 1, 1, i)$. Thus we want to move each player on a Type-I set (initially) to a corresponding Type-III set, an α^2 *increase* in cost. To this purpose, we will pass each such player through the first pump and move it on the Type-II set. This achieves the transition from $state_{2i}$ to $state_{2i+1}$. Since the player's cost is increased by an α factor (from $\frac{1}{\alpha^2}$ to $\frac{1}{\alpha}$), we can pass it through the second pump and then move it on the Type-III set. This achieves the transition from $state_{2i+1}$ to $state_{2i+2}$.

The social cost of our initial configuration is $W_0 = N \cdot \frac{1}{\alpha^2} + W_{P_1} + W_{P_2} \leq 2^\alpha + e \cdot \frac{1}{\alpha} 2^\alpha + e \cdot 2^\alpha \leq 7 \cdot 2^\alpha$. The final social cost (excluding the pumps) is at least $N = \alpha^2 2^\alpha$. Thus $PoU = \Omega(\alpha^2)$, and $\alpha = \Theta(\log n)$.

Finally, we note that the pump is constructed using $1 + 1/\epsilon$ sets $s_0, \ldots, s_{\frac{1}{\epsilon}}$ where set s_i has weight $W(1 + \epsilon)^i$. Additionally there are α 'storage' sets t_j with weight $2W/j$. In the initial configuration, the sets $s_1, \ldots, s_{2\alpha}$ are occupied, with set s_i having $\max(0, \alpha - \lceil \log_2 1 + i \rceil)$ players on it, for $i \geq 1$. s_0 also has $\alpha - 1$ players. The pump is activated by a chip of cost W/α moving onto set s_0. The chips then advance into a configuration where set s_i has precisely $\max(0, \alpha - \lceil \frac{1}{\epsilon} - i + 1 \rceil)$ chips. Note that this roughly doubles the cost of each chip. The chips on set s_1 then use the storage sets to capture their current cost. The chip in storage with cost $2W$ exits the pump, and the other chips in storage fill up set s_0. All the other chips can return to the initial configuration by making only cost decreasing moves. Note that the chip to leave the pump is not the same chip that entered. □

In the full version of the paper, we show how our pump gadget can be tweaked to provide a polylogarithmic lower bound on PoU for generalized set-covering games with *increasing* delay functions, as long as they have bounded jumps, i.e. if an additional user of a resource cannot increase its cost by more than a constant factor.

3.3 Lower Bound for Best-Response

We also show that a significant increase in costs is possible for a large range of ϵ even if players follow best-response dynamics with arbitrary orderings. This construction will use more sets than players, and so will not contradict Theorem 4

Theorem 6. $PoU_{BR}(\epsilon, \text{set-covering}) = \Omega(\epsilon n^{1/3}/\log n)$, for any $\epsilon = \Omega(n^{-1/3})$. This holds for any arbitrary ordering of the dynamics, i.e. no matter which player is given the opportunity to update at any time step.

Proof sketch. The proof has two stages. Our first step provides a construction which shows that in a set-covering game with best-response dynamics, the adversary can compel an increase of social cost that is polynomial (of fractional degree) in n. With our new construction, we then separately show that the adversary can cause this increase even when it cannot control which players update – we show that the adversary can cause only the relevant players to update. □

We note that previous work provided a stronger lower bound of $\Omega(\epsilon n^{1/2}/\log n)$ [3], but which only works for a specific ordering of the updates.

4 Open Questions

It would be interesting to close our gap on PoU for consensus games. It would also be interesting to study a model where the perturbations are not completely adversarial, but instead chosen from some distribution of bounded magnitude.

Acknowledgements. This work was supported by NSF grants CCF-0953192 and CCF-1101215, by ONR grant N00014-09-1-0751, and by a Microsoft Research Faculty Fellowship.

References

1. Anshelevich, E., Dasgupta, A., Kleinberg, J.M., Tardos, É., Wexler, T., Roughgarden, T.: The price of stability for network design with fair cost allocation. In: FOCS, pp. 295–304 (2004)
2. Awerbuch, B., Azar, Y., Epstein, A., Mirrokni, V.S., Skopalik, A.: Fast convergence to nearly optimal solutions in potential games. In: EC (2008)
3. Balcan, M.-F., Blum, A., Mansour, Y.: The price of uncertainty. In: EC (2009)
4. Balcan, M.-F., Constantin, F., Ehrlich, S.: The snowball effect of uncertainty in potential games (2011), www.cc.gatech.edu/~ninamf/papers/snowball-long.pdf
5. Buchbinder, N., Lewin-Eytan, L., Naor, J., Orda, A.: Non-Cooperative Cost Sharing Games Via Subsidies. In: Monien, B., Schroeder, U.-P. (eds.) SAGT 2008. LNCS, vol. 4997, pp. 337–349. Springer, Heidelberg (2008)
6. Christodoulou, G., Mirrokni, V.S., Sidiropoulos, A.: Convergence and Approximation in Potential Games. In: Durand, B., Thomas, W. (eds.) STACS 2006. LNCS, vol. 3884, pp. 349–360. Springer, Heidelberg (2006)
7. Monderer, D., Shapley, L.S.: Potential games. Games and Economic Behavior 14, 124–143 (1996)
8. Moscibroda, T., Schmid, S., Wattenhofer, R.: When selfish meets evil: byzantine players in a virus inoculation game. In: PODC, pp. 35–44 (2006)

Approximation Algorithm for Security Games with Costly Resources

Sayan Bhattacharya, Vincent Conitzer, and Kamesh Munagala

Department of Computer Science, Duke University
Durham, NC 27708, USA
{bsayan,conitzer,kamesh}@cs.duke.edu

Abstract. In recent years, algorithms for computing game-theoretic solutions have been developed for real-world security domains. These games are between a defender, who must allocate her resources to defend potential targets, and an attacker, who chooses a target to attack. Existing work has assumed the set of defender's resources to be fixed. This assumption precludes the effective use of approximation algorithms, since a slight change in the defender's allocation strategy can result in a massive change in her utility. In contrast, we consider a model where resources are obtained at a cost, initiating the study of the following optimization problem: Minimize the total cost of the purchased resources, given that every target has to be defended with at least a certain probability. We give an efficient logarithmic approximation algorithm for this problem.

1 Introduction

Taking a game as input and computing a solution of it is one of the core problems of algorithmic game theory. To be more precise, it is a collection of problems, one for each combination of a representation scheme and a solution concept. Perhaps the best-known example is the problem of computing a Nash equilibrium of a normal-form game, which is now known to be PPAD-complete for two-player games [2,3] and FIXP-complete for games with three or more players [5].

In contrast, this paper deals with an alternative solution concept, corresponding to a "Stackelberg" model. In this model, there are two players, a "leader" and a "follower". The leader first commits to a mixed strategy, and the follower responds only after observing the commitment. Recently, algorithms for computing optimal leader strategies have been developed for various real-world security domains, including US airports [11,12], the US Federal Air Marshals [13], and the US Coast Guard [1].

Motivated by such applications, Kiekintveld *et al.* [8] proposed a general model of security games. The players are a *defender* and an *attacker*. There is a set of *targets* that the attacker may want to attack (in the Federal Air Marshal example, these would be individual flights), and the defender has a set of *resources* (Federal Air Marshals). The defender can assign each resource to a *schedule*, which consists of a subset of the targets (for example, a tour of multiple flights that a single Federal Air Marshal can take). In general, not every resource can

N. Chen, E. Elkind, and E. Koutsoupias (Eds.): WINE 2011, LNCS 7090, pp. 13–24, 2011.

be assigned to every schedule. If a resource is assigned to a schedule, then it defends all the targets contained in that schedule.

A pure (resp., mixed) strategy for the defender is a deterministic (resp., randomized) assignment of the resources to schedules. The typical motivation given for the Stackelberg model, where the defender (the leader) commits to a mixed strategy and the attacker (the follower) subsequently best-responds, is as follows. Every day, the defender draws an assignment from her distribution. Over the course of time, the attacker observes the realized assignments and eventually learns the probabilities with which each target is defended on any given day. Then, the attacker decides to attack a single target that maximizes her expected utility. The utilities of both players usually depend on: (a) the target that is attacked, and (b) whether or not that target is defended on the day of the attack. In the existing literature on this topic, the typical problem is to find the optimal mixed strategy for the defender to commit to, that is, the one that will maximize the defender's expected utility.

In this paper, we explore the setting where the resources can be purchased by the defender at a cost. To be more specific, we consider the following problem. Some resources are to be (deterministically) purchased in advance. Next, they are to be randomly assigned to schedules, ensuring that each target is defended with at least a certain probability. The objective is to minimize the total cost of the purchased resources.

For example, suppose that the Federal Air Marshal Service can hire a set of Marshals on a contractual basis for one year, thereby incurring a certain cost. During each day of the following year, the hired Marshals are (randomly) assigned to the schedules, according to the mixed strategy chosen. It is important to emphasize that we require a set of resources to be purchased *deterministically*, after which these resources can be randomly assigned to schedules. It is impractical to randomize the resource supply itself each day, since, for example, security personnel cannot be hired as day laborers.

Existing literature [8,9] on security games assumes a fixed set of resources. In contrast, our model assumes that the resources are available in unlimited supply, but they can be purchased at a cost. There is a connection between these two settings: Given a fixed set of resources, there is an efficient algorithm to compute the optimal commitment strategy for a defender *if and only if* we can decide in polynomial time whether each target can be defended with a certain probability. The proof of this claim appears in the full version of our paper.

Our Results and Techniques. In Section 2, we formally state our problem and show that it is a generalization of Set Cover, and hence the problem is unlikely to admit an approximation ratio better than $O(\log |\mathcal{T}|)$, where $|\mathcal{T}|$ is the number of targets [6]. Section 3 gives an $O(\log |\mathcal{T}| + \log |\mathcal{S}|)$ approximation algorithm for this problem, where $|\mathcal{S}|$ is the number of schedules. The algorithm partitions the set of targets into two subsets, depending on whether their required coverage probabilities are *small* or *big*, and separately deals with these two subsets.

In Section 3.1, we show that the subproblem of defending the targets with big probabilities admits an LP relaxation (LP (3)), and it can be converted to the

LP relaxation for Set Cover (LP (6)) while losing at most a constant factor in the objective value. Thus, a greedy algorithm (Figure 1) gives a good approximation for this case. Unfortunately, this analysis cannot be extended to defend the targets with small probabilities, because the gap between the optimal objective value of LP (3) and the cost of the optimal solution can become arbitrarily large.

Section 3.2 considers the remaining subproblem—to defend the targets with small probabilities. For this, we present an exponentially sized covering *integer program* (IP (9)) that bounds the cost (Lemma 3) of the optimal solution. Interestingly, the LP relaxation of IP (9) can again be far away from the optimal solution, and thus, it is necessary to impose the integrality constraints.

IP (9) has the additional property that all of its constraint data are integral. Under such circumstances, a simple greedy heuristic (see Dobson [4]) gives a logarithmic approximation to the integer optimum (Lemma 4). The idea is to treat every variable as a set that *covers* part of the constraints, and during each iteration, increment the variable that gives maximum coverage per unit cost. However, IP (9) has exponentially many variables, and hence, we cannot directly apply Dobson's algorithm to solve it in polynomial time. Instead, we show that the subroutine of selecting the variable with maximum coverage per unit cost is exactly equivalent to maximizing a submodular function subject to a budget constraint. Hence, we can use another greedy algorithm [10] to implement this subroutine upto a constant factor approximation. Accordingly, we solve IP (9), and its outcome determines the resources we purchase and the way in which we randomly assign them to the schedules (Theorem 2).

Remark. All the missing proofs appear in the full version of this paper.

2   Notations and Preliminaries

There are a set of targets $\mathcal{T}$, a set of schedules $\mathcal{S}$, and a set of resource-types Θ. There is an *unlimited supply* of resources of each type. Any resource of type $\theta \in \Theta$ has cost $c(\theta)$. In addition, there is a subset $S(\theta) \subseteq \mathcal{S}$ such that any resource of type θ can be *assigned* to at most one schedule in $S(\theta)$. The type of a resource r is denoted by $\theta_r \in \Theta$. Whenever some resource is assigned to schedule $s \in \mathcal{S}$, it *defends* all targets in the subset $T(s) \subseteq \mathcal{T}$. Furthermore, each target $t \in \mathcal{T}$ has a *threshold requirement* $0 \leq q_t \leq 1$. In the SECURITY GAME problem, we want to (deterministically) purchase some resources, and randomly assign them to schedules so that every target $t \in \mathcal{T}$ is defended with probability at least q_t. We want to minimize the sum of the costs of the purchased resources.[1]

Consider an example. There are 4 targets, 3 schedules, and 2 different types of resources. Target t_1 (resp. t_4) needs to be defended with probability $2/3$

---

[1] Note that the unlimited availability of resources guarantees the existence of a feasible solution satisfying all the threshold requirements: simply purchase a sufficient number of resources of each type. This assertion holds provided each target can be defended by some schedule, and for each schedule, there is some resource that can be assigned to it. We will make these assumptions without any loss of generality.

(resp. $1/3$), whereas each of the targets in $\{t_2, t_3\}$ has a threshold requirement of 1. Any resource of type θ_1 costs 2, and any resource of type θ_2 costs 3. A resource of type θ_1 (resp. θ_2) can be assigned to at most one schedule in the set $\{s_1, s_2\}$ (resp. $\{s_2, s_3\}$). Finally, whenever it has some resource assigned to it, schedule s_1 defends target t_1; schedule s_2 defends both the targets t_2, t_3; and schedule s_3 defends both the targets t_3, t_4. In terms of notations, we have

$$\mathcal{T} = \{t_1, t_2, t_3, t_4\}, \ \mathcal{S} = \{s_1, s_2, s_3\}, \ \Theta = \{\theta_1, \theta_2\}$$
$$q_{t_1} = 2/3, \ q_{t_2} = q_{t_3} = 1, \ q_{t_4} = 1/3$$
$$c(\theta_1) = 2, \ c(\theta_2) = 3$$
$$S(\theta_1) = \{s_1, s_2\}, \ S(\theta_2) = \{s_2, s_3\}$$
$$T(s_1) = \{t_1\}, \ T(s_2) = \{t_2, t_3\}, \ T(s_3) = \{t_3, t_4\}$$

The optimal solution will purchase one resource r_1 of type θ_1 and one resource r_2 of type θ_2 so that $\theta_{r_1} = \theta_1$, $\theta_{r_2} = \theta_2$; thereby incurring a total cost of $2 + 3 = 5$. Next, with probability $2/3$, it will *simultaneously* assign resource r_1 to schedule s_1 and resource r_2 to schedule s_2; and with the remaining probability $1 - 2/3 = 1/3$, it will *simultaneously* assign resource r_1 to schedule s_2 and resource r_2 to schedule s_3. It is important to note how the optimal solution correlates the random assignments of the resources to schedules.

The SECURITY-GAME problem is a generalization of SET-COVER. Consider an instance of the SECURITY-GAME problem where the threshold requirements of all the targets are equal to 1, and there is only one resource-type. In this case, the task of finding the optimal solution is equivalent to finding a minimum cardinality subset of schedules to defend all the targets, which is exactly the SET-COVER problem. As a consequence, the SECURITY-GAME problem is NP-hard and unless NP has slightly superpolynomial time algorithms, we cannot even approximate it to a factor better than $O(\log |\mathcal{T}|)$ [6]. In the next section, we give an $O(\log |\mathcal{T}| + \log |\mathcal{S}|)$ approximation algorithm.

3   Approximation Algorithm

First, we partition the set of targets into two groups depending on their threshold requirements. Define

$$\mathcal{T}_{big} = \{t \in \mathcal{T} \ : \ 1/e < q_t \leq 1\} \tag{1}$$
$$\mathcal{T}_{small} = \{t \in \mathcal{T} \ : \ q_t \leq 1/e\} \tag{2}$$

We deal with the two subsets separately. In Section 3.1, we consider the subproblem where we need to defend *only* the subset $\mathcal{T}_{big}$ of targets, and give an $O(\log |\mathcal{T}|)$ approximation algorithm for this task. On the other hand, Section 3.2 deals with the subproblem where we have to defend *only* the targets $t \in \mathcal{T}_{small}$. For this task, we present an $O(\log |\mathcal{T}| + \log |\mathcal{S}|)$ approximation algorithm. Finally, we take the union of the two solutions, and this results in an $O(\log |\mathcal{T}| + \log |\mathcal{S}|)$ approximation for the SECURITY-GAME problem.

3.1 Targets with Big Threshold Requirements

Let OPT_{big} denote the minimum-cost solution that only defends the targets in the subset $\mathcal{T}_{big} \subseteq \mathcal{T}$ according to their threshold requirements, and ignores the remaining targets in $\mathcal{T}_{small} = \mathcal{T} \setminus \mathcal{T}_{big}$. We now derive an LP-relaxation.

$$\min \quad \sum_{\theta \in \Theta} c(\theta) \sum_{s \in S(\theta)} w(\theta, s) \tag{3}$$

$$\text{s.t.} \quad \sum_{\theta \in \Theta} \sum_{s \in S(\theta): t \in T(s)} w(\theta, s) \geq q_t, \quad \forall t \in \mathcal{T}_{big} \tag{4}$$

$$w(\theta, s) \geq 0 \qquad\qquad , \quad \forall \theta \in \Theta, s \in S(\theta) \tag{5}$$

Let $\mathcal{R}^*$ be the set of resources purchased by OPT_{big}. For all $r \in \mathcal{R}^*, s \in S(\theta_r)$, let y_{rs}^* be the probability that OPT_{big} assigns resource r to schedule s. Recall that the type of a resource r is denoted by θ_r. Define

$$w^*(\theta, s) = \sum_{r \in \mathcal{R}: \theta_r = \theta} y_{rs}^*$$

It is easy to verify that the $w^*(\theta, s)$ values are a feasible solution to LP (3). Constraint (4) holds since OPT_{big} defends every $t \in \mathcal{T}_{big}$ with probability at least q_t and by the union bound, the left hand side of the constraint is an overestimate of the probability with which target t is defended. The relaxation assumes that the resources can be purchased partially and hence, the objective value is at most the total cost incurred by OPT_{big}. This leads us to Lemma 1.

Lemma 1. *LP (3) gives a lower bound on the total cost incurred by* OPT_{big}.

Consider the following linear program (6). It replaces the right hand of Constraint (4) by 1. Recall that all targets in the subset $\mathcal{T}_{big} \subseteq \mathcal{T}$ have $q_t \geq 1/e$. Therefore, if we solve LP (3) optimally and multiply every $w(\theta, s)$ by e, then we get a feasible solution to LP (6). However, the objective value also increases by a factor of e. Combining this observation with Lemma 1, we obtain Fact 1.

$$\min \quad \sum_{\theta \in \Theta} c(\theta) \sum_{s \in S(\theta)} w(\theta, s) \tag{6}$$

$$\text{s.t.} \quad \sum_{\theta \in \Theta} \sum_{s \in S(\theta): t \in T(s)} w(\theta, s) \geq 1, \quad \forall t \in \mathcal{T}_{big} \tag{7}$$

$$w(\theta, s) \geq 0 \qquad\qquad , \quad \forall \theta \in \Theta, s \in S(\theta) \tag{8}$$

Fact 1. *The optimal objective value of LP (6) is at most* $O(1)$ *(specifically,* e*) times the cost incurred by* OPT_{big}.

We note that LP (6) is an LP-relaxation for the SET-COVER problem, where the targets $t \in \mathcal{T}_{big}$ behave like elements that have to be covered, and each pair (θ, s) acts like a set $T(s) \cap \mathcal{T}_{big}$ having a cost $c(\theta)$. Hence the greedy algorithm described in Figure 1 gives a $O(\log |\mathcal{T}|)$ approximation [7] to LP (6). Hence, Fact 1 implies the following Theorem 1.

Initialize $D \leftarrow \emptyset$, $F \leftarrow \emptyset$.
While $D \neq T_{big}$ do
        Find an ordered pair $(\theta', s') \in \arg\max_{\theta \in \Theta, s \in S(\theta)} |(T(s) \cap T_{big}) \setminus D|/c(\theta)$
        $F \leftarrow F \cup \{(\theta', s')\}$
        $D \leftarrow D \cup (T(s') \cap T_{big})$
For all $(\theta, s) \in F$ do
        Buy a resource of type θ, and *deterministically* assign it to schedule s.

Fig. 1. Greedy algorithm for targets with big threshold requirements. It defends every target $t \in T_{big}$ with probability 1.

Theorem 1. *The greedy algorithm described in Figure (1) gives an $O(\log |T|)$ approximation to OPT_{big}.*

3.2 Targets with Small Threshold Requirements

Let SG_{small} be the problem where we must defend each target $t \in T_{small}$ with probability at least q_t, but we are free to ignore the remaining targets in T_{big}. A *solution* to the SG_{small} problem purchases some resources and randomly assigns them to schedules so that each target $t \in T_{small}$ is defended with probability q_t.

Definition 1. *Given any solution to the SG_{small} problem, every purchased resource r can be associated with an assignment vector. It is a vector with $|S|$ components, where the value of component s equals y_{rs} if $s \in S(\theta_r)$ and is zero otherwise. Here, y_{rs} is the probability that resource r is assigned to schedule s.*

Lemma 2 shows that we can restrict our attention to a subset of feasible solutions to the SG_{small} problem. Specifically, we focus on those solutions where the values of all components of the assignment vectors come from a discrete set.

Lemma 2. *Let OPT be the minimum-cost solution to the SG_{small} problem. There exists a solution DISCRETE-OPT to the SG_{small} problem such that:*

1. *The cost incurred by DISCRETE-OPT is at most 2 times the cost of OPT.*
2. *For every resource purchased by DISCRETE-OPT, all the components of the corresponding assignment vector are integral multiples of $1/|S|^2$.*

Fix any target $t \in T_{small}$ with $0 < q_t < 1/|S|^2$. The solution DISCRETE-OPT defends this target according to its threshold requirement. Thus, DISCRETE-OPT purchases some resource of type θ, and assigns it with *non-zero* probability to some schedule $s \in S(\theta)$ such that $t \in T(s)$. However, the probability of assigning the resource to schedule s is an integral multiple of $1/|S|^2$ (Lemma 2), and hence the target t is defended with probability at least $1/|S|^2$.

Corollary 1. *The solution DISCRETE-OPT defends each target $t \in T_{small}$ with probability at least $\min(q_t, 1/|S|^2)$.*

Let $\mathcal{P}$ denote the set of all assignment vectors that have been discretized according to Lemma 2. More formally, the set $\mathcal{P}$ consists of all possible $|\mathcal{S}|$-tuples where the value of each component is an integral multiple of $1/|\mathcal{S}|^2$, and the sum of the values of all the components is at most 1. For all $\boldsymbol{p} \in \mathcal{P}, s \in \mathcal{S}$, let $\boldsymbol{p}(s)$ denote the component of the assignment vector $\boldsymbol{p}$ corresponding to schedule s.

Suppose that a resource r cannot be assigned to some schedule s, that is, $s \notin S(\theta_r)$, and a valid solution assigns the resource r to different schedules with probabilities that are given by the components of the vector $\boldsymbol{p} \in \mathcal{P}$. In this case, we must have $\boldsymbol{p}(s) = 0$. Definition 2 formalizes this concept.

Definition 2. *An assignment vector $\boldsymbol{p} \in \mathcal{P}$ is* feasible *for a resource-type $\theta \in \Theta$ if $\boldsymbol{p}(s) = 0$ for all schedules $s \in \mathcal{S} \setminus S(\theta)$. Define $\mathcal{P}_\theta$ to be the set of all feasible assignment vectors for resource type $\theta \in \Theta$.*

Now we present an *Integer* Program to bound the cost of DISCRETE-OPT.

$$\min \quad \sum_{\theta \in \Theta, \boldsymbol{p} \in \mathcal{P}_\theta} c(\theta) x(\theta, \boldsymbol{p}) \tag{9}$$

$$\text{s.t.} \quad \sum_{\theta \in \Theta, \boldsymbol{p} \in \mathcal{P}_\theta} \sum_{s:\, t \in T(s)} \eta(\boldsymbol{p}, s)\, x(\theta, \boldsymbol{p}) \geq \lambda_t, \quad \forall t \in \mathcal{T}_{small} \tag{10}$$

$$x(\theta, \boldsymbol{p}) \in \mathbb{N} \quad , \quad \forall \theta \in \Theta,\, \boldsymbol{p} \in \mathcal{P}_\theta \tag{11}$$

IP (9) introduces some new notation. In Constraint (11), the set of all nonnegative integers is denoted by $\mathbb{N}$. In Constraint (10), we have

$$\lambda_t = \lceil e \times q_t \times |\mathcal{S}|^2 \rceil \qquad \text{for all } t \in \mathcal{T}_{small} \tag{12}$$

$$\eta(\boldsymbol{p}, s) = \boldsymbol{p}(s) \times |\mathcal{S}|^2 \qquad \text{for all } \boldsymbol{p} \in \mathcal{P},\, s \in \mathcal{S} \tag{13}$$

By definition, each $\boldsymbol{p}(s) \in [0, 1]$ is an integral multiple of $1/|\mathcal{S}|^2$, and $q_t \in [0, 1/e]$ for all $t \in \mathcal{T}_{small}$. These observations lead to the the following fact.

Fact 2. *IP (9) is a covering integer program where all the coefficients in the constraint data, that is, all the values of $\eta(\boldsymbol{p}, s)$ and λ_t, are integers lying between 0 and $|\mathcal{S}|^2$.*

Lemma 3. *The optimal objective value of the Integer Program (9) is at most 4 times the cost incurred by DISCRETE-OPT.*

We now describe some intuitions behind IP (9). The variable $x(\theta, \boldsymbol{p})$ denotes the number of purchased resources that satisfy *both* of the following conditions.

1. The resource is of type $\theta \in \Theta$, *and*
2. For all $s \in \mathcal{S}$, the probability that the resource is assigned to schedule s is given by $\boldsymbol{p}(s)$.

Each resource of type θ costs $c(\theta)$. Therefore, summing over all possible resource types and feasible assignment vectors, we see that the total cost is given by the objective function. We now proceed towards verifying Constraint (10). Applying union-bound, we can show that the left hand side of Constraint (10) is at least $|\mathcal{S}|^2$ times the probability of defending target t. Glossing over some of the details, the constraint holds since, the right hand side is roughly equal to $|\mathcal{S}|^2$ times the probability of defending target t.[2]

---

1.   Let $\boldsymbol{\delta}$ denote a vector with $|\mathcal{T}_{small}|$ components, where $\boldsymbol{\delta}(t)$ gives
      the value of the component corresponding to target $t \in \mathcal{T}_{small}$.
2.   FOR ALL $t \in \mathcal{T}_{small}$:   Initialize $\boldsymbol{\delta}(t) \leftarrow \lambda_t$.
3.   FOR ALL $\theta \in \Theta$, $\boldsymbol{p} \in \mathcal{P}_\theta$:   Initialize $X(\theta, \boldsymbol{p}) \leftarrow 0$.
4.   WHILE $\boldsymbol{\delta} \neq 0$ DO
5.   FOR ALL $\boldsymbol{p} \in \mathcal{P}$, and $t \in \mathcal{T}_{small}$
          $$\Delta_{Cov}(\boldsymbol{p}, \boldsymbol{\delta}, t) \leftarrow \min\left(\boldsymbol{\delta}(t), \sum_{s:\, t \in T(s)} \boldsymbol{p}(s) \times |\mathcal{S}|^2\right).$$
6.   FOR ALL $\boldsymbol{p} \in \mathcal{P}$: $\Delta_{Cov}(\boldsymbol{p}, \boldsymbol{\delta}) \leftarrow \sum_{t \in \mathcal{T}_{small}} \Delta_{Cov}(\boldsymbol{p}, \boldsymbol{\delta}, t)$.
7.   Find some $(\theta, \boldsymbol{p}) \in \arg\max_{\theta \in \Theta, \boldsymbol{p} \in \mathcal{P}_\theta} \left\{\Delta_{Cov}(\boldsymbol{p}, \boldsymbol{\delta})/c(\theta)\right\}$.
8. $X(\theta, \boldsymbol{p}) \leftarrow X(\theta, \boldsymbol{p}) + 1$.
9.   FOR ALL $t \in \mathcal{T}_{small}$: $\boldsymbol{\delta}(t) \leftarrow \boldsymbol{\delta}(t) - \Delta_{Cov}(\boldsymbol{p}, \boldsymbol{\delta}, t)$.
10. Return the $X(\theta, \boldsymbol{p})$ values for all $\theta \in \Theta$, $\boldsymbol{p} \in\in \mathcal{P}_\theta$.

Fig. 2. Dobson's Algorithm applied to LP (9)

If a covering *integer* program has a constraint matrix with integral entries, then a simple greedy heuristic returns a logarithmic approximation to the *integral* optimum (Dobson [4]). Fact 2 tells us that we can apply Dobson's heuristic to IP (9). A simple intuition behind the algorithm (Figure 2) comes from an alternate way of viewing the problem: Each target t has to be *covered* by a threshold amount λ_t. The total *coverage* required is $\sum_{t \in \mathcal{T}_{small}} \lambda_t$. This coverage can be achieved by incrementing the *columns* $\{(\theta, \boldsymbol{p})\}$, where each column $(\theta, \boldsymbol{p})$ corresponds to the variable $x(\theta, \boldsymbol{p})$. We want to increment the columns so that the required coverage is attained at minimum cost.

At the beginning of a typical iteration of the WHILE loop (Steps 4–9), the value of $\boldsymbol{\delta}(t)$ equals the *remaining coverage* required for target t before we can attain its threshold λ_t. If we increment a column $(\theta, \boldsymbol{p})$ by 1, then the cost of our solution will increase by $c(\theta)$, and at the same time, the coverage of target t will increase (Step 5) by the amount $\Delta_{Cov}(\boldsymbol{p}, \boldsymbol{\delta}, t)$. Hence, the increase in total coverage of all the targets (Step 6) will be equal to $\Delta_{Cov}(\boldsymbol{p}, \boldsymbol{\delta})$. Let us term this quantity $\Delta_{Cov}(\boldsymbol{p}, \boldsymbol{\delta})$ as *marginal coverage*. The algorithm myopically selects the column that has the maximum marginal coverage to cost ratio (Step 7), and increments that column by 1 (Step 8). The remaining coverage required for all the targets are adjusted accordingly (Step 9). The WHILE loop terminates (Step 4) when $\boldsymbol{\delta} = 0$, that is, when all the targets in $\mathcal{T}_{small}$ have been covered up to their corresponding thresholds.

---

[2] To be more precise, the RHS is equal to $\lceil e \times q_t \times |\mathcal{S}|^2 \rceil$. While converting the IP solution to a feasible (random) assignment of resources to schedules, the final algorithm (Figure 3) looses a factor of e in the probability of defending any target $t \in \mathcal{T}_{small}$.

Note that IP (9) contains exponentially many variables $x(\theta, \boldsymbol{p})$. This follows from the fact that the set $\mathcal{P}$ of possible assignment vectors is exponential in size. Hence, we have to prove that Dobson's algorithm can be implemented in polynomial time. We also need to establish a bound on the approximation ratio.

Lemma 4. *Dobson's algorithm (Figure 2) can be used to solve the Integer Program (9). Although IP (9) contains exponentially many variables, an approximate version of Dobson's algorithm can be implemented in polynomial time. It returns a feasible solution to IP (9), where each variable $x(\theta, \boldsymbol{p})$ is assigned a nonnegative integral value $X(\theta, \boldsymbol{p})$. The solution satisfies two properties.*

1. *The objective value of the solution is at most $O(\log |\mathcal{T}| + \log |\mathcal{S}|)$ times the optimal objective value of IP (9).*
2. *The number of variables taking nonzero values are polynomially bounded.*

Proof (Sketch). The approximation ratio of Dobson's algorithm [4] grows logarithmically with the maximum column sum of the coefficient matrix. Recall that (Fact 2) each $\eta(\boldsymbol{p}, s)$ is an integer lying between 0 and $|\mathcal{S}|^2$, and the number of constraints in IP (9) is at most $|\mathcal{T}|$. Thus, in case of IP (9), the maximum column sum is upper bounded by $|\mathcal{S}|^2 \times |\mathcal{T}|$. Hence the approximation ratio is given by $O(\log(|\mathcal{S}|^2 |\mathcal{T}|)) = O(\log |\mathcal{S}| + \log |\mathcal{T}|)$. Next, we will show that an approximate version of Dobson's algorithm can be implemented in polynomial time, and asymptotically, it gives the same approximation ratio of $O(\log |\mathcal{S}| + \log |\mathcal{T}|)$.

It is sufficient to discuss the implementations of Step 3, the WHILE loop (Steps 4–9) and Step 10. We implement Step 3 by implicitly assuming that all the $X(\theta, \boldsymbol{p})$ values have been initialized to zero. During the course of the algorithm, we keep track of only those $X(\theta, \boldsymbol{p})$ values that have been incremented at least once. Since each λ_t is an integer lying between 0 and $|\mathcal{S}|^2$ (Fact 2), the total coverage required of all targets is at most $|\mathcal{T}| \times |\mathcal{S}|^2$. Every iteration of the WHILE loop (Steps 4–9) contributes at least 1 towards this total coverage, and it increments exactly one $X(\theta, \boldsymbol{p})$. Therefore, at the termination of the WHILE loop, the number of nonzero $X(\theta, \boldsymbol{p})$ values is upper bounded by the polynomial $|\mathcal{T}| \times |\mathcal{S}|^2$. In Step 10, the algorithm returns only these nonzero $X(\theta, \boldsymbol{p})$ values, and all other variables are implicitly set to zero.

Regarding the WHILE loop (Steps 4–9), note that the marginal coverage $\Delta_{Cov}(\boldsymbol{p}, \boldsymbol{\delta})$, as a function of the assignment vector $\boldsymbol{p}$, is monotone and submodular. To be more precise, fix some resource type θ. Next, take any two assignment vectors $\boldsymbol{p}, \boldsymbol{p}' \in \mathcal{P}_\theta$ such $\boldsymbol{p}$ is dominated by $\boldsymbol{p}'$, that is, $\boldsymbol{p}(s) \leq \boldsymbol{p}'(s)$ for all $s \in \mathcal{S}$. Furthermore, suppose that $\sum_{s \in \mathcal{S}} \boldsymbol{p}'(s) < 1$. Fix some schedule $s^* \in S(\theta)$ and consider two new assignment vectors $\boldsymbol{p}_1, \boldsymbol{p}'_1 \in \mathcal{P}_\theta$ with the following properties. For all $s \in \mathcal{S} \setminus \{s^*\}$, we have $\boldsymbol{p}_1(s) = \boldsymbol{p}(s)$, and $\boldsymbol{p}'_1(s) = \boldsymbol{p}'(s)$. We also have $\boldsymbol{p}_1(s^*) = \boldsymbol{p}(s^*) + 1/|\mathcal{S}|^2$, and $\boldsymbol{p}'_1(s^*) = \boldsymbol{p}'(s^*) + 1/|\mathcal{S}|^2$. Now, submodularity of marginal coverage means that the next inequality will always be satisfied.

$$\Delta_{Cov}(\boldsymbol{p}'_1, \boldsymbol{\delta}) - \Delta_{Cov}(\boldsymbol{p}', \boldsymbol{\delta}) \leq \Delta_{Cov}(\boldsymbol{p}_1, \boldsymbol{\delta}) - \Delta_{Cov}(\boldsymbol{p}, \boldsymbol{\delta})$$

We can exploit this submodularity condition as follows. While implementing Step 7, suppose we have correctly guessed the resource type θ that maximizes the marginal coverage to cost ratio.[3] All we need to do is to find an assignment vector $p \in \mathcal{P}_\theta$ with maximum marginal coverage, subject to the constraints that each component of the vector p is an integral multiple of $1/|\mathcal{S}|^2$, and the sum of all the components is at most *one* (Lemma 2). This is equivalent to maximizing a monotone submodular function subject to a budget constraint [10]. A simple greedy algorithm is known to give a $(1 - 1/e)$ approximation for this problem.

To summarize, in polynomial time we can obtain a column (θ, p) that gives a constant approximation to the optimal ratio of marginal coverage to cost. Going through Dobson's proof [4], it is easy to verify the following statement. Even if we implement Step 7 in this approximate fashion, the algorithm will have the same asymptotic approximation guarantee. This concludes the proof. □

We are now ready to present our algorithm for the SG_{small} problem (Figure 3). First, we solve IP (9) according to Lemma 4. Let B^* denote the set of columns of IP (9) where the corresponding variable is set to some nonzero value, that is, $B^* = \{(\theta, p) : X(\theta, p) \neq 0\}$. For each $(\theta, p) \in B^*$, we purchase $X(\theta, p)$ resources of type θ and tag them with the assignment vector p. Finally, each purchased resource is randomly assigned to some schedule according to its assignment vector; and this process occurs *independently* of all other resources.

Solve IP (9) according to Lemma 4.
Define $\mathcal{P}_\theta^* = \{p \in \mathcal{P}_\theta : X(\theta, p) \neq 0\}$ for all $\theta \in \Theta$.
Let $\mathcal{R}_\theta$ be the set of resources of type θ that will be purchased.
Let $\mathcal{R} = \bigcup_{\theta \in \Theta} \mathcal{R}_\theta$ be the set of *all* resources that will be purchased.
Let $\mathcal{R}_{\theta,p} \subseteq \mathcal{R}_\theta$ denote the resources in $\mathcal{R}_\theta$ whose
assignment probabilities are specified by $p \in \mathcal{P}_\theta^*$.
For all resource-types $\theta \in \Theta$
$\quad |\mathcal{R}_\theta| = \sum_{p \in \mathcal{P}_\theta^*} X(\theta, p).$
$\quad$ For all $p \in \mathcal{P}_\theta^*$: $\quad |\mathcal{R}_{\theta,p}| = X(\theta, p).$
Randomly assign each resource $r \in \mathcal{R}_{\theta,p}$ to schedules, according to the
assignment probabilities specified by p, independently of all other resources.

Fig. 3. Approximation Algorithm for the SG_{small} Problem

Theorem 2. *The algorithm described in Figure 3 gives an* $O(\log |\mathcal{T}| + \log |\mathcal{S}|)$ *approximation to the* SG_{small} *problem.*

Proof. If we purchase the resources according to Figure 3, then the total cost is equal to the objective value of the IP solution returned by Lemma 4. Now Lemma 2, Lemma 3 and Lemma 4 imply that this cost is at most $O(\log |\mathcal{T}| + \log |\mathcal{S}|)$ times the cost incurred by OPT. It remains to show that the solution defends all the targets in $\mathcal{T}_{small}$ according to their threshold requirements. Fix some target $t \in \mathcal{T}_{small}$ for the rest of this proof. Given any purchased resource $r \in \mathcal{R}$, let p_r be its assignment vector according to Figure 3. Since the $X(\theta, p)$ values

---

[3] Clearly, in $O(|\Theta|)$ time we can iterate over all possible resource types.

constitute a feasible solution to IP (9), we have that $\sum_{r \in \mathcal{R}} \sum_{s: t \in T(s)} \eta(\boldsymbol{p}_r, s) \geq \lambda_t$. Recall that $\eta(\boldsymbol{p}_r, s) = \boldsymbol{p}_r(s) |\mathcal{S}|^2$ and $\lambda_t = \lceil e \times q_t \times |\mathcal{S}|^2 \rceil \geq eq_t |\mathcal{S}|^2$. For all $r \in \mathcal{R}$, define $\phi_r(t) = \sum_{s: t \in T(s)} \boldsymbol{p}_r(s)$. Therefore, we get

$$\sum_{r \in \mathcal{R}} \phi_r(t) = \sum_{r \in \mathcal{R}} \sum_{s: t \in T(s)} \boldsymbol{p}_r(s) \geq eq_t \tag{14}$$

The probability that resource r does *not* defend target t is given by the expression $1 - \sum_{s: t \in T(s)} \boldsymbol{p}_r(s) = 1 - \phi_r(t)$. Since the event of assigning a resource to some (random) schedule occurs independently of other resources, the probability that *no resource* defends the target t is equal to $\prod_{r \in \mathcal{R}} (1 - \phi_r(t))$. Since $q_t \leq 1/e$,

$$\prod_{r \in \mathcal{R}} (1 - \phi_r(t)) \leq \prod_{r \in \mathcal{R}} \exp(-\phi_r(t)) = \exp\left(-\sum_{r \in \mathcal{R}} \phi_r(t)\right) \leq \exp(-eq_t)$$

Thus, the probability of defending target t is at least $1 - \exp(-eq_t) \geq q_t$.   □

Remark. We note that it is possible to devise a polynomial time algorithm that gives an $O(\log |\mathcal{T}|)$ approximation to the SG_{small} problem. We have to consider an exponential sized Linear Program that is similar to IP (9), solve it approximately using an approximate separation oracle for its dual, and then directly employ randomized rounding. However, the running time of such an algorithm might become prohibitive. We omit the details due to space constraints.

4   Conclusion

We investigated the security game problem when there is an unlimited supply of resources that can be purchased at a cost. We designed an algorithm for (deterministically) purchasing some resources at minimum cost, and then randomly assigning them to schedules so that each target is defended with at least a certain probability. The algorithm is efficient and gives a logarithmic approximation.

Since this problem is a generalization of SET-COVER, we cannot get a sublogarithmic approximation ratio. However, if each target has at most two schedules that are capable of defending it (a generalization of the VERTEX-COVER problem) and resources are homogeneous, then we can get a constant factor approximation algorithm. We omit the proof due to lack of space. We leave open the questions of exploring other settings with better approximation guarantees, and investigating the fixed parameter tractability of the problem.

Acknowledgements. The authors thank Dmytro Korzhyk and Ronald Parr for several helpful discussions. This research was supported by NSF under award numbers IIS-0812113, IIS-0953756, CCF-1101659, CCF-0745761, CCF-1008065, a gift from Cisco, by Conitzer's Alfred P. Sloan Research Fellowship, and by Munagala's Alfred P. Sloan Research Fellowship.

References

1. An, B., Pita, J., Shieh, E., Tambe, M., Kiekintveld, C., Marecki, J.: GUARDS and PROTECT: next generation applications of security games. ACM SIGecom Exchanges 10(1), 31–34 (2011)
2. Chen, X., Deng, X.: Settling the complexity of two-player Nash equilibrium. In: FOCS, pp. 261–272 (2006)
3. Daskalakis, C., Goldberg, P.W., Papadimitriou, C.H.: The complexity of computing a Nash equilibrium. In: STOC, pp. 71–78 (2006)
4. Dobson, G.: Worst-case analysis of greedy heuristics for integer programming with nonnegative data. Mathematics of Operations Research 7(4), 515–531 (1982)
5. Etessami, K., Yannakakis, M.: On the complexity of Nash equilibria and other fixed points. SIAM J. Comput. 39(6), 2531–2597 (2010)
6. Feige, U.: A threshold of $\ln n$ for approximating set-cover. Journal of the ACM 45(4), 634–652 (1998)
7. Hochbaum, D.: Approximation Algorithms for NP-hard Problems. PWS Publishing Company (1997)
8. Kiekintveld, C., Jain, M., Tsai, J., Pita, J., Ordóñez, F., Tambe, M.: Computing optimal randomized resource allocations for massive security games. In: AAMAS, pp. 689–696 (2009)
9. Korzhyk, D., Conitzer, V., Parr, R.: Complexity of computing optimal Stackelberg strategies in security resource allocation games. In: AAAI, pp. 805–810 (2010)
10. Nemhauser, G., Wolsey, L., Fisher, M.: An analysis of approximations for maximizing submodular set functions. Mathematical Programming 14(1), 265–294 (1978)
11. Pita, J., Jain, M., Ordóñez, F., Portway, C., Tambe, M., Western, C.: Using game theory for Los Angeles Airport security. AI Magazine 30(1), 43–57 (2009)
12. Pita, J., Jain, M., Western, C., Portway, C., Tambe, M., Ordóñez, F., Kraus, S., Parachuri, P.: Deployed ARMOR protection: The application of a game-theoretic model for security at the Los Angeles International Airport. In: AAMAS - Industry and Applications Track, pp. 125–132 (2008)
13. Tsai, J., Rathi, S., Kiekintveld, C., Ordóñez, F., Tambe, M.: IRIS - a tool for strategic security allocation in transportation. In: AAMAS - Industry Track, pp. 37–44 (2009)

On Allocations with Negative Externalities

Sayan Bhattacharya, Janardhan Kulkarni,
Kamesh Munagala, and Xiaoming Xu

Department of Computer Science
Duke University
Durham NC 27708-0129
{bsayan,kulkarni,kamesh,xiaoming}@cs.duke.edu

Abstract. We consider the problem of a monopolist seller who wants to sell some items to a set of buyers. The buyers are strategic, unit-demand, and connected by a social network. Furthermore, the utility of a buyer is a decreasing function of the number of neighbors who do not own the item. In other words, they exhibit negative externalities, deriving utility from being *unique* in their purchases. In this model, any fixed setting of the price induces a sub-game on the buyers. We show that it is an exact potential game which admits multiple pure Nash Equilibria. A natural problem is to compute those pure Nash equilibria that raise the most and least revenue for the seller. These correspond respectively to the most optimistic and most pessimistic revenues that can be raised.

We show that the revenues of *both* the best and worst equilibria are hard to approximate within sub-polynomial factors. Given this hardness, we consider a relaxed notion of pricing, where the price for the same item can vary within a constant factor for different buyers. We show a 4-approximation to the pessimistic revenue when the prices are relaxed by a factor of 4. The interesting aspect of this algorithm is that it uses a linear programming relaxation that only encodes part of the strategic behavior of the buyers in its constraints, and rounds this relaxation to obtain a starting configuration for performing relaxed Nash dynamics. Finally, for the maximum revenue Nash equilibrium, we show a 2-approximation for bipartite graphs (without price relaxation), and complement this result by showing that the problem is NP-Hard even on trees.

1   Introduction

This paper considers pricing and allocations over a social network, when buyers derive utility from being *unique* in their purchase. Such *negative externalities* arise in several consumer goods where buyers derive value from "showing off" the product to friends lacking it. Consider the following example. For many years, the DVD publication industry has utilized the so called "double-dipping" policy, a term for releasing multiple versions of the same movie on discs. A quick search for the movie "The Matrix" shows besides the original, there are "The Matrix Revisited", "The Matrix: Platinum Limited Edition Collector's Set", and "The Ultimate Matrix Collection". It is often the case that these

N. Chen, E. Elkind, and E. Koutsoupias (Eds.): WINE 2011, LNCS 7090, pp. 25–36, 2011.
© Springer-Verlag Berlin Heidelberg 2011

editions have the same core content (in this case, the movie), while they differ in the some "unique" different material that is packed with the disc, for instance, sound tracks of the music, toy character of the figures in the movie, etc. And it is often the case that the price discrepancy between these versions outweighs the "real value" or intrinsic value the extra material provides. An incentive as observed in [21] is that "..the extra 20 or 30 bucks ... is discreet enough to display without geek alarms flashing and whirling anytime I have friends and/or new people over". The same marketing policy exists in many other different industries, book publishing, expensive electronic gadgets to name a few.

The model for negative externalities we study in the paper is simple: The buyers are unit-demand and connected by a social network that we model as a graph; the edges in the graph represent friendships. There are two types of items, each with unlimited supply - a "cheap" item and an "expensive" item. Each user has an intrinsic value for each type of item; however, buyers of the expensive item also derive additional *extrinsic* utility from friends who only possess the cheaper item. We assume the edges in the graph are weighted, so that the extrinsic utility is the sum of the weights of edges leading to friends possessing the cheaper item.

In this model, a monopolist wishes to price the items to maximize revenue. Any fixed setting of prices induces a sub-game on the buyers where they decide which item to purchase in order to maximize their individual utility. If a buyer buys the cheaper item, then she gets its intrinsic valuation; else if she buys the more expensive item, she gets the sum of its intrinsic and extrinsic valuations. The whole process can be viewed as a strategic game occurring in two rounds: The seller commits to the two prices in advance, and then each buyer simultaneously decides which item to purchase. We term this the PRICING GAME, and investigate two natural questions for a fixed setting of prices:

- What is the *pessimistic* Nash equilibrium, *i.e.*, one that raises minimum revenue? This gives a guarantee on the revenue the seller raises regardless of the behavior of the buyers; a risk-averse seller will choose prices at which this revenue is as large as possible.
- What is the *best* Nash equilibrium, *i.e.*, one that raises largest revenue? This will give the most optimistic view of the buyers' behavior, and is appropriate when the seller can recommend which item to buy via targeted advertising.

Given efficient algorithms for the above problems, the seller can treat them as subroutines while iterating over all possible prices. This will help her set the prices in such a way that maximizes the pessimistic (resp. optimistic) revenue.

Sequential Pricing: Though our model is motivated by negative externalities, the same model arises in an entirely different context that is well-studied in economics. In *sequential pricing*, buyers derive positive utility from neighbors who bought the item earlier in time, and strategically decide when to buy the item to maximize their utility. The goal of the seller is to decide the prices to set for each stage, with later stages having higher prices, so that the resulting subgame among the buyers raises large revenue. Such a model of externality is motivated by buyers deciding to wait on a purchase if a lot of her friends buy now, since waiting will lead to more product feedback and hence raise utility. This is a well-studied problem: Exact and approximation algorithms are known for models with no network effects [11, 7, 9, 13, 20], non-atomic buyers [5, 2], or with non-strategizing buyers [14, 1, 3]. Our formulation exactly models two-stage sequential pricing for a single item with strategic, atomic buyers on a network: Buyers derive both an intrinsic utility from the item as well as linear utility from neighbors who have bought the item in the previous stage. To map this to our problem, simply note that the buyers in the second stage of sequential pricing correspond to buyers of the more expensive item, and those in the first stage, to the cheaper item. As we show next, the presence of atomic, strategizing buyers on a network causes the problem to become structurally different from previously considered versions.

1.1   Our Results

We first show that the PRICING GAME is an exact potential game, and hence admits to a pure Nash Equilibrium (Theorem 1). As a corollary, the Nash dynamics converges in poly-time when the edge-weights are polynomially bounded integers. It is straightforward to check that not only are there multiple pure Nash equilibria in this game, but also they have vastly different revenue properties for the seller. This motivates us to focus on computing equilibria with certain optimality properties. In particular, we show the following results:

- For the Pessimistic (minimum) Nash equilibrium problem, we show that its revenue is NP-HARD to approximate within a factor of $O(n^{1/3-\epsilon})$, where n is the number of nodes in the social network (Theorem 2). Interestingly, our hardness results hold even when the buyers derive significantly large intrinsic utility compared to the externalities.
- In view of the above lower bound, we focus on a δ-relaxed notion of Nash equilibria,[1] where the seller posts different prices to different buyers of the expensive item, but these prices are within a factor of δ of each other. We give an algorithm (Theorem 6) for computing a 4-relaxed NE that is a 4-approximation to the pessimistic revenue. The algorithm uses a linear program relaxation to bound the revenue of the pessimistic NE. It is interesting to note that this linear program only encodes the constraint that the buyers do not deviate from buying the cheaper item to the more expensive item, allowing the buyers of the more expensive item to deviate. Based on the ideas from vertex cover, we present a rounding scheme for this LP, and use the resulting solution as a starting point for performing relaxed Nash dynamics.
- We show (Theorem 7) that the Best Nash equilibrium problem is also hard to approximate within a factor of $O(n^{1/3-\epsilon})$ by reduction from the independent set problem. However, in contrast with independent set, our problem is NP-Hard even on trees (Theorem 8). We finally present a 2-approximation to maximum revenue when the underlying network is bipartite (Theorem 9).

---

[1] Refer to the discussion in the beginning of Section 3.1 for more details.

We emphasize that all the three hardness results (Theorems 2, 7, 8) hold even if the edge-weights are small integers so that Nash dynamics converges quickly to *some* Nash equilibrium. In contrast, both the algorithms presented in this paper - a 4 approximation with relaxed prices to the revenue of the Pessimistic Nash Equilibrium, and a 2 approximation to the revenue of the Best Nash Equilibrium on bipartite graphs - run in polynomial time for arbitrary edge-weights.

Remark. All the missing proofs appear in the full version of this paper.

1.2 Related Work

Viral Marketing: Marketing strategies in social networks have been studied extensively, starting with the seminal paper by Kempe *et. al.* [16]. Several recent papers [2, 14, 1, 3, 12] consider pricing and auction design in social networks when buyers exhibit *positive* network externalities, meaning they derive positive utility from neighbors possessing the product. Our main contribution is to show that the structure induced by uniqueness and negative externalities is very different from positive externalities. In particular, we show that computing Nash Equilibria with desirable revenue properties ends up being hard to approximate, and we need to consider relaxed notions of equilibria for positive results.

Potential Games: Our problem is a special case of potential (or congestion) games [18], which always admit to pure Nash equilibria. Typically, these games have been studied from two perspectives - how bad can the welfare of the resulting equilibria be compared to optimal social welfare (*price of anarchy*) [17, 19]; and how quickly does the Nash dynamics converge [10, 8, 4]. The literature on price of anarchy aims to find worst case loss in efficiency due to strategizing across all game instances. In contrast, our focus is on *computing* the best or worst equilibria for a *specific* input network. It is relatively easy to check that the *worst-case* revenue of these equilibria across all networks can be $\Omega(n)$ factor off from the optimal revenue that can be raised with non-strategizing buyers; and furthermore, the revenues of best and worst equilibria can be separated by a factor of $\Omega(n)$. For general potential games, computing a pure Nash equilibrium is PLS-complete [10], and most literature in this domain have tried to circumvent this hardness by considering the notion of *approximate* Nash equilibrium [6]. Our focus is not on PLS-completeness; in fact, our problem is interesting (and hard) even in the regime where the Nash dynamics converges in polynomial time.

2 Notations and Preliminaries

There is a *cheap* item and an *expensive* item, each of them being available in unlimited supply. The buyers are unit demand. Consider an undirected graph $G = (V, E)$ with a weight function over the edges $w : E \to \mathbb{R}^+$. Each node $i \in V$ denotes a buyer, there is an edge $(i, j) \in E$ if buyers i, j are friends, and the weight of their friendship is given by the quantity $w(i, j)$. The price of the cheap (resp. expensive) item is set at p_1 (resp. p_2). Naturally, the price of the cheap item should be less than that of the expensive one, that is, $p_1 \leq p_2$. The pricing is *uniform* in the sense that the same item is offered at the same price to all the nodes. Say that a node is *black* (resp. *white*) if she buys the cheap (resp. expensive) item, while paying an amount p_1 (resp. p_2).

Definition 1. *The* externality $Ext(i)$ *of node i is the total weight of the edges it shares with black nodes. Thus, we have $Ext(i) = \sum_{(i,j) \in E:\, j \text{ is black}} w(i,j)$.*

Definition 2. *The* weighted degree $\mathcal{D}(i)$ *of node i is the total weights of the edges incident to it. Thus, we have $\mathcal{D}(i) = \sum_{(i,j) \in E} w(i,j)$.*

Any buyer purchasing the cheap (resp. expensive) item gets an *intrinsic* valuation of a (resp. b), where $0 \leq a \leq b$, and a, b are publicly known and fixed in advance. In addition, any buyer purchasing the expensive item gets an *extrinsic* valuation that increases as more of her friends purchase the cheap item, and her total valuation equals the sum of intrinsic and extrinsic valuations. To be more specific, the valuation of every black node is equal to a; whereas the valuation of a white node i is given by the expression $b + Ext(i)$.

The buyers have *quasi-linear* utilities. If a node i is black, then her utility is $U(i) = a - p_1$. Else if the node i is white, then her utility is $U(i) = b + Ext(i) - p_2$. And if node i does not purchase any item, then she gets *zero* utility. We assume $p_1 \leq a$ and $p_2 \leq b$. These two inequalities guarantee that every buyer purchases the item, or equivalently, every node is colored either black or white.

In this paper, we consider the setting where we are given the values of p_1, p_2, a and b. This induces a normal form game between the buyers, and we term it as the Pricing Game. The strategy of each node consists of choosing whether to be colored black or white. A coloring of the nodes defines a strategy profile, and in a Nash equilibrium, the color chosen by each node is a best response to the colors chosen by other nodes. Thus, node i is colored black iff:

$$a - p_1 \geq b + Ext(i) - p_2 \Rightarrow Ext(i) \leq p_2 - p_1 + a - b = \Delta \text{ (say)}$$

Similarly, node i is colored white if and only if $Ext(i) \geq \Delta$.[2]

The game can therefore be summarized as follows.

Pricing Game. In any pure Nash equilibrium, every node is colored either black or white, and each black (resp. white) node has an externality of at most (resp. at least) Δ. Let $B_{\mathcal{C}}$ (resp. $W_{\mathcal{C}} = V \setminus B_{\mathcal{C}}$) denote the set of black (resp. white) nodes under coloring $\mathcal{C}$. Each black (resp. white) node mays a payment of p_1 (resp. p_2). Thus:

$$\text{Seller's Revenue } = p_1 \times |B_{\mathcal{C}}| + p_2 \times |W_{\mathcal{C}}| = (\Delta + b - a) \times |W_{\mathcal{C}}| + p_1 \times |V_{\mathcal{C}}|$$

In the above equation, the second equality holds since $p_2 = p_1 + \Delta + b - a$.

---

[2] If $\Delta < 0$, then it is a dominant strategy for each buyer to purchase the more expensive item, and the pure NE degenerates to the case where all nodes are colored white. Throughout the rest of the paper, we will consider the generic setting where $\Delta \geq 0$.

Our first theorem shows that the PRICING GAME is an exact potential game. Associate each coloring $\mathcal{C}$ with a potential $\phi(\mathcal{C})$, given by the total weight of black-black edges plus Δ times the number of white nodes. In terms of notations,

$$\phi(\mathcal{C}) = \Delta \times |W_{\mathcal{C}}| + \sum_{(i,j)\in E: i,j\in B_{\mathcal{C}}} w(i,j)$$

Theorem 1. *The PRICING GAME is an exact potential game with potential function $\phi()$, and hence it admits a pure Nash equilibrium. If a, b, p_1, p_2 and all the edge weights $w(i,j)$ take integral values in the range $\{1, \ldots, \mu\}$, then the Nash dynamics converges within $\Theta(\mu|V|^2)$ steps.*

In this paper, we are interested in estimating the maximum (resp. minimum) revenue the seller may obtain from any pure Nash equilibrium. Towards this end, we define the following problems.

Problem 1 (BNE: Best Nash Equilibrium). Given an instance of the PRICING GAME, find a pure Nash equilibrium that generates the maximum revenue.

Problem 2 (PNE: Pessimistic Nash Equilibrium). Given an instance of the PRICING GAME, find a pure Nash equilibrium that generates the minimum revenue.

Our focus is *not* on studying Nash dynamics. Instead we ask the question: *Even if* the edge-weights are small integers so that Nash dynamics converges in polynomial time (Theorem 1), how hard is it to compute a pure Nash equilibrium with specific revenue properties? All our hardness results (Theorems 2, 7, 8) hold under this very natural scenario of small integer edge-weights. Our hardness results also hold when intrinsic valuations are large compared to each individual externality (the reductions set $a = b = 1$), and the seller wants to obtain non-negligible revenue from each item (the reductions set $p_1 = 1$). However, both the positive results in this paper - a 4 approximation with relaxed prices to the revenue of the Pessimistic Nash Equilibrium (Theorem 6), and a 2 approximation to the revenue of the Best Nash Equilibrium on bipartite graphs (Theorem 9) - do not make *any* assumption on the input, and hold for arbitrary edge-weights.

3   The Pessimistic Nash Equilibrium

First, we derive a strong inapproximability result for the Pessimistic Nash Equilibrium (PNE) problem by reducing it from Minimum Maximal Independent Set (MMIS). In the MMIS problem, we are given a graph, and the objective to find an independent set of minimum size such that every node in the graph is adjacent to at least one node in the independent set. Kann [15] shows that it is NP-HARD to get a poly-time $O(n^{1-\epsilon})$ approximation for MMIS [15].

Theorem 2. *It is NP-HARD to compute in poly-time a pure Nash equilibrium whose revenue is a $O(n^{1/3-\epsilon})$ approximation to the minimum revenue, where n is the number of nodes in the social network.*

Proof. Given an instance $G = (V, E)$ of the minimum maximal independent set problem, we reduce it to an instance of PNE as outlined below. The resulting weighted graph will be denoted by $\mathcal{G}$.

- Set $a = b = 1$, $p_1 = 1$, and $p_2 = |V|^2$, implying that $\Delta = (p_2 - p_1 + a - b) = \Theta(|V|^2)$. Consequently, in any pure Nash equilibrium of the PNE instance we construct, every black (resp. white) node should have an externality at most (resp. at least) Δ. Further, every black node makes a payment of 1, whereas every white node contributes $\Theta(|V|^2)$ towards total revenue.
- Start with $G = (V, E)$ and assign a weight of 3 to each edge $(i, j) \in E$.
- For all nodes $i \in V$:

  - Create a new set of nodes T_i of cardinality $(3\Delta - 1)$.
  - Partition T_i into four disjoint subsets as L_i, R_i, S_i and $\{\alpha_i, \tilde{\alpha}_i, \beta_i, \gamma_i, \delta_i\}$, where $|L_i| = |R_i| = |S_i| = \Delta - 2$.
  - Create a cycle connecting the nodes of L_i.
  - For all $u \in L_i, v \in R_i$, create an edge (u, v). For all $u \in L_i \cup R_i$, create the edges $(u, \gamma_i), (u, \delta_i)$. For all $u \in L_i$, create an edge (u, β_i). For all $u \in S_i$, create the edges $(u, \alpha_i), (u, i)$.
  - Create the edges $(\beta_i, \gamma_i), (\beta_i, \delta_i), (\alpha_i, \beta_i), (i, \alpha_i)$ and $(\alpha_i, \tilde{\alpha}_i)$.
  - Assign a weight of 2 to each of the edges (α_i, β_i) and $(\alpha_i, \tilde{\alpha}_i)$. Every other newly created edge gets a weight of 1.

Since all nodes in $S_i \cup \{\tilde{\alpha}_i\}$ have a weighted degree of 2, and $\Delta > 2$, we have:

Observation 1. *For all $i \in V$, the set of nodes $S_i \cup \{\tilde{\alpha}_i\}$ must be colored black in any pure Nash equilibrium.*

The next three lemmas will be crucial in deriving the lower bound.

Lemma 3. *Consider any pure Nash equilibrium in graph $\mathcal{G}$. The nodes of the set V that are colored black form a maximal independent set in $G = (V, E)$.*

Lemma 4. *Consider the pure Nash equilibrium in graph $\mathcal{G}$ with minimum revenue. If a node $i \in V$ is colored white, then the revenue from the set T_i is $\Theta(\Delta)$.*

Lemma 5. *Consider any pure Nash equilibrium in graph $\mathcal{G}$. If a node $i \in V$ is colored black, then the revenue from the set T_i is $\Theta(\Delta^2)$.*

Let B^* (resp. $W^* = V \setminus B^*$) denote the set of black (resp. white) nodes from V in *the pure Nash equilibrium of graph $\mathcal{G}$ that minimizes revenue*. By Lemmas 4, 5, the revenue of the coloring is given by the expression $|B^*| \times \Theta(\Delta^2) + |W^*| \times \Theta(\Delta)$. Since $\Delta = \Theta(|V|^2)$ and $|W^*|$ can be at most $|V|$, the first term dominates the second one, and revenue equals $|B^*| \times \Theta(|V|^4)$. Since the coloring *minimizes* the quantity $|B^*| \times \Theta(|V|^4)$, Lemma 3 implies that, upto a constant factor, the set B^* is a minimum maximal independent set in the graph $G = (V, E)$.

Now, let B (resp. $W = V \setminus B$) denote the set of black (resp. white) nodes from V in any pure Nash equilibrium of $\mathcal{G}$. By Lemma 5, the revenue obtained is at least $|B| \times \Theta(\Delta^2) = |B| \times \Theta(|V|^4)$. By Lemma 3, the set B is also a maximal independent set in the graph $G = (V, E)$. Hence an approximation for the PNE instance will imply an approximation for the MMIS of graph $G = (V, E)$.

Finally, the graph $\mathcal{G}$ contains $\Theta(|V| \times \Delta) = \Theta(|V|^3) = n$ nodes. As a consequence, a $O(n^{1/3-\epsilon})$ approximation for the PNE instance will translate into a $O(|V|^{1-3\epsilon})$ approximation for the minimum maximal independent set of the graph $G = (V, E)$. This completes the reduction. □

3.1 Nash Equilibrium with Relaxed Prices

At first glance, it seems that a possible way to circumvent the hardness result (Theorem 2) is to focus on the relaxed notion of *approximate* Nash equilibrium, which is defined as a coloring where any node on flipping color can improve her utility by at most a constant factor. Unfortunately, such an approach is *not* going to work, for the following reason. In the proof of Theorem 2, we have $p_1 = a$, that is, the cheap item is offered at a price equal to its intrinsic valuation. Hence every black node gets *zero* utility. In this situation, any approximate Nash equilibrium must be an *exact* Nash equilibrium. As a result, the hardness lower bound carries over to the seemingly relaxed notion of approximate Nash equilibrium.

On a positive note, we provide a 4-approximation algorithm for the PNE problem when the constraint of uniform pricing is relaxed slightly, allowing the seller to offer different prices to different buyers purchasing the same item, under the condition that these prices are within a constant factor of each other. Our algorithm returns a coloring $\mathcal{C}$ satisfying the two properties described below.

- Every white node has an externality of at least $\Delta = (p_2 - p_1 + a - b)$, and every black node has an externality of at most 4Δ. It is easy to verify that such a coloring is a pure Nash equilibrium under the following *relaxed* pricing scheme: Each black (resp. white) node is offered the price $4p_2$ (resp. p_2) for the expensive item, whereas the cheap item is offered at the same price p_1 to every node.
- The revenue of the coloring $\mathcal{C}$ is *at most* 4 times the revenue of any pure Nash equilibrium where the cheap (resp. expensive) item is offered to all buyers at the same price p_1 (resp. p_2).

The algorithm is described in Figure 1. We give a LP relaxation for the PNE instance, encoding the constraint that every black node has an externality of at most Δ. Surprisingly, the LP does *not* enforce any lower bound on the externalities of the white nodes. Using a simple rounding scheme, we get an integral solution whose revenue is at most 4 times the minimum revenue. In the integral solution, every black node has an externality of at most 2Δ. Next, we employ an iterative improvement process that flips the colors of *bad* (see Definition 3) nodes. We show that the revenue does not increase during the iterative improvement process and that the process converges in polynomial time to a coloring that is a pure Nash equilibrium under the relaxed pricing scheme mentioned above.

We first write down the LP relaxation for the given PNE instance. Recall that the notation $\mathcal{D}(i)$ denotes the weighted degree of node i (Definition 2).

Consider any pure Nash equilibrium. For all $i \in V$, if node i is colored white (resp. black), set $x_i = 1$ (resp. $x_i = 0$). For all $(i, j) \in E$, set $y_{ij} = 1$ if edge (i, j) has both its end vertices colored black, otherwise set $y_{ij} = 0$. The revenue is given by the objective value of LP-MIN. Constraint (3) states that if a node has weighted degree less than Δ, then it must be colored black. Constraint (2) ensures $y_{ij} = 1$ if and only if both endpoints of the edge (i, j) are black. Finally, Constraint (1) requires that if a node is black, then she has an externality of at most Δ. Thus, any pure Nash equilibrium is a feasible solution to LP-MIN, and optimal objective value of the LP lower bounds the minimum revenue obtainable from any pure Nash equilibrium.

$$\text{Minimize} \quad (\Delta + b - a) \sum_i x_i + p_1 |V| \qquad \text{(LP-MIN)}$$

$$
\begin{aligned}
\sum_{j:(i,j)\in E} w(i,j) y_{ij} &\leq \Delta & \forall i : \mathcal{D}(i) \geq \Delta & \quad (1) \\
x_i + x_j + y_{ij} &\geq 1 & \forall (i,j) \in E & \quad (2) \\
x_i &= 0 & \forall i : \mathcal{D}(i) < \Delta & \quad (3) \\
x_i &\in [0,1] \; \forall i & & \quad (4)
\end{aligned}
$$

Definition 3. *A node is called* bad *if either it is white and has an externality less than Δ, or if it is black and has an externality more than 4Δ.*

RELAXED

INPUT: A graph $G = (V, E)$, and some $\Delta > 0$.
OUTPUT: A coloring of the nodes.

Let $\{x_i^*, y_{ij}^*\}$ denote the optimal solution to LP-MIN.
Rounding: For all nodes $i \in V$:
    If $x_i^* \geq 1/4$, then color node i as white; Else color node i as black.
Iterative Improvement:
    WHILE there exists a *bad* [a] node $i \in V$: *Flip* the color of node i.

---
[a] See Definition 3

Fig. 1. Approximation Algorithm for the Pessimistic Nash Equilibrium (PNE) problem with relaxed prices

Theorem 6. *Given an instance of the PNE problem, Algorithm RELAXED (Figure 1) returns a coloring of the nodes with the following properties.*

- *The algorithm terminates in polynomial time.*
- *The revenue of the returned coloring is a 4-approximation to the optimal objective value.*
- *It returns a coloring where each white node has an externality of at least Δ and each black node has an externality of at most 4Δ.*

Proof. Let $\mathcal{C}$ denote the coloring obtained from the rounding step. Since every white node has $x_i^* \geq 1/4$, the revenue of coloring $\mathcal{C}$ is at most 4 times the optimal objective value of LP-MIN. We now show that the revenue can only decrease during the iterative improvement process, which converges in polynomial time.

Consider a node i which is black under coloring $\mathcal{C}$. Consider any other black node j that is adjacent to node i. It follows that $x_i^*, x_j^* < 1/4$, and by constraint (2) of LP-MIN, we have $y_{ij}^* \geq 1/2$. Now constraint (1) implies that $Ext(i) \leq 2\Delta$. Hence, under coloring $\mathcal{C}$, every black node i has $Ext(i) \leq 2\Delta$.

At any instant of the iterative improvement process, let W (resp. $B = V \setminus W$) denote the set of white (resp. black) nodes. Associate with every node i a variable called $bank(i)$, and define the potential of the system at any instant as $\Phi = \Delta \times |W| + \sum_{i \in V} bank(i)$. Initially, all the bank variables are set to 0. Hence, the potential of the coloring $\mathcal{C}$ is given by Δ times the number of white nodes. As the process unfolds, we adjust the bank variables according to the following rules.

- If a node i is flipped from white to black, then for all black nodes j adjacent to i, increment $bank(j)$ by an amount $w(i,j)$. Furthermore, set $bank(i)$ to 0.
- If a node i is flipped from black to white, then set $bank(i)$ to 0.

First, we observe that when a node i is flipped from white to black, we have:

$$\text{Decrease in Potential} = \Delta - \sum_{j \in B:\, (i,j) \in E} w(i,j) = \Delta - Ext(i) > 0$$

The last inequality holds since any node flipped from white to black has an externality strictly less than Δ. Next, we note that whenever a black node i is flipped to white, then the decrease in potential is $-\Delta + bank(i)$.

The last time instant when node i was colored black, we had $Ext(i) \leq 2\Delta$ and $bank(i)$ was set to 0. At the present instant, when we are flipping node i to white, $Ext(i)$ is at least 4Δ. During this time period, node i has therefore gained an externality of 2Δ. During the same time period, whenever a friend j of node i has been flipped to black, both the externality of node i and the variable $bank(i)$ have increased by the same amount $w(i,j)$. Consequently, at the present time instant, the variable $bank(i)$ has a value at least 2Δ, and when we are flipping node i to white, we are decreasing the potential by at least Δ.

The preceding discussion shows that the iterative improvement process never increases the potential. The highest potential is obtained at the beginning of the process when all bank variables are zero, and consequently, this value is bounded from above by the expression $\Delta \times |V|$. And whenever a node is flipped from black to white, the potential decreases by at least Δ. Since a flip from black to white must occur at least once amongst any sequence of $|V| + 1$ consecutive flips, it follows that the process *decreases the potential* by an amount Δ within every $\Theta(|V|)$ steps. As a result, the process converges in $\Theta(|V|^2)$ steps.

Since the bank variables are initially set to zero and the potential decreases as the process unfolds, it implies that the number of white nodes (and hence revenue) in the final coloring must be lower than that of the coloring $\mathcal{C}$ obtained from the rounding step. Recall that the revenue of the coloring $\mathcal{C}$ is at most 4 times the minimum revenue, and the approximation guarantee follows.

The iterative improvement step is repeated till there are no bad nodes (Definition 3), and hence the final part of the theorem can be easily verified.

□

4   The Best Nash Equilibrium

This section begins with a strong inapproximability result for the BNE problem.

Theorem 7. *Unless $NP = ZPP$, it is not possible to compute in polynomial time a pure Nash equilibrium whose revenue is a $O(n^{1/3-\epsilon})$ approximation to maximum revenue, where n is the number of nodes in the social network.*

In view of the above hardness result on general graphs, we consider the BNE problem on bipartite graphs. Unfortunately, the next theorem shows that the problem is NP-HARD even on trees (and hence on general bipartite graphs).

Theorem 8. *It is NP-hard to optimally solve the BNE problem on trees.*

On the positive side, we can show a simple 2 approximation for the BNE problem on bipartite graphs. Consider a BNE instance on a bipartite graph with partite-sets L, R and set of edges $E \subseteq L \times R$. Suppose that $|L| \geq |R|$, and the weighted degree (Definition 2) of every node in the graph is at least Δ. In this case, it is easy to check that coloring all the nodes in L (resp. R) as white (resp. black) gives a 2 approximation to the BNE problem. This idea can be extended to graphs where nodes can have arbitrary weighted degrees.

Theorem 9. *There is an efficient algorithm that gives 2 approximation to the BNE problem on bipartite graphs.*

5   Conclusion

We considered the setting of monopoly pricing over a social network in presence of negative externalities. All of our algorithmic results, both for BNE and PNE, can easily be adapted to the scenario where different buyers have different intrinsic valuations for the same item. In addition, the algorithms work even if the extrinsic valuation for the expensive item increases linearly with the externality *only upto a certain threshold*, and then remains fixed at the threshold value.

It will be interesting to extend our results to pricing with more than two items, and investigate other solution concepts such as mixed Nash equilibrium. We also note that our work assumed a perfect information model where the seller knows the valuations of the buyers, and we leave open the question of studying the effects of imperfect information under a Bayesian setting.

Acknowledgments. We thank Vincent Conitzer and Aneesh Sharma for several helpful discussions. This research was supported by Munagala's Alfred P. Sloan Research Fellowship, a gift from Cisco, and by NSF via grants CCF-0745761 and CCF-1008065.

References

[1] Akhlaghpour, H., Ghodsi, M., Haghpanah, N., Mirrokni, V.S., Mahini, H., Nikzad, A.: Optimal Iterative Pricing over Social Networks (Extended Abstract). In: Saberi, A. (ed.) WINE 2010. LNCS, vol. 6484, pp. 415–423. Springer, Heidelberg (2010)

[2] Anari, N., Ehsani, S., Ghodsi, M., Haghpanah, N., Immorlica, N., Mahini, H., Mirrokni, V.S.: Equilibrium Pricing with Positive Externalities (Extended Abstract). In: Saberi, A. (ed.) WINE 2010. LNCS, vol. 6484, pp. 424–431. Springer, Heidelberg (2010)

[3] Arthur, D., Motwani, R., Sharma, A., Xu, Y.: Pricing Strategies for Viral Marketing on Social Networks. In: Leonardi, S. (ed.) WINE 2009. LNCS, vol. 5929, pp. 101–112. Springer, Heidelberg (2009)

[4] Awerbuch, B., Azar, Y., Epstein, A., Mirrokni, V., Skopalik, A.: Fast convergence to nearly optimal solutions in potential games. In: EC, pp. 264–273 (2008)

[5] Bensaid, B., Lesne, J.: Dynamic Monopoly Pricing with Network Externalities. International Journal of Industrial Organization 14(6), 837–855 (1996)

[6] Bhalgat, A., Chakraborty, T., Khanna, S.: Approximating pure nash equilibrium in cut, party affiliation, and satisfiability games. In: EC, pp. 73–82 (2010)

[7] Bulow, J.: Durable-goods Monopolists. The Journal of Political Economy 90(2), 314–332 (1982)

[8] Chien, S., Sinclair, A.: Convergence to approximate nash equilibria in congestion games. In: SODA, pp. 169–178 (2007)

[9] Coase, R.: Durability and Monopoly. Journal of Law and Economics, 143–149 (1972)

[10] Fabrikant, A., Papadimitriou, C., Talwar, K.: The complexity of pure nash equilibria. In: STOC, pp. 604–612 (2004)

[11] Gul, F., Sonnenschein, H., Wilson, R.: Foundations of dynamic monopoly and the coase conjecture. Journal of Economic Theory 39(1), 155–190 (1986)

[12] Haghpanah, N., Immorlica, N., Mirrokni, V.S., Munagala, K.: Optimal auctions with positive network externalities. In: EC, pp. 11–20 (2011)

[13] Hart, O., Tirole, J.: Contract Renegotiation and Coasian Dynamics. The Review of Economic Studies 55(4), 509–540 (1988)

[14] Hartline, J., Mirrokni, V., Sundararajan, M.: Optimal Marketing Strategies over Social Networks. In: WWW, pp. 189–198 (2008)

[15] Kann, V.: Polynomially bounded minimization problems that are hard to approximate. Nordic Journal of Computing 1(3), 317–331 (1994)

[16] Kempe, D., Kleinberg, J., Tardos, É.: Maximizing the Spread of Influence Through a Social Network. In: ACM SIGKDD, pp. 137–146 (2003)

[17] Koutsoupias, E., Papadimitriou, C.: Worst-Case Equilibria. In: Meinel, C., Tison, S. (eds.) STACS 1999. LNCS, vol. 1563, pp. 404–413. Springer, Heidelberg (1999)

[18] Monderer, D., Shapley, L.: Potential games. Games and Economic Behavior 14(1), 124–143 (1996)

[19] Roughgarden, T., Tardos, É.: How bad is selfish routing? In: FOCS, pp. 93–102 (2000)

[20] Thépot, J.: A Direct Proof of the Coase Conjecture. Journal of Mathematical Economics 29(1), 57–66 (1998)

[21] (December 2004),
http://www.washingtonpost.com/wp-dyn/articles/A55414-2004Dec10.html

An Improved 2-Agent Kidney Exchange Mechanism

Ioannis Caragiannis[1], Aris Filos-Ratsikas[1], and Ariel D. Procaccia[2,*]

[1] Department of Computer Engineering and Informatics,
University of Patras, 26504 Rio, Greece
[2] Computer Science Department,
Carnegie Mellon University, Pittsburgh, PA 15213, USA

Abstract. We study a mechanism design version of matching computation in graphs that models the game played by hospitals participating in pairwise kidney exchange programs. We present a new randomized matching mechanism for two agents which is truthful in expectation and has an approximation ratio of 3/2 to the maximum cardinality matching. This is an improvement over a recent upper bound of 2 [Ashlagi et al., EC 2010] and, furthermore, our mechanism beats for the first time the lower bound on the approximation ratio of deterministic truthful mechanisms. We complement our positive result with new lower bounds. Among other statements, we prove that the weaker incentive compatibility property of truthfulness in expectation in our mechanism is necessary; universally truthful mechanisms that have an inclusion-maximality property have an approximation ratio of at least 2.

1 Introduction

In an attempt to address the wide need for kidney transplantation and the scarcity of cadaver kidneys, several countries have launched, or are considering, national kidney exchange programs involving live donors [7,11,1,4]. Patients can enter such a program together with a member of their family or friend who is willing to donate them a kidney but cannot due to incompatibility. National kidney exchange programs aim to implement exchanges between two compatible patient-donor pairs u and v so that the donor of pair u donates her kidney to the patient of pair v and vice versa. This requires four simultaneous operations. More complicated exchanges involving more than two donor-patient pairs are also possible; however, we focus on pairwise exchanges since they are easier to perform in practice.

Donor-patient pairs approach a hospital in order to enroll into the national kidney exchange programs. In an ideal scenario, each hospital reports its donor-patient pairs to the program and a central authority runs an algorithm that decides which pairwise kidney exchanges will take place. In practice, strategic issues immediately arise. A hospital may prefer to not enroll some easy-to-match

---

* The third author is supported by gifts from Yahoo! and Google.

N. Chen, E. Elkind, and E. Koutsoupias (Eds.): WINE 2011, LNCS 7090, pp. 37–48, 2011.

donor-patient pairs to the program and instead match them and perform the kidney exchange operations internally. This may have an impact on patients of other hospitals who could have benefited if the hospital truthfully reported all its donor-patient pairs to the program. The current paper follows the line of research that seeks to design algorithms (or mechanisms) that discourage hospitals from behaving untruthfully. The main objective is to perform as many kidney exchanges as possible under this constraint. This is a *mechanism design* [5] problem, and in particular—because paying for organs is illegal in almost all countries—it falls within the scope of approximate mechanism design without money [6].

We can model the problem as a matching problem in graphs. The input consists of a graph in which the nodes represent donor-patient pairs and an edge connects two nodes u and v when the donor of pair u and the patient of pair v are compatible, and the donor of pair v and the patient of pair u are compatible. Each node of the graph is controlled by exactly one self-interested agent (a hospital). A *mechanism* takes the graph as input and returns a matching, i.e., a disjoint pair of edges indicating which pairwise kidney exchanges will take place. The *gain* of an agent is the number of nodes under her control that are matched. Clearly, an optimal solution is easy to find by a maximum matching computation. Unfortunately, a mechanism that returns such a solution may incentivize hospitals to behave untruthfully in the following sense. A hospital could hide some of its nodes from (i.e., not enroll them into) the system so that the mechanism is essentially applied on a graph that contains neither the hidden nodes nor the edges incident to them. Then, the gain of the hospital is the number of its nodes that are matched by the mechanism plus the number of nodes it managed to match internally. Such behavior can lead to fewer matched nodes compared to the best possible solution, i.e., fewer patients who receive kidneys. So, we seek mechanisms that guarantee that no agent has any incentive to deviate from truth-telling. Our goal is to design such mechanisms that also return matchings of high cardinality, i.e., high total gain.

The mechanisms can be deterministic or randomized. Given an instance of the problem, a deterministic mechanism returns a simple matching. A randomized mechanism returns a probability distribution over matchings. In the latter case, we distinguish between *universally truthful* mechanisms and mechanisms that are *truthful in expectation*. The former are induced by a probability distribution over truthful deterministic mechanisms, whereas the latter guarantee that no agent can deviate from truth-telling in order to increase her expected gain. The efficiency of truthful mechanisms is assessed through their *approximation ratio*, i.e., the maximum ratio over all possible instances of the problem of the size of the maximum cardinality matching over the expected size of the matching returned by the mechanism.

Early work on kidney exchange problems in Economics [8,9,10] has considered the incentives of incompatible donor-patient pairs. However, as national kidney exchange programs emerged, it has become apparent that such incentives are less important compared to the incentives of the hospitals [3]. The model considered in the current paper has also been studied in [2,3,12,13]. The fact that the maximum cardinality matching mechanism is not truthful was first observed by Sönmez and Ünver [12] (see also [3]). Ashlagi et al. [2] present a universally truthful randomized 2-approximation mechanism (called MIX-AND-MATCH) for arbitrarily many agents. MIX-AND-MATCH is based on a simple deterministic truthful 2-approximation mechanism for two agents, henceforth called MATCH. MATCH returns a matching that contains the maximum number of *internal edges* (where the nodes on both sides are controlled by the same agent), breaking ties in favor of the matching with maximum cardinality. A nice property of MATCH is *inclusion-maximality*; this translates to the requirement that a donor-patient pair does not participate in any kidney exchange only when all its compatible donor-patient pairs participate in some pairwise kidney exchange. A randomized mechanism has this property when it returns a probability distribution over inclusion-maximal matchings. On the negative side, there are lower bounds of 2 and 8/7 for deterministic truthful mechanisms and randomized mechanisms that are truthful in expectation, respectively [2,3]. Ashlagi et al. [2] also propose the mechanism FLIP-AND-MATCH for two agents. FLIP-AND-MATCH equiprobably selects among the outcome of MATCH and a maximum cardinality matching. They prove that this mechanism has approximation ratio 4/3 and leave open the question of whether it is truthful in expectation. Ashlagi and Roth [3] and Toulis and Parkes [13] consider weaker notions of truthfulness in random graph models that reflect the compatibility frequency among donors and patients from the human population. As in [2], no such information is required in our setting.

In an attempt to better understand the potential and limitations of randomized mechanisms, we consider the case of two agents. This case is of special interest because efficient mechanisms can enable cooperation between pairs of hospitals on an *ad-hoc* basis, in countries where a national kidney exchange program is not yet in place. Our main result is a randomized mechanism called WEIGHT-AND-MATCH for 2-agent pairwise kidney exchange that is truthful in expectation and has a tight approximation ratio of 3/2. This establishes, for the first time, a separation between the power of randomized mechanisms and deterministic mechanisms (for which there is a lower bound of 2).

WEIGHT-AND-MATCH is inspired by the mechanism FLIP-AND-MATCH proposed in [2]. Unfortunately, it turns out that FLIP-AND-MATCH is not truthful due to its use of maximum cardinality matchings. This observation is our starting point for the definition of the new mechanism. WEIGHT-AND-MATCH first assigns weights to the edges of the input graph and then selects equiprobably among two maximum-weight matchings: one with minimum cardinality (the particular weights assigned to the edges guarantee that this matching is identical to the one returned by MATCH) and one with maximum cardinality (which replaces the second matching used by FLIP-AND-MATCH). Informally, this definition guarantees that the bad incentives created by the second matching are canceled out by the outcome of MATCH.

We complement this result with new lower bounds on the approximation ratio of randomized mechanisms that are truthful in expectation or universally truthful, distinguishing between mechanisms that are inclusion-maximal and those that are not. Here we use the same 2-agent instance as in previous work [2,3,12] but our stronger analysis leads to improved bounds for inclusion-maximal and universally truthful mechanisms. Our general lower bound is 5/4. Interestingly, we prove a lower bound of 2 for inclusion-maximal universally truthful mechanisms that indicates that the weaker notion of truthfulness satisfied by WEIGHT-AND-MATCH (which is inclusion-maximal) is indeed necessary.

The rest of the paper is structured as follows. We warm up by showing that FLIP-AND-MATCH is not truthful in Section 2. Our mechanism and its analysis are presented in Section 3. The lower bounds are presented in Section 4. We conclude with a short discussion of open problems in Section 5.

2 An Unsuccessful Attempt: FLIP-AND-MATCH

Throughout the paper, we refer to the two agents as agent 1 and agent 2. We also call the nodes of agents 1 and 2 *white* and *gray* nodes, respectively.

Let us warm up by considering the mechanism FLIP-AND-MATCH proposed in [2]. FLIP-AND-MATCH selects equiprobably among the matching returned by MATCH and a maximum cardinality matching. In the original definition of [2], ties among maximum cardinality matchings are broken in favor of matchings that maximize the number of internal edges (i.e., edges between two nodes controlled by the same agent) and then arbitrarily. In our proof, we essentially show that any modification of the tie-breaking rule violates truthfulness.

Theorem 1. FLIP-AND-MATCH *is not truthful.*

Proof. Our proof uses the instance I and subinstances I_1 and I_2 of Figure 1. When applied to instance I, MATCH returns the matching $M_1 = \{(v_2, v_3), (v_4, v_5), (v_7, v_8)\}$. The gain of agent 1 is 4 while the gain of agent 2 is 2. Let M_2 be a maximum cardinality matching. It leaves exactly one node unmatched; this can be either a white or a gray node, i.e., M_2 matches either 4 white nodes and 4 gray nodes or 5 white nodes and 3 gray nodes. We distinguish between these two cases and show that, in both cases, some agent has an incentive to withhold nodes.

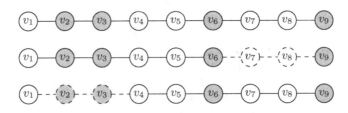

Fig. 1. The original instance I used in the proof of Lemma 1 and the two subinstances I_1 and I_2 used in cases 1 and 2, respectively. The dashed nodes and edges are not part of the instances I_1 and I_2 but are shown here in order to compare with instance I.

Case 1. M_2 matches 4 white nodes and 4 gray nodes and, hence, the expected gain of agent 1 from the application of FLIP-AND-MATCH on instance I is 4. Consider the instance I_1 in which agent 1 hides the white nodes v_7 and v_8 and matches them internally. In the new instance, MATCH returns the matching $\{(v_2, v_3), (v_4, v_5)\}$ that contains 2 matched white nodes while the maximum cardinality matching is $\{(v_1, v_2), (v_3, v_4), (v_5, v_6)\}$ that contains 3 matched white nodes. The expected gain of agent 1 (including the hidden nodes) is 4.5.

Case 2. M_2 matches 5 white nodes and 3 gray nodes and hence the expected gain of agent 2 from the application of FLIP-AND-MATCH to the original instance I is 2.5. Consider the instance I_2 in which agent 2 hides nodes v_2 and v_3 (and matches them internally). In the new instance, MATCH returns the matching $\{(v_4, v_5), (v_7, v_8)\}$ that contains no matched gray nodes while the maximum cardinality matching is $\{(v_4, v_5), (v_6, v_7), (v_8, v_9)\}$ that contains 2 matched gray nodes. The expected gain of agent 2 (including the hidden nodes) is 3.   □

3   Our Mechanism: WEIGHT-AND-MATCH

In this section, we present our new mechanism for two agents, which we call WEIGHT-AND-MATCH. The main idea behind it is similar to the one that led to FLIP-AND-MATCH: we try to combine mechanism MATCH with another mechanism that yields a higher gain. However, given the negative result for FLIP-AND-MATCH presented in the previous section, we should be careful with the definition of our mechanism. We can think of the following alternative definition for MATCH. We first assign weights to the edges of the input graph as follows. Internal edges have weight 1; edges between nodes of different agents have weight $1/2$. The matching returned by MATCH is then a maximum-weight matching on the weighted version of the input graph, where ties are broken in favor of the matching with minimum cardinality. Our mechanism WEIGHT-AND-MATCH also computes a maximum-weight maximum-cardinality matching on the weighted version of the input graph, and selects equiprobably among the two matchings. Note that WEIGHT-AND-MATCH is inclusion maximal. The rest of the section is devoted to proving the following statement.

Theorem 2. *Mechanism* WEIGHT-AND-MATCH *can be implemented in polynomial time, has approximation ratio* $3/2$*, and is truthful in expectation.*

Due to lack of space, we omit the proof that our mechanism can be implemented efficiently; we proceed with the proof of its approximation guarantee.

Lemma 1. WEIGHT-AND-MATCH *has an approximation ratio of* $3/2$*.*

Proof. Let M be a matching of maximum cardinality and let M_1 and M_2 be the maximum-weight matchings of minimum and maximum cardinality, respectively, that are used by WEIGHT-AND-MATCH. Consider the symmetric difference $M \triangle M_1 = (M \setminus M_1) \cup (M_1 \setminus M)$. It consists of several connected components which are either cycles (of even length), or paths with edges alternating between edges of M and edges of M_1. Let m_1 be the number of edges of M that either belong also to M_1 or belong to cycles or paths of $M \Delta M_1$ with even length. Let m_3 and m_5 be the edges of M that belong to paths of $M \Delta M_1$ with length exactly 3 and odd length at least 5, respectively. Clearly, $|M| = m_1 + m_3 + m_5$.

Note that the number of edges of M_1 that either belong also to M or belong to cycles or paths of $M \Delta M_1$ of even length is exactly m_1 as well. Also, since M has maximum cardinality, the first and the last edge in a path with odd length in $M \Delta M_1$ belong to M. So, M_1 contains exactly $m_3/2$ edges in paths of $M \Delta M_1$ of length 3 and at least $2m_5/3$ edges in paths of $M \Delta M_1$ of odd length at least 5. Hence, $|M_1| \geq m_1 + m_3/2 + 2m_5/3$.

We now show that M_2 (the maximum-weight matching of maximum cardinality) contains at least $m_1 + m_3 + 2m_5/3$ edges. Observe that, since M_1 is a maximum-weight matching, in any path with length 3 in $M \Delta M_1$, the edge of M_1 should have endpoints belonging to the same agent (and, hence, weight 1) and the two edges of M should have endpoints belonging to different agents (and, hence, weight $1/2$). Consider the edges of M_1 that do not belong to paths of length 3 of $M \Delta M_1$ and the edges of M that belong to paths of length 3 in $M \Delta M_1$. All these edges form a matching that has the same total weight as the edges of M_1, and their cardinality is at least $m_1 + m_3 + 2m_5/3$. Clearly, this is also a lower bound on the cardinality of M_2, i.e., $|M_2| \geq m_1 + m_3 + 2m_5/3$.

Hence, the expected cardinality of the mechanism's matching is

$$\frac{1}{2}(|M_1| + |M_2|) \geq m_1 + \frac{3m_3}{4} + \frac{2m_5}{3} \geq \frac{2}{3}(m_1 + m_3 + m_5) = \frac{2}{3}|M|. \quad \square$$

Fig. 2. An instance indicating that the analysis of Lemma 1 is tight. The maximum matching matches all 6 nodes but mechanism WEIGHT-AND-MATCH returns the matching that consists of edges (v_2, v_3) and (v_4, v_5). Note that here the symmetric difference is a path of length 5.

The bound obtained in Lemma 1 is tight through the example of Figure 2. We now turn to proving that our mechanism is truthful.

Lemma 2. WEIGHT-AND-MATCH *is truthful in expectation.*

Proof. We will show that agent 1 never has an incentive to deviate from truth-telling. The case of agent 2 is identical.

Let G be the input graph and consider the maximum-weight matchings M_1 and M_2 of minimum and maximum cardinality, respectively, that are used by WEIGHT-AND-MATCH. Also, assume that agent 1 hides some nodes and matches them internally. Then, the mechanism is applied to the subgraph G' of G which does not contain the hidden white nodes and edges incident to them. Let M_3 and

M_4 be the maximum-weight matchings of minimum and maximum cardinality computed by WEIGHT-AND-MATCH on input G', augmented by the edges used by agent 1 to match the hidden white nodes internally. Denote by $\mathtt{gain}(M)$ the gain of agent 1 from matching M and by $\mathtt{wgt}(M)$ the weight of matching M. Our proof will follow from the next two lemmas.

Lemma 3. $\mathtt{gain}(M_3) = \mathtt{gain}(M_1) - 2(\mathtt{wgt}(M_1) - \mathtt{wgt}(M_3))$.

Proof. Denote by $n_{\mathrm{ww}}(M)$, $n_{\mathrm{wg}}(M)$, and $n_{\mathrm{gg}}(M)$ the number of edges in matching M connecting two white nodes, two nodes belonging to different agents, and two gray nodes, respectively. We will first show that $n_{\mathrm{gg}}(M_1) = n_{\mathrm{gg}}(M_3)$. Consider the symmetric difference of the two matchings $M_1 \Delta M_3 = (M_1 \setminus M_3) \cup (M_3 \setminus M_1)$ and the subgraph of G induced by these edges. This subgraph consists of several connected components which can be cycles or paths (see Figure 3 for an example). Consider such a connected component C and let C_1 and C_3 be the sets of edges of M_1 and M_3 it contains, respectively.

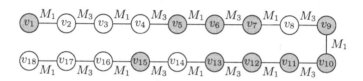

Fig. 3. A connected component of $M_1 \Delta M_3$ considered in the proof of Lemma 3. The sets of gray nodes $\{v_1\}$, $\{v_5, v_6, v_7\}$, $\{v_9, ..., v_{13}\}$, and $\{v_{15}\}$ form blocks. The main argument in the proof is that each block has an odd number of gray nodes.

In order to prove that $n_{\mathrm{gg}}(M_1) = n_{\mathrm{gg}}(M_3)$, it suffices to prove that $n_{\mathrm{gg}}(C_1) = n_{\mathrm{gg}}(C_3)$. This is clearly true if C is a cycle consisting of gray nodes only, since such a cycle should have an even number of edges, half of which belong to C_1 and half to C_3. Assume that C contains a block of t consecutive gray nodes $b_1, b_2, ..., b_t$ such that the first and the last have either degree 1 or are connected to another white node outside the block. We will show that t cannot be even. Assume that this was the case; then one of the two matchings (say M_1; the argument for M_3 is completely symmetric) would contain the $\frac{t}{2} - 1$ edges $(b_2, b_3), (b_4, b_5), \ldots, (b_{t-2}, b_{t-1})$ and the other (say M_3) would contain the $\frac{t}{2}$ edges $(b_1, b_2), (b_3, b_4), \ldots, (b_{t-1}, b_t)$. Then, by replacing the $\frac{t}{2} - 1$ edges of matching M_1 in the block as well as the edges of M_1 that are incident to nodes b_1 and b_t (if any) with the $\frac{t}{2}$ edges of M_3 in the block, we would obtain a matching that either has higher weight than M_1 (if some of nodes b_1 and b_t has degree 1) or the same weight as M_1 (recall that the edges connecting nodes b_1 and b_t to white nodes outside the block have weight 1/2) but smaller cardinality. Both cases contradict the fact that the matching M_1 is a minimum cardinality maximum-weight matching. Hence, every block has an odd number of nodes and an even number of edges between gray nodes that alternate between matchings M_1 and M_3. This implies that $n_{\mathrm{gg}}(C_1) = n_{\mathrm{gg}}(C_3)$. Consequently, by summing over all connected components of $M_1 \Delta M_3$ and the edges of $M_1 \cap M_3$ connecting gray nodes, we also have that $n_{gg}(M_1) = n_{gg}(M_3)$.

Next, observe that $\text{gain}(M) = 2n_{ww}(M) + n_{wg}(M)$ and $\text{wgt}(M) = n_{ww}(M) + n_{gg}(M) + n_{wg}(M)/2$. Hence, since $n_{gg}(M_1) = n_{gg}(M_3)$, we have

$$
\begin{aligned}
\text{gain}(M_3) &= 2n_{ww}(M_3) + n_{wg}(M_3) \\
&= 2n_{ww}(M_3) + n_{wg}(M_3) + 2n_{gg}(M_3) - 2n_{gg}(M_1) \\
&= \text{gain}(M_1) - 2(\text{wgt}(M_1) - \text{wgt}(M_3)),
\end{aligned}
$$

as desired.   □

Lemma 4. $\text{gain}(M_4) \leq \text{gain}(M_2) + 2(\text{wgt}(M_2) - \text{wgt}(M_4))$.

Proof. First consider each edge in $M_2 \cap M_4$ and observe that its contribution to $\text{gain}(M_4)$ equals its contribution to $\text{gain}(M_2) + 2(\text{wgt}(M_2) - \text{wgt}(M_4))$. We will now consider the symmetric difference of the two matchings $M_2 \Delta M_4 = (M_2 \setminus M_4) \cup (M_4 \setminus M_2)$ and the subgraph of G induced by these edges. Again, this subgraph consists of several connected components which can be cycles or paths. Consider such a connected component C and let C_2 and C_4 be the sets of edges of M_2 and M_4 it contains, respectively. We will complete the proof of the lemma by showing that

$$
\text{gain}(C_4) \leq \text{gain}(C_2) + 2(\text{wgt}(C_2) - \text{wgt}(C_4)). \tag{1}
$$

First, observe that since M_2 is a maximum-weight matching in G, it holds that $\text{wgt}(C_2) \geq \text{wgt}(C_4)$ (otherwise, we could replace the edges of C_2 with the edges of C_4 in M_2 and obtain a matching with higher weight). We now use a four-letter/number notation to classify the connected components of the subgraph of G induced by $M_2 \Delta M_4$ that are paths into different types: the first and last letters are w or g and denote whether the left and right endpoint of the connected component is a white or gray node, respectively. The second and third numbers are either 2 or 4 and denote whether the first and the last edge of the connected component belong to matching M_2 or M_4, respectively. Examples of paths of type w22w, w44g, and w44w are depicted in Figure 4. We distinguish between three main cases:

Case 1. If C is a cycle, or a path of type w22w, w24w, w42w, w22g, w24g, g22g, g24g, g42g, or g44g, we have $\text{gain}(C_4) \leq \text{gain}(C_2)$ and inequality (1) follows easily since $\text{wgt}(C_2) \geq \text{wgt}(C_4)$.

Case 2. If C is a path of type w42g or w44g, we claim that $\text{wgt}(C_2) + \text{wgt}(C_4)$ is non-integer. Indeed, since the first and the last node in the path belong to different agents, there is an odd number of external edges (between a white and a gray node) in C, and each such edge contributes $1/2$ to the sum $\text{wgt}(C_2) + \text{wgt}(C_4)$. Recall that $\text{wgt}(C_2) \geq \text{wgt}(C_4)$, and therefore $\text{wgt}(C_2) - \text{wgt}(C_4) \geq 1/2$. Inequality (1) follows by observing that $\text{gain}(C_2) = \text{gain}(C_4) - 1$ in this case.

Case 3. If C is of type w44w, observe that C_4 contains one more edge than C_2 and, hence, $\text{wgt}(C_2) > \text{wgt}(C_4)$ (otherwise, we could replace the edges of C_2 with

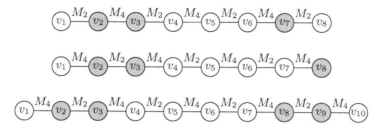

Fig. 4. Examples of connected components of $M_2 \triangle M_4$ considered in the proof of Lemma 4 (paths of type `w22w`, `w44g`, and `w44w`)

the edges of C_4 in M_2 in order to obtain a matching of the same weight but with higher cardinality). Also, observe that the number of external edges in C is even, and hence $\texttt{wgt}(C_2) + \texttt{wgt}(C_4)$ is integer. It follows that $\texttt{wgt}(C_2) \geq \texttt{wgt}(C_4) + 1$. Inequality (1) follows by further observing that $\texttt{gain}(C_2) = \texttt{gain}(C_4) - 2$. ☐

Since $\texttt{wgt}(M_1) = \texttt{wgt}(M_2)$ and $\texttt{wgt}(M_3) = \texttt{wgt}(M_4)$, by Lemmas 3 and 4 we have that the expected gain $\frac{1}{2}(\texttt{gain}(M_3) + \texttt{gain}(M_4))$ of agent 1 when she hides some white nodes and matches them internally is upper-bounded by the expected gain $\frac{1}{2}(\texttt{gain}(M_1) + \texttt{gain}(M_2))$ when she acts truthfully. ☐

4   Lower Bounds

Ashlagi et al. [2] and Ashlagi and Roth [3] provide a lower bound of 8/7 for truthful-in-expectation randomized mechanisms.[1] The proof of the next lemma starts with the same initial instance as [2,3] but uses a more detailed reasoning in order to prove lower bounds for randomized mechanisms that are either universally truthful or truthful in expectation, distinguishing between mechanisms that are inclusion-maximal and those that are not.

Theorem 3. *Let A be a randomized mechanism for 2-agent kidney exchange.*

(a) If A is truthful in expectation, then its approximation ratio is at least 5/4.
(b) If A is truthful in expectation and inclusion-maximal, then its approximation ratio is at least 4/3.
(c) If A is universally truthful, then its approximation ratio is at least 3/2.
(d) If A is universally truthful and inclusion-maximal, then its approximation ratio is at least 2.

Proof. Our proof uses the instances depicted in Figure 5. The starting point is instance I. We denote by I_1 the instance obtained by removing the white nodes

---

[1] Ashlagi et al. [2] actually claim a bound of 4/3 but this is inaccurate. In fact it is not hard to design an artificial mechanism (as a probability distribution over matchings) that is truthful in expectation and has approximation ratio at most 5/4 for the instances considered in their proof.

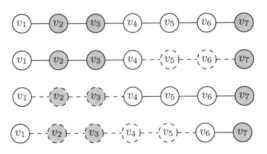

Fig. 5. The instances I, I_1, I_2, and I_3 used in the proof of Theorem 3. The dashed nodes and edges are not part of the instances I_1, I_2, and I_3 but are shown here in order to compare with instance I.

v_5 and v_6 and their incident edges from I, by I_2 the instance obtained from I by removing the nodes v_2 and v_3 and their incident edges, and by I_3 the instance obtained from I_2 by removing the nodes v_4 and v_5 and their incident edges.

(a) Consider the application of mechanism A to instance I. Observe that the maximum cardinality matching of this instance has size 3, i.e., the total gain of both agents from any matching is at most 6. So, assume that the expected gain of agents 1 and 2 from the matching returned by A is at most $4 - u$ and at most $2 + u$ respectively, for some $u \in [0, 1]$. Then, consider the application of mechanism A to instance I_1. The expected gain of agent 1 from the matching returned by A should be at most $2 - u$ (otherwise, in the original instance I, agent 1 would have an incentive to hide the white nodes v_5 and v_6 and match them internally). This means that, on input I_1, the probability that A returns a matching consisting of two edges is at most $1 - u/2$. Hence, the approximation ratio of mechanism A on instance I_1 is at least $\frac{4}{4-u}$.

Also, consider the application of mechanism A to instance I_2. The expected gain of agent 2 from the matching returned by A should be at most u (otherwise, in the original instance I, agent 2 would have an incentive to hide the gray nodes v_2 and v_3 and match them internally). This means that, on input I_2, the probability that A returns a matching consisting of two edges is at most u. Hence, the expected gain of agent 1 from instance I_2 is at most $2 + u$. Now, consider the application of A to instance I_3; A should return a non-empty matching with probability at most u (otherwise, agent 1 would have an incentive to hide nodes v_4 and v_5 from instance I_2 and match them internally). Hence, the approximation ratio of mechanism A on instance I_3 would be $1/u$.

We conclude that the approximation ratio of A is at least $\max\left\{\frac{4}{4-u}, \frac{1}{u}\right\}$ which is minimized to 5/4 for $u = 4/5$.

(b) From the analysis of (a), we have that A is inclusion-maximal only when $u = 1$ (otherwise, it would return an empty matching for instance I_3 with non-zero probability). In this case, the approximation ratio of A is at least 4/3.

(c) Since A is universally truthful, it uses a probability distribution over deterministic truthful mechanisms. We partition the set of truthful deterministic

mechanisms into two sets $\mathcal{A}_w$ and $\mathcal{A}_g$: the set $\mathcal{A}_w$ (respectively, $\mathcal{A}_g$) consists of mechanisms which, on input instance I, return a matching that leaves at least one white node (respectively, at least one gray node) unmatched. Any other truthful deterministic mechanism is arbitrarily put in one of the two sets.

Let A_w be a deterministic mechanism that belongs to $\mathcal{A}_w$. On input instance I_1, A_w should return a matching with just one edge. Otherwise, a matching with two edges would match the two white nodes v_1 and v_4 which means that agent 1 would have an incentive to hide nodes v_5 and v_6 from instance I and match them internally; this would violate the truthfulness of mechanism A_w. Hence, mechanism A_w returns matchings of size at most 1 on input instances I_1 and I_3.

Also, let A_g be a deterministic mechanism that belongs to $\mathcal{A}_g$. Consider the application of A_g to instance I_2. The matching it returns should not match node v_7 since otherwise agent 2 would have an incentive to hide nodes v_2 and v_3 in the original instance I and match them internally. Hence, only two white nodes are matched by mechanism A_g on input instance I_2. Now consider the application of A_g to the instance I_3. It should return an empty matching otherwise agent 1 would have an incentive to hide the white nodes v_4 and v_5 from instance I_2 and match them internally. Hence, the matchings returned by mechanism A_g on input instances I_2 and I_3 have size at most 2 and 0, respectively.

Next, let p be the probability that mechanism A runs a deterministic truthful mechanism from $\mathcal{A}_w$. Then, the expected size of the matching returned by A on input instances I_1 and I_3 is at most $2 - p$ and p, respectively, and its approximation ratio is at least $\max\left\{\frac{2}{2-p}, \frac{1}{p}\right\}$ which is minimized to $3/2$ for $p = 2/3$.

(d) In the proof of (c), the mechanisms in $\mathcal{A}_g$ are not inclusion-maximal. Hence, if A is universally truthful and inclusion-maximal, it should use only deterministic mechanisms from $\mathcal{A}_w$, i.e., $p = 1$. Following the analysis in the previous case for instance I_1, we obtain that A has approximation ratio at least 2.   □

Theorems 2 and 3(d) establish a separation between truthfulness in expectation and universal truthfulness with respect to inclusion-maximal mechanisms.

5   Discussion and Open Problems

Our work has shed some light on the efficiency of randomized truthful mechanisms for the 2-agent pairwise kidney exchange problem. Although the number of agents is restricted, we believe that this case is of special interest because 2-agent mechanisms can enable *ad-hoc* arrangements between hospitals in countries where national exchanges are not in place.

Clearly, the question of whether the upper bound of 2 of Ashlagi et al. [2] can be improved for instances with arbitrarily many agents remains wide open. Unfortunately, several extensions of WEIGHT-AND-MATCH that we have considered for this case have failed, and in fact it seems likely that this upper bound is tight for more than two agents. Still, the 2-agent case deserves some further investigation because there are gaps between our upper and lower bounds.

In this context, it is especially interesting to know whether a truthful in expectation, inclusion-maximal, 4/3-approximation mechanism exists. For the 2-agent case, we also believe that characterizations of truthful mechanisms would be very useful in order to complete the picture. Finally, Ashlagi et al. [2] were unable to provide a truthful deterministic mechanism for the case of more than two agents that gives any nontrivial approximation ratio. Providing such a mechanism, or proving a lower bound, remains an enigmatic open problem.

References

1. Abraham, D., Blum, A., Sandholm, T.: Clearing algorithms for barter exchange markets: enabling nationwide kidney exchanges. In: Proc. of the 8th ACM Conference on Electronic Commerce (EC), pp. 295–304 (2007)
2. Ashlagi, I., Fischer, F., Kash, I.A., Procaccia, A.D.: Mix and match. In: Proc. of the 11th ACM Conference on Electronic Commerce (EC), pp. 305–314 (2010)
3. Ashlagi, I., Roth, A.E.: Individual rationality and participation in large scale, multi-hospital kidney exchange. In: Proc. of the 12th ACM Conference on Electronic Commerce (EC), p. 321 (2011),
   http://web.mit.edu/iashlagi/www/papers/LargeScaleKidneyExchange_1_13.pdf
4. Biró, P., Manlove, D.F., Rizzi, R.: Maximum weight cycle packing in directed graphs, with application to kidney exchange programs. Discrete Mathematics, Algorithms, and Applications 1(4), 499–517 (2009)
5. Nisan, N.: Introduction to mechanism design for computer scientists. In: Algorithmic Game Theory, ch. 9, pp. 209–241. Cambridge University Press (2007)
6. Procaccia, A.D., Tennenholtz, M.: Approximate mechanism design without money. In: Proc. of the 10th ACM Conference on Electronic Commerce (EC), pp. 177–186 (2009)
7. Rees, M., Pelletier, R., Mulgaonkar, S., Laskow, D., Nibhanupudy, B., Kopke, J., Roth, A.E., Ünver, M.U., Sandholm, T., Rogers, J.: Report from a 60 transplant center multiregional kidney paired donation program. Transplantation 86(2S), 1 (2008)
8. Roth, A.E., Sönmez, T., Ünver, M.U.: Kidney exchange. Quarterly Journal of Economics 119, 457–488 (2004)
9. Roth, A.E., Sönmez, T., Ünver, M.U.: Pairwise kidney exchange. Journal of Economic Theory 125, 151–188 (2005)
10. Roth, A.E., Sönmez, T., Ünver, M.U.: Efficient kidney exchange: coincidence of wants in markets with compatibility-based preferences. American Economic Review 97, 828–851 (2007)
11. Saidman, S., Roth, A.E., Sönmez, T., Ünver, M.U., Delmonico, F.: Increasing the opportunity of live kidney donation by matching for two- and three-way exchanges. Transplantation 81(5), 773 (2006)
12. Sönmez, T., Ünver, M.U.: Market design for kidney exchange. In: Neeman, Z., Niederle, N., Vulkan, M. (eds.) Oxford Handbook of Market Design, Oxford University Press (to appear)
13. Toulis, P., Parkes, D.: A random graph model of kidney exchanges: efficiency, individual-rationality and incentives. In: Proc. of the 12th ACM Conference on Electronic Commerce (EC), pp. 323–332 (2011)

Optimal Pricing in Social Networks
with Incomplete Information

Wei Chen[1], Pinyan Lu[1], Xiaorui Sun[3,*], Bo Tang[2,*],
Yajun Wang[1], and Zeyuan Allen Zhu[4,*,**]

[1] Microsoft Research Asia
{weic,pinyanl,yajunw}@microsoft.com
[2] Shanghai Jiaotong University
tangbo1@sjtu.edu.cn
[3] Columbia University
xiaoruisun@cs.columbia.edu
[4] MIT CSAIL
zeyuan@csail.mit.edu

Abstract. In revenue maximization of selling a digital product in a so-cial network, the utility of an agent is often considered to have two parts: a private valuation, and linearly additive influences from other agents. We study the incomplete information case where agents know a common distribution about others' private valuations, and make decisions simul-taneously. The "rational behavior" of agents in this case is captured by the well-known Bayesian Nash equilibrium.

Two challenging questions arise: how to *compute* an equilibrium and how to *optimize* a pricing strategy accordingly to maximize the revenue assuming agents follow the equilibrium? In this paper, we mainly focus on the natural model where the private valuation of each agent is sampled from a uniform distribution, which turns out to be already challenging.

Our main result is a polynomial-time algorithm that can *exactly* com-pute the equilibrium and the optimal price, when pairwise influences are non-negative. If negative influences are allowed, computing any equilib-rium even approximately is PPAD-hard. Our algorithm can also be used to design an FPTAS for optimizing discriminative price profile.

1 Introduction

In this paper, we study the problem of selling a digital product to agents in a social network. To incorporate social influence, we assume each agent's utility of having the product is the summation of two parts: the private intrinsic valuation and the overall influence from her friends who also have the product. In this paper, we study the linear influence case, i.e., the overall influence is simply the summation of influence values from her friends who have the product.

---

* Part of this work was done while the authors were visiting Microsoft Research Asia.
** A preliminary version of this work has appeared as a part of the B.Sci thesis of this author [18].

N. Chen, E. Elkind, and E. Koutsoupias (Eds.): WINE 2011, LNCS 7090, pp. 49–60, 2011.
© Springer-Verlag Berlin Heidelberg 2011

Given such assumption, the purchasing decision of one agent is not solely made based on her own valuation, but also on information about her friends' purchasing decisions. However, a typical agent does not have complete information about others' private valuations, and thus might make the decision based on her belief of other agents' valuations.

We study the case when this belief forms a public distribution, and rely on the solution concept of Bayesian Nash equilibrium [8]. Specifically, each agent knows her own private valuation (also referred to as her *type*); in addition, there is a distribution of this private valuation, publicly known by everyone in the network as well as the seller. We assume that the joint distribution is a product of uniform distributions, and the valuations for all agents are sampled from it.

Computing the Equilibria. Usually, there exist multiple equilibria in this game. We first study the case when all influences are *non-negative*. We show that there exist two special ones: the *pessimistic equilibrium* and the *optimistic equilibrium*, and all other equilibria are between these two. We then design a polynomial time algorithm to compute the pessimistic (resp. optimistic) equilibrium *exactly*.

The overall idea is to utilize the fact that the pessimistic (resp. optimistic) equilibrium is "monotonically increasing" when the price increases. However, the iterative method requires exponential number of steps to converge, just like many potential games which may well be PLS-hard. Our algorithm is based on the line sweep paradigms, by increasing the price p and computing the equilibrium on the way. There are several challenges we have to address to implement the line sweep algorithm. See Section 3.1 for more discussions on the difficulties.

On the negative side, when there exist negative influences among agents, the monotone property of the equilibria does not hold. In fact, we show that computing an approximate equilibrium is PPAD-hard for a given price, by a reduction from the two player Nash equilibrium problem.

Optimal Pricing Strategy. When the seller considers offering a uniform price, our proposed line sweep algorithm calculates the equilibrium as a function of the price. This closed form allows us to find the price for the optimal revenue.

We also discuss the extensions to discriminative pricing setting: agents are partitioned into k groups and the seller can offer different prices to different groups. Depending on whether the algorithm can choose the partition or not, we discuss the hardness and approximation algorithms of these extensions.

1.1 Related Work

Pricing with equilibrium models. When there is social influence, a large stream of literature is focusing on simultaneous games. This is also known as the "two-stage" game where the seller sets the price in the first stage, and agents play a one-shot game in their purchasing decisions. Agents' rational behavior in this case is captured by Nash equilibrium (or Bayesian Nash equilibrium).

The concept and existence of pessimistic and optimistic equilibria is not new. For instance, in analogous problems with externalities, Milgrom and Roberts [12] and Vives [17] have witnessed the existence of such equilibria in the *complete information* setting. Notice that our pricing problem, when restricted to complete information, can be trivially solved by an iterative method.

In incomplete information setting, Vives and Van Zandt [16] prove a similar existential result using iterative methods. However, they do not provide any convergence guarantee. In our setting, such type of iterative methods may take exponential time to converge. (See the full version of this paper for an example.) Our proposed algorithm instead *exactly* computes the equilibrium, through a much move involved (but constructive) method. In parallel to this work, Sundararajan [15] also discover the monotonicity of the equilibria, but for symmetry and limited knowledge of the structure (only the degree distribution is known).

It is worth noting that those works above have considered non-linear influences. Though our paper focuses on linear influences, our monotonicity results for equilibria do easily extend to non-linear ones. See Section 2.

When the influence is linear, Candogan, Bimpikis and Ozdaglar [4] study the problem with (uniform) pricing model for a divisible good on sale. It differs from our paper in the model: they are in complete information and divisible good setting; more over, they have relied on a diagonal dominant assumption, which simplifies the problem and ensures the uniqueness of the equilibrium.

Another paper for linear influence is by Bloch and Querou [3], which also studies the uniform pricing model. When the influence is small, they approximate the influence matrix by taking the first 3 layers of influence, and then an equilibrium can be easily computed. They also provide experiments to show that the approximation is numerically good for random inputs.

Pricing with cascading models. In contrast to the simultaneous-move game considered by us (and many others), another stream of work focuses on the cascading models with social influence.

Hartline, Mirrokni and Sundararajan [9] study the *explore and exploit* framework. In their model the seller offers the product to the agents in a sequential manner, and assumes all agents are *myopic*, i.e., each agent is making the decision based on the known results of the previous agents in the sequence. As they have pointed out, if the pricing strategy of the seller and the private value distributions of the subsequent agents are publicly known, the agents can make more "informed" decisions than the myopic ones. In contrast to them, we consider "perfect rational" agents in the simultaneous-move game, where agents make decisions *in anticipation* of what others may do given their beliefs to other agents' valuations.

Arthur et al. [2] also use the explore and exploit framework, and study a similar problem; potential buyers do not arrive sequentially as in [9], but can choose to buy the product with some probability only if being recommended by friends.

Recently, Akhlaghpour et al. [1] consider the multi-stage model that the seller sets different prices for each stage. In contrast to [9], within each stage, agents are "perfectly rational", which is characterized by the pessimistic equilibrium in our setting with *complete information*. As mentioned in [1], they did not consider the case where a rational agent may defer her decision to later stages in order to improve the utility.

Other works. Another notable body of work in computer science is the *optimal seeding* problem (e.g. Kempe et al. [11] and Chen et al. [5]), in which a set of k seeds are selected to maximize the total influence according to some stochastic propagation model. If the value of the product does not exhibit social influence, the seller can maximize the revenue following the optimal auction process by the seminal work of Myerson [13]. Truthful auction mechanisms have also been studied for digital goods, where one can achieve constant ratio of the profit with optimal fixed price [7,10]. On computing equilibria for problems that guarantees to find an equilibrium through iterative methods, most of them, for instance the famous congestion game, is proved to be PLS-hard [6].

2   Model and Solution Concept

We consider the sale of one digital product by a seller with zero cost, to the set of agents $V = [n] = \{1, 2, \ldots, n\}$ in a social network. The network is modeled as a simple *directed* graphs $G = (V, E)$ with no self-loops.

- **Valuation:** Agent i has a private value $v_i \geq 0$ for the product. We assume v_i is sampled from a uniform distribution with interval $[a_i, b_i]$ for $0 \leq a_i < b_i$, which we denote as $U(a_i, b_i)$. The values a_i and b_i are common knowledge.
- **Price:** We consider the seller offering the product at a uniform price p.
- **Revenue:** Let $\mathbf{d} = \{d_1, \ldots, d_n\} \in \{0, 1\}^n$ be the decision vector the agents make, i.e., $d_i = 1$ if agent i buys the product and 0 otherwise. The revenue of the seller is defined as $\sum_i p \cdot d_i$. When the decisions are random variables, the revenue is defined as the expected payments received from the users.
- **Influence:** Let matrix $T = (T_{j,i})$ with $T_{j,i} \in \mathbb{R}$ and $i, j \in V$ represent the influences among agents, with $T_{j,i} = 0$ for all $(j, i) \notin E$. In particular, $T_{j,i}$ is the utility that agent i receives from agent j, if both of them buy the product. Except for the hardness result, we consider $T_{j,i}$ to be non-negative.
- **Utility:** Let $\mathbf{d}_{-i}$ be the decision vector of the agents other than agent i. For convenience, we denote $\langle d_i', \mathbf{d}_{-i} \rangle$ the vector by replacing the i-th entry of $\mathbf{d}$ by d_i'. In particular, given the influence matrix T, the utility is defined as:

$$u_i(\langle d_i, \mathbf{d}_{-i} \rangle, v_i, p) = \begin{cases} v_i - p + \sum_{j \in [n]} d_j \cdot T_{j,i}, & \text{if } d_i = 1 \\ 0, & \text{if } d_i = 0 \end{cases} \quad (1)$$

Remark 2.1. In our algorithm later, the requirement $a_i < b_i$ is only for ease of presentation. It can be relaxed to $a_i \leq b_i$ to handle fixed value case as well.

We study the agents' *rational behavior* using the concept Bayesian Nash equilibrium (BNE).[1]

---

[1] Given equilibrium $\mathbf{q}$ in our definition, the strategy profile that each agent i "buys the product iff her valuation $v_i \geq p - \sum_{j \neq i} T_{j,i} q_j$" is a BNE. See the full version of this paper for details.

Definition 2.2. *The probability vector* $\mathbf{q} = (q_1, q_2, ..., q_n) \in [0, 1]^n$ *is an equilibrium at price* p, *if for all* $i \in [n]$: *(where med is the median function)*

$$q_i = \Pr_{v_i \sim U(a_i, b_i)} \left[v_i - p + \sum_{j \in [n]} T_{j,i} \cdot q_j \geq 0 \right] = \text{med} \left\{ 0, 1, \frac{b_i - p + \sum_{j \in [n]} T_{j,i} q_j}{b_i - a_i} \right\}. \quad (2)$$

Eq.(2) can be also defined in the language of a transfer function, which we will extensively reply on in the rest of the paper.

Definition 2.3 (Transfer function). *Given price* p, *we define the transfer function* $f_p : [0, 1]^n \to [0, 1]^n$ *as*

$$[f_p(\mathbf{q})]_i = \text{med}\{0, 1, [g_p(\mathbf{q})]_i\} \quad (3)$$

in which

$$[g_p(\mathbf{q})]_i = \frac{b_i - p + \sum_{j \in [n]} T_{j,i} q_j}{b_i - a_i}.$$

Notice that $\mathbf{q}$ *is an equilibrium at price* p *if and only* $f_p(\mathbf{q}) = \mathbf{q}$.

Using Brouwer fixed point theorem, the existence of BNE is not surprising, even when influences are negative. However, we will show in Section 4 that computing BNE will be PPAD-hard with negative influences. We now define the pessimistic and optimistic equilibria based on the transfer function.

Definition 2.4. *Let* $f_p^{(1)} = f_p$, *and* $f_p^{(m)}(\mathbf{q}) = f_p(f_p^{(m-1)}(\mathbf{q}))$ *for* $m \geq 2$. *When all influences are non-negative, we define*

- **Pessimistic equilibrium:** $\underline{\mathbf{q}}(p) = \lim_{m \to \infty} f_p^{(m)}(\mathbf{0})$;
- **Optimistic equilibrium:** $\overline{\mathbf{q}}(p) = \lim_{m \to \infty} f_p^{(m)}(\mathbf{1})$.

We remark that both limits exist by monotonicity of f (see Fact 2.5 below), when all influences are non-negative. In addition, $\underline{\mathbf{q}}(p)$ and $\overline{\mathbf{q}}(p)$ are both equilibria themselves, because $f_p(\underline{\mathbf{q}}(p)) = \underline{\mathbf{q}}(p)$ and $f_p(\overline{\mathbf{q}}(p)) = \overline{\mathbf{q}}(p)$. We later show that $\underline{\mathbf{q}}(p)$ and $\overline{\mathbf{q}}(p)$ are the lower bound and upper bound for any equilibrium at price p respectively. Now we state some properties of equilibria, which we will use extensively later. See the full version of this paper for proofs.

For two vectors $\mathbf{v}_1, \mathbf{v}_2 \in \mathbb{R}^n$, we write $\mathbf{v}_1 \geq \mathbf{v}_2$ if $\forall i \in [n]$, $[\mathbf{v}_1]_i \geq [\mathbf{v}_2]_i$ and we write $\mathbf{v}_1 > \mathbf{v}_2$ if $\mathbf{v}_1 \geq \mathbf{v}_2$ and $\mathbf{v}_1 \neq \mathbf{v}_2$.

Fact 2.5. *When all influences are non-negative, given* $p_1 \leq p_2, \mathbf{q}^1 \leq \mathbf{q}^2$, *the transfer function satisfies* $f_{p_2}(\mathbf{q}^1) \leq f_{p_1}(\mathbf{q}^1) \leq f_{p_1}(\mathbf{q}^2)$.

Lemma 2.6. *When all influences are non-negative, equilibria satisfy the following properties:*

a) *For any equilibrium* $\mathbf{q}$ *at price* p, *we have* $\underline{\mathbf{q}}(p) \leq \mathbf{q} \leq \overline{\mathbf{q}}(p)$.
b) *Given price* p, *for any vector* $\mathbf{q} \leq \underline{\mathbf{q}}(p)$, *we have* $f_p^{(\infty)}(\mathbf{0}) = \underline{\mathbf{q}}(p) = f_p^{(\infty)}(\mathbf{q})$.
c) *Given price* $p_1 \leq p_2$, *we have* $\underline{\mathbf{q}}(p_1) \geq \underline{\mathbf{q}}(p_2)$ *and* $\overline{\mathbf{q}}(p_1) \geq \overline{\mathbf{q}}(p_2)$.
d) $\underline{\mathbf{q}}(p) = \lim_{\varepsilon \to 0+} \underline{\mathbf{q}}(p + \varepsilon)$ *and* $\overline{\mathbf{q}}(p) = \lim_{\varepsilon \to 0-} \underline{\mathbf{q}}(p + \varepsilon)$.

In this paper, we consider the problem that whether we can exactly calculate the pessimistic (resp. optimistic) equilibrium, and whether we can maximize the revenue. The latter is formally defined as follows:

Definition 2.7 (Revenue maximization problem)
Assume the value of agent i is sampled from $U(a_i, b_i)$ and the influence matrix T is given. The revenue maximization problem is to compute an optimal price with respect to the pessimistic *equilibrium (resp.* optimistic *equilibrium):*

$$\arg\max_{p>0} \sum_{i \in [n]} p \cdot [\underline{q}(p)]_i \quad (resp. \ \arg\max_{p>0} \sum_{i \in [n]} p \cdot [\overline{q}(p)]_i \ ).$$

Notice that the optimal revenue with respect to the pessimistic equilibrium is robust against equilibrium selection. By Lemma 2.6(a), no matter which equilibrium the agents choose, this revenue is a minimal guarantee from the seller's perspective. The revenue guarantees for pessimistic and optimistic equilibria is an important objective to study; see for instance the *price of anarchy* and the *price of stability* in [14] for details.

3   The Main Algorithm

When all influences are non-negative, can we calculate $\underline{q}(p)$ and $\overline{q}(p)$ in polynomial time? We answer this question positively in this section by providing an efficient algorithm. Notice that it is possible to iteratively apply the transfer function Eq.(3) to reach the equilibria, but this may take exponential time. See the full version of this paper for a counter example.

3.1   Outline of Our Line Sweep Algorithm

We start to introduce our algorithm with the easy case where valuations of agents are fixed. Consider the pessimistic decision vector as a function of p. By monotonicity, there are at most $O(n)$ different such vectors when p varies from $+\infty$ to 0. In particular, at each price p, if we decrease p gradually to some threshold value, one more agent would change his decision to buy the product. Such kind of process can be casted in the "line sweep algorithm" paradigm.

When the private valuations of the agents are sampled from uniform distributions, the line sweep algorithm is much more complicated. We now introduce the algorithm to obtain the pessimistic equilibrium $\underline{q}(p)$, while the method to obtain $\overline{q}(p)$ is similar.[2] The essence of the line sweep algorithm is processing the events corresponding to some structural changes. We define the possible structures of a probability vector as follows.

Definition 3.1. *Given $q \in [0, 1]^n$, we define the structure function $S : [0, 1]^n \rightarrow \{0, \star, 1\}^n$ satisfying:*

$$[S(\mathbf{q})]_i = \begin{cases} 0, & q_i = 0 \\ \star, & q_i \in (0, 1) \\ 1, & q_i = 1. \end{cases} \tag{4}$$

---

[2] We sweep the price from $+\infty$ to 0 to compute the pessimistic equilibrium, but we need to sweep from 0 to $+\infty$ for the optimistic one.

Our line sweep algorithm is based on the following fact: when p is sufficiently large, obviously $\mathbf{q}(p) = \mathbf{0}$; with the decreasing of p, at some point $p = p_1$ the pessimistic equilibrium $\underline{\mathbf{q}}(p)$ becomes non-zero, and there exists some *structural change* at this moment. Due to the monotonicity of $\underline{\mathbf{q}}(p)$ in Lemma 2.6, such structural changes can happen at most $2n$ times. (Each agent i can contribute to at most two changes: $0 \rightarrow \star$ and $\star \rightarrow 1$.) Therefore, there exist threshold prices $p_1 > p_2 > \cdots > p_m$ for $m \leq 2n$ such that within two consecutive prices, the structure of the pessimistic equilibrium remains unchanged and $\underline{\mathbf{q}}(p)$ is a linear function of p. This indicates that the total revenue, i.e., $p \cdot \sum_i [\underline{\mathbf{q}}(p)]_i$, and its maximum value is easy to obtain. If we can compute the threshold prices and the corresponding pessimistic equilibrium $\underline{\mathbf{q}}(p)$ as a function of p, it will be straightforward to determine the optimal price p.

There are several difficulties to address in this line sweep algorithm.

- First, degeneracies, i.e., more than one structural changes in one event, are intrinsic in our problem. Unlike geometric problems where degeneracies can often be eliminated by perturbations, the degeneracies in our problem are persistent to small perturbations.
- Second, to deal with degeneracies, we need to identify the next structural change, which is related to the eigenvector corresponding to the largest eigenvalue of a linear operator. By a careful inspection, we avoid solving eigen systems so that our algorithm can be implemented by pure algebraic computations.
- Third, after the next change is identified, the usual method of pushing the sweeping line further does not work directly in our case. Instead, we recursively solve a subproblem and combine the solution of the subproblem with the current one to a global solution. The polynomial complexity of our algorithm is guaranteed by the monotonicity of the structures.

We first design a line sweep algorithm for the problem with a diagonal dominant condition, which will not contain degenerate cases, in Section 3.2. Then we describe techniques to deal with the unrestricted case in Section 3.3.

3.2 Diagonal Dominant Case

Definition 3.2 (Diagonal dominant condition)
Let $L_{i,j} = T_{j,i}/(b_i - a_i)$ and $L_{i,i} = T_{i,i} = 0$. The matrix $I - L$ is strictly diagonal dominant, if $\sum_j L_{i,j} = \sum_j T_{j,i}/(b_i - a_i) < 1$.

This condition has some natural interpretation on the buying behavior of the agents. It means that the decision of any agent cannot be solely determined by the decisions of her friends. In particular, the following two situations cannot occur *simultaneously* for any agent i and price p: a) agent i will not buy the product regardless of her own valuation when none of her friends bought the product($p \geq b_i$), and b) agent i will always buy the product regardless of her own valuation when all her friends bought the product ($\sum_j T_{j,i} + a_i \geq p$).

In our line sweep algorithm, we maintain a *partition* $Z \cup W \cup O = V = [n]$, and name Z the *zero set*, W the *working set* and O the *one set*. This corresponds to the structure $\mathbf{s} \in \{0, \star, 1\}^n$ as follows:

$$s_i = 0 \ (\forall i \in Z), \quad s_i = \star \ (\forall i \in W), \quad s_i = 1 \ (\forall i \in O).$$

We use $\mathbf{x}_W$ to denote the restriction of vector $\mathbf{x}$ on set W, and for simplicity we write $\langle \mathbf{x}_Z, \mathbf{x}_W, \mathbf{x}_O \rangle = \mathbf{x}$. Let $L_{W \times W}$ be the projection of matrix L to $W \times W$.

We start from the price $p = +\infty$ where the structure of the pessimistic equilibrium $\underline{\mathbf{q}}(p)$ is $\mathbf{s}^0 = \mathbf{0}$, i.e., $Z = [n]$ and $W = O = \emptyset$. The first event happens when p drops to $p_1 = \max_i b_i$ and $\underline{\mathbf{q}}(p)$ starts to become non-zero.

Assume now we have reached threshold price p_t, the current pessimistic equilibrium is $\mathbf{q}^t = \underline{\mathbf{q}}(p_t)$, and the structure in interval (p_t, p_{t-1}) (or $(p_t, +\infty)$ if $t = 1$) is $\mathbf{s}^{t-1}$. We define

$$\mathbf{x} = \left(\frac{b_1 - p_t}{b_1 - a_1}, \frac{b_2 - p_t}{b_2 - a_2}, \ldots, \frac{b_n - p_t}{b_n - a_n} \right)^T, \text{ and } \mathbf{y} = \left(\frac{1}{b_1 - a_1}, \frac{1}{b_2 - a_2}, \ldots, \frac{1}{b_n - a_n} \right)^T.$$

To analyze the pessimistic equilibrium in the next price interval, for price $p = p_t - \varepsilon$ where $\varepsilon > 0$, we write function $g_p(\cdot)$ (recall Eq.(3)) as:

$$g_{p_t - \varepsilon}(\mathbf{q}) = \mathbf{x} + \varepsilon \mathbf{y} + L\mathbf{q}.$$

For $p \in (p_t, p_{t-1})$, let partition $Z \cup W \cup O = [n]$ be consistent with the structure $\mathbf{s}^{t-1}$. According to Def. 3.1 and the right continuity $\mathbf{q}^t = \lim_{p \to p_t+} \underline{\mathbf{q}}(p)$ (see Lemma 2.6d), we have

$$\begin{aligned}
\forall i \in Z, \ [g_{p_t}(\mathbf{q}^t)]_i &= [\mathbf{x} + L\mathbf{q}^t]_i \leq 0 \\
\forall i \in W, \ [g_{p_t}(\mathbf{q}^t)]_i &= [\mathbf{x} + L\mathbf{q}^t]_i \in (0, 1] \\
\forall i \in O, \ [g_{p_t}(\mathbf{q}^t)]_i &= [\mathbf{x} + L\mathbf{q}^t]_i \geq 1
\end{aligned} \tag{5}$$

Step 1: For any $i \in Z$, if $[\mathbf{x} + L\mathbf{q}^t]_i = 0$, move i from zero set Z to working set W; for any $i \in W$, if $[\mathbf{x} + L\mathbf{q}^t]_i = 1$, move i from working set W to one set O.

Notice that the structural changes we apply in Step 1 are exactly the changes defining the threshold price p_t. We will see in a moment that after the process in Step 1, the new partition will be the next structure $\mathbf{s}^t$ for $p \in (p_{t+1}, p_t)$. In other words, there is no more structural change at price p_t.

In the next two steps, we calculate the next threshold price p_{t+1}. For notation simplicity, we assume Z, W and O remain unchanged in these two steps. When p decreases by ε, we show that the probability vector of agents in W, $[\underline{\mathbf{q}}(p)]_W$, increases linearly with respect to ε. (See $\mathbf{r}_W(\varepsilon)$ below.) However, this linearity holds until we reach some point, where the next structural change takes place.

Step 2: Define the vector $\mathbf{r}(\varepsilon) \in \mathbb{R}^n$, and let:

$$\begin{aligned}
\mathbf{r}_W(\varepsilon) &= \varepsilon(I - L_{W \times W})^{-1} \mathbf{y}_W + \mathbf{q}_W^t \\
&= \varepsilon(I - L_{W \times W})^{-1} \mathbf{y}_W + [\mathbf{x} + L\mathbf{q}^t]_W \\
\mathbf{r}_Z(\varepsilon) &= \mathbf{x}_Z + \varepsilon \mathbf{y}_Z + L_{Z \times W} \mathbf{r}_W(\varepsilon) + L_{Z \times O} \mathbf{1}_O \\
&= \varepsilon(\mathbf{y}_Z + L_{Z \times W}(I - L_{W \times W})^{-1} \mathbf{y}_W) + [\mathbf{x} + L\mathbf{q}^t]_Z \\
\mathbf{r}_O(\varepsilon) &= \mathbf{x}_O + \varepsilon \mathbf{y}_O + L_{O \times W} \mathbf{r}_W(\varepsilon) + L_{O \times O} \mathbf{1}_O \\
&= \varepsilon(\mathbf{y}_O + L_{O \times W}(I - L_{W \times W})^{-1} \mathbf{y}_W) + [\mathbf{x} + L\mathbf{q}^t]_O
\end{aligned} \tag{6}$$

Clearly, $\mathbf{r}(\varepsilon)$ is linear to ε and we write $\mathbf{r}(\varepsilon) = \varepsilon\boldsymbol{\ell} + (\mathbf{x} + L\mathbf{q}^t)$ where $\boldsymbol{\ell} = \langle \ell_1, \ell_2, \ldots, \ell_n \rangle \in \mathbb{R}^n$ is the linear coefficient derived from Eq.(6). When $I - L$ is strictly diagonal dominant, the largest eigenvalue of $L_{W \times W}$ is smaller than 1. Using this property one can verify (see full version) that $\boldsymbol{\ell}$ is strictly positive.

Step 3

$$\varepsilon_{min} = \min \left\{ \min_{i \in Z} \left\{ \frac{0 - [\mathbf{x} + L\mathbf{q}^t]_i}{\ell_i} \right\}, \min_{i \in W} \left\{ \frac{1 - [\mathbf{x} + L\mathbf{q}^t]_i}{\ell_i} \right\} \right\} \tag{7}$$

Using the positiveness of vector $\boldsymbol{\ell}$ one can verify that $\varepsilon_{min} > 0$. Also, the next threshold price $p_{t+1} = p_t - \varepsilon_{min}$. (See full version for proofs.)

Lemma 3.3. $\forall 0 < \varepsilon \leq \varepsilon_{min}$, $\underline{\mathbf{q}}(p_t - \varepsilon) = \langle \mathbf{0}_Z, \mathbf{r}_W(\varepsilon), \mathbf{1}_O \rangle$.

We remark here that the above lemma has confirmed that our structural adjustments in Step 1 are correct and complete. Now we let $p_{t+1} = p_t - \varepsilon_{min}$, $\mathbf{q}^{t+1} = \langle \mathbf{0}_Z, \mathbf{r}_W(\varepsilon_{min}), \mathbf{1}_O \rangle$. The next structural change will take place at $p = p_{t+1}$. This is because according to the definition of ε_{min} (Eq.(7)), there must be some

$$i \in W \wedge \left[\mathbf{x} + \varepsilon_{min}\mathbf{y} + L\mathbf{q}^{t+1} \right]_i = 1, \quad \text{or} \quad i \in Z \wedge \left[\mathbf{x} + \varepsilon_{min}\mathbf{y} + L\mathbf{q}^{t+1} \right]_i = 0.$$

One can see that in the next iteration, this i will move to one set O or working set W accordingly. Therefore, we can iteratively execute the above three steps by sweeping the price further down.

The return value of our constrained line sweep method is a function $\underline{\mathbf{q}}$ which gives the pessimistic equilibrium for any price $p \in \mathbb{R}$, and $\underline{\mathbf{q}}(p)$ is a piecewise linear function of p with no more than $2n + 1$ pieces. All three steps in our algorithm can be done in polynomial time. Since there are only $O(n)$ threshold prices, we have the following result.

Theorem 3.4. *When the matrix $I - L$ is strictly diagonal dominant, we can calculate the pessimistic equilibrium $\underline{\mathbf{q}}(p)$ (resp. $\overline{\mathbf{q}}(p)$) for any given price p in polynomial time, together with the optimal revenue.*

3.3   General Case

After relaxing the diagonal dominance condition, the algorithm becomes more complicated. This can be seen from this simple scenario. There are 2 agents, with $[a_1, b_1] = [a_2, b_2] = [0, 1]$, and $T_{1,2} = T_{2,1} = 2$. One can verify that $\underline{\mathbf{q}}(p) = (0, 0)^T$ when $p \geq 1$; $\underline{\mathbf{q}}(p) = (1, 1)^T$ when $p < 1$.

In this example, there is an *equilibrium jump* at price $p = 1$, i.e., $\underline{\mathbf{q}}(1) \neq \lim_{p \to 1-} \underline{\mathbf{q}}(p)$. Our previous algorithm essentially requires that both the left and the right continuity of $\underline{\mathbf{q}}(p)$. However, only the right continuity is unconditional by Lemma 2.6d. More importantly, degeneracies may occur: the new structure $\mathbf{s}^t$ when $p = p_t$ cannot be determined all in once in Step 1. When p goes from $p_t + \varepsilon$ to $p_t - \varepsilon$, there might take place even two-stage jumps: some index i might leave Z for O, without being in the intermediate state.

Let $\rho(L)$ be the largest norm of the eigenvalues in matrix L. The ultimate reason for such degeneracies, is $\rho(L_{W \times W}) \geq 1$ and $(I - L_{W \times W})^{-1} \neq \lim_{m \to \infty}(I + L_{W \times W} + \cdots + L_{W \times W}^{m-1})$. We will prove shortly in such cases, those structural changes in Step 1 are *incomplete*, that is, as p sweeps across p_t, at least one more structural change will take place. We derive a method to identify one *pivot*, i.e. an additional structural change, in polynomial time. Afterwards, we recursively solve a subproblem with set O taken out, and combine the solution from the subproblem with the current one. The follow lemma shows that whether $\rho(L) < 1$ can be determined efficiently.

Lemma 3.5. *Given non-negative matrix M, if $I - M$ is reversible and $(I - M)^{-1}$ is also non-negative, then $\rho(M) < 1$; on the contrary, if $I - M$ is degenerate or if $(I - M)^{-1}$ contains negative entries, $\rho(M) \geq 1$.*

Finding the pivot. When $\rho(L_{W \times W}) < 1$ for the new working set W, one can find the next threshold price p_{t+1} following Step 2 and 3 in the previous subsection. Now, we deal with the case that $\rho(L_{W \times W}) \geq 1$ by showing that there must exists some additional agent $i \in W$ such that $[\mathbf{q}(p)]_i = 1$ for any p smaller than the current price. We call such agent a *pivot*.

Since $\rho(L_{W \times W}) \geq 1$, we can always find a non-empty set $W_1 \subset W$ and $W_2 = W_1 \cup \{w\} \subset W$, satisfying $\rho(L_{W_1 \times W_1}) < 1$ but $\rho(L_{W_2 \times W_2}) \geq 1$. The pair (W_1, W_2) can be found by ordering the elements in W and add them to W_1 one by one. We now show that there is a pivot in W_2.

As $L_{W_2 \times W_2}$ is a non-negative matrix, based on knowledge from spectral theory, exists a non-zero eigenvector $\mathbf{u}_{W_2} \geq \mathbf{0}_{W_2}$ such that $L_{W_2 \times W_2} \mathbf{u}_{W_2} = \lambda \mathbf{u}_{W_2}$ and $\lambda = \rho(L_{W_2 \times W_2}) \geq 1$. $\mathbf{u}_{W_2}$ can be extended to $[n]$ by defining $\mathbf{u}_{[n] \backslash W_2} = \mathbf{0}_{[n] \backslash W_2}$. Let

$$k = \underset{k \in W_2, u_k \neq 0}{\arg\min} \frac{1 - q_k^t}{u_k} = \underset{k \in [n], u_k \neq 0}{\arg\min} \frac{1 - q_k^t}{u_k} \tag{8}$$

Now we prove that k is a pivot. Intuitively, if we slightly increase the probability vector $\mathbf{q}_{W_2}^t$ by $\delta \mathbf{u}_{W_2}$, where δ is a small constant, by performing the transfer function only on agents in W m times, their probability will increase by $\delta(1 + \lambda + .. + \lambda^m)\mathbf{u}_{W_2}$, while $\lambda \geq 1$. Therefore, after performing the transfer function sufficiently many times, agent $k \in W_2$'s probability will hit 1 first.

Lemma 3.6. $\forall W_2 \subset W$ *s.t.* $\rho(L_{W_2 \times W_2}) \geq 1$, *we have* $\forall \varepsilon > 0$, $[\mathbf{q}(p_t - \varepsilon)]_k = 1$.

We remark that if we can exactly estimate the eigenvector (which may be irrational), then the above lemma has already determined that the k defined in Eq.(8) is a pivot. To avoid the eigenvalue computation, we find a quasi-eigenvector $\mathbf{u}$ in the following manner.

$$\mathbf{u} = \begin{cases} \mathbf{u}_{W_1} = (I - L_{W_1 \times W_1})^{-1} L_{W_1 \times \{w\}}; \\ u_w = 1; \\ \mathbf{u}_{Z \cup O \cup W \backslash W_2} = \mathbf{0}_{Z \cup O \cup W \backslash W_2}. \end{cases} \tag{9}$$

The meaning of the above vector is as follows. If we raise agent w's probability by δ, those probabilities of agents in W_1 increase proportionally to $L_{W_1 \times \{w\}} \delta$.

Assuming that we ignore the probability changes outside W_2 (which will even increase the probabilities in W_2), the probability of agents in W_1 will eventually converge to $(I + L_{W_1 \times W_1} + L_{W_1 \times W_1}^2 + ...)L_{W_1 \times \{w\}}\delta = (I - L_{W_1 \times W_1})^{-1}L_{W_1 \times \{w\}}\delta$.

We will see that the real probability vector increases at least "as much as if we increase in the direction of $\mathbf{u}$". In other words, we pick a pivot in the same way as Eq.(8). The following is the critical lemma to support our result.

Lemma 3.7. *For $\mathbf{u}$ in Eq.(9) and k in Eq.(8), we have $\forall \varepsilon > 0, [\mathbf{q}(p_t - \varepsilon)]_k = 1$.*

Recursion on the subproblem. Let $W' = W \setminus \{k\}$, $O' = O \cup \{k\}$, and we consider a subproblem with $n' = n - |O'| < n$ agents, where k is the pivot identified in the previous section. This subproblem is a projection of the original one, assuming that the agents in O' always tend to buy the product.

$$\forall i \in Z \cup W', \quad [a_i', b_i'] = [a_i + \textstyle\sum_{j \in O'} T_{j,i}, b_i + \sum_{j \in O'} T_{j,i}]. \tag{10}$$

By recursively solving this new instance, we can solve the pessimistic equilibrium of the subproblem for any given price p. This recursive procedure will eventually terminate because every invocation reduces the number of agents by at least 1. The following lemma tells us that for any $p < p_t$, the pessimistic equilibrium of the original problem and the subproblem are one-to-one.

Lemma 3.8. *Let $\underline{\mathbf{q}}'(p)$ be the pessimistic equilibrium function in the subproblem. We have:*

$$\forall p < p_t, \underline{\mathbf{q}}(p) = \langle \underline{\mathbf{q}}'(p), \mathbf{1}_{O'} \rangle.$$

At this moment we have solved the pessimistic equilibrium $\underline{\mathbf{q}}(p)$ for $p < p_t$, and thus solved the original problem. Again $\underline{\mathbf{q}}(p)$ is a piecewise linear function of p with no more than $2n + 1$ pieces.

Theorem 3.9. *For matrix T satisfying $T_{i,i} = 0$ and $T_{i,j} \geq 0$, in polynomial time we can calculate the pessimistic equilibrium $\underline{\mathbf{q}}(p)$ (resp. $\overline{\mathbf{q}}(p)$) at any price p, together with the optimal revenue.*

4 Extensions

In the full version of this paper, we also prove the following theorems. When the influence values can be negative, it is actually PPAD-hard to compute an *approximate* equilibrium. We define a probability vector $\mathbf{q}$ to be an ε-approximate equilibrium for price p if:

$$q_i \in (q_i' - \varepsilon, q_i' + \varepsilon),$$

where $q_i' = \mathrm{med}\left\{0, 1, \frac{b_i - p + \sum_{j \in [n]} T_{j,i} q_j}{b_i - a_i}\right\}$. We have the following theorem:

Theorem 4.1. *It is PPAD-hard to compute an n^{-c}-approximate equilibrium of our pricing system for any $c > 1$ when influences can be negative.*

In discriminative pricing setting, we study the revenue maximization problem in two natural models. We assume the agents are partitioned into k groups.

The seller can offer different prices to different groups. The first model we consider is the fixed partition model, i.e., the partition is predefined. In the second model, we allow the seller to partition the agents into k groups and offer prices to the groups respectively. We have the following two theorems:

Theorem 4.2. *There is an FPTAS for the discriminative pricing problem in the fixed partition case with constant k.*

Theorem 4.3. *It is NP-hard to compute the optimal pessimistic discriminative pricing equilibrium in the choosing partition case.*

References

1. Akhlaghpour, H., Ghodsi, M., Haghpanah, N., Mirrokni, V.S., Mahini, H., Nikzad, A.: Optimal Iterative Pricing over Social Networks. In: Saberi, A. (ed.) WINE 2010. LNCS, vol. 6484, pp. 415–423. Springer, Heidelberg (2010)
2. Arthur, D., Motwani, R., Sharma, A., Xu, Y.: Pricing Strategies for Viral Marketing on Social Networks. In: Leonardi, S. (ed.) WINE 2009. LNCS, vol. 5929, pp. 101–112. Springer, Heidelberg (2009)
3. Bloch, F., Quérou, N.: Pricing with local network externalities. Technical report (July 2009)
4. Candogan, O., Bimpikis, K., Ozdaglar, A.E.: Optimal Pricing in the Presence of Local Network Effects. In: Saberi, A. (ed.) WINE 2010. LNCS, vol. 6484, pp. 118–132. Springer, Heidelberg (2010)
5. Chen, W., Wang, Y., Yang, S.: Efficient influence maximization in social networks. In: SIGKDD 2009, pp. 199–208 (2009)
6. Fabrikant, A., Papadimitriou, C., Talwar, K.: The complexity of pure nash equilibria. In: STOC 2004, pp. 604–612 (2004)
7. Goldberg, A.V., Hartline, J.D., Karlin, A.R., Saks, M., Wright, A.: Competitive auctions. Games and Economic Behavior 55(2), 242–269 (2006)
8. Harsanyi, J.C.: Games with incomplete information played by "bayesian" players, i-iii. part i. the basic model. Management Science 14(3), 159–182 (1967)
9. Hartline, J., Mirrokni, V., Sundararajan, M.: Optimal marketing strategies over social networks. In: WWW 2008, pp. 189–198 (2008)
10. Hartline, J.D., McGrew, R.: From optimal limited to unlimited supply auctions. In: ACM-EC 2005, pp. 175–182 (2005)
11. Kempe, D., Kleinberg, J., Tardos, É.: Maximizing the spread of influence through a social network. In: SIGKDD 2003, pp. 137–146 (2003)
12. Milgrom, P., Roberts, J.: Rationalizability, learning, and equilibrium in games with strategic complementarities. Econometrica 58(6), 1255–1277 (1990)
13. Myerson, R.B.: Optimal auction design. Math. Oper. Res. 6(1), 58 (1981)
14. Nisan, N., Roughgarden, T., Tardos, É., Vazirani, V.V.: Algorithmic game theory. Cambridge University Press (2007)
15. Sundararajan, A.: Local network effects and complex network structure. The B.E. Journal of Theoretical Economics 7(1) (2008)
16. Van Zandt, T., Vives, X.: Monotone equilibria in bayesian games of strategic complementarities. Journal of Economic Theory 134(1), 339–360 (2007)
17. Vives, X.: Nash equilibrium with strategic complementarities. Journal of Mathematical Economics 19(3), 305–321 (1990)
18. Zhu, Z.A.: Two Topics on Nash Equilibrium in Algorithmic Game Theory. Bach- elor's thesis. Tsinghua University (June 2010)

On the Approximation Ratio of k-Lookahead Auction

Xue Chen[1,*], Guangda Hu[2,*], Pinyan Lu[3], and Lei Wang[4]

[1] Department of Computer Science, University of Texas at Austin
xchen@cs.utexas.edu
[2] Department of Computer Science, Princeton University
guangdah@cs.princeton.edu
[3] Microsoft Research Asia
pinyanl@microsoft.com
[4] Georgia Institute of Technology
lwang@cc.gatech.edu

Abstract. We consider the problem of designing a profit-maximizing single-item auction, where the valuations of bidders are correlated. We revisit the k-lookahead auction introduced by Ronen [6] and recently further developed by Dobzinski, Fu and Kleinberg [2]. By a more delicate analysis, we show that the k-lookahead auction can guarantee at least $\frac{e^{1-1/k}}{e^{1-1/k}+1}$ of the optimal revenue, improving the previous best results of $\frac{2k-1}{3k-1}$ in [2]. The 2-lookahead auction is of particular interest since it can be derandomized [2, 5]. Therefore, our result implies a polynomial time deterministic truthful mechanism with a ratio of $\frac{\sqrt{e}}{\sqrt{e}+1} \approx 0.622$ for any single-item correlated-bids auction, improving the previous best ratio of 0.6. Interestingly, we can show that our analysis for 2-lookahead is tight. As a byproduct, a theoretical implication of our result is that the gap between the revenues of the optimal deterministically truthful and truthful-in-expectation mechanisms is at most a factor of $\frac{1+\sqrt{e}}{\sqrt{e}}$. This improves the previous best factor of $\frac{5}{3}$ in [2].

1 Introduction

Optimal auction design is an important subject that has been heavily studied in both economics and theoretical computer science. Among the accomplished research in this area, a solid part is focused on *single-item auction*, which serves as a basic that provides insight to other more complicated problems. In the seminal paper [4], Myerson gave a complete characterization of the optimal single-item auction in the setting where bidders' valuations are drawn from independent distributions. However, the design of optimal auction with correlated bidders was left open.

From the economics aspect, a natural attempt for solving this problem is to generalize Myerson's characterization. Unfortunately, most results obtained via this approach are for restricted special cases, see [3] for a survey. One exception is [1] by Cremer and McLean where they relax the *individually rational* constraint and obtain mechanisms

---

[*] This work was done when the authors were undergraduate students at Tsinghua university and intern students at Microsoft Research Asia.

N. Chen, E. Elkind, and E. Koutsoupias (Eds.): WINE 2011, LNCS 7090, pp. 61–71, 2011.

that extract the full social welfare. On the other hand, from a computer science aspect, two research directions (see [2]) were suggested.

The first one is the introduction of approximation algorithms into optimal auction design. In other word, instead of providing a characterization of the optimal auction, which might not even exist, one would look for efficient algorithms that guarantee the approximate optimality.

Along this direction, two computational models were considered-the *explicit model* [5] and the *oracle model* [6]. In the explicit model, the running time of an algorithm has to be polynomial in the support size of the distribution. However, in the oracle model, the algorithm is only allowed to make polynomial in the number of bidders queries to an oracle that returns the conditional distribution of a set of bidders given the values of the remaining ones. Ronen [6] gave the first efficient mechanism in the oracle model called 1-lookahead that 2-approximates the optimal revenue. In [7], Ronen and Saberi further proved that no deterministic efficient *ascending auction* can do better than $\frac{3}{4}$. On the other hand, in the explicit model, Papadimitriou and Pierrakos [5] showed that although there is an optimal deterministic auction among optimal truthful-in-expectation auctions for two bidders and this auction can be computed efficiently, it is NP-hard to find the optimal deterministic one for more than three bidders. The understanding the approximability of the optimal auction remains as a major challenge.

The second direction suggested is to relax the solution concept to *truthfulness-in-expectation*. One advantage of such relaxation is that the optimal truthful-in-expectation auction can be described as a linear program [2, 5] whose size is polynomial in the support of the distribution, hence can be computed efficiently in the explicit model.

Based on this observation, Dobzinski et.al. [2] studied a class of truthful-in-expectation mechanisms called *k-lookahead*. To be precise, for any fixed constant k, the k-lookahead mechanism runs the linear program among the k bidders with the highest bids, conditioning on the remaining bidders. Since k is a constant, the linear program can be solved efficiently in the oracle model.

In [2], the authors showed that the k-lookahead mechanism has approximation ratio $\frac{2k-1}{3k-1}$. As usual in computer science, improving this approximation ratio would be an important issue in this direction. Furthermore, a question that is of theoretical interest itself is the task of evaluating the gap between truthful-in-expectation and deterministically truthful mechanisms. Obviously, one would expect truthful-in-expectation mechanisms to achieve more revenue than the deterministic ones. Dobzinski et.al. [2] showed the gap is existed by providing an example of truthful-in-expectation mechanism that cannot be implemented as an universally truthful mechanism. At the same time, Papadimitriou et.al. [5] and Dobzinski et.al. [2] showed that there is an elegant derandomization of the 2-lookahead mechanism. In [2], Dobzinski showed that the gap is at most a factor of 5/3 between truthful-in-expectation and the optimal deterministically truthful. Closing the gap further requires either better truthful-in-expectation mechanisms that can be derandomized, or simply tighter analysis of the 2-lookahead mechanism.

Our results. In this paper, we contribute to both research directions mentioned earlier by providing more delicate analysis of the k-lookahead mechanisms in the oracle model. We show that the approximation ratio of k-lookahead mechanism is at least $\frac{e^{1-1/k}}{1+e^{1-1/k}}$, which improves the ratio given in [2]. In particular, our result implies that 2-lookahead mechanism is at least $\frac{\sqrt{e}}{1+\sqrt{e}}$-approximate and interestingly, we prove that our analysis is tight by showing an example in which 2-lookahead mechanism obtains exactly $\frac{\sqrt{e}}{1+\sqrt{e}}$ fraction of the optimal revenue.

Our analysis is based on the clever idea from [2] of comparing the revenue obtained by k-lookahead mechanism to the *t-fixed-price* and *t-pivot* auctions. The novelty of our approach is that instead of picking only one *threshold* t, we consider a series of thresholds $t_1, \ldots, t_m$ and choose the best series. Apparently, our analysis will lead to better ratio but become more complicated. Therefore, new idea and technique will be introduced for our analysis.

2 Preliminary

In this section, we formally define our problem and provide some useful facts that will be needed in the future discussion.

In a single-item auction, a seller wishes to sell one item to a group of n self-interested bidders. Each bidder has a private valuation $v_i \in \mathbb{R}^+$. We assume that there is a publicly known distribution $\mathcal{D}$ on the valuation space of the bidders. In this paper, we make no assumption on the distribution. In particular, bidders' valuations could be correlated. Since we only consider truthful mechanisms in this paper, we will equalize the notions of *bid* and *valuation*.

An auction M is a mechanism that takes a bid vector v and then decides who wins the item and for what price. We use $(\mathbf{x}, \boldsymbol{p})$ to denote the allocation and payment where $x_i(v)$ is the probability that bidder i gets the item and $p_i(v)$ is her expected payment. Here, the goal of each bidder i is to maximize her own *utility* defined as $x_i v_i - p_i$.

A mechanism is *deterministically truthful* if reporting the true valuation is a dominant strategy for each agent and $x_i(v) \in \{0, 1\}$ for every bidder i and every bid vector v, and we say that a randomized mechanism is *universally truthful* if the mechanism is a probability distribution over deterministically truthful mechanisms. At last, *truthful-in-expectation* is a weaker notion in which an agent maximizes her expected utility by being truthful. It is easy to see that every deterministically truthful mechanism is universally truthful and every universally truthful mechanism is truthful in expectation.

In this paper, we are interested in designing truthful-in-expectation mechanisms. From now on, without particular specification, we will simply say a mechanism is *truthful* if it is truthful-in-expectation and an *optimal auction* is referred to a truthful-in-expectation mechanism that maximizes the seller's expected profit $E_{\mathcal{D}}[M] = E_{v \sim \mathcal{D}}(\sum_{i=1}^{n} p_i(v))$ on input distribution $\mathcal{D}$.

An useful observation is that the optimal auction can be described as a linear program [2, 5] that its size is polynomial in the support size of distribution. Therefore we can obtain an optimal auction in polynomial time in the size of distribution, which implies that the optimal auction can be computed efficiently in the explicit model. But the linear program is not generally efficient in the oracle model unless the number of bidders is a constant. This motivates the study of k-lookahead mechanisms [6, 2]. Due to the lack of space, we omit this linear program.

In a k-lookahead mechanism, we find the k bidders with the highest values. We denote the set of these k bidders by K. Next we get the conditional distribution D_K on $v_i \geq \max\{v_j | j \notin K\}$ for $i \in K$ and v_j is fixed for all $j \notin K$. Then we reject the bidders not in K and use the mentioned linear program for distribution D_K to get the allocation vector $\mathbf{x}_K$ and payment vector $\mathbf{p}_K$.

In this paper, we will investigate the approximation ratio of the k-lookahead mechanism. Here, we say an auction M is a c-approximation mechanism if $\frac{E_D[M]}{E_D[OPT]} \geq c$ where OPT is the revenue-maximizing valid auction on distribution $\mathcal{D}$.

Finally, the following theorem provides a characterization of deterministic mechanisms for single item auctions, which will be useful in the analysis of 2-lookahead mechanism.

Theorem 1. *[4] A deterministic mechanism, with allocation and payment rule q, p respectively, is truthful if and only if for each bidder i and each v_{-i}, the following conditions hold:*

1. *Monotone Allocation: $q_i(v_i, v_{-i}) \leq q_i(v_i', v_{-i})$ for all $v_i \leq v_i'$;*
2. *Threshold Payment: There exists a threshold $t_i(v_{-i})$ such that $p_i(v_i, v_{-i}) = t_i(v_{-i}) \cdot q_i(v_i, v_{-i})$.*

3 The Approximation Ratio

In this section, we present our main result. From now on, we fix a constant k and let K be the agents with the highest k bids. Let D_K be the conditional distribution of bidders in K conditioned on the remaining bidders. We show that the approximation ratio of k-lookahead mechanism is at least $\frac{e^{1-1/k}}{1+e^{1-1/k}}$.

Our high-level idea is to partition the optimal revenue into different components. Then we design several auctions that only sell the item in K and each of them approximately realizes part of the components. The revenues of these auctions provide a lower bound on the revenue of k-lookahead since it is the optimal auction that only sells the item to bidders in K. Without lose of generality, we assume $K = \{1, \cdots, k\}$ and v_{k+1} is the highest valuation not in K.

In the following, we always assume that the optimal revenue is 1. Now we consider the expected revenue of the optimal auction. As we mentioned before, we first partition the optimal revenue into four parts.

Definition 1. *Fix the optimal auction, for any $t > 1$, we define $L(t), \tilde{L}(t), M(t), H(t)$ as follows:*

1. *$L(t)$:the expected revenue from bidders in $N \backslash K$ for instances where no bidder in K has value at least $t \cdot v_{k+1}$.*
2. *$\tilde{L}(t)$:the expected revenue from bidders in $N \backslash K$ for instances where there are some bidders in K whose valuations are at least $t \cdot v_{k+1}$.*
3. *$M(t)$:the expected revenue from bidders in K for instances where no bidder in K has value at least $t \cdot v_{k+1}$.*
4. *$H(t)$:the expected revenue from bidders in K for instances where there are some bidders in K whose valuations are at least $t \cdot v_{k+1}$.*

Let the expected revenue from K in the optimal auction be $\alpha(\alpha \leq 1)$. By our definition, $M(t) + H(t) = \alpha$ and $L(t) + \tilde{L}(t) = 1 - \alpha$ for all $t \geq 1$.

Lemma 1. *The expected revenue of k-lookahead auction is at least α.*

Proof. Consider the following auction: If the optimal auction sells the item to bidder i in K with probability x_i and p_i, we still sell the item to i with probability x_i and ask for a payment p_i. Otherwise no one gets the item. This mechanism might not be truthful because it is possible that some bidders in $N\backslash K$ raises her bid so that she becomes a bidder in K and has a chance to get the item. To make this mechanism truthful, we raise the expected payment of each bidder i by $\max\{0, (v_{k+1} - p_i(v_{k+1}, v_{-i})) \cdot \frac{x_i(v_{k+1}, v_{-i})}{x_i(v)}\}$. This is then a truthful mechanism with expected revenue at least α. Furthermore, one can see that the mechanism only sells the item to bidders in K, therefore, the expected revenue of k-lookahead auction is at least α. $\square$

The above lemma provides a lower bound on the revenue of k-lookahead auction related to the components of M and H in the optimal auction. To get more such bounds, we need the following auctions first introduced by Dobzinski, Fu and Kleinberg [2]. Suppose there is a threshold $t \geq 1$:

t**-Fixed Price Auction:** Select a bidder j uniform from K at random. If any bidders in $K\backslash\{j\}$ have valuations no less than $t \cdot v_{k+1}$ then he gets the item with payment $t \cdot v_{k+1}$. If there are several bidders satisfy this condition, break ties arbitrary. Otherwise, bidder j gets the item with payment v_{k+1}.

t**-Pivot Auction:** Select a bidder j uniform from K at random. If any bidders in $K\backslash\{j\}$ have valuations no less than $t \cdot v_{k+1}$, we choose the bidder i with the smallest index. We run the *k-lookahead auction* on the conditional distribution D'_k that fix the valuations of bidders not in K, and require $v'_l \geq v_{k+1}(l \in K)$ and $v'_i \geq t \cdot v_{k+1}$. Otherwise, we allocate the item to bidder j with a payment v_{k+1}.

It is easy to verify that t-Fixed Price Auction is truthful. To check that t-Pivot Auction is truthful, the only case we should be careful is that a bidder i raises her valuation and let the mechanism run the k-lookahead auction. However, bidder i must be the only bidder whose valuation is not less than $t \cdot v_{k+1}$ in this case. So k-lookahead auction runs under the conditional distribution that $v_i \geq t \cdot v_{k+1}$. As a result, her payment must be at least $t \cdot v_{k+1}$, which exceeds her actual valuation. Therefore, t-Pivot auction is truthful.

In the following, we will choose a series of s threshold values $t_1 < t_2 < \cdots < t_s$ (whose values will be determined later) and relate the revenue of each t_i-Fixed Price Auction and t_i-Pivot Auction to the four components of the optimal revenue defined earlier.

To be precise, assume that $t_0 = 1$ and we define $M_i = M(t_i) - M(t_{i-1}) \geq 0$ which is the revenue from K in the optimal auction when the highest valuation is in $[t_{i-1} \cdot v_{k+1}, t_i \cdot v_{k+1})$.

Lemma 2. *The expected revenue of t_i-Fixed Price Auction is at least $P_i = L(t_i) + \sum_{j=1}^{i} \frac{M_j}{t_j} + (\frac{k-1}{k}t_i + \frac{1}{k})(\tilde{L}(t_i) + \sum_{j=i+1}^{s} \frac{M_j}{t_j})$.*

Proof. We consider two cases. In the first case, there is no bidder in K whose valuation is greater or equal than $t_i \cdot v_{k+1}$. So the auction allocates the item to the selected bidder j with payment v_{k+1}. The corresponding expected revenue in the optimal auction is $L(t_i) + \sum_{j=1}^{i} M_j$. From the definition of M_i, the revenue of our auction is at least $L(t_i) + \sum_{j=1}^{i} \frac{M_j}{t_j}$.

In the second case, there are some bidders whose valuations are at least $t_i \cdot v_{k+1}$. In our auction, the auction will obtain $t_i \cdot v_{k+1}$ with probability at least $\frac{k-1}{k}$. Otherwise the auction will obtain at least v_{k+1}. Therefore the expected revenue of this auction is at least $(\frac{k-1}{k}t_i + \frac{1}{k})\tilde{L}(t_i)$ when the optimal auction allocates the item to bidders not in K. At the same time, the expected revenue of this auction is at least $(\frac{k-1}{k}t_i + \frac{1}{k})\sum_{j=i+1}^{s} \frac{M_j}{t_j}$ when the optimal auction allocates the item to K.

From all discussion above, the expected revenue of t_i-Fixed Price Auction is at least $P_i = L(t_i) + \sum_{j=1}^{i} \frac{M_j}{t_j} + (\frac{k-1}{k}t_i + \frac{1}{k})\tilde{L}(t_i) + (\frac{k-1}{k}t_i + \frac{1}{k})\sum_{j=i+1}^{s} \frac{M_j}{t_j}$.

Similarly, we can prove the following:

Lemma 3. *The expected revenue of t_i-Pivot Auction is at least* $Q_i = L(t_i) + \sum_{j=1}^{i} \frac{M_j}{t_j} + \frac{k-1}{k}H(t_i) + \frac{1}{k}(\tilde{L}(t_i) + \sum_{j=i+1}^{s} \frac{M_j}{t_j})$.

Let $R_i = \max\{P_i, Q_i\}$ and we can see that $\max_{1 \le i \le s} R_i$ is a lower bound on the revenue of k-lookahead. From the above lemma, this lower bound is explicitly related to the components M, H, L and $\tilde{L}$. In the following, we will choose s large enough and $t_1, \ldots, t_s$ appropriately to obtain a lower bound on $\max_{1 \le i \le s} R_i$ that is only related to α. Together with Lemma 1, we will get the desired approximation ratio. Now we prove this lower bound:

Lemma 4. $\max_{1 \le i \le s} R_i \ge 1 - e^{-(1-1/k)}\alpha$.

Proof. To prove this lemma, we need to eliminate the explicit dependency of $\max_{1 \le i \le s} R_i$ to the components of M, H, L and $\tilde{L}$.

First of all, for each t_i, we can replace $\tilde{L}(t_i), H(t_i)$ by $1 - \alpha - L(t_i), \alpha - M(t_i)$ and simplify P_i, Q_i as:

$$P_i = (t_i + \frac{1}{k})(1 - \alpha) - (\frac{k-1}{k}t_i + \frac{1}{k} - 1)L(t_i) + \sum_{j=1}^{i} \frac{M_j}{t_j} + \sum_{j=i+1}^{s} (\frac{k-1}{k}t_i + \frac{1}{k})\frac{M_j}{t_j}$$

$$Q_i = \alpha + \frac{1}{k}(1 - \alpha) + \frac{k-1}{k}L(t_i) + \sum_{j=1}^{i}(\frac{1}{t_j} - \frac{k-1}{k})M_j + \sum_{j=i+1}^{s} \frac{1}{k}\frac{M_j}{t_j}$$

Now we are ready to eliminate $L(t_i)$. Since $R_i = \max\{P_i, Q_i\}$, we have

$$R_i \ge \frac{1}{t_i}P_i + \frac{t_i - 1}{t_i}Q_i = 1 - (\frac{k-1}{kt_i} + \frac{1}{k})\alpha + \sum_{j=1}^{s} \frac{M_j}{t_j} - \frac{k-1}{k}(1 - \frac{1}{t_i})\sum_{j=1}^{i} M_j.$$

At last, we will eliminate M_j for all j. Observe that $\max_{1 \le i \le s} R_i$ is lower bounded by the average, we have the following:

$$\max_{1 \le i \le s} R_i \ge \sum_{i=1}^{s} R_i \ge s - \sum_{i=1}^{s} \left(\frac{k-1}{kt_i} + \frac{1}{k} \right) \alpha + \sum_{j=1}^{s} \left[\frac{s}{t_j} - \sum_{i=j}^{s} \frac{k-1}{k} \left(1 - \frac{1}{t_i} \right) \right] M_j$$

Therefore, in order to eliminate M_j for all j, we only need to choose numbers $t_1, \ldots, t_s$ such that

$$\frac{s}{t_j} - \sum_{i=j}^{s} \frac{k-1}{k} \left(1 - \frac{1}{t_i} \right) = 0, \quad \text{for all } 1 \le j \le s. \tag{1}$$

As long as we can find such $t_1, \ldots, t_s$, we have:

$$\max_{1 \le i \le s} R_i \ge 1 - \left[\frac{k-1}{k} \cdot \frac{1}{s} \sum_{i=1}^{s} \frac{1}{t_i} + \frac{1}{k} \right] \alpha. \tag{2}$$

At first, we use β to denote $\frac{k-1}{k}$ and have $\frac{s}{t_s} - \beta(1 - \frac{1}{t_s}) = 0$ when $j = s$. So $1 - \frac{1}{t_s} = \frac{s}{s+\beta}$. Then, comparing the equations of j and $j+1$, we obtain $(1 - \frac{1}{t_j}) = (\frac{s}{s+\beta})^{s-j+1}$. Therefore $\sum_{i=1}^{s} \frac{1}{t_i} = s - \sum_{i=1}^{s} (\frac{s}{s+\beta})^i = s - \frac{s}{\beta}(1 - (\frac{s}{s+\beta})^s)$. At the same time, we know $\lim_{s \to \infty} (\frac{s}{s+\beta})^s = e^{-\beta} = e^{-\frac{k-1}{k}}$

Together with (2), we have $\max_{1 \le i \le s} R_i \ge 1 - e^{-(1-1/k)} \alpha$.

Finally, we are ready to prove:

Theorem 2. *The approximation ratio of k-lookahead mechanism is at least $\frac{e^{1-1/k}}{1+e^{1-1/k}}$.*

Proof. Let rev_k be the revenue of the k-lookahead mechanism. From Lemma 4, we know that $\text{rev}_k \ge 1 - e^{-(1-1/k)} \alpha$. Together with Lemma 1, we have

$$\text{rev}_k \ge \max\{\alpha, 1 - e^{-(1-1/k)} \alpha\}.$$

Simple calculation shows that for all positive value x, $\max\{\alpha, 1 - x\alpha\} \ge \frac{1}{1+x}$. Therefore, we have $\text{rev}_k \ge \frac{e^{1-1/k}}{1+e^{1-1/k}}$. This completes our proof.

4   Tightness of Analysis

In the previous section, we showed that the approximation ratio of k-lookahead is $\frac{e^{1-1/k}}{1+e^{1-1/k}}$. In particular, the 2-lookahead mechanism, which is of special interest, has an approximation ratio of at least $\frac{\sqrt{e}}{1+\sqrt{e}}$. In this section, we design an example to show that our analysis for 2-lookahead is tight.

First of all, we need some definitions. Because 2-lookahead auction can be made deterministically [2, 5], it either allocates the item, or does not allocate to anyone. So we only consider the 2-lookahead mechanisms that are deterministic from now on. We will consider the *empty instances* that a 2-lookahead mechanism doesn't allocate the item. We use empt(D) to denote the maximal *empty probability* over 2-lookahead mechanism that empty instances occur on a distribution D. In the following, we will use

$E_2(D)$ to denote the 2-lookahead mechanism with the maximum empty probability for distribution D.

In a setting where there are only three bidders, we say that a distribution D is *valid*, if the third bidder always has valuation $v_3 = 1$ and the valuations of the other two bidders are at least 1.

We first prove a property of valid distributions. Let $\mathsf{rev}_2(D)$ and $\mathsf{opt}(D)$ denote the revenue of the 2-lookahead and the optimal auction for a distribution D respectively.

Lemma 5. *Let D be a valid distribution on three bidders, then* $\mathsf{opt}(D) \geq \mathsf{rev}_2(D) + \mathsf{empt}(D)$.

Proof. Consider this auction $\mathcal{A}$: run the 2-lookahead auction $E_2(D)$ and if it allocates the item to bidder i in $K = \{1, 2\}$ with payment p, we still allocate the item to i with payment p. Otherwise we allocate the item to bidder 3 with payment 1. It is easy to see that $\mathcal{A}$ is truthful, and its revenue is $\mathsf{rev}_2(D) + \mathsf{empt}(D)$. Therefore, $\mathsf{opt}(D) \geq \mathsf{rev}_2(D) + \mathsf{empt}(D)$.

The above lemma provides a lower bound of opt. In the following, we will explicitly construct a valid distribution D such that $\mathsf{empt}(D) \geq \frac{\mathsf{rev}_2(D)}{\sqrt{e}}$, hence prove our desired ratio.

In our example, there are three bidders and the third bidder's valuation is always 1. Now we construct the distribution D_2 for the first two bidders explicitly. We assume that there are m possible valuations $p_0, p_1, \cdots, p_m$ (m is an *odd number*). Then we define $x_0 = 1$ and $x_i = (1+p)^i - (1+p)^{i-1} = p(1+p)^{i-1}$ for $1 \leq i \leq m$ where p is a parameter. We will set the value of p and choose $p_1, \ldots, p_m$ later. One can see that a property of our construction is $\sum_{0 \leq i < j} x_i = (1+p)^j$ for all $j \leq m$.

Now we consider this following distribution D_2 where $D_2(i, j)$ denotes the probability of $v_1 = p_i, v_2 = p_j$:

$$
D_2(i, j) = \begin{cases}
x_i x_j & i + j < m \\
x_i (\sum_{j \leq k \leq m} x_k) & (i+j = m) \text{ and } (i < j) \\
(\sum_{i \leq k \leq m} x_k) x_j & (i+j = m) \text{ and } (i > j) \\
0 & i + j > m
\end{cases}
$$

In fact, D_2 should be normalized to become a distribution. However, since we only care about the ratio between $\mathsf{empt}(D)$ and $\mathsf{rev}_2(D)$, we will simply use D_2 as the distribution without normalizing. From now on, we will simply use E_2, rev_2 and empt to denote $E_2(D), \mathsf{rev}_2(D)$ and $\mathsf{empt}(D)$.

Now we choose $p_0 = 1$ and $p_i = \sum_{0 \leq j \leq m} x_j / \sum_{i \leq j \leq m} x_j$ for all $1 \leq i \leq m$. Therefore, we have $p_0 \leq p_1 \leq \cdots \leq p_m$. Furthermore, we obtain the following characterization of the event that E_2 allocates the item:

Lemma 6. *Let p_i, p_j be the bid of bidder 1 and 2 respectively, then E_2 allocates the item if and only if $i + j = m$.*

Proof. First of all, by our choice, it is easy to verify the following:

Property 1. If $i < m/2$, then: $p_0(\sum_{k=0}^{m-i} D_2(i, k)) = \cdots = p_j(\sum_{k=j}^{m-i} D_2(i, k)) = \cdots = p_{m-i} D_2(i, m - i) = x_i \sum_l x_l$.

Basically, this property can be interpreted as follows: fix $v_1 = p_i$, the expected revenue obtained by offering bidder 2 a threshold price p_j is a constant when $0 \leq j \leq m - i$. As a result, recall that by Theorem 1, the winner in a single item auction pays the threshold price, we have:

Corollary 1. *In $E_2, t_2(v_1) \geq p_{m-i}$ when $v_1 = p_i$ for all $i < m/2$. Similarly, $t_1(v_2) \geq p_{m-j}$ when $v_2 = p_j$ for all $j < m/2$.*

The proof of the corollary is straightforward: Suppose $v_1 = p_i$ for some $i < m/2$. If $t_2(v_1) < p_{m-i}$, then we can always increase the threshold price to p_{m-i} without decreasing the revenue. By doing this, we only increase the empty probability. This is a contradiction to our assumption that E_2 maximizes the empty probability.

Now we are ready to prove the lemma. If it is not true, suppose $i + j < m$ but E_2 allocates the item to either bidder 1 or 2. Consider the smallest sum of $i + j$ that satisfies the above. Without lose of generality, we may assume $i < m/2$. From Corollary 1, since $i + j < m$, we know that bidder 2 can not get the item. Therefore, bidder 1 gets the item when $v_1 = p_i$ and $v_2 = p_j$. At the same time, $j > m/2$ otherwise we can get a contradiction from Corollary 1. So bidder 1 still gets the item when $v_1 = p_{m-j} > p_i$ and $v_2 = p_j$.

Now we show that we can modify the allocation of E_2 when $v_2 = p_j$ to get more empty probability and the expected revenue of modified auction is not less than the original one. Let E_2' be an auction as follows: (1) it performs exactly the same as E_2 when $v_2 \neq p_j$ and (2) when $v_2 = p_j$, E_2' allocates the item to bidder 2 only when $v_1 = p_{m-j}$ and otherwise allocates nothing.

Obviously, E_2' has a larger empty probability than E_2. To get a contradiction, we only need to prove that its expected revenue is at least as large as E_2. In other words, we want to show:

$$p_j D_K(m - j, j) \geq p_i \sum_{k=i}^{m-j} D_K(k, j) \tag{3}$$

By our construction, simple calculation shows that (3) is equivalent to the following

$$p(1 + p)^{m-j-1}\left((1 + p)^{j-1} - \sum_{l=0}^{i-1} x_l\right) \geq p(1 + p)^{j-1}\left((1 + p)^{m-j-1} - \sum_{l=0}^{i-1} x_l\right),$$

which always holds when $j > m/2$. This is a contradiction.

By the above characterization, we can easily calculate rev_2 and empt. We will show that by choosing the parameter p appropriately, $rev_2 \leq \sqrt{e} \cdot$ empt, which implies:

Theorem 3. *The approximation ratio of 2-lookahead auction is at most $\frac{\sqrt{e}}{\sqrt{e}+1}$.*

Proof. By Lemma 6 and our construction, we first estimate rev_2 as follows:

$$rev_2 \leq \sum_{i,j:i+j=m} \max\{p_i, p_j\} D_2(i, j) = 2\left(\sum_{0 \leq i < m/2} x_i\right)\left(\sum_{l=0}^{m} x_l\right) = 2(1 + p)^{3m/2}.$$

Now we compute the empty probability empt. Again, by Lemma 6, we have empt $= \sum_{i,j:i+j<m} x_i x_j$, which can be calculate as follows:

$$\sum_{i,j:i+j<m} x_i x_j = \sum_{j=0}^{m-1} x_j + \sum_{i=1}^{m-1} \sum_{j=0}^{m-1-i} x_i x_j$$
$$= (1+p)^{m-1} + (m-1)p(1+p)^{m-2}.$$

We set $p = 1/m$ and let $m \to \infty$, we have $\text{rev}_2 \leq 2e^{3/2}$ and empt $\geq 2e$. Therefore, $\text{rev}_2 \leq \sqrt{e} \cdot$ empt. Therefore, by our previous argument, the approximation ratio of 2-lookahead auction is at most $\frac{\sqrt{e}}{\sqrt{e}+1}$.

5   Discussion and Open Questions

Perhaps the first question that every theoretical computer scientist would ask here is whether the analysis of the k-lookahead mechanism can be improved in general. An important open problem is whether the approximation ratio of k-lookahead mechanism tends to 1 when k tends to positive infinity. A nature attempt for this question from the negative aspect is to generalize our tight instance for 2-lookahead mechanism in section 4 to the k-lookahead mechanism for general k. In particular, one might consider the following distribution $D_K(i_1, ..., i_k)$ for the set K of the highest k bidders:

1. $D_K(i_1, \cdots, i_k) = 0$: there exists $p, q \in [k]$ such that $p \neq q$ and $i_p + i_q > m$.
2. $D_K(i_1, \cdots, i_k) = \prod_{j \in K} x_{i_j}$: for all $p, q \in [k](p \neq q)$, we have $i_p + i_q < m$.
3. $D_K(i_1, \cdots, i_k) = \prod_{j \in K \setminus \{l\}} x_{i_j} \cdot \sum_{j=i_l}^{m} x_j$: $\max_{p,q \in [k](p \neq q)} \{i_p + i_q\} = m$, where $i_l = \max\{i_1, \cdots, i_k\}$.

Again, we assume that the highest bid outside K is $v_{k+1} = 1$. Similar to the analysis for 2-lookahead, we can prove that k-lookahead allocates to some bidder in K if and only if $i_1, i_2, \cdots, i_k$ is such that $\max_{p,q \in [k](p \neq q)} \{i_p + i_q\} = m$. However, simple calculation implies that the ratio between the empty probability and the revenue of the k-lookahead is at most $2/k$ when k tends to infinity. This only implies a $\frac{k}{k+2}$ upper bound on the approximation ratio of the k-lookahead mechanism. Therefore, to obtain better upper bound, if possible, one might need new ideas and techniques.

From the positive aspect, one might improve the analysis via the following approach: Instead comparing the revenue of k-lookahead to t-fixed price and t-pivot auctions, we could compare to more delicate auctions such as a hybrid of t_1-fixed price and t_2-pivot auctions for distinct values of t_1, t_2.

Another interesting open question is to further close the gap between the revenues of the optimal deterministically truthful and truthful-in-expectation mechanisms. Our analysis of 2-lookahead implies that the gap is at most a factor of $\frac{1+\sqrt{e}}{\sqrt{e}}$. As we mentioned, our analysis is tight, hence closing the gap further requires better truthful-in-expectation mechanisms which can be derandomized.

References

1. Cremer, J., McLean, R.P.: Optimal selling strategies under uncertainty for a discriminating monopolist when demands are interdependent. Econometrica 53(2), 345–361 (1985)
2. Dobzinski, S., Fu, H., Kleinberg, R.D.: Optimal auctions with correlated bidders are easy. In: Proceedings of the 43rd Annual ACM Symposium on Theory of Computing, STOC 2011, pp. 129–138 (2011)
3. Klemperer, P.: Auction theory: A guide to the literature. Microeconomics, EconWPA (March 1999)
4. Myerson, R.B.: Optimal acution design. Mathematics of Operations Research 6(1), 58–73 (1981)
5. Papadimitriou, C.H., Pierrakos, G.: On optimal single-item auctions. In: Proceedings of the 43rd Annual ACM Symposium on Theory of Computing, STOC 2011, pp. 119–128 (2011)
6. Ronen, A.: On approximating optimal auctions. In: Proceedings of the 3rd ACM Conference on Electronic Commerce, EC 2001, pp. 11–17 (2001)
7. Ronen, A., Saberi, A.: On the hardness of optimal auctions. In: Proceedings of the 43rd Symposium on Foundations of Computer Science, FOCS 2002, pp. 396–405 (2002)

Decision Markets with Good Incentives*

Yiling Chen, Ian Kash, Mike Ruberry, and Victor Shnayder

Harvard University

Abstract. Decision markets both predict and decide the future. They allow experts to predict the effects of each of a set of possible actions, and after reviewing these predictions a decision maker selects an action to perform. When the future is independent of the market, strictly proper scoring rules myopically incentivize experts to predict consistent with their beliefs, but this is not generally true when a decision is to be made. When deciding, only predictions for the chosen action can be evaluated for their accuracy since the other predictions become counterfactuals. This limitation can make some actions more valuable than others for an expert, incentivizing the expert to mislead the decision maker. We construct and characterize decision markets that are – like prediction markets using strictly proper scoring rules – myopic incentive compatible. These markets require the decision maker always risk taking every available action, and reducing this risk increases the decision maker's worst-case loss. We also show a correspondence between strictly proper decision markets and strictly proper sets of prediction markets, creating a formal connection between the incentives of prediction and decision markets.

1 Introduction

To make an informed decision a decision maker must understand the likely consequences of their actions. Hanson proposed a "decision market" to directly predict these consequences [11]. His proposal consists of a set of conditional prediction markets, one for each possible action. After the markets close the decision maker could evaluate each action's predicted effect on the set of possible outcomes, and choose the most preferred action. Conditional markets for actions not taken are voided.

Consider, for example, a project manager deciding between two developers, A and B. The manager prefers to hire the candidate more likely to complete a project on time, so it runs two conditional prediction markets. One determines the likelihood A will finish on time, conditional on A being hired, and the latter does the same for B. If the project manager has access to knowledgeable experts and these markets reflect the experts' information then the manager can make an informed hiring decision.

---

* This material is based upon work supported by NSF Grant No. CCF-0915016. Any opinions, findings, conclusions, or recommendations expressed in this material are those of the authors alone.

Using a prediction market to make a decision is natural and previous work has demonstrated they can produce accurate forecasts [1,2,18,3,8]. However, while a prediction market using a strictly proper scoring rule is myopic incentive compatible, Hanson's proposed decision market is not. That is, in a prediction market an expert maximizes its score for a prediction by predicting consistent with its beliefs, but the same is not true when a decision is to be made [16,6].

We return to our hypothetical project manager and the design of its two prediction markets. The manager would like to reward experts for improving either market's accuracy, but only one market's condition will ever be realized since only one developer will be hired. The other market's predictions will become unscored counterfactuals. If an expert has improved one market's prediction more than the other's, it has an incentive to convince the project manager to hire the associated developer regardless of how poor an employee that developer may be!

More concretely, if the markets currently predict developer A has a 60% and developer B a 80% chance of finishing the project on time, and an expert believes the correct likelihoods are 70% and 80%, respectively, truthful reporting can only improve the market's accuracy for developer A. If developer B is hired this expert will receive a score of zero, but if A is hired they expect to score for a 10% improvement. Instead of being honest, then, the expert can pretend B is incompetent, lowering the market's prediction for the likelihood B will finish on time to less than 70%, cause A to be hired instead, and enjoy the profits.

We address this manipulation and construct and characterize decision markets that are myopic incentive compatible, like prediction markets. Instead of a scoring rule, these markets use a decision scoring rule that can account for the likelihood actions are taken when scoring predictions. When a decision maker risks taking an action at random, these decision scoring rules allow the scores of unlikely actions to be amplified while the scores of likely actions are comparatively reduced, making risk neutral experts indifferent to their affect on the decision. We show this risk of taking an action at random is a requirement for myopic incentive compatible decision markets, and reducing this risk increases the decision maker's worst-case loss. We also show that, for risk-neutral experts, every myopic incentive compatible decision market describes a game equivalent to that described by a myopic incentive compatible set of prediction markets, creating a formal connection between decision and prediction markets.

The rest of the paper is organized as follows. Section 1.1 describes previous work on prediction and decision markets. Section 2 provides a formal description of prediction markets in our notation, and Section 3 describes our decision market model. Section 4 presents our construction and characterization results. Section 5 extends these results, describing optimal behavior for a risk neutral decision maker and connecting prediction and decision markets. Finally, Section 6 discusses further research challenges and concludes.

1.1 Related Work

Strictly proper scoring rules have long been understood to be able to truthfully elicit a single risk-neutral expert's beliefs over the outcome of an uncertain event

[14,17,10]. Hanson [12,13] showed these same rules could be used to myopically incentivize any number of experts to be honest in a prediction market, and described an extension of scoring rules – market scoring rules – that prevent the market maker's worst-case loss from growing with the number of experts. Importantly, all strictly proper scoring rules require the market maker correctly observe the event's outcome. We formally describe these rules in Section 2.

When making a decision, some outcomes are not observed, and strictly proper scoring rules do not generally myopically incentivize an expert to be truthful. Othman and Sandholm [16] first formalized this incentive problem. They considered a single expert predicting and a decision maker picking their most preferred action based on the expert's predictions, and they showed the expert can be incentivized to honestly reveal the decision maker's most preferred action. They describe this decision rule as MAX—simply picking the best action from what's available. Chen and Kash [6] also considered a single expert but allowed both deterministic and stochastic decision rules. Given a decision rule they characterized all scoring rules incentivizing a single risk-neutral expert to predict truthfully.

But while strictly proper scoring rules can be used for a single expert and extend to prediction markets, these scoring and decision pairs do not have a such a natural extension . In a prediction market, an expert's expected score for a prediction is fixed once the prediction is made, and the same is true when a single expert is informing a decision. In a decision market, however, a prediction's score may not be fixed until the market closes, creating new strategic complexities. In fact, Othman and Sandholm showed that no scoring rule can myopically incentivize experts to predict honestly in a decision market using their MAX rule, and we extend this result in Section 4. We also describe myopic incentive compatible decision markets for the first time.

Recently, manipulation in the presence of outside incentives has been studied [9,5]. In this paper we do not consider outside payoffs. The decision maker's choice of action may affect an expert's utility, but not because of any inherent preferences over actions that expert may have.

2   Prediction Markets: Background and Notation

This section presents the standard market scoring rule model of a prediction market, first described by Hanson [12,13], and defines our notation.

Let $\mathcal{O}$ be a finite, mutually exclusive, and exhaustive set of outcomes. A *prediction market* is a sequential game played by any number of risk-neutral, expected-value–maximizing *experts* predicting the likelihood of these outcomes. The market opens at round zero with some initial prediction $p^0 \in \Delta(\mathcal{O})$, where $\Delta(\mathcal{O})$ is the set of probability distribution over outcomes. At each round after the market opens, an arbitrarily chosen expert makes a prediction $p \in \Delta(\mathcal{O})$, and we let p^t be the prediction made in round t. The market closes at some round $\bar{t}$, after which an outcome o^* is observed and experts are scored for each prediction by a *scoring rule*,

$$s : \mathcal{O} \times \Delta(\mathcal{O}) \rightarrow \mathbb{R} \cup \{-\infty\},$$

where $\mathbb{R}$ is the set of real numbers. We write $s_o(p) \equiv s(o, p)$ as a shorthand, and an expert's payment for a prediction is the difference between the scores of its and the immediately preceding prediction. Letting $\mathcal{T}$ be the set of rounds when an expert made a prediction, its total payoff is

$$\sum_{t \in \mathcal{T}} s_{o^*}(p^t) - s_{o^*}(p^{t-1}).$$

Markets with this sequential difference payoff structure are described as *market scoring rule* markets.

Scoring rules are *regular* when only predictions assigning zero likelihood to the observed event are scored negative infinity, and *proper* when a risk-neutral expert's expected score for a prediction is maximized when predicting consistent with its belief. Formally, a rule is proper if for all beliefs $q \in \Delta(\mathcal{O})$ over the likelihood of outcomes and predictions p

$$\sum_{o \in \mathcal{O}} q_o s_o(q) \geq \sum_{o \in \mathcal{O}} q_o s_o(p),$$

where q_o is the believed likelihood of outcome o. A rule is *strictly proper* when the inequality is strict unless $q = p$, uniquely maximizing an expert's score when they predict consistent with their beliefs. An example of a strictly proper scoring rule is $s_o(p) = a_o + b \log p_o$ with $a_o \in \mathbb{R}$ and $b > 0$. When experts uniquely maximize their score for a prediction by predicting consistent with their beliefs we describe the mechanism as *myopically incentive compatible*.

In aggregate, experts receive a payoff of $\Sigma_{t=1}^{\bar{t}} s_{o^*}(p^t) - s_{o^*}(p^{t-1}) = s_{o^*}(p^{\bar{t}}) - s_{o^*}(p^0)$, so the market institution's worst-case loss is

$$\max_{o^*, p^{\bar{t}}} s_{o^*}(p^0) - s_{o^*}(p^{\bar{t}}).$$

Note that the market institution's payment is bounded and independent of the number of experts. In practice, a market institution's budget must be at least their worst-case loss.

Ideally, the final prediction is an accurate consensus of experts' beliefs. Bayesian experts, for example, update their beliefs as they observe other's predictions. However, while a market using a strictly proper scoring rule is myopic incentive compatible, it is not incentive compatible in general. An expert participating in multiple rounds may provide a prediction inconsistent with its belief, with the intention to mislead other experts and later capitalize on their mistakes [4]. Despite such possible manipulations by forward-looking Bayesian experts, previous work has shown that under certain conditions prediction markets that are myopic incentive compatible can fully aggregate information in finite rounds [4] or in the limit [15]. In this paper, however, we do not restrict experts to be Bayesian but allow arbitrary – or free – beliefs, as is typical when working with scoring rules.

3   Decision Market Model

A prediction market is a special case of a decision market. Both use the same sequential market structure, but a decision market uses a decision rule to pick from a set of actions before the outcome is observed, and which action is chosen may affect the likelihood an outcome occurs. Unlike previously proposed models of decision markets, we score experts using a decision scoring rule instead of a standard scoring rule. This more general function is necessary to recreate the myopic incentive compatibility of a prediction market for the broadest possible class of decision markets.

Let $\mathcal{A}$ be a finite set of actions, and $\mathcal{O}$ a set of outcomes as before. Without loss of generality and for notational convenience we assume the outcomes for every action are the same. As in a prediction market, a decision market opens with an initial prediction, but instead of a single probability distribution it is a set of conditional distributions, one for each action, denoted $P^0 \in \Delta(\mathcal{O})^{|\mathcal{A}|}$. Experts still report sequentially, and we let P^t be the prediction made in round t, P_a^t that prediction's distribution over outcomes given action a is chosen, and $P_{a,o}^t$ be that conditional distribution's likelihood for outcome o.

After the market closes, the decision maker selects an action using a *decision rule*

$$D : \Delta(\mathcal{O})^{|\mathcal{A}|} \to \Delta(\mathcal{A}),$$

applied to the final report $P^{\bar{t}}$, drawing an action a^* from $\mathcal{A}$ according to the distribution $D(P^{\bar{t}})$. We say that a decision rule has *full support* if it only maps to distributions with full support. As a shorthand we write d for a distribution over actions and d_a the likelihood action a is drawn from the set.

Once the action is selected, an outcome o^* is revealed, and experts are scored for each prediction by a *decision scoring rule*

$$S : \mathcal{A} \times \mathcal{O} \times \Delta(\mathcal{A}) \times \Delta(\mathcal{O})^{|\mathcal{A}|} \to \mathbb{R} \cup \{-\infty\},$$

written $S_{a,o}(d, P) \equiv S(a, o, d, P)$. Paralleling scoring rules, we describe decision scoring rules as *regular* when only predictions assigning zero likelihood to the observed event are scored negative infinity.

Letting $\mathcal{T}$ again be rounds where an expert made a prediction, its total payoff is

$$\sum_{t \in \mathcal{T}} S_{a^*,o^*}(d, P^t) - S_{a^*,o^*}(d, P^{t-1}),$$

so the market institution's worst-case loss is

$$\max_{P^{\bar{t}}, a \in \bar{\mathcal{A}}, o \in \mathcal{O}} S_{a,o}(d, P^0) - S_{a,o}(d, P^{\bar{t}}), \tag{1}$$

where $\bar{\mathcal{A}}$ is the support of $D(P^{\bar{t}})$. Previous work on decision markets used a similar model, but with a *conditional scoring rule*

$$s_c : \mathcal{A} \times \mathcal{O} \times \Delta(\mathcal{O})^{|\mathcal{A}|} \to \mathbb{R} \cup \{-\infty\},$$

instead of a decision scoring rule.

As we show in the next section, however, considering the likelihood an action is selected is necessary to create the same myopic incentive compatibility as in prediction markets.

4   Decision Market Incentives

In a prediction market, a strictly proper scoring rule uniquely maximizes an expert's score for a prediction when they predict consistent with their beliefs. The same is not always true in a decision market. While both markets can reward improvements over a prior prediction, a decision market only observes and scores the improvement in the prediction for the chosen action. Since this action is a function of the market's final prediction, experts may have an incentive to change this prediction (either directly or by manipulating other experts) to create a distribution over actions more likely to score their greatest improvement.

In this section we extend the myopic incentives of prediction markets to decision markets, demonstrating how to construct myopic incentive compatible decision markets, and characterizing some of their properties. While myopic incentive compatibility does not guarantee that an expert who participates in multiple rounds will predict consistent with its beliefs in *every* round, previous work has shown that myopic incentives are sufficient to aggregate experts' private information at perfect Bayesian equilibria under certain conditions [4,15].

4.1   Myopic Incentive Compatibility

We first provide a formal treatment of myopic incentive compatibility for decision markets. Recall, for a prediction market, myopic incentive compatibility requires an expert always maximize their score for a prediction when they predict consistent with their beliefs. Assume a decision market uses decision rule D and decision scoring rule S. Then an expert with beliefs Q over the conditional outcomes who expects that d will be the final distribution over actions has an expected score for a prediction P of

$$\sum_{a \in \mathcal{A}} d_a \sum_{o \in \mathcal{O}} Q_{a,o} S_{a,o}(d, P).$$

And a myopic incentive compatible decision market must account not only for an expert's prediction, but also the likelihood each action is taken.

Definition 1. *A decision market (D, S) with a regular decision scoring rule S is proper if*

$$\sum_{a \in \mathcal{A}} d_a \sum_{o \in \mathcal{O}} Q_{a,o} S_{a,o}(d, Q) \geq \sum_{a \in \mathcal{A}} d'_a \sum_{o \in \mathcal{O}} Q_{a,o} S_{a,o}(d', P),$$

for all beliefs Q, distributions d and d' in the codomain of D and predictions P. The market is strictly proper if the inequality is strict unless $P = Q$.

If the market is strictly proper we also describe it as myopic incentive compatible, analogous to our treatment of prediction markets. When a decision market is not strictly proper there exist final predictions and beliefs such that experts maximize their score for a prediction by misrepresenting their beliefs.

4.2   A Simple Construction for Strictly Proper Decision Markets

Given a decision rule with full support, a simple construction can extend any strictly proper scoring rule into a decision scoring rule that makes the resulting decision market strictly proper, too.

Theorem 1. *Let D be a decision rule with full support. Then there exists a decision scoring rule S such that (D, S) is strictly proper.*

Proof. The proof is by construction. Let s be any strictly proper scoring rule. Construct

$$S_{a,o}(d, P) = \frac{1}{d_a} s_o(P_a). \tag{2}$$

(D, S) is strictly proper: an expert's expected score for a prediction is

$$\sum_{a \in \mathcal{A}} d_a \sum_{o \in \mathcal{O}} Q_{a,o} \frac{1}{d_a} s_o(P_a) = \sum_{a \in \mathcal{A}} \sum_{o \in \mathcal{O}} Q_{a,o} s_o(P_a),$$

the sum of the expected scores of the same prediction in a set of prediction markets, one for each action, using a strictly proper scoring rule. Since predicting consistent with beliefs maximizes the expected score of the expert in each market, it maximizes the sum of the expected scores.   □

This constructive result positively answers Chen and Kash's open question whether it is possible to construct decision markets with good incentives [6].

4.3   Strictly Proper Decision Markets Have Full Support

The construction in Theorem 1 requires a decision rule have full support, and makes experts' expected scores independent of future reports while their actual scores vary inversely with the likelihood an action is chosen. Surprisingly, every strictly proper decision market with a differentiable decision scoring rule has these properties. We prove the necessity of full support before characterizing all strictly proper decision market with differentiable decision scoring rules.

Theorem 2. *Let D be a decision rule. A decision scoring rule S that makes (D, S) strictly proper exists if and only if D has full support.*

Proof. First we prove that if a decision rule D does not have full support, there is no decision scoring rule S such that (D, S) is strictly proper. We proceed by contradiction. Let D be a decision rule without full support, choose a final report

P so that $d = D(P)$ has $d_k = 0$ for some $k \in \mathcal{A}$, and let S be a decision scoring rule such that (D, S) is strictly proper. Let $Q, Q' \in \Delta(\mathcal{O})^{|\mathcal{A}|}$ be two beliefs differing only on action k: for all $a \neq k$ and all o, $Q_{a,o} = Q'_{a,o}$; $\exists o\ Q_{k,o} \neq Q'_{k,o}$. Consider the expected utility of an expert with each of these beliefs reporting truthfully, while the final report remains P. One of these utilities must be weakly greater than the other. Without loss of generality, let

$$\sum_{a \in \mathcal{A}} \sum_{o \in \mathcal{O}} d_a Q_{a,o} S_{a,o}(d, Q) \geq \sum_{a \in \mathcal{A}} \sum_{o \in \mathcal{O}} d_a Q'_{a,o} S_{a,o}(d, Q'), \tag{3}$$

Because Q and Q' only differ on action k, and $d_k = 0$,

$$\sum_{a \in \mathcal{A}} \sum_{o \in \mathcal{O}} d_a Q_{a,o} S_{a,o}(d, Q) = \sum_{a \in \mathcal{A}} \sum_{o \in \mathcal{O}} d_a Q'_{a,o} S_{a,o}(d, Q), \tag{4}$$

Combining lines (3) and (4) contradicts strict properness with respect to Q'.

The other direction, which shows how to construct a strictly proper decision market for any decision rule with full support, follows by the construction in the proof of Theorem 1. □

This result extends Othman and Sandholm's impossibility result for deterministic decision markets [16] to the more general class of decision markets without full support. The theorem does not apply to non-strictly proper decision markets, however; for example, all constant decision scoring rules are proper for all decision rules.

4.4   Decision Markets with Good Incentives

While Theorem 1 provided a simple construction to create a strictly proper decision market, we now characterize all strictly proper decision and decision scoring rule pairs. The proof of Theorem 3 parallels similar characterizations of proper scoring rules given by Gneiting and Raftery [10] and of strictly proper pairs for a single expert [6], and appears in the full version of the paper[1].

Theorem 3. *A decision market (D, S), where S is regular and D has full support, is (strictly) proper if*

$$S_{a,o}(d, P) = \frac{1}{d_a |\mathcal{A}|} \left(G(P) - G'(P) : P + |\mathcal{A}| G'_{a,o}(P) \right) \tag{5}$$

where $G : \Delta(\mathcal{O})^{|\mathcal{A}|} \to \mathbb{R}$ is a (strictly) convex function, $G'(P)$ is a subgradient of G at P and : denotes the Frobenius inner product. Conversely, if S is differentiable in P and (D, S) is (strictly) proper, then S can be written in the form of (5) for some (strictly) convex G.

---

[1] Available from the authors' personal webpages.

Like the construction of Theorem 1, the characterization shown in Theorem 3 requires an expert's expected score to be independent of the final report, and that the realized score vary inversely with the likelihood an action is taken. It does, however, allow more complicated constructions than the normalized strictly proper scoring rules used in Theorem 1. For example, given a decision rule D with full support, defining

$$S_{a,o}(d, P) = \frac{1}{d_a|\mathcal{A}|}(2|\mathcal{A}|P_{a,o} - \sum_{i,j} P_{i,j}^2), \tag{6}$$

makes (D, S) a strictly proper decision market.

Theorem 3 also illustrates that our expansion of the payment rule in decision markets from scoring rules to decision scoring rules is necessary to obtain myopic incentive compatibility, because scoring rules do not allow a dependence on d. Scoring rules function properly in the special case of a prediction market because for any constant decision rule a strictly proper scoring rule is sufficient to create myopic incentive compatibility.

5   Extensions

In this section we discuss how a decision maker can approximate a deterministic rule, and what the optimal decision rule for a risk-neutral decision maker would be. We also demonstrate a correspondence between any strictly proper decision market and a set of strictly proper prediction markets, suggesting a framework for applying previous prediction market results to decision markets.

5.1   Approximating Deterministic Decisions

Deterministic decision rules, like MAX, are natural. Unfortunately, no strictly proper decision market can use a deterministic decision rule. It is possible, however, to approximate deterministic decision rules with stochastic ones, but better approximations of a deterministic decision rule increase the decision maker's worst case loss.

Corollary 1. *Every strictly proper decision market (D, S) where* $\inf_{P \in \Delta(\mathcal{O})^{|\mathcal{A}|}} D_a(P) = 0$ *for some action a has unbounded worst-case loss.*

We omit the proof as it follows directly from the inverse relationship between scores and the likelihood of actions required by Theorem 3.

5.2   Expected Utility Maximizing Decision Rules

A natural question is how a decision maker should design a strictly proper decision market to maximize their expected utility. A decision maker's utility is the payoff they receive for the observed outcome minus the cost of paying experts. Since the expected payment to experts is independent of the decision rule used, an expected utility maximizing decision rule maximizes the likelihood the most preferred action is taken, subject to the decision maker's budget constraint. We call this decision rule APPROX-MAX, and since picking a decision scoring rule is analogous to picking a scoring rule for a prediction market, we take it as given when defining the decision rule.

Given a budget b and decision scoring rule S, and the final reports $P^{\bar{t}}$, we compute a minimal feasible probability for each action a,

$$p_a = \max_o \frac{S_{a,o}([1]^{|a|}, P^{\bar{t}}) - S_{a,o}([1]^{|a|}, P^0)}{b},$$

where $[1]^{|a|}$ is a vector of ones. This expression computes the decision maker's worst-case payment to experts for each action, unweighted, then divides that value by the budget to find a feasible inverse multiplier for the decision scoring rule, which is equal to the minimal feasible probability. If $\sum_{a \in \mathcal{A}} p_a > 1$ then no decision rule fits the decision maker's budget, but otherwise a "probability surplus" of $1 - \sum_{a \in \mathcal{A}} p_a$ can be assigned arbitrarily. APPROX-MAX adds this surplus to the minimal feasible probability of the most preferred action to maximize the decision maker's expected utility.

5.3 A Correspondence between Decision Markets and Prediction Markets

Strictly proper decision markets constructed using the technique in Theorem 1 have an expected score for a prediction P, given beliefs Q, of

$$\sum_{a \in \mathcal{A}} \sum_{o \in \mathcal{O}} Q_{a,o} s_o(P_a),$$

where s is a strictly proper scoring rule. This is also equal to an expert's expected score for a set of predictions in a set of independent prediction markets, one for each action in $\mathcal{A}$. This equivalence holds more generally: every strictly proper decision market has a corresponding set of prediction markets. Theorem 4 states this correspondence formally.

Theorem 4. *Every strictly proper decision market (D, S), where S is differentiable, has a corresponding strictly proper set of prediction markets. This correspondence implies that when the previous prediction in both settings is the same, the expected score for a new prediction, given any beliefs, is also the same.*

Informally, this theorem implies risk-neutral experts are indifferent to participating in a strictly proper decision market or the corresponding set of strictly proper prediction markets. Their available predictions and expected scores for each prediction are the same in both settings. This correspondence suggests that results applying to sets of prediction markets may also apply directly to decision markets.

6   Conclusion

We extended the myopic incentive compatibility of prediction markets to decision markets. We proved that this extension requires the decision maker use a decision rule with full support, and showed how to construct a strictly proper decision market for any such decision rule, answering an open question posed by Chen and Kash [6]. We characterized the set of myopic incentive compatible decision markets, and show that it is possible to approximate any deterministic decision rule with a stochastic decision rule, although better approximations cause higher worst-case loss for the decision maker. We also showed a correspondence between strictly proper decision and sets of prediction markets, suggesting a unifying technique to apply results to both types of markets.

There remain many interesting research questions involving decision markets. Requiring decision makers commit to a randomized decision rule poses an important practical challenge. Returning to our example from the introduction, the project manager must be willing to risk hiring a slower developer for the privilege of running a myopic incentive compatible decision market. This is simply not credible behavior—managers prefer to hire faster developers. Designing a more credible mechanism is likely to be a prerequisite for the deployment of decision markets in practice.

Another practical concern is extending our decision market results to a cost function framework. Instead of requiring experts provide an entire probability distribution, cost function based prediction markets allow traders to buy and sell contracts associated with particular outcomes [7]. The price of each contract is expected to represent the likelihood that outcome occurs. These interfaces are similar to that provided by stock markets, and there is an equivalence between scoring rule and cost function markets. The same equivalence holds for decision markets, but our decision scoring rules require contracts with variable payouts or large upfront costs, both undesirable features. Designing a more natural contract structure for a decision may be of considerable practical value.

References

1. Berg, J.E., Forsythe, R., Nelson, F.D., Rietz, T.A.: Results from a dozen years of election futures markets research. In: Plott, C.A., Smith, V. (eds.) Handbook of Experimental Economic Results (2001)
2. Berg, J.E., Rietz, T.A.: Prediction markets as decision support systems. Information Systems Frontier 5, 79–93 (2003)
3. Chen, K.-Y., Plott, C.R.: Information aggregation mechanisms: Concept, design and implementation for a sales forecasting problem. Working paper No. 1131, California Institute of Technology, Division of the Humanities and Social Sciences (2002)
4. Chen, Y., Dimitrov, S., Sami, R., Reeves, D.M., Pennock, D.M., Hanson, R.D., Fortnow, L., Gonen, R.: Gaming prediction markets: Equilibrium strategies with a market maker. Algorithmica 58(4), 930–969 (2010)

5. Chen, Y., Gao, X.A., Goldstein, R., Kash, I.A.: Market manipulation with outside incentives. In: AAAI 2011: Proceedings of the 25th Conference on Artificial Intelligence (2011)
6. Chen, Y., Kash, I.A.: Information elicitation for decision making. In: AAMAS 2011: Proceedings of the 10th International Conference on Autonomous Agents and Multiagent Systems (2011)
7. Chen, Y., Pennock, D.M.: A utility framework for bounded-loss market makers. In: UAI 2007: Proceedings of the 23rd Conference on Uncertainty in Artificial Intelligence, pp. 49–56 (2007)
8. Debnath, S., Pennock, D.M., Giles, C.L., Lawrence, S.: Information incorporation in online in-game sports betting markets. In: EC 2003: Proceedings of the 4th ACM Conference on Electronic Commerce, pp. 258–259. ACM, New York (2003)
9. Dimitrov, S., Sami, R.: Composition of markets with conflicting incentives. In: EC 2010: Proceedings of the 11th ACM Conference on Electronic Commerce, pp. 53–62. ACM, New York (2010)
10. Gneiting, T., Raftery, A.E.: Strictly proper scoring rules, prediction, and estimation. Journal of the American Statistical Association 102(477), 359–378 (2007)
11. Hanson, R.: Decision markets. IEEE Intelligent Systems 14(3), 16–19 (1999)
12. Hanson, R.D.: Combinatorial information market design. Information Systems Frontiers 5(1), 107–119 (2003)
13. Hanson, R.D.: Logarithmic market scoring rules for modular combinatorial information aggregation. Journal of Prediction Markets 1(1), 1–15 (2007)
14. McCarthy, J.: Measures of the value of information. PNAS: Proceedings of the National Academy of Sciences of the United States of America 42(9), 654–655 (1956)
15. Ostrovsky, M.: Information aggregation in dynamic markets with strategic traders. In: EC 2009: Proceedings of the Tenth ACM Conference on Electronic Commerce, p. 253. ACM, New York (2009)
16. Othman, A., Sandholm, T.: Decision rules and decision markets. In: Proceedings of the 9th International Conference on Autonomous Agents and Multiagent Systems (AAMAS), pp. 625–632 (2010)
17. Savage, L.J.: Elicitation of personal probabilities and expectations. Journal of the American Statistical Association 66(336), 783–801 (1971)
18. Wolfers, J., Zitzewitz, E.: Prediction markets. Journal of Economic Perspective 18(2), 107–126 (2004)

A Global Characterization of Envy-Free Truthful Scheduling of Two Tasks

George Christodoulou[1] and Annamária Kovács[2,*]

[1] University of Liverpool, Liverpool, UK
gchristo@liv.ac.uk
[2] Department of Informatics, Goethe University, Frankfurt M., Germany
panni@cs.uni-frankfurt.de

Abstract. We study envy-free and truthful mechanisms for domains with additive valuations, like the ones that arise in scheduling on unrelated machines. We investigate the allocation functions that are both weakly monotone (truthful) and locally efficient (envy-free), in the case of only two tasks, but *many* players. We show that the only allocation functions that satisfy both conditions are affine minimizers, with strong restrictions on the parameters of the affine minimizer. As a further result, we provide a common payment function, i.e., a single mechanism that is both truthful and envy-free.

For additive combinatorial auctions our approach leads us (only) to a non- affine maximizer similar to the counterexample of Lavi et al. [26]. Thus our result demonstrates the inherent difference between the scheduling and the auctions domain, and inspires new questions related to the classic problem of characterizing truthfulness in additive domains.

1 Introduction

We are interested in characterizing the class of *deterministic* mechanisms that are both *incentive-compatible* and *envy-free* for domains with *additive* valuations. Such valuations arise naturally in many interesting problems, like for instance scheduling on unrelated machines, and combinatorial auctions with additive bidders. We describe the whole setting as a scheduling problem. There are n machines (agents) and m tasks, and the processing time needed for a task j to run on machine i is t_{ij}, and is *privately* known only to the agent that owns the machine. Incentive-compatibility assures that no player can gain by misreporting her true values, while envy-freeness that no individual is envious of the combination of tasks and payments given to other players.

Incentive-compatibility. The scheduling setting was originally proposed by Nisan and Ronen, in their seminal paper [32] that pioneered the field of Algorithmic Mechanism Design, as a vehicle to explore the potentiality/limitations of truthful mechanisms in optimization problems. It was demonstrated that not all objectives can be truthfully optimized, *even by non polynomial-time algorithms*.

* Research supported by the German Research Foundation (DFG) grant KO 4083/1-1.

N. Chen, E. Elkind, and E. Koutsoupias (Eds.): WINE 2011, LNCS 7090, pp. 84–96, 2011.
© Springer-Verlag Berlin Heidelberg 2011

In particular, a standard performance criterion in the scheduling literature is makespan minimization (i.e. minimizing the maximum completion time of a machine), which is radically different than the common, well-studied social welfare maximization objective in economics. Nisan and Ronen showed that it is impossible to design deterministic truthful mechanisms with approximation guarantee better than 2, and they conjectured that VCG [36,12,20] (that achieves the rather unattractive ratio of n) is optimal among truthful mechanisms. The conjecture still remains open; the constant lower bound has been slightly improved to 2.41 for three machines [9], and later to 2.61 for n machines [25].

One of the reasons that make the problem particularly difficult, is the lack of a useful characterization of the allocation functions used by incentive-compatible mechanisms for restricted domains. There are two types of characterizations that dominate the literature of Mechanism Design. Characterizations of the first type, like Weak Monotonicity [30,26,35] or Cycle Monotonicity [34], describe the implementable allocations in a *local* fashion. Roughly, these are properties that describe the restrictions imposed on a *single* player's possible allocations with respect to his declarations. The second type characterizes the implementable allocations in a more *global* fashion. The most important result of this type is due to Roberts who showed that for *unrestricted* domains (where all possible valuations over outcomes are allowed) the only implementable social choice rules are a simple generalization of VCG mechanisms, namely *affine maximizers* [24].

Since we follow the scheduling notation, and the players are cost minimizers instead of utility maximizers, it will be useful to define affine minimizers.

Definition 1 (Affine Minimizers). *We say that an allocation function is an affine minimizer if there exist nonnegative constants λ_i, one for each player $i = 1, \ldots, n$, and γ_a one for each allocation a, such that the mechanism selects the allocation a, that minimizes*

$$\sum_{i=1}^{n} \sum_{j=1}^{m} \lambda_i \cdot a_{ij} \cdot t_{ij} + \gamma_a,$$

where a_{ij} is 1 if player i gets task j according to a, and 0 otherwise.

Both characterizations of the first type have been proved very useful in the design of truthful mechanisms, but only in domains that are very restricted; there exist *non-VCG* deterministic monotone algorithms with optimal performance for single-parameter[1] valuation domains (e.g. scheduling on related machines [1,11], single-minded combinatorial auctions [28,7]), and cycle-monotone algorithms for multi-dimensional domains with only two possible values [27]. However, for more general multi-dimensional domains such characterizations did not seem to be informative so far. A global, Roberts-like characterization would be much more

---

[1] We refer the reader to Chapters 9 and 12 of part II of [31] for basic definitions and discussion about valuation domains.

useful. Unfortunately, Roberts' requirement of unrestricted valuations does not apply to many realistic setups with richer structure, like combinatorial auctions and scheduling where inherently, externalities[2] make not much sense. Only for the special cases of two players, when *all items/tasks must be allocated*[3], is a global characterization known [18,10].

Finally, since the problem has remained open for so long, there have been efforts to impose extra conditions on top of incentive-compatibility, in order to restrict further the class of possible mechanisms, and try to make the problem easier to attack. Ashlagi et al. [2] consider the natural restriction of anonymity, i.e. the allocation should not depend on the identities of the players. They succeeded in proving the Nisan-Ronen conjecture for that case.[4] Lavi et al. [26] showed that assuming a restriction analogous to the Arrowian Independence of Irrelevant Alternatives, the only truthful mechanisms (in order based domains) are so-called "almost-" affine maximizers.

Envy-Freeness. Envy-freeness has traditionally been considered a very important fairness criterion in Economics and Political Science in settings without money and with infinitely divisible goods [5,33]. While generally in settings with indivisible goods, envy-free allocations do not always exist, if we allow payments, in the standard quasilinear utility setting, envy-free outcomes do exist. Money is used to compensate envious players. Formally, a mechanism is envy-free for the scheduling setting, if for every player $i \in [n]$, and for every other player $h \neq i$,

$$\sum_{j=1}^{m} a_{ij} t_{ij} - p^i \leq \sum_{j=1}^{m} a_{hj} t_{ij} - p^h,$$

where p^i, and p^h are the respective payments for the players.

Haake et al. [22] characterized the class of allocations that can be implemented in an envy-free manner, in terms of a property that is called local efficiency in [29]. This requires that the allocation must maximize the social welfare over allocations permuting the same bundles, and is necessary and sufficient for envy-free implementations. For our setting the definition is the following.

Definition 2 (Local Efficiency). *We say that a mechanism is* locally efficient *if the mechanism selects an allocation a, such that for all $t = (t_1, \ldots, t_n)$, and all permutations π of $[n]$, it satisfies*

$$\sum_{i=1}^{n} \sum_{j=1}^{m} a_{ij} \cdot t_{ij} \leq \sum_{i=1}^{n} \sum_{j=1}^{m} a_{\pi(i)j} \cdot t_{ij}.$$

---

[2] The valuation of a player i in such settings is a function of the bundle of items (or set of tasks) that i gets, and not of the other players' bundles.

[3] For settings that allow partial allocations, there exist positive results that escape those characterizations [4,17].

[4] The lower bound proof in [2] did not need a characterization of anonymous truthful mechanisms, which still remains a major open problem.

There have been many papers that considered envy-free pricing for revenue maximization problems [21,8,6,3,23], while hardness results have been shown in [16]. Mu'alem [29], and later Cohen et al. [13] considered bounding the performance of (non-truthful) envy-free mechanisms for makespan minimization.

Our Contribution. We study envy-free and truthful mechanisms for domains with additive valuations, like scheduling. It is known [15] that this class is non-empty, as VCG with Clarke payments satisfy both conditions. Cohen et al. [14] have characterized this class in terms of a Rochet-like cycle monotonicity. In [15] the same authors studied a variation where each agent has a capacity that determines the maximum number of items that she can be assigned. They focus on VCG mechanisms, and they seek for payments that are both truthful and envy-free. Very recently, Fleischer and Wang [19] showed that for the case of *two related* machines, the only mechanism that is truthful, envy-free, scalable, anonymous, and individually rational is the VCG. Our domain is multi-dimensional *(unrelated machines)*, our results hold for many players (for two items), and we require only envy-freeness on top of truthfulness.

- We investigate the allocation functions that are both weakly monotone (truthful) and locally efficient, in the case of only two tasks, but many players. We are interested in a global, Roberts-like characterization. For the sake of more generality and simple exposition, in the technical part we allow that the t_{ij} take arbitrary real values. We show that if equal bids for the same task are excluded, then the only allocation functions in this class are affine minimizers with all $\lambda_i = 1$, and further strong restrictions on the parameters γ_a (see Theorem 1). We complete the theorem by showing a simple non affine-minimizer mechanism with singularities for three players, if equal bids of different players for the same task are allowed in the input.
- Surprisingly, we found that our proof methods and results carry over to the scheduling domain (i.e., when all t_{ij} are positive), while they do *not* carry over to additive combinatorial auctions with two items (equivalent to our model with every t_{ij} negative)! This fact is especially interesting, given that so far the two problems have been treated as "almost" equivalent. For combinatorial auctions, we present a new non affine-minimizer mechanism for three or more players, that is continuous, truthful and envy-free.
- Since the affine minimizers of the characterization theorem are both monotone, and locally efficient, they admit a truthful payment scheme, and a (possibly different) envy-free payment scheme. We provide a *common* payment function, i.e., a single mechanism that is both truthful and envy-free.[5]

It should be emphasized that this is a genuinely multi-parameter setting. To the best of our knowledge, this is the first time that a global characterization has been proven for a scheduling-type multi-dimensional domain for more than two players. Even for the simple case of three players and two tasks, no global

---

[5] For results on payments see the full version at
http://www.csc.liv.ac.uk/~gchristo/

characterization of incentive-compatible (non-envy-free) mechanisms is known, which is considered a very important open problem. Our primary goal has been to purify the general problem with the envy-freeness constraints, so that a new, structural approach to characterization becomes feasible.

Open Problems. The most important question here is, whether the non-envy-free problem, or other problem variants can be tackled by generalizing our methods. Similar results for two tasks in the non- envy-free case could possibly serve as cornerstones for the general many-tasks problem [32], as has been the case in the two-player setting [10].

Taking an opposite view, we ask the following: The counterexample of Section 3.3 turns out to be of similar flavor as the non affine-maximizer auction of Lavi et al. ([26] Example 4.). Note that this kind of example exists despite the envy-freeness restriction, whereas no counterexample exists for scheduling. Are there nontrivial[6] counterexamples for scheduling, (or for the unbounded domain) in the truthful, *non envy-free* case? Is the orientation of the domain crucial?

Notation and Basic Geometry of Truthfulness. The allocation of tasks to player i is denoted by a_i, and can take the values $a_i \in \{11, 10, 01, 00\}$; the allocation a to all the players is the vector $a = (a_1, a_2, \ldots, a_n)$. Further, we denote by a^{ij} the allocation giving task 1 to player i and task 2 to player j.

In a truthful mechanism, the payment of player i depends on the bid matrix t_{-i} of the *other* players, and on the allocation a_i of player i. Let $p^i_{a_i}(t_{-i})$ denote this payment. We introduce the following functions:

$$f_i(t_{-i}) = p^i_{11}(t_{-i}) - p^i_{01}(t_{-i})$$
$$f'_i(t_{-i}) = p^i_{10}(t_{-i}) - p^i_{00}(t_{-i})$$
$$g_i(t_{-i}) = p^i_{11}(t_{-i}) - p^i_{10}(t_{-i})$$
$$g'_i(t_{-i}) = p^i_{01}(t_{-i}) - p^i_{00}(t_{-i})$$

Most of the time we will apply the short notation f_i, f'_i, g_i, g'_i, and for player $i = 1$ we omit the subscript, using only f, f', g, g', for the respective values. It is well known [10] that in any truthful mechanism, for fixed t_{-i} the allocation of player i as a function of (t_{i1}, t_{i2}) has a geometrical representation of one of three possible shapes – see Figure 2 –, where the two vertical boundaries are on the lines $t_{i1} = f_i$ and $t_{i1} = f'_i$, and the horizontal ones are on the lines $t_{i2} = g_i$ and $t_{i2} = g'_i$. Furthermore, $f'_i - f_i = g'_i - g_i$ holds. We call the 45° boundary 10/01 or 11\00 the *flipping boundary*, since there the allocation of both tasks gets flipped (the flipping boundary may happen to be a single point). Our proofs are based on this type of representation.

Let t^m be the point (in general not a single bid) with coordinates $t^m_1 = \min_{k \neq 1,2} t_{k1}$ and $t^m_2 = \min_{k \neq 1,2} t_{k2}$. Furthermore, let $M = \min_{i \neq 1} \{t_{i1} + t_{i2}\}$.

---

[6] The known non affine-minimizers are 'practically' task-independent mechanisms [10].

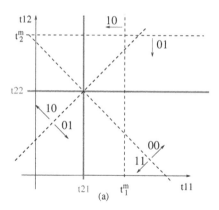 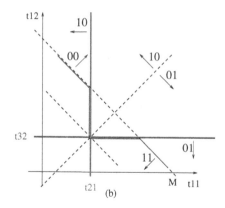

Fig. 1. Envy-freeness constraints on the allocations, in case of minimum bids by a single agent (a) and two different agents (b)

2 Constraints Due to Envy-Freeness

We start by investigating the (geometric) restrictions that envy-freeness imposes on the possible allocations. Without loss of generality, we consider the allocation figure of player 1. In the next propositions we deal with the cases, when in t_{-1} a single player (assume wlog. player 2) bids minimum, respectively when different players (say players 2, and 3) bid minimum for the two tasks.

Proposition 1. *Assume that $t_{22} < t_{i2}$, and $t_{21} < t_{i1}$ for every player $i \neq 1,2$. The following restrictions are implied by local efficiency (see Figure 1 (a)). If the allocation of player 1 is*

(a) (11) then $t_{11} + t_{12} \leq t_{21} + t_{22}$;
(b) (10) then $t_{11} + t_{22} \leq t_{12} + t_{21}$, and $t_{11} \leq t_1^m$;
(c) (01) then $t_{11} + t_{22} \geq t_{12} + t_{21}$, and $t_{12} \leq t_2^m$;
(d) (00) then $t_{11} + t_{12} \geq t_{21} + t_{22}$.

Proposition 2. *Assume that $t_{21} \leq t_{i1}$, and $t_{32} \leq t_{i2}$ for every player $i \neq 1$. The following restrictions are implied by local efficiency (see Figure 1 (b)). If the allocation of player 1 is*

(a) (11) then $t_{11} + t_{12} \leq M$;
(b) (10) then $t_{11} + t_{32} \leq t_{12} + t_{21}$, and $t_{11} \leq t_{21}$;
(c) (01) then $t_{11} + t_{32} \geq t_{12} + t_{21}$, and $t_{12} \leq t_{32}$;
(d) (00) then $t_{11} + t_{12} \geq M$, or ($t_{11} \geq t_{21}$ and $t_{12} \geq t_{32}$).

The geometric implications for envy-free allocations are summarized by Corollary 1 below. They admit allocations of types shown in Figures 2 and 3.

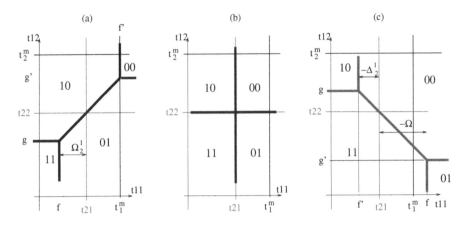

Fig. 2. Possible forms of allocations when a single player bids minimum for both tasks

Corollary 1. *For the allocation of player 1 in a truthful and envy-free mechanism the following hold. If $t_{21} < t_1^m$, and $t_{22} < t_2^m$, then the point t_2 is on the flipping boundary, furthermore $f' \le t_1^m$ and $g' \le t_2^m$. If in t_{-1} players 2 and 3 bid minimum for tasks 1 and 2 respectively, then either $f' = t_{21}$ and $g' = t_{32}$, OR $f' \le t_{21}$ and $g' \le t_{32}$, and the flipping boundary (11\00) is on the line $t_{11} + t_{12} = M$.*

3   Characterization of Envy-Free Truthful Mechanisms

The characterization has two major steps. First, focusing on the case of minimum bids by a single player (say, player 2), we prove that the distances $f' - t_{21}$, $t_{21} - f$, $g' - t_{22}$, and $t_{22} - g$ are independent of t_2, that is, by moving t_2 the allocation figure moves along with t_2 while keeping its shape (cf. Figure 2). To be precise, this holds as long as $f' < t_1^m$, $g' < t_2^m$ and $t_{21} + t_{22}$ is minimum, i.e., $t_{21} + t_{22} = M$. Therefore, the first two lemmas consider an extended domain for t_2 (as compared to $(-\infty, t_1^m) \times (-\infty, t_2^m)$, where player 2 bids minimum).

Second, by looking at the case of minimum bids by different players (Figure 3), it becomes clear that many of these constant distances must be equal, further implying that they are even independent of all other bids (e.g., $f' - t_{21}$, is independent not only of t_2 but even of t_{-12}, the input of all players other than 1 and 2). Therewith the allocation rule turns out to be identical to that of an affine minimizer, given that all payment functions are continuous. If arbitrary functions are allowed, then it is an affine minimizer over inputs with pairwise different bids, with possible singularities when some bids are equal.

We start with the observation that whenever player 1 is sure to exchange the first task with player 2, the functions f and f' are non-decreasing in t_{21} *independently* of t_{22}.[7] Note that the conditions of the next Lemma ensure exactly

---

[7] In a more special form, the same was observed in [10].

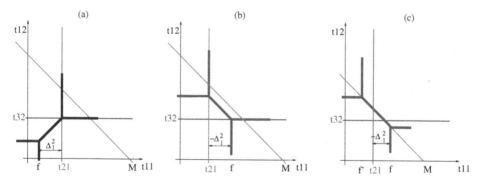

Fig. 3. Possible forms of allocations when different players bid minimum for both tasks

that the task gets exchanged with player 2, and not with some other player. Below we omit the fixed constant t_{-12} from the argument of these functions.

Lemma 1. *(a) For any t_2, t_2' s.t. $t_{21} < t_{21}' < t_1^m$, it holds that $f(t_2) \leq f(t_2')$.*
(b) Let t_2 be so that $t_{22} < t_{i2}$ and $t_{21} + t_{22} < t_{i1} + t_{i2}$ for $i \geq 3$, and let the same
hold for t_2'. If $t_{21} < t_{21}'$, and $g'(t_2') < t_2^m$, then $f'(t_2) \leq f'(t_2')$.
Analogous statements hold for g and g'.

The next lemma completes the first main step of the characterization.

Lemma 2. *In every truthful envy-free mechanism, for fixed t_{-12} there exist constants $\Delta = \Delta(t_{-12})$, and $\Omega = \Omega(t_{-12})$ s.t. for every t_2 such that $t_{21} + t_{22} = M$,*

(a) $f(t_2) = t_{21} - \Omega$ if $t_{21} < t_1^m$ and $t_{22} \leq \max\{t_2^m, t_2^m - \Omega\}$, and
(b) $f'(t_2) = \min\{t_{21} + \Delta, t_1^m\}$ whenever $t_{22} \leq \min\{t_2^m, t_2^m - \Delta\}$.

Furthermore, if Δ is positive (negative) then Ω is non-negative (non-positive).

Analogous statements hold for g and g'. In particular, given that t_2 is on the flipping boundary, we obtain that if Ω and Δ are non-negative, then $g = t_{22} - \Omega$, and $g' = \min\{t_{22} + \Delta, t_2^m\}$; if they are non-positive, then $g = t_{22} - \Delta$, and $g' = \min\{t_{22} + \Omega, t_2^m\} = t_{22} + \Omega$.

An intuitive proof of the lemma is the following. By Lemma 1, $f(t_{21}, t_{22})$ is a monotone function of t_{21}, and therefore it is continuous in almost all t_{21} (say, for fixed t_{22}). Moreover, since f is monotone in t_{21} *regardless* of t_{22}, it is necessarily independent of t_{22} (that is, $f(t_{21}, t_{22}) = f(t_{21}, t_{22}')$), whenever it is continuous in t_{21}. Assume e.g., that $f(t_{21}, t_{22}) < t_{21}$. Then the fact that t_2 is on the flipping boundary (cf. Corollary 1), and $f(t_{21}, t_{22})$ is independent of t_{22} implies that $t_{22} - g(t_{21}, t_{22}) = t_{21} - f(t_{21}, t_{22})$ is a constant for fixed t_{21}. We name this constant Ω, and obtain that $g(t_{21}, t_{22}) = t_{22} - \Omega$ must hold for *all* t_{21}, since g is monotone in t_{22} regardless of t_{21}.

Notation. Let $\Delta_2^1 = \Delta_2^1(t_{-12})$, and $\Omega_2^1 = \Omega_2^1(t_{-12})$ denote the constants obtained in Lemma 2. For arbitrary two players $i \neq j$ we define $\Delta_j^i(t_{-ij})$, and $\Omega_j^i(t_{-ij})$ analogously. Note that Δ_j^i and Ω_j^i appear in the allocation figure of player i, when player j has minimum bids in t_{-i}.

Observation 1. *For any $i \neq j$, for fixed t_{-ij} we have $\Delta_j^i = \Omega_i^j$.*

In the second part we prove that the $\Delta_j^i(t_{-ij}) = \Omega_i^j(t_{-ij})$ values are independent of (t_{-ij}). We complete the characterization by showing that many of these constants are equal, and either all Δ are non-negative, or all are non-positive.

Lemma 3. *Let t_{-1} be fixed so that $t_{21} < t_{i1}$ for all $i \neq 1, 2$, and $t_{32} < t_{i2}$ for all $i \neq 1, 3$ (see Figure 3). Then for the allocation of player 1 $f = t_{21} - \Delta_1^2(t_{-12})$. Furthermore, $g = t_{32} - \Delta_1^3(t_{-13})$ if $g < t_{32}$, and $g = t_{32} - \Omega_1^3(t_{-13})$ if $g > t_{32}$, and one of these two holds if $g = t_{32}$.*

Lemma 4. *Let $t_{-12} = (t_3, t_4, \ldots, t_n)$, and $t'_{-12} = (t'_3, t_4, \ldots, t_n)$ be such that $\max\{t_{32}, t'_{32}\} < t_{i2}$ (resp. $\max\{t_{31}, t'_{31}\} < t_{i1}$) for $i \neq 1, 3$. Then $\Delta_1^2(t_{-12}) = \Delta_1^2(t'_{-12})$.*

Intuitively, the lemma implies that given the set of points $\{t_3, t_4, \ldots, t_n\}$ in the plane, we can move around the point in the lowermost (leftmost) position, the $\Delta_1^2(t_{-12})$ does not change as long as the point remains in the lowermost (leftmost) position. Next we show that an arbitrary array of points $(t_3, t_4, \ldots, t_n)$ can be transformed to another arbitrary array $(t'_3, t'_4, \ldots, t'_n)$ using only such movements. Consequently, $\Delta_1^2(t_{-12})$ is independent of t_{-12} (at least for t_{-12} where the points have pairwise different coordinates). This holds obviously for arbitrary pair of players $i \neq j$.

Lemma 5. *Let $t_{-12} = (t_3, t_4, \ldots, t_n)$ and $t'_{-12} = (t'_3, t'_4, \ldots, t'_n)$ be arbitrary such that the first (or second) coordinates of the points are pairwise different in t_{-12} and similarly in t'_{-12}. Then $\Delta_1^2(t_{-12}) = \Delta_1^2(t'_{-12})$.*

Having the Δ_j^i values constant, it is straightforward to verify that restricted to inputs having pairwise different bids for each task, the allocation of the mechanism is identical to that of an affine minimizer, where $\Delta_i^j = \gamma_{a^{ii}} - \gamma_{a^{ij}}$. As a concluding step, we investigate the question, to what extent the constants Δ_j^i determine each other.

If $\Delta_j^i = 0$ for all $i \neq j$, then the allocation is trivially the VCG allocation. Assume now that there exist two players $h \neq k$, such that $\Delta_k^h \neq 0$. We have the following corollaries of Lemmas 3 and 5.

Corollary 2. *If $\Delta_k^h > 0$, then $\Delta_k^i = \Delta_k^h$ for every player i.*

Corollary 3. *If $\Delta_k^h < 0$, then for each pair i, j of different players $\Delta_j^i = \Omega_i^j \leq 0$. Furthermore, for $n \geq 3$, all of the $\Delta_j^i = \Omega_i^j$ values are equal.*

It needs a straightforward verification that the obtained types of allocations are, in fact, locally efficient. We summarize our results in terms of affine minimizers. Notice that if we assume continuous payment functions, then all f and g functions are also continuous, and therefore the characterization can be extended to the whole domain. (Observe also that we require a rather weak form of continuity.)

Theorem 1. *For domains with additive valuations with two tasks (items), and any number of players, excluding equal bids of different players, the allocations that admit both truthful and envy-free mechanisms are affine minimizers with $\lambda_i = 1$, and of one of the following types:*

(1) $\gamma_{a^{ii}} \geq 0$, and $\gamma_{a^{ij}} = 0$ for $i \neq j$; or
(2) $\gamma_{a^{ij}} \geq 0$, and $\gamma_{a^{ii}} = 0$. Furthermore, for $n \geq 3$ all $\gamma_{a^{ij}}$ $(i \neq j)$ are equal.

Assuming that (fixing the rest of the input) the payments are continuous functions of every bid t_{ij}, and equal bids are allowed, the allocation is an affine minimizer over the whole domain.

3.1   Counterexample with Singularity

Some kind of restriction of the domain to pairwise different bids, or the continuity requirement is really necessary for the theorem to hold. Here we show a simple mechanism with singularity that is not an affine minimizer.

Consider the following simple allocation rule for $n \geq 3$ players. Let A be the allocation of an affine minimizer with $\gamma_{a^{ii}} = 1$ for all i, and $\gamma_{a^{ij}} = 0$ for $i \neq j$ (i.e., case $\Delta > 0$), and define the allocation rule $a()$ to be $a(t) = A(t)$ if $t_1 \neq t_2$, and $a(t)$ be the VCG allocation if $t_1 = t_2$.[8] Moreover, if t_1 and t_2 have equal, and minimum coordinates, then players 1 and 2 must both get a job. For players $i \neq 1, 2$, for fixed t_{-i} the allocation looks either like A or like VCG, and is truthful and envy-free. For player 1 (and similarly for player 2), for fixed t_{-1}, the allocation figure is that of A. We only need to perform a straightforward check – assuming different relative positions of t_2 –, that in the single point $t_1 = t_2$, the allocation of player 1 is consistent with this figure.

3.2   Task Scheduling

Our setting models the problem of (envy-free) unrelated scheduling mechanisms, if we restrict the t_{ij} to positive values.[9] The difficulty with allowing t_i points only in the positive orthant is that Lemma 4, and thus also Lemma 5 do not necessarily hold if the t_1 and t_2 coordinates are bounded. In fact, for this reason, the characterization result fails for $t_{ij} \in R_-$. Rather surprisingly, for positive t_{ij} Lemmas 4 and 5 can be 'saved' in some modified form, and we obtain Theorem 2.[10] Despite that the unrelated scheduling problem and the additive

---

[8] We could have used any affine minimizer with $\gamma_{a^{ii}} < 1$ instead of VCG; however $\gamma_{a^{ii}} > 1$ would not work.

[9] For simplicity we exclude $t_{ij} = 0$, since our results hold for continuous mechanisms, or for inputs with pairwise different coordinates.

[10] Notice that on R_+^2 our mechanisms are not decisive (i.e., a single player cannot force any outcome for himself, by bidding properly), except for the VCG mechanism.

combinatorial auctions problem look very similar, our results demonstrate that they do not exhibit symmetric behaviour, and are by no means equivalent problems.

Theorem 2. *Restrict the domain of bids to $t_i \in R_+^2$, and consider only inputs where the t_{i1} are pairwise different, and similarly for the t_{i2}, and for the $t_{i1} + t_{i2}$. The truthful and envy-free mechanisms $m = 2$ are exactly the same mechanisms as in Theorem 1.*

3.3   Additive Combinatorial Auctions

In additive combinatorial auctions each player i has a positive value v_{ij} for every item j to be sold. As opposed to the cost model (scheduling), players with higher v_{ij} tend to get the item j. By using $t_{ij} = -v_{ij}$, the problem becomes equivalent to the setting used in the paper, with the t_{ij} restricted to take *negative* values. The next example shows that Theorem 1 does *not* carry over to additive combinatorial auctions. (In particular, Lemma 4 does not hold. We conjecture, though, that all counterexamples for envy-free additive auctions are variants of this one.) Assume that $t_{i1} + t_{i2} \le t_{j1} + t_{j2} \le t_{k1} + t_{k2} \le \dots$ are the three smallest sums of bids over all players (break ties by player indices). Then allocate the two jobs to players i and j, according to an affine minimizer with (negative) $\Delta = t_{k1} + t_{k2}$. This mechanism is well-defined, and checking the allocation figures shows that restricted to the negative orthant, it is also truthful and envy-free.[11]

Acknowledgements. We would like to thank Elias Koutsoupias, Amos Fiat, and Angelina Vidali for fruitful discussions.

References

1. Archer, A., Tardos, É.: Truthful mechanisms for one-parameter agents. In: FOCS, pp. 482–491 (2001)
2. Ashlagi, I., Dobzinski, S., Lavi, R.: An optimal lower bound for anonymous scheduling mechanisms. In: EC, pp. 169–176 (2009)
3. Bansal, N., Chen, N., Cherniavsky, N., Rudra, A., Schieber, B., Sviridenko, M.: Dynamic pricing for impatient bidders. ACM Transactions on Alg. 6(2) (2010)
4. Bartal, Y., Gonen, R., Nisan, N.: Incentive compatible multi unit combinatorial auctions. In: TARK, pp. 72–87 (2003)
5. Brams, S., Taylor, D.: Fair division: from cake cutting to dispute resolution. Cambridge University Press (1996)
6. Briest, P.: Uniform Budgets and the Envy-Free Pricing Problem. In: Aceto, L., Damgård, I., Goldberg, L.A., Halldórsson, M.M., Ingólfsdóttir, A., Walukiewicz, I. (eds.) ICALP 2008, Part I. LNCS, vol. 5125, pp. 808–819. Springer, Heidelberg (2008)

---

[11] Note that if $t_{i1} + t_{i2} \le t_{k1} + t_{k2}$ then for player j the area $t_{k1} + t_{k2} < t_{j1} + t_{j2}$ must be part of the 00 allocation of j. This area is bounded in the negative orthant, but not bounded if t_{ij} can take positive values. Therefore, in the scheduling or unbounded domain only degenerate affine minimizers with $\Delta = -\infty$ fulfil this requirement.

7. Briest, P., Krysta, P., Vöcking, B.: Approximation techniques for utilitarian mechanism design. In: STOC, pp. 39–48 (2005)

8. Chen, N., Ghosh, A., Vassilvitskii, S.: Optimal envy-free pricing with metric substitutability. In: EC, pp. 60–69 (2008)

9. Christodoulou, G., Koutsoupias, E., Vidali, A.: A lower bound for scheduling mechanisms. In: SODA, pp. 1163–1170 (2007)

10. Christodoulou, G., Koutsoupias, E., Vidali, A.: A Characterization of 2-Player Mechanisms for Scheduling. In: Halperin, D., Mehlhorn, K. (eds.) ESA 2008. LNCS, vol. 5193, pp. 297–307. Springer, Heidelberg (2008)

11. Christodoulou, G., Kovács, A.: A deterministic truthful ptas for scheduling related machines. In: SODA, pp. 1005–1016 (2010)

12. Clark, E.H.: Multipart pricing of public goods. Public Choice 11, 17–33 (1971)

13. Cohen, E., Feldman, M., Fiat, A., Kaplan, H., Olonetsky, S.: Envy-free makespan approximation: extended abstract. In: EC, pp. 159–166 (2010)

14. Cohen, E., Feldman, M., Fiat, A., Kaplan, H., Olonetsky, S.: On the interplay between incentive compatibility and envy freeness. CoRR, abs/1003.5328 (2010)

15. Cohen, E., Feldman, M., Fiat, A., Kaplan, H., Olonetsky, S.: Truth and envy in capacitated allocation games. CoRR, abs/1003.5326 (2010)

16. Demaine, E.D., Feige, U., Hajiaghayi, M., Salavatipour, M.R.: Combination can be hard: Approximability of the unique coverage problem. SIAM J. Comp. 38(4), 1464–1483 (2008)

17. Dobzinski, S., Nisan, N.: Multi-unit auctions: Beyond roberts. In: EC (2011)

18. Dobzinski, S., Sundararajan, M.: On characterizations of truthful mechanisms for combinatorial auctions and scheduling. In: EC, pp. 38–47 (2008)

19. Fleischer, L., Wang, Z.: Lower Bound for Envy-Free and Truthful Makespan Approximation on Related Machines. In: Persiano, G. (ed.) SAGT 2011. LNCS, vol. 6982, pp. 166–177. Springer, Heidelberg (2011)

20. Groves, T.: Incentives in teams. Econometrica 41, 617–663 (1973)

21. Guruswami, V., Hartline, J.D., Karlin, A.R., Kempe, D., Kenyon, C., McSherry, F.: On profit-maximizing envy-free pricing. In: SODA, pp. 1164–1173 (2005)

22. Haake, C.-J., Raith, M.G., Su, F.E.: Bidding for envyfreeness: A procedural approach to n-player fair division problems. Social Choice and Welfare 19 (2002)

23. Kempe, D., Mu'alem, A., Salek, M.: Envy-Free Allocations for Budgeted Bidders. In: Leonardi, S. (ed.) WINE 2009. LNCS, vol. 5929, pp. 537–544. Springer, Heidelberg (2009)

24. Roberts, K.: The characterization of implementable choice rules. Aggregation and Revelation of Preferences, pp. 321–348 (1979)

25. Koutsoupias, E., Vidali, A.: A Lower Bound of 1+Phi for Truthful Scheduling Mechanisms. In: Kučera, L., Kučera, A. (eds.) MFCS 2007. LNCS, vol. 4708, pp. 454–464. Springer, Heidelberg (2007)

26. Lavi, R., Mu'alem, A., Nisan, N.: Towards a characterization of truthful combinatorial auctions. In: FOCS, pp. 574–583 (2003)

27. Lavi, R., Swamy, C.: Truthful mechanism design for multi-dimensional scheduling via cycle monotonicity. In: EC (2007)

28. Lehmann, D.J., O'Callaghan, L., Shoham, Y.: Truth revelation in approximately efficient combinatorial auctions. J. ACM 49(5), 577–602 (2002)

29. Mu'alem, A.: On Multi-Dimensional Envy-Free Mechanisms. In: Rossi, F., Tsoukias, A. (eds.) ADT 2009. LNCS, vol. 5783, pp. 120–131. Springer, Heidelberg (2009)

30. Myerson, R.B.: Optimal auction design. Math. of Op. Res. 6(1), 58–73 (1981)

31. Nisan, N., Roughgarden, T., Tardos, E., Vazirani, V.V.: Algorithmic Game Theory. Cambridge University Press (2007)
32. Nisan, N., Ronen, A.: Algorithmic mechanism design. Games and Economic Behavior 35, 166–196 (2001)
33. Robertson, J.M., Webb, W.A.: Cake-cutting algorithms: Be fair if you can. A.K. Peters (1998)
34. Rochet, J.-C.: A necessary and sufficient condition for rationalizability in a quasi-linear context. Journal of Mathematical Economics 16, 191–200 (1987)
35. Saks, M.E., Yu, L.: Weak monotonicity suffices for truthfulness on convex domains. In: 6th ACM Conference on Electronic Commerce (EC), pp. 286–293 (2005)
36. Vickrey, W.: Counterspeculation, Auctions and Competitive Sealed Tenders. Journal of Finance, 8–37 (1961)

Truth, Envy, and Truthful Market Clearing Bundle Pricing

Edith Cohen[1], Michal Feldman[2], Amos Fiat[3], Haim Kaplan[3], and Svetlana Olonetsky[3]

[1] AT&T Labs-Research, 180 Park Avenue, Florham Park, NJ
[2] School of Business Administration, The Hebrew University of Jerusalem
[3] The Blavatnik School of Computer Science, Tel Aviv University

Abstract. We give a non-trivial class of valuation functions for which we give auctions that are efficient, truthful and envy-free.

We give interesting classes of valuations for which one can design such auctions. Surprisingly, we also show that minor modifications to these valuations lead to impossibility results, the most surprising of which is that for a natural class of valuations, one cannot achieve efficiency, truthfulness, envy freeness, individual rationality, and no positive transfers.

We also show that such auctions also imply a truthful mechanism for computing bundle prices ("shrink wrapped" bundles of items), that clear the market. This extends the class of valuations for which truthful market clearing prices mechanisms exist.

1 Introduction

In this paper we consider auctions that are

1. Efficient — the mechanism maximizes the sum of the valuations of the agents. Alternately, efficient mechanisms are said to maximize social welfare.
2. Incentive compatible (IC) — it is a dominant strategy for agents to report their private information [11].
3. Envy-free (EF) - no agent wishes to exchange her outcome with that of another [6,7,22,15,16,24].
4. Make no positive transfers (NPT)— the payments of all agents are non-negative.
5. Individually rational (IR) — no agent gets negative utility.

We argue that such auctions are natural and interesting for a variety of reasons:

- It is not clear how to obtain efficiency without truthfullness.
- An auction that is not envy-free discriminates between bidders, moreover — it is relatively easy for bidders to realize that they are being discriminated against. Experiments suggest that human subjects prefer degraded performance over discrimination, e.g., ([21]).
- Posted prices that clear the market are inherently envy-free, but computing such Walrasian pricing, even if it exists, is itself not necessarily truthful. Auctions that are both truthful and envy-free can be interpreted as a mechanism for computing market clearing prices where the posted price is associated with bundles of items rather than individual items. We present a natural subset of gross substitute valuations for which we show that

N. Chen, E. Elkind, and E. Koutsoupias (Eds.): WINE 2011, LNCS 7090, pp. 97–108, 2011.
© Springer-Verlag Berlin Heidelberg 2011

- There is no incentive compatible mechanism to compute Walrasian prices.
- We give a truthful mechanism for computing market clearing *bundle* prices.

— Auctions that are not individually rational or make positive transfers represent situations when one forces the bidders to participate against their will or subsidizes the auction. That said, we show a class of valuation functions (capacitated valuations with unequal capacities), for which Walrasian prices exist, but no auction exists that is incentive compatible, envy-free, makes no positive transfers, and is individually rational. Moreover, for a subset of this class, we give an efficient, incentive compatible, envy-free and individually rational auction (albeit — with positive transfers).

We consider a specific class of additive valuations where agents have a limit on the number of goods they may receive. We refer to such valuations as *capacitated* valuations and seek mechanisms that maximize social welfare and are simultaneously incentive compatible, envy-free, individually rational, and have no positive transfers. Capacitated valuations are a special case of gross substitute valuations (Kelso and Crawford [12]) and they are a natural generalization of the unit demand valuation. One may view the capacity of an agent as the size of the market basket, an agent with capacity c may carry no more than c items. If such a bidder gets more than c items, (say, in a shrink wrapped bundle containing $2c$ goods, of various types), she can discard any excess items.

If capacities are infinite, then sequentially repeating the 2nd price Vickrey auction meets these requirements. In 1983, Leonard showed that for unit capacities, VCG with Clarke pivot payments is also envy-free[1]. In this paper we consider generalizations of the setting considered by Leonard. For homogeneous capacities (all capacities equal) we show that VCG with Clarke pivot payments is envy-free (VCG with Clarke pivot payments is always efficient, incentive compatible, individually rational, and has no positive transfers). Also, we show that there is no incentive compatible mechanism to compute Walrasian prices for capacities > 1. For heterogeneous capacities, we show that there is no mechanism with all 5 properties, but at least in some cases, one can achieve both incentive compatibility and envy freeness.

Let $[s] = \{1, \ldots, s\}$ be the set of goods to be allocated amongst n agents with private valuations. An agent's valuation function is a mapping from every subset of the goods into the non negative reals. A *mechanism* receives the valuations of the agents as input, and determines an allocation a_i and a payment p_i for every agent. We assume that agents have quasi-linear utilities; that is, the utility of agent i is the difference between her valuation for the bundle allocated to her and her payment.

Given an efficient, incentive compatible, envy-free auction, with no positive transfers and individually rational, we can convert the allocations a_i and associated prices p_i into market clearing bundle prices. As the auction is efficient, we can assume that all items are allocated. (In capacitated valuations, there may be goods later discarded simply because the agents do not have the capacity to accept them). For every agent i we create a bundle of all items in a_i, and attach the price p_i to this shrink-wrapped bundle of goods. Of all these bundles, the bundle a_i, and it's associated price, p_i, maximize the utility for agent z, the bundle a_j and it's associated price, p_j, maximize the utility for agent j, etc.

---

[1] Lehmann, Lehmann, and Nisam, [13], show that computing VCG in the case of gross substitutes is poly time.

Most of our results concern the class of capacitated valuations: every agent i has an associated capacity c_i, and her value is additive up to the capacity, *i.e.*, for every set $S \subseteq [s]$,

$$v_i(S) = \max \left\{ \sum_{j \in T} v_i(j) \Big| T \subseteq S, |T| = c_i \right\},$$

where $v_i(j)$ denotes the agent i's valuation for good j.

Consider the following classes of valuation functions:

1. Gross substitutes: good x is said to be a gross substitute of good y if the demand for x is monotonically non-decreasing with the price of y, *i.e.*,

$$\partial(\text{demand } x)/\partial(\text{price } y) \geq 0 .$$

   A valuation function is said to obey the gross substitutes condition if for every pair of goods x and y, good x is a gross substitute of good y.
2. Subadditive valuations: A valuation $v : 2^{[s]} \rightarrow \mathbb{R}_{\geq 0}$ is said to be subadditive if for every two disjoint subsets $S, T \subseteq [s]$, $v(S) + v(T) \geq v(S \cup T)$.
3. Superadditive valuations: A valuation $v : 2^{[s]} \rightarrow \mathbb{R}_{\geq 0}$ is said to be superadditive if for every two disjoint subsets $S, T \subseteq [s]$, $v(S) + v(T) \leq v(S \cup T)$.

Capacitated valuations are a subset of gross substitutes, which are themselves a subset of subadditive valuations.

In a Walrasian equilibrium (See [12,10]), prices are *item prices*, that is, prices are assigned to *individual goods* so that every agent chooses a bundle that maximizes her utility and the market clears. Thus, Walrasian prices automatically lead to an envy-free allocation. Every Walrasian pricing gives a mechanism that is efficient and envy-free, has no positive transfers, and is individually rational [13].

We remark that while Walrasian pricing $\Rightarrow$ EF, NPT, IR, the converse is not true. Even a mechanism that is EF, NPT, IR, *and* IC does not imply Walrasian prices. Note that envy-free prices may be assigned to bundles of goods which cannot necessarily be interpreted as item prices. It is well known that in many economic settings, bundle prices are more powerful than item prices [1,19]. [12] showed that gross substitutes imply the existence of Walrasian equilibrium, Gul and Stacchetti [10] show that this is necessarily the case.

As capacitated valuations are also gross substitutes (see Theorem 2.4 in Section 2.2), it follows that capacitated valuations always have a Walrasian equilibrium. Walrasian prices, however, may not be incentive compatible. In fact, we show that even with 2 agents with capacities 2 and 3 goods, there is *no incentive compatible* mechanism that produces Walrasian prices.

For superadditive valuations it is known that Walrasian equilibrium may not exist. Pápai [18] has characterized the family of mechanisms that are simultaneously EF and IC under superadditive valuations. In particular, VCG with Clarke pivot payments satisfies these conditions. However, Pápai's result for superadditive valuations does not hold for subadditive valuations. Moreover, Clarke pivot payments do not satisfy envy freeness even for the more restricted family of capacitated valuations, as demonstrated in the following example:

Example 1.1. Consider an allocation problem with two agents, $\{1, 2\}$, and two goods, $\{a, b\}$. Agent 1 has capacity $c_1 = 1$ and valuation $v_1(a) = v_1(b) = 2$, and agent 2 has capacity $c_2 = 2$ and valuation $v_2(a) = 1$, $v_2(b) = 2$. According to VCG with Clarke pivot payments, agent 1 is given a and pays 1, while agent 2 is given b and pays nothing (as he imposes no externality on agent 1). Agent 1 would rather switch with agent 2's allocation and payment (in which case, her utility grows by 1), therefore, the mechanism is not envy-free.

Two extremal cases of capacitated valuations are "no capacity constraints", or, all capacities are equal to one. If capacities are infinite, running a Vickrey 2nd price auction [23] for every good, independently, meets all requirements (IC + Walrasian $\Rightarrow$ efficient, IC, EF, NPT, IR). If all agent capacities are one, [14] shows that VCG with Clarke pivot payments is envy-free, and it is easy to see that it also meets the stronger notion of an incentive compatible Walrasian equilibrium. For arbitrary capacities (not only all ∞ or all ones), we distinguish between *homogeneous* capacities, where all agent capacities are equal, and *heterogeneous* capacities, where agent capacities are arbitrary.

When considering incentive compatible and heterogeneous capacities, we distinguish between capacitated valuations with *public* or *private* capacities: being incentive compatible with respect to private capacities and valuation is a more difficult task than incentive compatible with respect to valuation, where capacities are public. In this paper, we primarily consider public capacities.

The main technical results of this paper (which are also summarized in Figure 1) are as follows:

- For arbitrary homogeneous capacities c, such that
  $(c \equiv c_1 = c_2 = \cdots = c_n)$:
  - VCG with Clarke pivot payments is efficient, IC, NPT, IR, and EF.
  - However, there is no incentive compatible mechanism that produces Walrasian prices, even for $c = 2$.
- For arbitrary heterogeneous capacities
  $c = (c_1, c_2, \ldots, c_n)$:
  - Under the VCG mechanism with Clarke pivot payments (public capacities), a higher capacity agent will never envy a lower capacity agent.
  - In the full version we also show that
    * There is no mechanism that is IC, NPT, and EF (for public and hence also for private capacities).
    * 2 agents, public capacities - there exist mechanisms that are IC, IR, and EF.
    * 2 agents, 2 goods - there exist mechanisms that are IC, IR, and EF for every subadditive valuation.

2   Model and Preliminaries

Let $[s] = \{1, \ldots, s\}$ be a set of goods to be allocated to a set $[n] = \{1, \ldots, n\}$ of agents.

	Subadditive	Gross substitutes	capacitated - heterogeneous	capacitated - homogeneous
Walras.	NO [10]	YES [10]	($\rightarrow$) YES	($\rightarrow$) YES
Walras.+IC	NO ($\leftarrow$)	NO ($\leftarrow$)	NO ($\leftarrow$)	NO*
EF + IC	? YES* for $m = 2, n = 2$	? ($\rightarrow$) YES for $m = 2, n = 2$	? YES* for $m = 2$	YES ($\uparrow$)
EF + IC + NPT	NO ($\leftarrow$)	NO ($\leftarrow$)	NO*	YES*

Fig. 1. This table specifies the existence of a particular type of mechanism (rows) for various families of valuation functions (columns). Efficiency is required in all entries. The valuation families satisfy capacitated homogeneous $\subset$ capacitated heterogeneous $\subset$ gross substitutes $\subset$ subadditive. Wherever results are implied from other table entries, this is specified with corresponding arrows. We note that for the family of additive valuations (no capacities), all entries are positive, as the Clarke pivot mechanism satisfies all properties. * Appears in the full version only.

An allocation $a = (a_1, a_2, \ldots, a_n)$ assigns agent i the bundle $a_i \subseteq [s]$ and is such that $\bigcup_i a_i \subseteq [s]$ and $a_i \cap a_j = \emptyset$ for $i \neq j$. We use $\mathcal{L}$ to denote the set of all possible allocations.

For $S \subseteq [s]$, let $v_i(S)$ be the valuation of agent i for set S. Let $v = (v_1, v_2, \ldots, v_n)$, where v_i is the valuation function for agent i.

Let V_i be the domain of all valuation functions for agent $i \in [n]$, and let $V = V_1 \times V_2 \times \cdots \times V_n$.

An allocation function $a : V$ maps $v \in V$ into an allocation

$$a(v) = (a_1(v), a_2(v), \ldots, a_n(v)) .$$

A payment function $p : V$ maps $v \in V$ to $\mathbb{R}^n_{\geq 0}$: $p(v) = (p_1(v), p_2(v), \ldots, p_n(v))$, where $p_i(v) \in \mathbb{R}_{\geq 0}$ is the payment of agent i. Payments are from the agent to the mechanism (if the payment is negative then this means that the transfer is from the mechanism to the agent).

A mechanism is a pair of functions, $\langle a, p \rangle$, where a is an allocation function, and p is a payment function. For a valuation v, the utility to agent i in a mechanism $\langle a, p \rangle$ is defined as $v_i(a_i(v)) - p_i(v)$. Such a utility function is known as quasi-linear.

For a valuation v, we define (v'_i, v_{-i}) to be the valuation obtained by substituting v'_i for v_i, i.e.,

$$(v'_i, v_{-i}) = (v_1, \ldots, v_{i-1}, v'_i, v_{i+1}, \ldots, v_n).$$

A mechanism is incentive compatible (IC) if for all i, v, and v'_i:

$$v_i(a_i(v)) - p_i(v) \geq v_i(a_i(v'_i, v_{-i})) - p_i(v'_i, v_{-i});$$

this holds if and only if

$$p_i(v) \leq p_i(v', v_{-i}) + \Big(v_i(a_i(v)) - v_i(a_i(v'_i, v_{-i}))\Big). \tag{1}$$

A mechanism is envy-free (EF) if for all $i, j \in [n]$ and all v:

$$v_i(a_i(v)) - p_i(v) \geq v_i(a_j(v)) - p_j(v);$$

this holds if and only if

$$p_i(v) \leq p_j(v) + \Big(v_i(a_i(v)) - v_i(a_j(v))\Big). \tag{2}$$

Given valuation functions $v = (v_1, v_2, \ldots, v_n)$, a social optimum Opt is an allocation that maximizes the sum of valuations

$$\text{Opt} \in \arg\max_{a \in \mathcal{L}} \sum_{i=1}^{n} v_i(a_i).$$

Likewise, the social optimum when agent i is missing, Opt^{-i}, is the allocation

$$\text{Opt}^{-i} \in \arg\max_{a \in \mathcal{L}} \sum_{j \in [n]\setminus\{i\}} v_j(a_j).$$

2.1   VCG Mechanisms

A mechanism $\langle a, p \rangle$ is called a VCG mechanism [4,23] if:

- $a(v) = \text{Opt}$, and
- $p_i(v) = h_i(v_{-i}) - \sum_{j \neq i} v_j(a_j(v))$, where h_i does not depend on v_i, $i \in [n]$.

For connected domains, the only efficient incentive compatible mechanism is VCG (See Theorem 9.37 in [17]). Since capacitated valuations induce a connected domain, we get the following proposition.

Proposition 2.1. *With capacitated valuations, a mechanism is efficient and IC if and only if it is VCG.*

VCG with *Clarke pivot payments* has

$$h_i(v_{-i}) = \max_{a \in \mathcal{L}} \sum_{j \neq i} v_j(a) \qquad \Big(= \sum_{j \neq i} v_j(\text{Opt}_j^{-i})\Big).$$

Agent valuations for bundles of goods are non negative. The only mechanism that is efficient, incentive compatible, individually rational, and with no positive transfers is VCG with Clarke pivot payments.

The following proposition, which appears in [18], provides a criterion for the envy freeness of a VCG mechanism.

Proposition 2.2. *[18] Given a VCG mechanism, specified by functions $\{h_i\}_{i \in [n]}$, agent i does not envy agent j iff for every v,*

$$h_i(v_{-i}) - h_j(v_{-j}) \leq v_j(\text{Opt}_j) - v_i(\text{Opt}_j).$$

2.2 Gross Substitutes and Capacitated Valuations

We define the notion of gross substitute valuations and show that every capacitated valuation (i.e., additive up to the capacity) has the gross substitutes property. As this discussion refers to a valuation function of a single agent, we omit the index of the agent.

Fix an agent and let $D(p)$ be the collection of all sets of goods that maximize the utility of the agent under the price vector p, $D(p) = \arg\max_{S \subseteq [s]} \{v(S) - \sum_{j \in S} p_j\}$.

Definition 2.3. [10] A valuation function $v : 2^{[s]} \to \mathbb{R}_{\geq 0}$ satisfies the gross substitutes condition if the following holds: Let $p = (p_1, \ldots, p_s)$ and $q = (q_1, \ldots, q_s)$ be two price vectors such that the price for good j is no less under q than under p: i.e., $q_j \geq p_j$, for all j. Consider the set of all items whose price is the same under p and q, $E(p, q) = \{1 \leq j \leq s \mid p_j = q_j\}$, then for any $S^p \in D(p)$ there exists some $S^q \in D(q)$ such that $S^p \cap E(p, q) \subseteq S^q \cap E(p, q)$.

Theorem 2.4. Every capacitated valuation function (additive up to the capacity) obeys the gross substitutes condition.

As a corollary, we get that capacitated valuations admit a Walrasian equilibrium. However, not necessarily within an IC mechanism.

3 Envy-Free and Incentive Compatible Assignments with Capacities

The main result of this section is that Clarke pivot payments are envy-free when capacities are homogeneous. This follows from a stronger result, which we establish for heterogeneous capacities, showing that with Clarke pivot payments, no agent envies a lower-capacity agent.

In full version we show that one cannot aim for an incentive compatible mechanism with Walrasian prices (if this was possible then envy freeness would follow immediately).

The following theorem establishes a general result for capacitated valuations: in a VCG mechanism with Clarke pivot payments, no agent will ever envy a lower-capacity agent.

Theorem 3.1. If we apply the VCG mechanism with Clarke pivot payments on the assignment problem with capacitated valuations, then

- The mechanism is incentive compatible, individually rational, and makes no positive transfers (follows from VCG with CPP).
- No agent of higher capacity envies an agent of lower or equal capacity.

The input to the VCG mechanism consists of capacities and valuations. The capacity of agent i, $c_i \geq 0$, is publicly known. The number of units of good j, $q_j \geq 0$ is also a public knowledge. The valuations $v_i(j)$ are private.

The b-Matching Graph. Given capacities c_i, q_j, and a valuation matrix v, we construct an edge-weighted bipartite graph G as follows:

- We associate a vertex with every agent $i \in [n]$ on the left, let $\mathcal{A}$ be the set of these vertices.
- We associate a vertex with every good $j \in [s]$ on the right, let $\mathcal{I}$ be the set of these vertices.
- Edge (i, j), $i \in \mathcal{A}$, $j \in \mathcal{I}$, has weight $v_i(j)$.
- Vertex $i \in \mathcal{A}$ (associated with agent i) has *degree constraint* c_i.
- Vertex $j \in \mathcal{I}$ (associated with good j) has degree constraint q_j.

We seek an allocation a $(= a(v))$ where a_{ij} is the number of units of good j allocated to agent i. The value of the allocation is $v(a) = \sum_{ij} a_{ij} v_i(j)$. We seek an allocation of maximal value that meets the degree constraints: $\sum_j a_{ij} \leq c_i$, $\sum_i a_{ij} \leq q_j$, this is known as a b-matching problem and has an integral solution if all constraints are integral, see [20]. Let $a_i = (a_{i1}, a_{i2}, \ldots, a_{in})$ denote the i'th row of a, which corresponds to the bundle allocated to agent i.

Let $v_k(a_i) = \sum_{j \in [s]} a_{ij} v_k(j)$ denote the value to agent k of bundle a_i. Let M denote some allocation that attains the maximum social value, $M \in \arg\max_a v(a)$. Finally, let G^{-i} be the graph derived from G by removing the vertex associated with agent i and all its incident edges, and let M^{-i} be a matching of maximum social value with agent i removed.

Specializing the Clarke pivot rule to our setting, the payment of agent k is

$$p_k = v(M^{-k}) - v(M) + v_k(M_k) . \tag{3}$$

In the special case of permutation games (the number of agents and goods is equal, and every agent can receive at most one good), the social optimum corresponds to a maximum weighted matching in G. Such "permutation games" were first studied by [14] who showed that Clarke pivot payments are envy-free. However, the shadow variables technique used in this proof does not seem to generalize for larger capacities.

Proof sketch of Theorem 3.1: Let agent 1 and agent 2 be two arbitrary agents such that $c_1 \geq c_2$. Agent 1 does not envy agent 2 if and only if

$$v_1(M_1) - p_1 \geq v_1(M_2) - p_2$$

By substituting the Clarke pivot payments (3) and rearranging, this is true if and only if

$$v(M^{-2}) \geq v(M^{-1}) + v_1(M_2) - v_2(M_2). \tag{4}$$

Thus in order to prove the theorem we need to establish (4).

We construct a new allocation D^{-2} on G^{-2} (from the allocations M and M^{-1}) such that

$$v(D^{-2}) \geq v(M^{-1}) + v_1(M_2) - v_2(M_2) . \tag{5}$$

From the optimality of M^{-2}, it must hold that $v(M^{-2}) \geq v(D^{-2})$. Combining this with (5) shall establish (4), as required.

In what follows we make several preparations for the construction of the allocation D^{-2}. Given M and M^{-1}, we construct a directed bipartite graph G_f on $\mathcal{A} \cup \mathcal{I}$ coupled with a flow f as follows. For every pair of vertices $i \in \mathcal{A}$ and $j \in \mathcal{I}$,

- If $M_{ij} - M_{ij}^{-1} > 0$, then G_f includes arc $i \to j$ with flow $f_{i \to j} = M_{ij} - M_{ij}^{-1}$.
- If $M_{ij} - M_{ij}^{-1} < 0$, then G_f includes arc $j \to i$ with flow $f_{j \to i} = M_{ij}^{-1} - M_{ij}$.
- If $M_{ij} = M_{ij}^{-1}$, then G_f contains neither arc $i \to j$ nor arc $j \to i$.

We define a vertex to be a *source* vertex if the difference between the amount of flow flowing out of the vertex and the amount of flow flowing into the vertex is positive, and define vertex to be a *target* vertex otherwise.

Using the flow decomposition theorem, we can decompose the flow f into simple paths and cycles, where each path connects a source to a target. Associated with each path and cycle T is a positive flow value $f(T) > 0$. Given an arc $x \to y$, $f_{x \to y}$ is obtained by summing up the values $f(T)$ of all paths and cycles T that contain $x \to y$. Notice that $M_{1j}^{-1} = 0$ for all j and therefore $f_{1 \to j} \geq 0$ for all j. It follows that there are no arcs of the form $j \to 1$ in G_f.

We define the *value* of a path or a cycle $T = u_1, u_2, \ldots, u_t$ in G_f, to be

$$v(P) = \sum_{\substack{u_i \in \mathcal{A}, \\ u_{i+1} \in \mathcal{I}}} v_{u_i}(u_{i+1}) - \sum_{\substack{u_i \in \mathcal{I}, \\ u_{i+1} \in \mathcal{A}}} v_{u_{i+1}}(u_i).$$

It is easy to verify that $\sum_T f(T) \cdot v(T) = v(M) - v(M^{-1})$, where we sum over all paths and cycles T in our decomposition.

The proofs of the following Lemmata are omitted.

Lemma 3.2. *Let $T = u_1, u_2, \ldots, u_t$ be a cycle in G_f or a path in the flow decomposition of G_f, and let ϵ be the minimal flow along any arc of T. We construct an allocation $\widehat{M} (= \widehat{M}(T))$ from M by canceling the flow along T, start with $\widehat{M} = M$ and then for each $(u_i, u_{i+1}) \in T$ set:*

$$\widehat{M}_{u_i u_{i+1}} = M_{u_i u_{i+1}} - \epsilon \qquad u_i \in \mathcal{A}, u_{i+1} \in \mathcal{I}$$
$$\widehat{M}_{u_{i+1} u_i} = M_{u_{i+1} u_i} + \epsilon \qquad u_i \in \mathcal{I}, u_{i+1} \in \mathcal{A}.$$

Alternatively, we construct $\widehat{M}^{-1} (= \widehat{M}^{-1}(T))$ from M^{-1}, starting from $\widehat{M}^{-1} = M^{-1}$ and then for each $(u_i, u_{i+1}) \in T$ set

$$\widehat{M}_{u_i u_{i+1}}^{-1} = M_{u_i u_{i+1}}^{-1} + \epsilon \qquad u_i \in \mathcal{A}, u_{i+1} \in \mathcal{I}, \quad ,$$
$$\widehat{M}_{u_{i+1} u_i}^{-1} = M_{u_{i+1} u_i}^{-1} - \epsilon \qquad u_i \in \mathcal{I}, u_{i+1} \in \mathcal{A}.$$

The allocations $\widehat{M}, \widehat{M}^{-1}$ are valid (do not violate capacity constraints).

Lemma 3.3. *The graph G_f does not contain a cycle. The vertex that corresponds to agent 1 is the unique source vertex.*

In particular, Lemma 3.3 implies that there are no cycles in our flow decomposition. and that all the paths in our flow decomposition originate at agent 1. We are now ready to describe the construction of the allocation D^{-2}:

1. Stage I: initially, $D^{-2} := M^{-1}$.
2. Stage II: for every good j, let $x = \min\{M_{2j}, M_{2j}^{-1}\}$, and set $D_{2j}^{-2} := M_{2j}^{-1} - x$ and $D_{1j}^{-2} := x$.
3. Stage III: for every flow path P in the flow decomposition of G_f that contains agent 2, let $\hat{P}$ be the prefix of P up to agent 2. For every agent to good arc $(i \to j) \in \hat{P}$ set $D_{ij}^{-2} := D_{ij}^{-2} + f(P)$, and for every good to agent arc $(j \to i) \in \hat{P}$ set $D_{ij}^{-2} := D_{ij}^{-2} - f(P)$.

It is easy to verify that D^{-2} indeed does not allocate any good to agent 2. Also, the allocation to agent 1 in D^{-2} is of the same size as the allocation to agent 2 in M^{-1}. Since $c_1 \geq c_2$, D^{-2} is a valid allocation. To conclude the proof of Theorem 3.1 we show that:

Lemma 3.4. *Allocation D^{-2} satisfies (5).*

Proof. Rearranging (5), we obtain

$$v(D^{-2}) \geq v(M^{-1}) \tag{6}$$

$$+ \sum_{j=1}^{s} (v_1(j) - v_2(j)) \cdot \min(M_{2j}, M_{2j}^{-1}) \tag{7}$$

$$+ \sum_{j: M_{2j} > M_{2j}^{-1}} (v_1(j) - v_2(j)) (M_{2j} - M_{2j}^{-1}). \tag{8}$$

At the end of stage I, we have $D^{-2} = M^{-1}$ and so the inequality above at line (6) (excluding expressions (7) and (8)) holds trivially. It is also easy to verify that at the end of stage II, the inequality above that spans expressions (6) and (7) (and excludes expression (8)) holds. What we show next is that at the end of stage III, the full inequality above will hold.

Consider a good j such that $M_{2j} > M_{2j}^{-1}$. In G_f we have an arc $2 \to j$ such that $f_{2 \to j} = M_{2j} - M_{2j}^{-1}$, therefore in the flow decomposition we must have paths $P_1, \ldots, P_\ell$, all containing the arc $2 \to j$, such that

$$\sum_{k=1}^{\ell} f(P_k) = f_{2 \to j} = M_{2j} - M_{2j}^{-1}. \tag{9}$$

For every $k = 1, \ldots, \ell$, let $\hat{P}_k$ denote the prefix of P_k up to agent 2. Consider the cycle C consisting of $\hat{P}_k$ followed by arcs $2 \to j$ and $j \to 1$. We claim that the value of this cycle is non-negative.

Consider the allocation $\widehat{M}(C)$ which is a valid allocation from Lemma 3.2. Observe that $v(\widehat{M}) = v(M) - \epsilon v(C) > v(M)$. This now contradicts the assumption that M maximizes v over all allocations. We obtain $v(\hat{P}_k) + v_2(j) - v_1(j) \geq 0$. Rearranging and multiplying by $f(P_k)$, it follows that $f(P_k)v(\hat{P}_k) \geq f(P_k)(v_1(j) - v_2(j))$. Summing over all paths $k = 1, \ldots, \ell$, we get

$$\sum_{k=1}^{\ell} \left(f(P_k)v(\hat{P}_k) \right) \geq (v_1(j) - v_2(j)) \sum_{k=1}^{\ell} f(P_k).$$

Substituting (9) in the last inequality establishes the following inequality:

$$\sum_{k=1}^{\ell} \left(f(P_k)v(\widehat{P}_k) \right) \geq \left(v_1(j) - v_2(j) \right) \left(M_{2j} - M_{2j}^{-1} \right). \tag{10}$$

The left hand side of (10) is exactly the gain in value of the allocation when applying stage III to the paths $\widehat{P}_1, \ldots, \widehat{P}_\ell$ during the construction of D^{-2} above. The right hand side is the term which we add in (8).

To conclude the proof of Lemma 3.4, we note that stage III may also deal with other paths that start at agent 1 and terminate at agent 2. This part of the proof appears in the full version.

The following is a direct corollary of Theorem 3.1.

Corollary 3.5. *If all agent capacities are equal, then the VCG allocation with Clarke pivot payments is EF.*

4   Discussion and Open Problems

This work initiates the study of efficient, incentive compatible, and envy-free mechanisms for capacitated valuations.

Our work suggests a host of problems for future research on heterogeneous capacitated valuations and generalizations thereof.

We know that, generally, there may be no mechanism that is both IC and EF even if we allow positive transfers (example in full version).

We conclude by posing the following open problems, from the very concrete to the more general:

- Is there a mechanism for k agents, $k > 2$, with heterogenous capacitated allocations, that is efficient, IC, and EF ? We conjecture that such mechanisms do exist for any combinatorial auction with subadditive valuations. We know they exist for two agents and public capacities, and for subadditive valuations with two agents and two goods.
- We have focused on efficient mechanisms; *i.e.*, that maximize social welfare. A natural question is how well the optimal social welfare can be *approximated* by a mechanism that is IC, EF, and NPT.
- In [8], Fleischer and Wang consider lower bounds for envy-free and truthful mechanisms for makespan minimization on related machines. Ergo, one can ask these questions not only in the context of efficiency but also in other contexts. This is yet another step in the most general problem of all (see below).
- And, the most general problem of all: can one characterize the set of truthful and envy-free mechanisms? There have been some attempts, including a characterization due to the authors that generalizes Rochet's cyclic monotonicity characterization for truthfulness to a full characterization for truthfulness and envy freeness (See [5]). However, like cyclic monotonicity itself, this characterization is hardly satisfactory.

References

1. Ausubel, L., Milgrom, P.: Ascending auctions with package bidding. Frontiers of Theoretical Economics 1, 1–42 (2002)
2. Blumrosen, L., Nisan, N.: Combinatorial auctions. In: Tardos, E., Vazirani, V., Nisan, N., Roughgarden, T. (eds.) Algorithmic Game Theory. Cambridge University Press (2007)
3. Blumrosen, L., Nisan, N.: Informational limitations of ascending combinatorial auctions. Journal of Economic Theory 145, 1203–1223 (2001)
4. Clarke, E.: Multipart Pricing of Public Goods. Public Choice 1, 17–33 (1971)
5. Cohen, E., Feldman, M., Fiat, A., Kaplan, H., Olonetsky, S.: On the Interplay between Incentive Compatibility and Envy Freeness, http://arxiv.org/abs/1003.5328
6. Dubins, L.E., Spanier, E.H.: How to cut a cake fairly. American Mathematical Monthly (1961)
7. Foley, D.: Resource allocation and the public sector. Yale Economic Essays 7, 45–98 (1967)
8. Fleischer, L., Wang, Z.: Lower Bound for Envy-Free and Truthful Makespan Approximation on Related Machines. In: Persiano, G. (ed.) SAGT 2011. LNCS, vol. 6982, pp. 166–177. Springer, Heidelberg (2011)
9. Demange, G., Gale, D., Sotomayor, M.: Multi-Item Auctions. Journal of Political Economy (1986)
10. Gul, F., Stacchetti, E.: Walrasian equilibrium with gross substitutes. Journal of Economic Theory 87, 95–124 (1999)
11. Hurwicz, L.: Optimality and informational efficiency in resource allocation processes. In: Arrow, K.J., Karlin, S., Suppes, P. (eds.) Mathematical Methods in the Social Sciences (1960)
12. Kelso, A., Crawford, V.: Job Matching, Coalition Formation, and Gross Substitutes. Econometrica (1982)
13. Lehmann, B., Lehmann, D.J., Nisan, N.: Combinatorial Auctions with Decreasing Marginal Utilities. In: ACM Conference on Electronic Commerce (2001)
14. Leonard, H.B.: Elicitation of honest preferences for the assignment of individuals to positions. The Journal of Political Economy 91(3), 461–479 (1983)
15. Maskin, E.S.: On the fair allocation of indivisible goods. In: Feiwel, G. (ed.) Arrow and the Foundations of the Theory of Economic Policy (essays in honor of Kenneth Arrow) (1987)
16. Moulin, H.: Fair Division and Collective Welfare. MIT Press (2004)
17. Nisan, N.: Introduction to mechanism design. In: Nisan, N., Roughgarden, T., Tardos, E., Vazirani, V. (eds.) Algorithmic Game Theory. Cambridge University Press (2007)
18. Pápai, S.: Groves sealed bid auctions of heterogeneous objects with fair prices. Social choice and Welfare 20, 371–385 (2003)
19. Parkes, D.: Iterative combinatorial auctions: Achieving economic and computational efficiency. Ph.D. Thesis, Department of Computer and Information Science, University of Pennsylvania (2001)
20. Pulleyblank, W.: Dual integrality in b-matching problems. In: Cottle, R.W., et al. (eds.) Combinatorial Optimization. Mathematical Programming Studies, vol. 12 (1980)
21. Raz, D., Levy, H., Avi-Itzhak, B.: A resource-allocation queueing fairness measure. In: SIGMETRICS (2004)
22. Svensson, L.G.: On the existence of fair allocations. Journal of Economics (1983)
23. Vickrey, W.: Counterspeculation, Auctions, and Competitive Sealed Tenders. Journal of Finance (1961)
24. Young, H.P.: Equity: In Theory and Practice. Princeton University Press (1995)

Simple, Optimal and Efficient Auctions

Constantinos Daskalakis[1,*] and George Pierrakos[2,**]

[1] MIT, EECS, CSAIL
costis@csail.mit.edu
[2] UC Berkeley, EECS
georgios@cs.berkeley.edu

Abstract. We study the extent to which simple auctions can simultaneously achieve good revenue and efficiency guarantees in single-item settings. Motivated by the optimality of the second price auction with monopoly reserves when the bidders' values are drawn i.i.d. from regular distributions [12], and its approximate optimality when they are drawn from independent regular distributions [11], we focus our attention to the second price auction with general (not necessarily monopoly) reserve prices, arguably one of the simplest and most intuitive auction formats. As our main result, we show that for a carefully chosen set of reserve prices this auction guarantees at least 20% of both the optimal welfare and the optimal revenue, when the bidders' values are distributed according to independent, not necessarily identical, regular distributions. We also prove a similar guarantee, when the values are drawn i.i.d. from a—possibly irregular—distribution.

1 Introduction

Social welfare and revenue are without doubt the two most important objectives in mechanism design. They are both well-studied, and extremely well-understood when there is a single item for sale. Not only do the Vickrey and Myerson auctions optimize these objectives in isolation, but there also exist (typically randomized) mechanisms that simultaneously optimize for both objectives, in the sense of maximizing revenue subject to a lower bound on social welfare, or vice-versa [13]. Interestingly, when the bidders' values are independently and identically distributed according to some regular distribution,[1] the Vickrey and Myerson mechanisms behave very much alike: Myerson's auction is just Vickrey's auction with an additional reserve price. Motivated by this astonishing similarity (and the somewhat peculiar format of Myerson's auction in more general settings), Hartline and Roughgarden [11] showed that a Vickrey auction with appropriately chosen reserve prices can approximate the revenue of the optimal auction in more general settings. Inspired by their result, and the fact that

---

* Supported by a Sloan Foundation Fellowship and NSF Award CCF-0953960 (CAREER) and CCF-1101491.
** This work was done while the author was visiting Microsoft Research, New England.
[1] See Sec. 2 for a definition.

N. Chen, E. Elkind, and E. Koutsoupias (Eds.): WINE 2011, LNCS 7090, pp. 109–121, 2011.

the mechanism of [13] is at least as complicated as Myerson's mechanism and potentially randomized, in this paper we ask the question of whether one can design *simple and deterministic* mechanisms that achieve *approximately-optimal* guarantees for both objectives *simultaneously*.

At first glance it is not obvious why such simple auctions should even exist. Indeed, despite the fact that Vickrey's auction achieves at least half of the optimal revenue, when the values are drawn i.i.d. from regular distributions (see e.g. [9]), this is no longer the case when the values are independent but drawn from different regular distributions. In particular, it is easy to see that the revenue of Vickrey's auction can be arbitrarily far from the optimal revenue: just consider $n - 1$ bidders distributed independently and uniformly in $[0, 1]$, and a single bidder distributed uniformly in $[h, h + 1]$, for some large $h > 1$. The situation does not become any better if we resort to the mechanism of [11], i.e. running Vickrey with a different reserve price for every bidder, taken to be Myerson's monopoly reserve price for that bidder. The auction now can be arbitrarily inefficient even for a single bidder whose value is distributed according to a regular distribution: consider the (almost) equal revenue distribution, where the bidder's value is supported on $\{1 - \epsilon, 2 - \epsilon, \ldots, h - \epsilon\}$, for some $\epsilon \in (0, 1)$ and $h > 1$, and the probability that it is larger than or equal to $i - \epsilon$ is exactly $1/i$, for $i = 1 \ldots h$. In this paper, we show that by appropriately tweaking the reserve price of each bidder, we can fix this inefficiency:

Main Result (Th. 1 of Sec. 3): *In every single-item setting with n bidders whose values are distributed according to independent (possibly non-identical) regular distributions and for every $p \in [0, 1]$, there exists a Vickrey auction with (generally non-anonymous) reserve prices that simultaneously achieves a p-fraction of the optimal social welfare and a $\left(\frac{1-p}{4}\right)$-fraction of the optimal revenue. In particular, there exists a Vickrey auction with reserve prices that achieves at least a 20% of the optimal social welfare and revenue.*

We can use our techniques to prove a similar approximation guarantee for non-identical distributions satisfying the monotone hazard rate condition (which has already been obtained by [7]), and we also show that a Vickrey auction with an anonymous reserve simultaneously approximates both objectives for general (possibly non-regular) distributions, as long as all values are i.i.d (Th. 3). We summarize our results together with already known welfare and revenue guarantees for various settings in Table 1.

Table 1. (α, β) stands for α-approximation for welfare and β-approximation for revenue. Notice that our result for regular distributions gets a handle on the whole Pareto boundary achieved by the Vickrey auction with non-anonymous reserve prices.

	i.i.d.	independent
mhr	$\left(1, \frac{1}{e}\right)$ and $\left(\frac{1}{e}, 1\right)$ [2]	$\left(\frac{1}{e}, \frac{1}{2}\right)$ [7]
regular	$\left(1, \frac{1}{2}\right)$ [9]	$\left(\frac{1}{5}, \frac{1}{5}\right)$ and $\left(p, \frac{1-p}{4}\right)$, for all $p \in [0, 1]$ [this work]
non-regular	$\left(\frac{1}{2}, \frac{1}{2}\right)$ [this work]	?

Two questions left open are whether one can extend our results to the setting of n bidders distributed according to independent but not necessarily identical and possibly irregular distributions, and to general single-dimensional settings.

1.1 Related Work

The work closer in spirit to ours is that of [11], where the authors show that for a variety of single-dimensional settings, second price auctions with carefully chosen reserve prices are approximately revenue-optimal. In particular, when the bidders' values are independently drawn from (possibly different) regular distributions, they show that Vickrey's auction with monopoly reserve prices (see Sec. 2 for a definition) achieves at least half of the optimal revenue. Moreover they show that Vickrey's auction with an anonymous reserve achieves a factor 4 approximation to the optimal revenue.

In an unpublished manuscript [8], the authors study the problem of designing *deterministic* mechanisms that optimize for both objectives, as a multi-objective optimization problem. They show that, even though exactly optimizing the trade-off curve is an NP-hard problem, there exists a polynomial-time deterministic mechanism that approximates within arbitrary precision any point on the trade-off curve of those two objectives, when there are 2 bidders with arbitrarily correlated values. Their mechanism, despite being deterministic, is far from simple; this work complements theirs by showing that, if one is willing to settle for less than an arbitrarily small approximation factor, simple mechanisms are possible, even when the number of bidders is large. Moreover, the existence of an auction that simultaneously achieves a constant factor approximation to both objectives, characterizes the "knee" of the Pareto curve, a structural result which is of independent interest.

A different type of result relating the two objectives is that of Bulow and Klemperer [4], where it is shown that in a single-item setting the revenue benefits of adding an extra bidder and running the efficiency-maximizing (Vickrey) auction surpass those of running the revenue-maximizing (Myerson) auction without adding the extra bidder, when the bidders' values are i.i.d. according to a regular distribution. In [2] the authors show that for values drawn i.i.d. from a monotone hazard rate distribution, an analogous theorem holds for efficiency: by adding $\Theta(\log n)$ extra bidders and running Myerson's auction, one gets at least the efficiency of Vickrey's auction. Finally, [11] extends Bulow and Klemperer's result to more general single-dimensional settings, as follows: they show that by duplicating all bidders (whose values are drawn independently from not necessarily identical, regular distributions), and then running the VCG auction, one can guarantee at least half of the optimal revenue (while being optimal with respect to welfare). Our result shows that in single-item settings with independent (but not necessarily i.i.d.) bidders, one can simultaneously achieve constant factor approximations to both optimal revenue and welfare *without adding any extra bidders* via the use of a Vickrey auction with appropriate (non-anonymous) reserve prices.

There has also been substantial work studying the revenue and welfare guarantees of welfare-optimizing and revenue-optimizing auctions respectively. In [2] the authors show that, for values drawn independently from the same monotone hazard rate distribution, both the welfare and revenue ratios of Vickrey and Myerson's auctions are bounded by $1/e$ (see the top-left square of Table 1). Similar kinds of revenue and welfare ratios are also studied in [9] for keyword auctions, in [14] for single-item English auctions, and in [1], where the authors present bounds on the efficiency loss of revenue-optimal mechanisms in single-item settings with i.i.d. bidders of finite support. Moreover, in [9] and [7] the authors present simple auctions that simultaneously achieve constant factor approximations to both objectives in single-item settings where bidders' values are i.i.d. from a regular distribution (see the middle-left square of Table 1), and independently (but not necessarily identically) distributed according to a monotone hazard rate distribution (see the top-right square of Table 1). Some of their results also hold for more general single-dimensional settings, namely when the feasibility constraints form a matroid.

Finally, despite our different motivation, methodologically our paper is somewhat related to [3]: in that paper the goal is to provide a general reduction from the mechanism design problem for many bidders, to that of a single bidder, while preserving the value of a separable objective (such as welfare or revenue) within a constant factor. In Lem. 3 and 5 we establish analogous many-to-one reductions; however, our goal is not only to preserve the approximation factor, but also for the resulting auction to be of a specific simple format, in contrast to the much more generic reduction of [3].

2   Preliminaries

Our auction setting is that of a single item for sale and n interested bidders, each with a value v_i for the item, which is distributed independently according to some distribution F_i. The distributions $\{F_i\}_i$ are not necessarily identical. For simplicity we assume that all F_i's in this paper are differentiable. So we can define the corresponding probability density functions as follows $f_i(x) = F_i'(x)$.

A single-item auction $\mathcal{A}$ consists of an allocation rule $\mathbf{x}$ and a payment rule $\mathbf{p}$; an allocation rule is a function from bid vectors to $[0, 1]^n$, encoding the probability by which every bidder receives the item, while a payment rule is a function from bid vectors to n-vectors of non-negative payments. We want from our auctions to satisfy the two standard constraints of ex-post incentive compatibility (IC) and individual rationality (IR) [12], so that the terms "bid" and "value" can be used interchangeably. We are interested in the objectives of revenue and welfare, defined as follows:

$$\text{Rev}[\mathcal{A}] = \mathbb{E}\left[\sum_{i=1}^n p_i(v_1, \ldots, v_n)\right] \quad \text{and} \quad \text{SW}[\mathcal{A}] = \mathbb{E}\left[\sum_{i=1}^n v_i \cdot x_i(v_1, \ldots, v_n)\right],$$

where the above expectations are with respect to value vectors $v = (v_1, \ldots, v_n)$ drawn from the product distribution $\times_i F_i$. For convenience, we sometimes write $\mathcal{R}_{\mathcal{A}} = \sum_{i=1}^n p_i(v_1, \ldots, v_n)$, so that $\text{Rev}[\mathcal{A}] = \mathbb{E}[\mathcal{R}_{\mathcal{A}}]$.

We say that an auction $\mathcal{A}$ is an α-approximation for welfare (resp. revenue) if $\text{SW}[\mathcal{A}] \geq \alpha \cdot \text{SW}[\text{Vic}]$ (resp. $\text{Rev}[\mathcal{A}] \geq \alpha \cdot \text{Rev}[\text{Mye}]$), where Vic denotes the Vickrey auction and Mye denotes Myerson's auction. We say that an auction is an (α, β)-approximation if it is simultaneously an α-approximation for welfare and a β-approximation for revenue. Also, given an auction $\mathcal{A}$, and a set $B \subseteq \{1, \ldots, n\}$, we may write $\mathcal{A}(B)$ to denote the auction $\mathcal{A}$ run only on the subset B of bidders. When we use this notation it will be clear from context how the "projected" auction operates.

In [12] Myerson introduced the notion of a bidder's *virtual valuation function* ϕ_i, defined as follows:

$$\phi_i(v_i) = v_i - \frac{1 - F_i(v_i)}{f_i(v_i)}.$$

In terms of this notion, we say a distribution F_i is *regular* if the virtual value function ϕ_i is non-decreasing, and that it satisfies the *monotone hazard rate condition* if the ratio $\frac{1 - F_i(x)}{f_i(x)}$ is non-increasing. For distributions that are non-regular, Myerson's ironing technique can be used to get the corresponding *ironed* virtual valuation function $\hat{\phi}_i(v_i)$. The following result is central to Myerson's analysis, and we also use it in the present paper:

Proposition 1. [Myerson's Lemma] *For any truthful mechanism $(\mathbf{x}, \mathbf{p})$, where all F_i are regular distributions, we can express the expected payment of bidder i as follows, where the expectation is over the players' values:*

$$\mathbb{E}\left[p_i(v_1, \ldots, v_n)\right] = \mathbb{E}\left[\phi_i(v_i) \cdot x_i(v_1, \ldots, v_n)\right].$$

We are interested in the following (family) of auction(s):

Definition 1. *The Vickrey auction with reserve prices $\mathbf{r} = (r_1, \ldots, r_n)$, denoted $\text{Vic}_{\mathbf{r}}$, is the following mechanism:*

1. *Reject all bidders whose values are $v_i < r_i$.*
2. *Allocate the item to the highest valued of the remaining bidders (or to none if no one clears their reserve in Step 1).*
3. *Charge the winner the maximum of the second highest bidder (among those who were not eliminated in Step 1) and her reserve price.*

Tie-break lexicographically if there are multiple highest bidders in Step 2.

Two cases of particular interest are the *Vickrey auction with an anonymous reserve*, where a common reserve r is used for all bidders, and *the Vickrey auction with monopoly reserves*, denoted by $\text{Vic}_{\mathbf{m}}$, where $m_i = \phi_i^{-1}(0)$, the *monopoly reserve* of bidder i.

3   The Regular, Independent Case

In this section we focus on the setting of n bidders whose values are distributed according to regular, but not necessarily identical, distributions. We start with a couple of probabilistic lemmas –not requiring regularity– whose easy proofs are postponed to the full version of the paper.

Lemma 1. *Let X and Y be independent random variables and $g : \mathbb{R} \to \mathbb{R}$ a (weakly) increasing function. Then, for any constant $c \in \mathbb{R}$,*

$$\Pr[X \geq Y \mid g(X) \geq c] \geq \Pr[X \geq Y \mid g(X) \leq c].$$

Lemma 2. *Let X and Y be independent random variables and $g : \mathbb{R} \to \mathbb{R}$ a (weakly) increasing function. Then*

$$\mathbb{E}[g(X)] \leq \mathbb{E}[g(X) \mid X \geq Y].$$

Our next lemma shows that if we take the Vickrey auction and add a reserve price for each bidder, such that the probability of any single bidder's value exceeding her reserve price is at least p, then the resulting welfare is at least a p fraction of Vickrey's (optimal) social welfare $\mathbb{E}[\max_i\{v_i\}]$. The proof of this lemma is relatively straightforward and is deferred to the full version of our paper as well. In what follows we use $\mathbb{I}_{(.)}$ to denote the indicator function.

Lemma 3. [Many-to-One Reduction—Welfare] *Suppose that $X_1,\ldots,X_n$ are independent, non-negative random variables (possibly non-identically distributed), $t_1,\ldots,t_n$ are (possibly different) thresholds, and $p \in [0,1]$. If it holds that $\Pr[X_i \geq t_i] \geq p$, for all $i = 1\ldots n$, then:*

$$\mathbb{E}\left[\max_i\{X_i \cdot \mathbb{I}_{(X_i \geq t_i)}\}\right] \geq p \cdot \mathbb{E}\left[\max_i\{X_i\}\right].$$

Lemma 3 immediately implies the following corollary, already known from [7].

Corollary 1. [mhr, independent] *In every single-item setting with n bidders whose values are distributed according to independent (possibly non-identical) distributions that satisfy the monotone hazard rate condition, the Vickrey auction with monopoly reserves is a $(1/e, 1/2)$-approximation.*

Proof. It is known from [11] that, if $\mathbf{m}$ is the vector of monopoly reserve prices, then $\text{Vic}_{\mathbf{m}}$ (the Vickrey auction with monopoly reserves) is a $1/2$-approximation to the optimal revenue. The welfare guarantee follows from Lem. 3 and the following fact from [2]: if v is drawn from a monotone hazard rate distribution, then $\Pr[v \geq \phi^{-1}(0)] \geq 1/e$.   □

Unfortunately, as discussed in Sec. 1, the Vickrey auction with monopoly reserve prices may be arbitrarily inefficient when we allow for regular distributions; in particular we cannot employ Lem. 3 directly as the probability of any single bidder being above her monopoly reserve may be arbitrarily small. To fix this, we recall a lemma for regular distributions from [5]. For a single bidder setting, this lemma guarantees that there is always a reserve price r (which generally needs to be smaller than the monopoly reserve) that achieves a constant factor of the optimal revenue, while at the same time is smaller than the bidder's value with constant probability.

Lemma 4 ([5]). *Let F be a regular distribution, and let $R_F(x) = x \cdot F^{-1}(1-x)$, for all $x \in [0,1]$,[2] be the revenue curve in quantile space. Then, for all $0 < \tilde{q} \le q \le p < 1$,*

$$R_F(\tilde{q}) \le \frac{1}{1-p} R_F(q).$$

If we try to use Lem. 4 to generalize Cor. 1 to regular distributions, we run into an additional difficulty. Indeed, if we lower the bidders' reserve prices to some vector $\mathbf{r} \le \mathbf{m}$ below their monopoly reserves and run $\text{Vic}_\mathbf{r}$, the bidders will start contributing negative virtual values to the expected virtual welfare of the auction (i.e. its expected revenue). So we need to control the absolute value of the overall negative contribution to the expected virtual social welfare. This is not straightforward and is established in the following lemma, *which alongside our main result is one of the main contributions of this paper.*

Before providing its proof, it is worth noting that the obvious approach of decomposing the auction's virtual welfare into every bidder's contribution, using the law of total expectation, and then comparing each bidder's contribution under different reserve prices poses technical challenges. In particular, the terms of the decomposition cannot be directly compared as each of these terms depends on the probabilistic experiment that determines the winner of the auction, and this experiment depends on the reserves in ways that makes it hard to find a useful coupling that enables term-by-term comparisons. Our technique tries to disentangle the contribution of each bidder to the virtual welfare of the auction from the competition among the bidders, enabling us to first relate the revenue of $\text{Vic}_\mathbf{r}$ with the revenue of a *hybrid* auction, instead of $\text{Vic}_\mathbf{m}$ (for which we have good revenue guarantees from [11]). Our hybrid auction uses the tweaked reserves $\mathbf{r}$ to truncate the bidders' values, but only gives the item to the winner of $\text{Vic}_\mathbf{r}$ if the winner also meets her monopoly reserve. Next we relate the revenue of our hybrid auction to $\text{Vic}_\mathbf{m}$. This is quite more challenging and involves a calculation that matches events where the hybrid auction makes no sale while $\text{Vic}_\mathbf{m}$ makes a sale to events where both auctions make a sale, establishing a factor 2 approximation. We expect our technique to find broader use in mechanism design.

Lemma 5. [Many-to-One Reduction—Revenue] *Consider a single-item setting with n bidders whose values are distributed according to independent (possibly non-identical) regular distributions. Let also $\mathbf{r} = (r_1, \ldots, r_n)$ be a vector of reserve prices such that, for all $i \in \{1, \ldots, n\}$, $r_i \le \phi_i^{-1}(0)$ (i.e. r_i is no larger than the monopoly reserve for bidder i) and $\text{Rev}[\text{Vic}_{r_i}(\{i\})] \ge (1-p) \cdot \text{Rev}[\text{Mye}(\{i\})]$, for some $p \in (0,1)$. (That is, if bidder i were considered in isolation then the Vickrey auction with reserve price r_i would achieve a $(1-p)$-fraction of the optimal revenue.) Then it holds that $\text{Rev}[\text{Vic}_\mathbf{r}] \ge \frac{1-p}{4} \cdot \text{Rev}[\text{Mye}]$.*

---

[2] See the discussion in [5] for why F^{-1} is a well-defined function for a differentiable regular distribution.

Proof. Let $\mathcal{E}_i$ denote the event that i is the winner of the Vikrey auction with reserves $\mathbf{r}$, i.e. $i = \arg\max_j\{v_j \cdot \mathbb{I}_{(v_j \geq r_j)}\}^3$ and $v_i \geq r_i$. Using Prop. 1 we can write $\mathrm{Rev}[\mathrm{Vic}_\mathbf{r}]$ in terms of the bidders' virtual values as follows:

$$\mathrm{Rev}[\mathrm{Vic}_\mathbf{r}] = \sum_{i=1}^{n} \mathbb{E}\left[\phi_i(v_i) \mid \mathcal{E}_i, \phi_i(v_i) \in [\phi_i(r_i), 0]\right] \Pr\left[\mathcal{E}_i, \phi_i(v_i) \in [\phi_i(r_i), 0]\right]$$
$$+ \mathbb{E}\left[\phi_i(v_i) \mid \mathcal{E}_i, \phi_i(v_i) \geq 0\right] \Pr\left[\mathcal{E}_i, \phi_i(v_i) \geq 0\right]. \quad (1)$$

In the course of the proof, we use the following inequalities:

$$\mathbb{E}\left[\phi_i(v_i) | \phi_i(v_i) \in [\phi_i(r_i), 0]\right] \leq \mathbb{E}\left[\phi_i(v_i) \mid \mathcal{E}_i, \phi_i(v_i) \in [\phi_i(r_i), 0]\right] \ (\leq 0) \quad (2)$$

$$(0 \leq) \ \mathbb{E}\left[\phi_i(v_i) | \phi_i(v_i) \geq 0\right] \leq \mathbb{E}\left[\phi_i(v_i) \mid \mathcal{E}_i, \phi_i(v_i) \geq 0\right] \quad (3)$$

$$\left|\mathbb{E}\left[\phi_i(v_i) \mid \phi_i(v_i) \in [\phi_i(r_i), 0]\right]\right| \cdot \Pr\left[\phi_i(v_i) \in [\phi_i(r_i), 0]\right] \leq$$
$$p \cdot \mathbb{E}\left[\phi_i(v_i) \mid \phi_i(v_i) \geq 0\right] \cdot \Pr\left[\phi_i(v_i) \geq 0\right] \quad (4)$$

Inequalities (2) and (3) follow from Lem. 2 when g is ϕ_i and $Y = \max_{j \neq i}\{v_j \cdot \mathbb{I}_{v_j \geq r_j}\}$. Inequality (4) involves a single bidder, and follows immediately from our assumption $\mathrm{Rev}[\mathrm{Vic}_{r_i}(\{i\})] \geq (1 - p) \cdot \mathrm{Rev}[\mathrm{Mye}(\{i\})]$ and noting that

$$\mathrm{Rev}[\mathrm{Vic}_{r_i}(\{i\})] = \mathbb{E}\left[\phi_i(v_i) \mid \phi_i(v_i) \in [\phi_i(r_i), 0]\right] \cdot \Pr\left[\phi_i(v_i) \in [\phi_i(r_i), 0]\right]$$
$$+ \mathbb{E}\left[\phi_i(v_i) \mid \phi_i(v_i) \geq 0\right] \cdot \Pr\left[\phi_i(v_i) \geq 0\right];$$
$$\mathrm{Rev}[\mathrm{Mye}(\{i\})] = \mathbb{E}\left[\phi_i(v_i) \mid \phi_i(v_i) \geq 0\right] \cdot \Pr\left[\phi_i(v_i) \geq 0\right].$$

Using (2), (3) and (4), we can bound the terms of the negative contribution to the expected revenue (1) as follows:

$$\left|\mathbb{E}\left[\phi_i(v_i) \mid \mathcal{E}_i, \phi_i(v_i) \in [\phi_i(r_i), 0]\right]\right| \cdot \Pr\left[\mathcal{E}_i, \phi_i(v_i) \in [\phi_i(r_i), 0]\right]$$
$$\leq \left|\mathbb{E}\left[\phi_i(v_i) \mid \phi_i(v_i) \in [\phi_i(r_i), 0]\right]\right| \left|\Pr\left[\phi_i(v_i) \in [\phi_i(r_i), 0]\right]\right. \Pr\left[\mathcal{E}_i \mid \phi_i(v_i) \in [\phi_i(r_i), 0]\right]$$

$$\leq p \cdot \mathbb{E}\left[\phi_i(v_i) \mid \phi_i(v_i) \geq 0\right] \cdot \Pr\left[\phi_i(v_i) \geq 0\right] \cdot \Pr\left[\mathcal{E}_i \mid \phi_i(v_i) \geq 0\right]$$
$$\leq p \cdot \mathbb{E}\left[\phi_i(v_i) \mid \mathcal{E}_i, \phi_i(v_i) \geq 0\right] \cdot \Pr\left[\mathcal{E}_i, \phi_i(v_i) \geq 0\right]$$

where for the first inequality we used (2) (and the fact that both sides of the inequality are non-positive), for the second inequality we used (4) and Lem. 1 taking g equal to ϕ_i, $X = v_i$ (conditioned on $X \geq r_i$), $Y = \max_{j \neq i}\{v_j \cdot \mathbb{I}_{v_j \geq r_j}\}$ and $c = 0$, and in the third inequality we used (3). We can now bound the revenue as follows:

$$\mathrm{Rev}[\mathrm{Vic}_\mathbf{r}] \geq (1 - p) \cdot \sum_{i=1}^{n} \mathbb{E}\left[\phi_i(v_i) \mid \mathcal{E}_i, \phi_i(v_i) \geq 0\right] \cdot \Pr\left[\mathcal{E}_i, \phi_i(v_i) \geq 0\right]. \quad (5)$$

---

[3] Throughout the proof we assume that all maximizations have a unique maximizer. This is ok, since we consider continuous distributions so this happens with prob. 1.

To continue, we observe that the summation on the right-hand-side of (5) can be interpreted as the revenue of the following *hybrid* auction, $\mathcal{H}$, which lies between Vic_r and Vic_m: $\mathcal{H}$ truncates all bidders at their respective reserve prices r_i; among the surviving bidders it identifies the larger bidder i^* as a potential winner, but only allocates the item to i^* if she clears her *monopoly* reserve m_{i^*}; if this happens, i^* pays the maximum of her reserve price m_{i^*} and $\max_{j \neq i^*} \{v_j \cdot \mathbb{I}_{(v_j \geq r_j)}\}$. We can clearly lower bound the expected payment of bidder i in the hybrid auction by the following expression:

$$\int_{x=0}^{m_i} \Pr\left[\max_{j \neq i}\{v_j \cdot \mathbb{I}_{(v_j \geq r_j)}\} = x\right] \cdot m_i \cdot \Pr\left[v_i \geq m_i\right] \, dx.$$

Hence:

$$\mathbb{E}\left[\mathcal{R}_\mathcal{H}\right] \geq \sum_{i=1}^{n} \int_{x=0}^{m_i} \Pr\left[\max_{j \neq i}\{v_j \cdot \mathbb{I}_{(v_j \geq r_j)}\} = x\right] \cdot m_i \cdot \Pr\left[v_i \geq m_i\right] \, dx. \qquad (6)$$

Next we compare the revenue of $\mathcal{H}$ to that of the Vickrey auction with monopoly reserves Vic_m. Our first observation is that whenever (i.e. for any value vector for which) $\mathcal{H}$ sells to some bidder i, Vic_m also sells to the same bidder i; moreover, the payment of bidder i in $\mathcal{H}$ is at least as large as her payment in Vic_m.[4] So the contribution of bidder i to the revenue from the event where she gets the item in both auctions is larger in the hybrid auction. This implies that the revenue in the event that both $\mathcal{H}$ and Vic_m sell the item is larger in $\mathcal{H}$ than Vic_m. Let us call this event the *good event* $\mathcal{G}$. We have just argued that

$$\mathbb{E}\left[\mathcal{R}_\mathcal{H} \mid \mathcal{G}\right] \cdot \Pr[\mathcal{G}] \geq \mathbb{E}\left[\mathcal{R}_{\text{Vic}_m} \mid \mathcal{G}\right] \cdot \Pr[\mathcal{G}]. \qquad (7)$$

So it suffices to bound the revenue of Vic_m under the event that Vic_m sells to some bidder, but the hybrid auction does not sell to any bidder. Let us call this event the *bad event*, $\mathcal{B}$. We claim that the bad event is contained in the union of the following disjoint events:

$$B_i = \left\{v_i \cdot \mathbb{I}_{(v_i \geq r_i)} = \max_j v_j \cdot \mathbb{I}_{(v_j \geq r_j)} \text{ and } v_i \leq m_i\right\}, \text{ for all } i.$$

Indeed, if the bad event happens it must be that the winner j^* of Vic_m does not satisfy $v_{j^*} \cdot \mathbb{I}_{(v_{j^*} \geq r_{j^*})} = \max_j\{v_j \cdot \mathbb{I}_{(v_j \geq r_j)}\}$. Suppose instead that $v_i \cdot \mathbb{I}_{(v_i \geq r_i)} = \max_j\{v_j \cdot \mathbb{I}_{(v_j \geq r_j)}\}$. For i not to be the winner in the hybrid auction it must be that $v_i \leq m_i$. Hence B_i is satisfied.

---

[4] The reason for this is that $\mathcal{H}$ uses lower reserves to truncate the bidders' values. So if i wins in $\mathcal{H}$ her value is larger than her monopoly reserve as well as all other bidders' values truncated at the reserves $\mathbf{r}$. So her value must also be larger than the other bidders' values truncated at the (higher) monopoly reserves $\mathbf{m}$. By the same token, the second highest truncated value will be higher if truncation happens at $\mathbf{r}$ than if it happens at $\mathbf{m}$.

Now, in event B_i, the maximum possible revenue that any auction (and hence $\text{Vic}_\mathbf{m}$) could be making is $\max_{j \neq i} v_j \cdot \mathbb{I}_{(v_j \geq r_j)}$. Hence, the revenue of $\text{Vic}_\mathbf{m}$ from the event B_i can be upper bounded as:

$$\mathbb{E}\left[\mathcal{R}_{\text{Vic}_\mathbf{m}} \mid B_i\right] \cdot \Pr[B_i] \leq \int_{x=0}^{m_i} \Pr\left[\max_{j \neq i}\{v_j \cdot \mathbb{I}_{(v_j \geq r_j)}\} = x\right] \cdot x \cdot \Pr\left[x \leq v_i \leq m_i\right] \, dx$$

$$\leq \int_{x=0}^{m_i} \Pr\left[\max_{j \neq i}\{v_j \cdot \mathbb{I}_{(v_j \geq r_j)}\} = x\right] \cdot x \cdot \Pr\left[v_i \geq x\right] \, dx$$

$$\leq \int_{x=0}^{m_i} \Pr\left[\max_{j \neq i}\{v_j \cdot \mathbb{I}_{(v_j \geq r_j)}\} = x\right] \cdot m_i \cdot \Pr\left[v_i \geq m_i\right] \, dx \qquad (8)$$

where the last inequality follows from the definition of the monopoly reserve m_i.

Hence, the revenue of $\text{Vic}_\mathbf{m}$ from the bad event $\mathcal{B}$ can be upper bounded as:

$$\mathbb{E}\left[\mathcal{R}_{\text{Vic}_\mathbf{m}} \mid \mathcal{B}\right] \cdot \Pr[\mathcal{B}] \leq \sum_{i=1}^{n} \mathbb{E}\left[\mathcal{R}_{\text{Vic}_\mathbf{m}} \mid B_i\right] \cdot \Pr[B_i] \leq \mathbb{E}\left[\mathcal{R}_\mathcal{H}\right], \qquad (9)$$

where for the first inequality we used that $\mathcal{B} \subseteq \cup_i B_i$, and for the second inequality we combined (8) and (6). Combining (9) and (7) we obtain:

$$\text{Rev}[\mathcal{H}] \geq \frac{1}{2} \cdot \text{Rev}[\text{Vic}_\mathbf{m}]. \qquad (10)$$

The lemma follows by combining (5), (10) and noticing that the revenue of $\text{Vic}_\mathbf{m}$ is known by [11] to be a $1/2$-approximation to the optimal revenue, i.e. $\text{Rev}[\text{Vic}_\mathbf{m}] \geq \frac{1}{2} \cdot \text{Rev}[\text{Mye}]$. $\qquad \square$

We are now ready to prove our main theorem:

Theorem 1 (Main). *For every single-item setting with n bidders whose values are distributed according to independent (possibly non-identical) regular distributions, and any $p \in [0,1]$, there is a vector of reserve prices $\mathbf{r} = (r_1, \ldots, r_n)$ such that $\text{Vic}_\mathbf{r}$ is a $(p, (1-p)/4)$-approximation.*

Proof. We argue that, for all i, there exists a price r_i such that the following are satisfied:

$$\Pr[v_i \geq r_i] \geq p; \text{ and}$$

$$\text{Rev}[\text{Vic}_{r_i}(\{i\})] \geq (1-p) \cdot \text{Rev}[\text{Mye}(\{i\})].$$

Indeed, we distinguish two cases. If $1 - F(\phi_i^{-1}(0)) \geq p$, we take $r_i = \phi_i^{-1}(0)$ and the above are satisfied automatically. Otherwise, the existence of a reserve with the above properties is implied by Lem. 4. Given reserves $r_1, \ldots, r_n$ as above, the theorem follows immediately from Lem. 3 and 5. $\qquad \square$

Picking $p = 1/5$ we obtain a $(1/5, 1/5)$-approximate mechanism for regular distributions.

Corollary 2. [regular, independent] *For every single-item setting with n bidders whose values are distributed according to independent (possibly non-identical) regular distributions, there exist reserve prices $\mathbf{r}$ such that $\text{Vic}_\mathbf{r}$ achieves a $(1/5, 1/5)$-approximation.*

4   The Non-regular, i.i.d. Case

In this section we show that the Vickrey auction with an anonymous reserve price achieves a constant factor approximation to both objectives for general distributions, when the bidders' values are distributed independently but identically. We will follow the approach of [6], which makes use of *prophet inequalities* [15] to show that this auction achieves a 1/2-approximation to the optimal revenue.

We first describe prophet inequalities. Imagine a gambler facing a series of n games in a casino, one on each of n days. Game i has a prize associated with it, whose value is distributed according to some distribution F_i. The distributions of the prize values are known to the gambler in advance, but their exact realization is not known in advance, and neither is the order of the games. On day i a game is chosen by an adversary trying to minimize the gambler's profit and its prize value is drawn from the corresponding distribution; the gambler needs to decide whether to pick the prize and leave the casino, or ignore it and keep playing. Clearly the gambler's optimal strategy can be computed using backwards induction; on the other hand, there exists a simple *threshold* strategy that guarantees the gambler at least half of the expected value of the maximum prize. A threshold strategy is a single value t, such that the gambler accepts the first prize i with $v_i \geq t$; the proof of the following theorem can be found in [15,10].

Theorem 2. *There exists a threshold t such that, independently of the order the games are played, the expected prize of the gambler is at least half of the expected value of the maximum prize, and the probability that the gambler receives a prize is exactly 1/2.*

In [6] they leverage this theorem to show that the Vickrey auction with an anonymous reserve price achieves at least half of the optimal revenue. We can easily extend this to show a guarantee for both social welfare and revenue.

Theorem 3. *In every single-item setting with n bidders whose values are drawn independently from the same (possibly non-regular) distribution, a Vickrey auction with an anonymous reserve price achieves a 1/2-approximation to both optimal revenue and welfare.*

Proof. For the sake of completeness we first sketch the proof for revenue. (For full details we refer the reader to [10].) Observe that the problem a revenue-optimizing auctioneer faces is similar to the gambler's problem described above, if prizes are taken to be the bidders' *ironed* virtual values (assuming that the gambler's strategy treats all values in every flat region of the ironed virtual valuation functions the same). Indeed, let t be the threshold that is guaranteed by Th. 2, and pick the reserve price to be $p = \hat{\phi}^{-1}(t)$, where $\hat{\phi}$ denotes the ironed virtual valuation of the bidders. If there are multiple p's mapped to t by $\hat{\phi}$ pick the smallest such p. Given this tie-breaking, observe that the Vickrey auction with reserve price p treats all flat regions in the ironed virtual valuation function the same; hence its revenue is equal to the expected ironed virtual value of the winner (prize picked), which by Th. 2 is at least 1/2 of the optimal expected ironed virtual surplus (expected maximum prize). Since the latter is an upper bound to the optimal revenue, the revenue of the Vickrey auction with reserve p is a 1/2-approximation to the optimal revenue. Moreover, Th. 2 guarantees that a prize will be picked with probability at least 1/2, i.e.

$$\Pr\left[\max_i\{v_i\} \geq p\right] \geq 1/2 \geq \Pr\left[\max_i\{v_i\} \leq p\right]. \tag{11}$$

Note that the way we defined our tie-breaking rule is important for this to hold. Next we show that this auction achieves at least half of the optimal social welfare:

$$
\begin{aligned}
\mathbb{E}\left[\max_i\{v_i\}\right] &= \int_0^p x \cdot \Pr\left[\max_i\{v_i\} = x\right] dx + \int_p^\infty x \cdot \Pr\left[\max_i\{v_i\} = x\right] dx \\
&\leq p \cdot \int_0^p \Pr\left[\max_i\{v_i\} = x\right] dx + \int_p^\infty x \cdot \Pr\left[\max_i\{v_i\} = x\right] dx \\
&\overset{(11)}{\leq} p \cdot \int_p^\infty \Pr\left[\max_i\{v_i\} = x\right] dx + \int_p^\infty x \cdot \Pr\left[\max_i\{v_i\} = x\right] dx \\
&\leq \int_p^\infty x \cdot \Pr\left[\max_i\{v_i\} = x\right] dx + \int_p^\infty x \cdot \Pr\left[\max_i\{v_i\} = x\right] dx \\
&= 2 \cdot \mathbb{E}\left[\max_i\{v_i \cdot \mathbb{I}_{v_i \geq p}\}\right] \qquad \square
\end{aligned}
$$

References

1. Abhishek, V., Hajek, B.E.: Efficiency loss in revenue optimal auctions. In: CDC, pp. 1082–1087 (2010)
2. Aggarwal, G., Goel, G., Mehta, A.: Efficiency of (revenue-)optimal mechanisms. In: ACM Conference on Electronic Commerce, pp. 235–242 (2009)
3. Alaei, S.: Bayesian combinatorial auctions: Expanding single buyer mechanisms to many buyers. In: FOCS (2011)
4. Bulow, J., Klemperer, P.: Auctions versus negotiations. American Economic Review 86(1), 180–194 (1996)
5. Cai, Y., Daskalakis, C.: Extreme-value theorems for optimal multidimensional pricing. In: FOCS (2011)
6. Chawla, S., Hartline, J.D., Malec, D.L., Sivan, B.: Multi-parameter mechanism design and sequential posted pricing. In: STOC, pp. 311–320 (2010)
7. Dhangwatnotai, P., Roughgarden, T., Yan, Q.: Revenue maximization with a single sample. In: ACM Conference on Electronic Commerce, pp. 129–138 (2010)
8. Diakonikolas, I., Papadimitriou, C., Pierrakos, G., Singer, Y.: Complexity of efficiency-revenue trade-offs in bayesian auctions
9. Dughmi, S., Roughgarden, T., Sundararajan, M.: Revenue submodularity. In: ACM Conference on Electronic Commerce, pp. 243–252 (2009)
10. Hartline, J.: Lectures on approximation in mechanism design. Lecture notes. Northwestern University (2010)
11. Hartline, J.D., Roughgarden, T.: Simple versus optimal mechanisms. In: ACM Conference on Electronic Commerce, pp. 225–234 (2009)

12. Myerson, R.B.: Optimal auction design. Mathematics of Operations Research 6, 58–73 (1981)
13. Myerson, R.B., Satterthwaite, M.A.: Efficient mechanisms for bilateral trading. Journal of Economic Theory 29(2), 265–281 (1983)
14. Neeman, Z.: The effectiveness of english auctions. Games and Economic Behavior 43(2), 214–238 (2003)
15. Samuel-Cahn, E.: Comparison of threshold stop rules and maximum for independent nonnegative random variables. The Annals of Probability 12(4), 1213–1216 (1984)

Prior-Independent Multi-parameter Mechanism Design

Nikhil Devanur[1], Jason Hartline[2,*], Anna Karlin[3], and Thach Nguyen[3]

[1] Microsoft Research
nikdev@microsoft.com
[2] Northwestern University
hartline@eecs.northwestern.edu
[3] University of Washington[**]
{karlin,ncthach}@cs.washington.edu

Abstract. In a unit-demand multi-unit multi-item auction, an auction-eer is selling a collection of different items to a set of agents each inter-ested in buying at most unit. Each agent has a different private value for each of the items. We consider the problem of designing a truthful auc-tion that maximizes the auctioneer's profit in this setting. Previously, there has been progress on this problem in the setting in which each value is drawn from a known prior distribution. Specifically, it has been shown how to design auctions tailored to these priors that achieve a constant factor approximation ratio [2, 5]. In this paper, we present a prior-independent auction for this setting. This auction is guaranteed to achieve a constant fraction of the optimal expected profit for a large class of, so called, "regular" distributions, without specific knowledge of the distributions.

1 Introduction

In a unit-demand multi-unit multi-item auction (UMMA), there are n agents and a seller selling a set of m items. The seller has a supply of m_j units of each item j. Each agent, say the i-th, has a private value v_{ij} for item j, and is only interested in purchasing one unit. The seller runs an auction to determine whom to sell to and at what prices. The auction (or mechanism) takes as input a bid b_{ij} from each agent, and based on the collection of bids, determines a feasible[1] allocation of items to agents and a price to charge each agent. The question we consider here is how to design a truthful auction for this unit-demand setting that maximizes the seller's profit.

This is an example of a *multi-parameter* mechanism design problem. While single parameter truthful mechanism design is reasonably well-understood, the

---

[*] Supported in part by NSF CAREER Award CCF-0846113.

[**] Part of this work was done while the authors were visiting the IAS, Hebrew Univer-sity, Israel.

[1] An allocation is feasible if each agent is allocated at most one item and if no more than m_j items of type j are sold.

N. Chen, E. Elkind, and E. Koutsoupias (Eds.): WINE 2011, LNCS 7090, pp. 122–133, 2011.

understanding of truthful multi-parameter mechanism design is still very much in its infancy. In particular, when the objective of the mechanism designer is something other than maximizing social welfare, we know very little.

Among multi-parameter mechanism design problems, the problem of designing profit maximizing mechanisms for UMMAs has received the most attention [2,5] and has yielded the greatest breakthroughs so far. The main results in this area so far concern Bayesian mechanism design, in which each agent's values v_{ij} are drawn from *known* prior distributions F_{ij}. In this setting, the goal is, given knowledge of the priors, to design a truthful mechanism which maximizes the seller's expected profit, where the expectation is taken over the random draws from the prior distributions. For example, Chawla, Hartline, Malec and Sivan [5], and independently (in somewhat different settings) Bhattacharya, Goel, Gollapudi and Munagala [2] have shown how to design truthful mechanisms which are guaranteed to obtain a constant fraction of the optimal expected profit. In addition, Cai, Daskalakis and Weinberg [4] have recently shown how to design PTASes for some special cases of the problem. For a large class of Bayesian combinatorial auction settings, where the priors are known to the mechanism designer, Alaei [1] gives a general framework for approximately reducing the mechanism design problem for multiple buyers to single buyer subproblems, which applies to revenue problems such as the one we consider here.

Inspired by [8], we present a "prior-independent" mechanisms for this problem. By prior-independent, we mean two things: first, that there exist prior distributions from which the agents' values are drawn, and, second, that the mechanism designer has no knowledge of these priors. Thus, the mechanism has to work well, that is, guarantee a constant fraction of the expected profit achieved by the optimal mechanism tailored to the particular prior distributions, without any knowledge of these priors, and no matter what they happen to be, as long as the distributions satisfy a relevant "regularity" condition. In an independent and and contemporaneous work, Roughgarden, Talgam-Cohen and Yan [12] shows that, for settings that are very similar to ours, a simple "welfare maximization with supply reduction" mechanism is also a prior-independent constant approximation mechanism.

Our main theorem is the following:

Theorem. *Consider a UMMA setting where for each item j, v_{ij} is drawn independently from an arbitrary regular distribution F_j. There is an efficiently implementable, truthful mechanism $\mathcal{M}$ that, with no knowledge of the F_j's, achieves*

$$\mathbb{E}_{\mathbf{v}}\left[\mathcal{M}(\mathbf{v})\right] \geq \tfrac{1}{8}(\mathbb{E}_{\mathbf{v}}\left[\mathsf{OPT}(\mathbf{v})\right]).$$

Here OPT *is the optimal deterministic mechanism tailored to the priors F_j.*[2]

---

[2] In [6], Chawla, Malec and Sivan show that for UMMA, the profit of the optimal randomized and deterministic mechanisms are within a constant factor of each other. Therefore, our mechanism also obtain a constant factor of the profit of the optimal randomized mechanism.

To prove this theorem, we build on a number of of ideas from previous works. First, we take advantage of a reduction from Chawla et al [5] that shows that the optimal expected profit achievable in the unit-demand auction setting is upper bounded by the optimal expected profit achievable in a certain single-parameter variant of the problem. We then show how to design a prior-independent mechanism for this related single-parameter variant and, also, how to convert this mechanism back to a multi-parameter mechanism. To design a prior-independent mechanism for the single-parameter variant, we use three ideas. First, we take advantage of our understanding of the optimal mechanism in single-parameter settings, namely the Myerson mechanism [11]. Second, we relax the unit-demand constraint and instead design a mechanism for a relaxed global supply constraint. The effect of this relaxation is to convert the feasibility constraint on the subset of simultaneously served agents from a matroid intersection constraint to the much easier to handle matroid constraint. Finally, we use a Bulow-Klemperer [3] style result due to Hartline and Roughgarden [10] that shows that in single-parameter matroid settings, if each agent is duplicated, and only one of each pair of duplicates is served in any allocation, then VCG is a 2-approximation to Myerson's optimal mechanism. Putting these ideas together, we are able to design a prior-independent mechanism for the single-parameter variant of the problem. The conversion back to a multi-parameter mechanism consists of an attempt to simulate the single-parameter mechanism by offering each agent a menu of prices and letting the agent choose his favorite item. However, this simulation is not (and cannot be) faithful because of the differences between the single parameter and multi-parameter setting. Thus we need to show that in expectation not too much revenue is lost, which we prove by taking advantage of the interchangeability of the random variables v_{ij}, $1 \leq i \leq n$ (that is a consequence of the fact that they are independent draws from the same distribution).

Other Results. We present a simpler mechanism that obtain a constant factor approximation to the optimal mechanism in the case where there is exactly one unit of each item. Depends on the number of agents and items, the approximation factor of this mechanism can be better than the mechanism in our main result.

We also obtain Bulow-Klemperer type result for some special cases of the problem. In particular, we show that VCG with duplicates approximates the optimal mechanism when either there is no constraint on the supplies of the items, or the distributions of the values satisfy the "monotone hazard rate condition". For these results, we allow each value v_{ij} to be drawn from a different distribution. However, due to the space limit, these results, as well as most proofs, are deferred to the full paper.

2 Preliminaries

2.1 Settings and Definitions

We define UMMA environments. In such an environment, a seller has k items, with m_j units of item j. Each agent i has, as her private information, a valuation

v_{ij} for obtaining each item j and would like to buy at most one unit. We will assume that the values v_{ij} are drawn independently from underlying distributions F_{ij}. Formally,

Definition 1. *A UMMA environment E is a tuple $(N, M, \mathcal{S}, \mathbf{F})$ where*

- $N = \{1, 2, \ldots n\}$ *is the set of bidders*
- $M = \{1, 2, \ldots m\}$ *is the set of items; there are m_j units of item type j.*
- $\mathcal{S}$ *is the collection of possible allocations given the supply constraints and the unit-demand constraint. Each $S \in \mathcal{S}$ is a set of pairs (i, j), which represents the assignment of items to agents. Thus, for each bidder i, each set S contains at most one pair containing i, and for each item j, each set S contains at most m_j pairs containing j.*
- $\mathbf{F} = \prod_{i \in N, j \in M} F_{ij}$ *where F_{ij} is the distribution of v_{ij}. Defining $\mathbf{F}$ to be a product distribution is equivalent to assuming that all values are independent.*

In this paper, we will assume that $F_{ij} = F_j$ for all i. For $\mathbf{v}$ that is a valuation profile drawn from $\mathbf{F}$, we call the tuple $(N, M, \mathcal{S}, \mathbf{v})$, sometimes abbreviated to $(E, \mathbf{v})$, an instance *of E.*

Similarly, a *single-parameter environment* is defined by a tuple $(N, \mathcal{S}, \mathbf{F})$ where N is the set of agents, $\mathcal{S}$ is the feasible set system (i.e., the subsets of agents N that can be simultaneously served) and $\mathbf{F}$ is the distribution of the valuation profiles. This environment corresponds to the scenario where a seller is offering a service (or goods), and the subsets in $\mathcal{S}$ are the sets of agents that can feasibly be served simultaneously. (For example, if we are describing a t-unit auction, then $\mathcal{S}$ consists of all subsets of agents of size at most t.) We will only consider scenarios where the set system is downward-closed, that is, every subset of a feasible set is also feasible. In this setting, each agent $i \in N$, has a value v_i for being served, where v_i is drawn from prior distribution F_i.

A mechanism takes as input a set of bids from the agents, where in the UMMA environment b_{ij} is agent i's bid for item j, and in the single-parameter setting b_i is agent i's bid for service. The mechanism then outputs an allocation and payments. The outcome of a (deterministic) mechanism on a UMMA instance consists of a set of (item, bidder) pairs represented by an 0/1 allocation vector $\mathbf{x}$ and a payment vector $\mathbf{p}$. Here $x_{ij} = 1$ if and only if item j is assigned to i and p_i represents the amount bidder i has to pay. Similarly for a single-parameter mechanism, the outcome is a set of winning agents and a payment vector. Again we will use the allocation vector $\mathbf{x}$ as an indicator for which agents are served: $x_i = 1$ if agent $i \in N$ is served and, again, p_i is agent i's payment.

Given the outcome of a mechanism, the utility of a single-parameter agent i is defined by $u_i = x_i v_i - p_i$, while the utility of a unit-demand multi-parameter agent i is $u_i = \sum_j x_{ij} v_{ij} - p_i$. We assume that agents act to maximize their utility, and we will focus on the design of truthful mechanisms.

Definition 2. *A mechanism is* truthful *if each bidder i maximizes her utility by bidding her true values, no matter what other agents do.*

As we have already discussed, our goal will be to design truthful mechanisms for the UMMA environment that maximize the expected profit of the auctioneer without knowledge of the priors from which agent's values are drawn. The fact that a prior-independent mechanism does not use information about the priors means that for any two environments E and E' that differ only on the distributions from which the agents' values are drawn, a prior-independent mechanism does not distinguish between $(E, \mathbf{v})$ and $(E', \mathbf{v})$ for any valuation profile $\mathbf{v}$ which is in the support of both distributions.

In the rest of this section, we review a number of important prior results that we will be using.

2.2   VCG

The VCG mechanism [7, 9, 13] is a truthful mechanism for maximizing social welfare in the various environments we consider in this paper. (It also applies much more generally.)

In the single-parameter environments we are discussing, the VCG mechanism takes as input a vector of bids $\mathbf{b}$ and chooses as its output the feasible set S that maximizes social welfare, i.e. $\sum_{i \in S} b_i$. The payment of an agent is its threshold bid, the minimum value it could have bid and still been part of the winning set.

2.3   Myerson's Optimal Mechanism for Single-Parameter Environments

We will rely heavily on Myerson's optimal mechanism [11] for profit maximization in single-parameter environments with known priors. This result assumes that agent's valuations are drawn from a product distribution $\mathbf{F} = F_1 \times F_2 \times \cdots \times F_n$. Thus, the agents' values are independently (but not identically) distributed.

Given a value v_i drawn from the distribution F_i, the *virtual value* corresponding to v_i, denoted by $\phi_{F_i}(v_i)$ is defined by

$$\phi_i(v_i) = v_i - \frac{1 - F_i(v_i)}{f_i(v_i)}.$$

F_i is *regular* if the function ϕ_i is monotone nondecreasing, and a product distribution $\mathbf{F} = \prod_i F_i$ is *regular* if each F_i is regular. The class of regular distributions is very large and includes many common distributions such as exponential and normal distributions.

Myerson's result is then the following.

Theorem 1 ([11]). *Let* $(N, \mathcal{S}, \mathbf{F})$ *be a single-parameter environment, where* $\mathbf{F}$ *is a regular product distribution. For any truthful mechanism for this environment, characterized by an allocation and a payment rules* $\mathbf{x}$ *and* $\mathbf{p}$*, we have*

$$\mathbb{E}_\mathbf{v} \left[\sum_i p_i(\mathbf{v}) \right] = \mathbb{E}_\mathbf{v} \left[\sum_i \phi_i(v_i) x_i(\mathbf{v}) \right].$$

The Myerson mechanism Mye *for regular distributions is a truthful mechanism that optimizes the quantity inside the expectation on the right hand side point-wise. In other words, given a set of bids* **b** *as input, Myerson selects as winners the feasible subset S such that $\sum_{i \in S} \phi_{F_i}(b_i)$ is maximized. This mechanism max-imizes the expected profit among truthful mechanisms.*

2.4   Reduction from UMMA Environments to Single-Parameter Environments

Definition 3. *Given a UMMA environment $E = (N, M, \mathcal{S}, \mathbf{F})$, the representa-tive environment* Rep (E) *of E is a single-parameter environment represented by the tuple $(N', \mathcal{S}', \mathbf{F})$ where*

- $N' = \{ij : i \in N, j \in M\}$,
- *Each set $S' \in \mathcal{S}'$ is constructed by taking a set $S \in \mathcal{S}$ and replacing each pair (i, j) by the agent ij.*

Each single-parameter agent ij is a representative *of the unit-demand agent i.*

Chawla et al [5] show that for UMMA, the optimal revenue in the representative environment upper bounds the optimal revenue of the original environment.

Lemma 1 (Corollary of Lemma 5 in [5]). *Let $E = (N, M, \mathcal{S}, \mathbf{F})$ be a UMMA environment and let* Rep (E) *be its representative environment. Also, let* OPT *be the optimal deterministic mechanism for E. We have*

$$\mathbb{E}_{\mathbf{v} \sim \mathbf{F}} [\mathsf{OPT}(E, \mathbf{v})] \leq \mathbb{E}_{\mathbf{v} \sim \mathbf{F}} [\mathsf{Mye} (\mathsf{Rep}(E), \mathbf{v})]$$

2.5   Bulow-Klemperer Type Results

We first review the concept of duplicates.

Definition 4. *Given a single parameter environment E, the environment with duplicates* Dup (E) *is obtained by adding a new agent i' for each agent i in E such that:*

- *The value of i and i' are drawn from the same distribution.*
- *A feasible set in* Dup (E) *is constructed by taking a feasible set in E and replace some of the agents by their duplicates.*

Hartline and Roughgarden [10] prove the following results:

Lemma 2 (Theorem 4.4 in [10]). *Suppose $E = (N, \mathcal{S}, \mathbf{F})$ is a single-parameter environment where $\mathcal{S}$ is a matroid set system (the feasible sets are independent sets in a matroid on N) and $\mathbf{F}$ is a regular product distribution. Then the ex-pected revenue of* VCG *on* Dup (E) *is at least $1/2$ the expected revenue of* Mye *on E, i.e.*

$$\mathbb{E}_{\mathbf{u} \sim \mathbf{F} \times \mathbf{F}} [\mathsf{VCG} (\mathsf{Dup}(E), \mathbf{u})] \geq \tfrac{1}{2} \mathbb{E}_{\mathbf{v} \sim \mathbf{F}} [\mathsf{Mye}(E, \mathbf{v})].$$

3 Prior-Independent Mechanism for UMMA

We design a prior-independent mechanism that approximates, in expectation, the revenue of the optimal mechanism for UMMA. Since there is no known characterization of the optimal mechanism for UMMA, we will make use of Lemma 1 and design a mechanism that approximates the revenue of Myerson's optimal auction on the representative environment, via a sequence of reductions using a few intermediate environments. To introduce the elements of this process, we start by considering a very simple special case.

3.1 Unit-Demand Multi-item Auction with Unit Supply

As a warm-up, we consider the case where there is exactly one unit of each item, and m, the number of items, is at most the number of agents, n. Let E be the original unit-demand environment, $\mathsf{Rep}\,(E)$ be the representative environment of E. An important intermediate environment in our reduction is obtained by relaxing the unit-demand constraint in $\mathsf{Rep}\,(E)$.

Definition 5. *Let E be a single-parameter environment where the agents can be partitioned into t groups such that at most one agent in each group can be served. Then $\mathsf{Global}\,(E)$ is the environment where this constraint is replaced by the constraint that overall, at most t agents can be served.*

In particular, the environment $\mathsf{Global}\,(\mathsf{Rep}\,(E))$, or in short, $\mathsf{G.R}(E)$, is obtained by replacing the unit-demand constraint in $\mathsf{Rep}\,(E)$ by the constraint that in total, at most n representatives can be served.

In the special case where $m < n$, the global constraint that at most n representatives can be served is subsumed by the supply constraint. Therefore, $\mathsf{G.R}(E)$ is equivalent to a combination of m independent single-unit auctions. For each single-unit auction, Bulow and Klemperer [3] show that the second-price auction obtains at least $\frac{n-1}{n}$ times the expected profit of Mye. This implies $\sum_j \mathsf{SPA}_j(\mathsf{G.R}(E)) \geq \frac{n-1}{n}\mathsf{Mye}\,(\mathsf{G.R}(E))$, where SPA_j is second price auction on the representatives interested in item j. Hence, it suffices to design a mechanism that simulates these second price auctions.

The straightforward approach is to offer to sell to each agent every item at a price equal to the highest bid of other agents for that item and ask her to choose her favorite one, as described in Fig. 1

Mechanism $\mathcal{M}_1$ for unit-demand multi-item auctions with unit supply

Offer each agent i a price menu $\mathbf{p}_i$ where $p_{ij} = \max_{i' \neq i} v_{i'j}$, and ask her to choose her favorite item.

Fig. 1. A mechanism for unit-demand multi-item auction where there is exactly one unit of each item

It is immediate that $\mathcal{M}_1$ is truthful and outputs a feasible allocation. To analyze the the revenue of $\mathcal{M}_1$, let p_j be the second highest bid for item j and i_j be the highest bidder for item j. Moreover, let ξ_j be the event that $i_\ell \neq i_j$ for all $\ell \neq j$. Then if ξ_j happens, $\mathcal{M}_1$ gets at least p_j from item j. Therefore,

$$\mathbb{E}\left[\mathcal{M}_1(E)\right] \geq \sum_j p_j \mathbf{Pr}\left[\xi_j\right] = \sum_j p_j \left(\tfrac{n-1}{n}\right)^{m-1}$$

$$= \left(\tfrac{n-1}{n}\right)^{m-1} \sum_j \mathsf{SPA}_j(\mathsf{G.R}(E)) = \left(\tfrac{n-1}{n}\right)^m \mathsf{Mye}\left(\mathsf{G.R}(E)\right)$$

$$\geq \left(\tfrac{n-1}{n}\right)^m \mathsf{Mye}\left(\mathsf{Rep}\left(E\right)\right) \geq \left(\tfrac{n-1}{n}\right)^m \mathsf{OPT}\left(E\right)$$

While we start with the assumption that $m \leq n$ to motivate the decompose of $\mathsf{G.R}(E)$ into m single-item auctions, the above analysis is independent of this assumption. This yields the following theorem.

Theorem 2. *For unit-supply unit-demand multi-item auction with m items and n agents, $\mathcal{M}_1$ approximates the revenue of the optimal mechanism within a factor of $\left(\tfrac{n}{n-1}\right)^m$.*

In particular, when $m = O(n)$, $\mathcal{M}_1$ is a constant approximation to the optimal mechanism. When $m \leq n$, the approximation ratio is at most 4 for $n \geq 2$, and converges to e when n tends to ∞.

Remark 1. There is a mechanism, which is a combination of $\mathcal{M}_1$ and the mechanism $\mathcal{M}$ described in the next section, which approximates the expected revenue of the optimal mechanism for unit-supply unit-demand multi-item auction within a factor of $2(\tfrac{n}{n-1})^{n+1}$, even when m is much larger than n. This approximation ratio is worse than that of $\mathcal{M}$ when n is small. However, as n tends to ∞, it approaches $2e$.

3.2   Unit-Demand Multi-unit Multi-item Auction

We turn to the general case where there are more than one unit of each items. For the ease of representation, we will assume that n, the number of agents, is even. (If the number of agents is odd, we can simply discard one agent at a small loss in revenue.)

We use the same approach as in the previous section: simulating a prior-independent mechanism for the single-parameter environment by offering a price menu to each agent. To this end, we have to construct a prior-independent single-parameter mechanism for $\mathsf{Rep}\left(E\right)$. Lemma 2 gives us a starting point. To use this result, we introduce duplicates. We restrict $\mathsf{Rep}\left(E\right)$ to half of the agents and use the remaining agents as duplicates. Moreover, since the resulting environment is not a matroid environment, we will relax the unit-demand constraint to a global constraint, as discussed in the previous section, to transform it into a matroid environment.

Formally, we make use of the following intermediate environments:

- $\mathsf{H.R}(E)$ (an abbreviation of $\mathsf{Half}\left(\mathsf{Rep}\left(E\right)\right)$) is the environment obtained by restricting $\mathsf{Rep}\left(E\right)$ to the set of representatives $\{ij : 1 \leq i \leq (i + n/2)\}$.

- G.H.R(E) (an abbreviation of Global (Half (Rep (E)))) is the environment obtained by relaxing the unit-demand constraint in H.R(E) to the constraint that, overall, at most $n/2$ representatives can be served.
- D.G.H.R(E) (an abbreviation of Dup (Global (Half (Rep (E)))))) is the environment obtained by adding a duplicate for each representative in G.H.R(E). We use the representatives discarded by Half as the duplicates, i.e., D.G.H.R(E) contains the representatives of all agents in N, and for each $i \in \{1, 2, \ldots, n/2\}$ and each j, ij and $(i + n/2)j$ are duplicates of each other.

We will show that VCG (D.G.H.R(E)) is a good approximation of OPT (E) and then design a multi-parameter mechanism $\mathcal{M}$ that approximates VCG (D.G.H.R(E)). The chain of reductions is summarized as follows

$$\mathbb{E}\left[\text{OPT}\left(E\right)\right] \leq \mathbb{E}\left[\text{Mye}\left(\text{Rep}\left(E\right)\right)\right] \leq 2\mathbb{E}\left[\text{Mye}\left(\text{H.R}(E)\right)\right] \leq 2\mathbb{E}\left[\text{Mye}\left(\text{G.H.R}(E)\right)\right]$$
$$\leq 4\mathbb{E}\left[\text{VCG}\left(\text{D.G.H.R}(E)\right)\right] \leq 8\mathbb{E}\left[\mathcal{M}(E)\right] \tag{1}$$

This chain is the proof of our main theorem.

Lemma 1 already gives us the first inequality.

Intuitively, the revenue Mye gets from Rep (E) is at most the revenue it gets from H.R(E) and the environment obtained by restricting Rep (E) to the other half of the representatives. Since these two restricted environment are identical, the optimal revenue in Rep (E) is at most twice the optimal revenue in H.R(E). Hence the second inequality holds.

The third inequality follows from the fact that Mye (G.H.R(E)) optimizes the virtual surplus over a relaxed set of constraints as compared to Mye (H.R(E)).

The fourth inequality follows from Lemma 2 and the fact that D.G.H.R(E) is a matroid environment.

It remains to describe $\mathcal{M}$ and prove the last inequality. As discussed, $\mathcal{M}$ would offer each agent i a price menu $\mathbf{p}_i$ and ask her to choose her favorite item. The question is how to determine p_{ij} for each i and j. Since we would like to simulate VCG (D.G.H.R(E)), the straightforward answer is to set p_{ij} to the VCG price of representative ij.

However, this straightforward approach does not work, as the VCG price of ij may be determined by the value of another representative of i; hence the menu offered to i is not independent of her bid. This complication stems from the fact that the VCG price of ij is computed by comparing the welfare of other representatives when (i) ij is included and (ii) ij is excluded from the environment. The important observation is that if in (ii), instead of excluding only ij, we excluded all representatives of i, the price menu would be independent of i's bids. In another word, p_{ij} should be the externality that i would impose on other representatives of D.G.H.R(E) by taking item j. This leads to our mechanism $\mathcal{M}$, detailed in Fig. 2.

$\mathcal{M}$ is clearly truthful. The following two lemmas complete (1) and the proof of our main theorem.

Lemma 3. *$\mathcal{M}$ outputs a feasible allocation.*

Mechanism $\mathcal{M}$ for UMMA [a]

For each agent i, do the following

1. Compute a price menu $\mathbf{p}_i$, where p_{ij} is the externality i would impose on other representatives in D.G.H.R(E) by taking item j. In another word, let E_{-i} be D.G.H.R(E) with all representatives of i removed, then p_{ij} is the maximum of three quantities:
   - the value of ij's duplicate,
   - the value of m_j-th winner of item j in E_{-i}, i.e. the smallest value among the winners if we are to sell m_j unit of item j and nothing else.
   - the value of the $\frac{n}{2}$-th winner in E_{-i}.
2. Offer $\mathbf{p}_i$ to agent i and ask her to choose her favorite item.

---

[a] This description of the mechanism assumes the absence of ties. When ties are present, extra steps are required to make sure that $\mathcal{M}$ and VCG (D.G.H.R(E)) break ties in the same way. The detailed mechanism is deferred to the full paper.

Fig. 2. A prior-independent, truthful mechanism for UMMA

Proof. The unit-demand constraint is automatically satisfied because each agent is asked to choose one item. On the other hand, $\mathcal{M}$ offers each item j to at most m_j agents at prices smaller than their bids for it. Therefore, at most m_j agents would buy j and the supply constraint is satisfied.   □

Lemma 4. *The expected revenue of $\mathcal{M}$ is at least $1/2$ the expected revenue of* VCG (D.G.H.R(E)).

To prove this lemma, we first give a condition so that $\mathcal{M}$ and VCG (D.G.H.R(E)) get the same revenue from an agent.

Lemma 5. *Consider welfare maximization in environment* D.G.H.R(E) *with and without representative ij. If representative ij is served in the former and none of i's representatives ij' are served in the latter then the payment of representative ij in* VCG (D.G.H.R(E)) *equals that of agent i in $\mathcal{M}$.*

Proof. First, in $\mathcal{M}$, $p_{ij'} > v_{ij'}$ for any $j' \neq j$. This is because surplus maximization without ij failed to assign an item to ij', so the externality from serving ij' (thus, the payment i must make for item j) must be greater than $v_{ij'}$. Hence, i would not buy any item $j' \neq j$.

Second, the payment of i for item j in $\mathcal{M}$ is the externality when all of i's representatives are removed, whereas the payment in VCG for ij is the externality when just ij is removed. By the assumption of the lemma, even when we just remove ij, surplus maximization chooses not to serve another representative of i, so these externalities are the same.   □

Based on this lemma, we can now prove Lemma 4.

Proof (of Lemma 4). Let us condition on the set of values drawn from each distribution and the pairing of values given by the duplicates, i.e., from each distribution F_j draw $n/2$ pairs of values, but defer the decision of which representatives belong to which agents until later. Given this conditioning, VCG (D.G.H.R(E)) is deterministic, i.e., both the winning representatives and their payments are fixed[3].

We argue that $\mathcal{M}$'s revenue from each item j is at least $1/2$ of VCG (D.G.H.R(E))'s revenue from it. To this end, fix the representatives that win copies of item j and let representative ij be one of them. Now consider (as in the statement of Lemma 5) finding the surplus maximizing allocation in D.G.H.R(E) with ij removed. Since D.G.H.R(E) is a matroid environment, ij will be replaced by some other representative $i'j'$ and all of VCG's other winners will remain winners. This process allocates at most $n/2$ units of items other than item j to representatives of at most $n/2$ distinct agents. While we have conditioned on the representatives that win units of item j, the agents whose representatives win the other items have not yet been fixed. We now consider realizing the assignment of these other representatives to agents. The probability that agent i is assigned one of these (at most) $n/2$ representatives is at most $1/2$. Hence, the assumption of Lemma 5 holds for representative ij with probability at least $1/2$. Therefore, ij's expected contribution to $\mathcal{M}$'s revenue is at least half its contribution to VCG's revenue.

The lemma follows.   □

References

1. Alaei, S.: Bayesian combinatorial auctions: Expanding single buyer mechanisms to many buyers. In: Proc. 52nd IEEE Symp. on Foundations of Computer Science (2011)
2. Bhattacharya, S., Goel, G., Gollapudi, S., Munagala, K.: Budget constrained auctions with heterogeneous items. In: Proc. 41st ACM Symp. on Theory of Computing (2010)
3. Bulow, J., Klemperer, P.: Auctions versus negotiations. American Economic Review 86, 180–194 (1996)
4. Cai, Y., Daskalakis, C., Matthew Weinberg, S.: On optimal multidimensional mechanism design. SIGecom Exchanges 10(2), 29–33 (2011)
5. Chawla, S., Hartline, J., Malec, D., Sivan, B.: Sequential posted pricing and multiparameter mechanism design. In: Proc. 41st ACM Symp. on Theory of Computing (2010)
6. Chawla, S., Malec, D., Sivan, B.: The power of randomness in bayesian optimal mechanism design. In: ACM Conference on Electronic Commerce, pp. 149–158 (2010)
7. Clarke, E.H.: Multipart pricing of public goods. Public Choice 11, 17–33 (1971)
8. Dhangwatnotai, P., Roughgarden, T.: Qiqi Yan. Revenue maximization with a single sample. In: Proc. 12th ACM Conf. on Electronic Commerce (2010)

---

[3] For example, the winning set can be calculated as follows: (i) choose the highest-valued representative in each pair; (ii) among the representatives chosen in (i), choose the m_j highest-valued ones for each item j; and (iii) among the representatives chosen in (ii), chose the $n/2$ highest-valued ones. The payments can be calculated by keeping track of the minimum value a representative needs to have so that she is not discarded in each step.

9. Groves, T.: Incentives in teams. Econometrica 41, 617–631 (1973)
10. Hartline, J., Roughgarden, T.: Simple versus optimal mechanisms. In: Proc. 11th ACM Conf. on Electronic Commerce (2009)
11. Myerson, R.: Optimal auction design. Mathematics of Operations Research 6, 58–73 (1981)
12. Roughgarden, T., Talgam-Cohen, I., Yan, Q.: Prior-independence without sampling (2011) (manuscript)
13. Vickrey, W.: Counterspeculation, auctions, and competitive sealed tenders. J. of Finance 16, 8–37 (1961)

Discrete Choice Models of Bidder Behavior in Sponsored Search

Quang Duong[1] and Sébastien Lahaie[2]

[1] University of Michigan
Ann Arbor, MI 48109
qduong@umich.edu
[2] Yahoo! Research
New York, NY 10018
lahaies@yahoo-inc.com

Abstract. There are two kinds of bidders in sponsored search: most keep their bids static for long periods of time, but some do actively manage their bids. In this work we develop a model of bidder behavior in sponsored search that applies to both active and inactive bidders. Our observations on real keyword auction data show that advertisers see substantial variation in rank, even if their bids are static. This motivates a discrete choice approach that bypasses bids and directly models an advertiser's (perhaps passive) choice of rank. Our model's value per click estimates are consistent with basic theory which states that bids should not exceed values, even though bids are not directly used to fit the model. An empirical evaluation confirms that our model performs well in terms of predicting realized ranks and clicks.

1   Introduction

A major portion of the revenue of search engines such as Google and Bing comes from advertising next to search results. Advertisers bid for placement on keywords relevant to their business, a practice known as sponsored search. A central problem in empirical modeling of sponsored search is to infer bidder values from their observed bidding behavior. Information on bidder values can inform virtually all aspects of the keyword auction design, including changes to the ranking rule to improve revenue [7]; reserve pricing policies [9]; and the impact of improved click-through rate models on efficiency [6]. With bidder values at hand, counterfactual experiments can be performed to evaluate the effect of changes to auction parameters before live testing, and even to compare the current design to more classical auctions such as VCG [1].

In this work we develop a model of bidder behavior in sponsored search that applies to both active and inactive bidders. We observe that most advertisers in Yahoo's sponsored search market keep their bids essentially constant for long periods of time (e.g., several weeks), as others have noticed independently in Bing data [10]. On the other hand, some advertisers on competitive, high-volume keywords do actively manage their bids [3]. We propose a discrete choice model

N. Chen, E. Elkind, and E. Koutsoupias (Eds.): WINE 2011, LNCS 7090, pp. 134–145, 2011.

of bidder decisions that can identify values under both kinds of behavior. Our key insight is that even though an advertiser's bids may show little variation, its *rank* typically varies considerably because of exogenous changes in the auction's parameters, such as the weights (related to click-through rates) placed on bids for ranking.

Our approach is to bypass bids and instead directly model an advertiser's (perhaps passive) choice of rank across auctions. Because there are only a small number of ad slots—no more than twelve—available on a search results page, an advertiser's choice of rank lends itself well to discrete choice modeling [11]. Besides value per click, our model also provides a useful estimate of the advertiser's *regret* variance, which captures how consistent its behavior is with a single value per click. We evaluate our model in terms of its ability to predict advertiser rank and realized clicks in future auctions, against both simple baselines assuming constant rank and click-through rates, and the recent stochastic variability model of Pin and Key [10].

Related Work. The earliest empirical estimates of advertiser values in sponsored search appear in the work of Varian [12]. His approach is to develop an equilibrium concept to model bidding, and jointly estimate bidder values on individual auction instances (i.e., on single queries) by minimizing deviation from equilibrium. This method, however, does not extend easily to several auction instances over time [1]. In another early work Borgers et al. [2] estimate values using a revealed preference approach: an advertiser's bid updates imply bounds on its value, assuming best-response, and with enough observations the bounds can pin down the value. However, this approach is ineffective if advertisers do not update their bids often, which is very common as previously mentioned.

More recently, Athey and Nekipelov [1] have developed an approach tailored to advertisers with static bids. By modeling the distribution over an advertiser's opponent bids, they derive a marginal cost, or "incremental cost per click" curve, and obtain a value based on where the advertiser bids along the curve. (A rational agent sets marginal cost equal to marginal value.) Pin and Key [10] have developed a simplified version of this approach that yields very similar predictions but is much more scalable. For advertisers that update their bids often, their method must estimate a separate value corresponding to each bid, which may be problematic unless there is good reason to believe their value per click indeed changes with each bid update.

2   The Model

In this section we provide the necessary background on sponsored search needed to understand advertisers' decision problems. We describe the basic model of sponsored search introduced in [4,12]; for a survey of the literature see [8]. We then present our discrete choice logit model of bidder behavior; for a full treatment of logit and other discrete choice models see the monograph [11].

2.1 Sponsored Search

We first focus on a single search query for a given keyword. When the query is issued, an auction is run to allocate the ad slots on the search results page among advertisers bidding on the keyword. Let K be the number of slots and N be the number of agents, where $N > K$. The core of the current auction mechanism (ranking and pricing) used by major search engines is known as the *generalized second-price auction* (GSP) [4]. Each agent i places a bid b_i, and the search engine assigns weights w_i that depend on the ad's past click-through rates. The ads are then ranked in descending order of their *score* $w_i b_i$. Without loss of generality, we can re-index the agents so that $w_1 b_1 \geq w_2 b_2 \geq \ldots \geq w_N b_N$.

Agents are charged only when a click is received. In the GSP, payment follows a second-price rule: an agent is charged the lowest bid it could have placed while maintaining its position. In particular, to maintain its position, agent i must bid so that $w_i b_i \geq w_{i+1} b_{i+1}$, and so its price per click (PPC) is $w_{i+1} b_{i+1}/w_i$. In practice search engines also set a *reserve score* s for each keyword, so that the minimum PPC i can pay is $r_i = s/w_i$. If i bids below r_i, its ad is not shown.

The click-through rate (CTR) of ad i in position j is denoted c_{ij}. We assume that CTRs are *separable* into an advertiser effect a_i and a position effect x_j, meaning that they factor according to $c_{ij} = a_i x_j$. Although separability is only an approximation to actual CTR patterns [1], search engines still estimate ad-specific and position-specific parameters because the ad effect a_i is a key input into the ad's weight w_i. We assume that each agent i has a value per click v_i and that its utility is quasi-linear, meaning that if it obtains slot j at a PPC of p_j then its (expected) utility is:

$$V_j = (v_i - p_j)c_{ij}. \tag{1}$$

In practice the bid space is discretized into increments (e.g., 10 cents), but these are fine enough relative to the range of allowed bids that the bid space can be viewed as continuous. However, note that an agent's utility only depends on the particular position selected, holding the other agents' bids fixed. In a single auction scenario, we can therefore view an agent's bidding decision as a discrete choice problem of selecting which position to appear in. The bid confers no more information about the agent's value beyond the position selected.

2.2 Discrete Choice

From the perspective just developed, we can model an agent's collective rank decisions across the auctions it participates in by using methods of discrete choice analysis from econometrics [11]. The basis of discrete choice analysis is the random utility model. In our context, this model posits that an agent i's utility for slot j decomposes into $U_j = V_j + \epsilon_j$, where ϵ_j is a random error, and V_j is the *representative utility* given by (1), derived from observable features of the chosen alternative—in our case, simply the position effect. Agents act rationally in that they choose the slot j with highest utility U_j. Under the random utility model, an agent's choice of rank can change from auction to auction even if the others' bids are held fixed, as the error terms vary across auctions. In each auction, the random utility induces a distribution over the agent's choice of position.

We use a maximum likelihood approach to fit the representative utility's parameters. In discrete choice modeling the observations take the form $U_{\sigma(t)} \geq U_j$ for $j = 1, \ldots, K$ and $t = 1, \ldots, T$, where t indexes the auctions and $\sigma(t)$ is the slot chosen at auction t. We emphasize that the observations, and therefore the model, do not take into account the bids placed, only the ranks obtained at each auction instance. The actual parametric model we fit is of the form:

$$U_j = \beta_v x_j + \beta_p x_j p_j + \epsilon_j. \tag{2}$$

Because utility can be normalized to any scale, we have dropped the leading a_i term from the equations, and the error variance can also be normalized to some convenient constant C. This follows from the fact that only differences in utility matter when making a choice—we refer to [11] for the technical and conceptual details. Once we fit the model to data, the coefficient $-\beta_p$ corresponds to the marginal utility of money, and hence $-\beta_v/\beta_p$ gives an estimate of v_i. The error variance, which was normalized to C, is proportional to β_p^{-2} on the money scale.

Error Interpretation. A common interpretation of the error term is that it captures unobserved features of the alternatives that impact utility. However, under our value-per-click model, nothing differentiates slots besides their position effects. Instead, we find it more appropriate to interpret the error terms as capturing an agent's *regret*, defined as the amount of foregone utility from choosing one slot over another. If the agent chooses slot j over k, for instance, then $V_k - V_j$ is its regret and $\rho_{kj} = \epsilon_k - \epsilon_j \geq V_k - V_j$ is a bound on this regret (which is binding when errors are minimized). The error distribution induces a distribution over regret. Note that regret can be negative, in which case it indicates the amount by which the chosen slot is preferred over the alternative.

As we will see in Section 3, agents typically hold their bids constant for long periods of time. In this case, variation in rank across auctions comes from exogenous changes such as updates to the advertiser effects, the number of opponents, or the reserve score [10]. The distribution of an agent's regret from keeping its bid fixed is therefore induced by these exogenous changes. Nevertheless, we find it fair to characterize an agent's distribution over ranks and regret as its "behavior", even if it holds its bid fixed, because the distribution captures the extent to which the agent manages its bid to maximize utility. Indeed, the regret (equivalently, error) variance in discrete choice models is sometimes interpreted as a measure of "bounded rationality" [5].

Error Distribution. To complete the model specification we need to detail the error distribution. In this work we assume that errors are independently and identically distributed according to an extreme value distribution with mean 0, which implies that regrets are distributed according to a logistic distribution with mean 0. This is known as the *logit* model, and with this specification there is a closed-form formula for the choice probabilities of different slots given their representative utilities [11]. Once we have fit $v_i = -\beta_v/\beta_p$ and therefore obtain

$V_1, \ldots, V_K$ for agent i in a given auction, and the *inverse* estimated error variance is $\lambda = \beta_p^{-2}/C$, then the choice probabilities are given by the familiar logit formula:

$$\Pr(i, j) = \frac{e^{\lambda V_j}}{\sum_{k=1}^{K} e^{\lambda V_k}}.$$

Observe that as λ increases (error variance decreases), the choice probabilities put increasing mass on the slot with highest representative utility, and the agent is utility maximizing in the limit. As λ decreases (error variance increases), the choice probabilities become increasingly uniform over slots, and the agent's choices of ranks across auctions are less consistent with a fixed value per click.

3   Data Description

Our empirical analysis is based on Yahoo's sponsored search logs for a one month period in the first half of 2010. We randomly sampled 20 keywords from each of the top 5 keyword deciles by volume. These 100 keywords together yield a data set of nearly three million auctions that involve 15699 unique advertisers. We used the first three weeks of data for training and the last week for testing. As our study only examines advertisers present in both the training and testing data sets, the number of included advertisers drops to 2603.

To estimate our logit model as specified in (2) we used the `mlogit` package in R [13]. We fit a model to each advertiser separately. The construction of the logit model entails computing the PPC p_j of each position j that ad i sees as an option, for every auction. Because ad i only occupied a certain position j, yielding one p_j value, we computed p_k for every $k \neq j$, by applying the second-price rule described in Section 2.1 and using available data on its opponents' weights and bids, as well as its reserve price.[1]

We needed to further filter the data in order to avoid regression problems such as collinearity. Estimating the PPC of the last position requires information about the ad immediately below the last ranked ad, which is unavailable in our data—we only have records on ads that were shown. Therefore we discarded the final slot in each auction as an alternative. We also filtered out advertisers that were almost always charged their reserve price, and consequently appeared mainly in the bottom positions, because such ads saw the same PPC for multiple slots (i.e., the reserve price) which created singularity issues for the regression. Although the amount of discarded data due to this latter issue accounts for more than 20% of the remaining data, these advertisers' rank decisions, which overwhelmingly focus on getting the bottom slots, would provide little information about bidders' behavior in general. Moreover, as the inclusion of ads displayed

---

[1] Yahoo maintains two different reserve scores: one for the mainline (ads shown at the top) and the sidebar (ads shown on the right). The mainline reserve price was not available for this analysis. However, we found that using the second-price rule together with the sidebar reserve price alone was enough to reproduce observed PPC's to within 0.02% accuracy on average.

infrequently would add significant noise to our analysis, while providing few insights about bidding behavior, we further removed more than 1500 ads that were shown less than once a day on average, leaving us with a dataset of 197 ads.

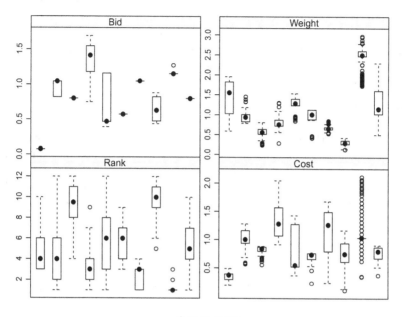

Sample of Ads

Fig. 1. Variation in bid, rank, weight (closely related to ad effect), and cost (i.e., PPC) for a representative set of 10 ads sampled uniformly at random from our dataset. The center dot gives the median; the box gives the lower and upper quartiles; the whiskers give the minimum and maximum; and any remaining dots indicate outliers. The ads' bids, weights, and costs have been normalized by the mean bid, weight, and cost for the ads' respective keywords to enable variation comparisons across panels.

Preliminary Analysis. Figure 1 presents a simple summary of bidding behavior for a representative sample of 10 ads from our dataset of 197. In this figure an ad's bids were normalized by the average bid (over opponents) of its associated keyword; we also normalized weight and cost (i.e., PPC) in the same way. A normalized bid of 1 means that it matches the average bid on the keyword.

We observe that six out of the ten ads barely vary their bid at all, and only the fourth and fifth substantially vary their bid. Pin and Key [10] also found that bids changed very little in Microsoft's sponsored search market, so this is a general feature of sponsored search and not just Yahoo's market. On the other hand, note that there is substantial variation in rank for all ads. In particular, the first ad takes on positions between 3 and 10, and the sixth ad between 3 and 9, even though their bids stay constant. For the first ad, some of the rank variation can be attributed to its changing weight (i.e., the estimate of its ad effect changes across auctions). There is less weight variation for the sixth ad, but other factors can change the rank such as variation in the number of opponents.

The variation in rank and cost here makes it possible for our discrete choice approach to identify values for the advertisers. Revealed preference approaches based on bid changes, as in [2], would fare poorly because of the dearth of bid update observations for most ads. On the other hand, for those ads whose bids do vary a lot, such as the third and fourth, approaches based on static bids need to estimate a different value for each bid, while our logit model also handles this case and estimates a single value for each advertiser.

4   Regression Results

We first report on the regression coefficients of the fitted models for each ad to confirm that they take on sensible values. Among the 197 ads we examine, 178 (or 90%) have nonzero β_v and β_p coefficients significant at the 5% level. All of these 178 ads have positive β_v coefficients and negative β_p coefficients, implying that utility is increasing in CTR and decreasing in PPC, as expected.

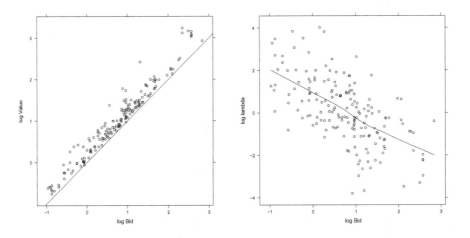

Fig. 2. Regressions results (value and error variance) against average bid (over the testing period) over 178 ads. Axes have been arbitrarily normalized for confidentiality reasons. In the left panel we see that bids lie uniformly below values. The Loess curve in the right panel confirms that inverse error variance λ decreases with bid in general.

Figure 2 provides a more detailed look at the estimated values per click v_i and inverse error variances λ_i, derived from the regression coefficients as explained in Section 2.2. According to the theory on sponsored search [4,12], bidding above one's value is a dominated strategy, so we would expect estimated values to exceed bids. This is corroborated in the figure, where values uniformly lie above the agents' average bids over the training period. We find this result striking because our model imposes no such constraints on agents' values, and indeed with little training data (one week rather than three) we do observe a few estimated values falling below bids. In fact, recall that bids are not an input to our model: we only rely on the observed position effects, PPC's, and an agent's rank at each auction. According to these estimated values agents shade their bids 20% below their value, on average (in terms of median and mean). The agents' return on investment (ROI), defined as profit per click over PPC, had a median of 48% and a mean of 95%, indicating a skewed distribution.

In Figure 2 we also see how the inverse error variances λ_i correlate with average bids. Recall that a high λ_i suggests that i is behaving more 'rationally', in the sense that its choice of slots across auctions is almost consistent with a fixed value per click, whereas a low λ_i indicates a more 'irrational' agent because its choices imply a high regret variance no matter what value per click is ascribed. According to the figure bid is negatively correlated with 'rationality', or stated more formally, correlated with high regret variance. We see the following possible reason. Low bidders tend to compete either on low-competition keywords or for low-ranked slots, and in those cases the slots are similar in terms of both CTR and PPC. Therefore even if the agent's position varies among these bottom slots (as it holds it bid fixed), its regret stays low and varies little. The situation is the opposite for high bidders that appear on high-competition keywords, where the top slots are highly differentiated in terms of CTR and PPC.

5   Model Evaluation

In this section we first describe the baseline models against which we compare our logit model, and then proceed to evaluate their performance in predicting future ranks and realized clicks.

5.1   Baseline Models

We first compare our logit model, denoted as M_{logit}, against two simple baseline models that provide predictions about bidders' positions and number of clicks using empirical distributions constructed directly from training data. The first simple baseline model, the *constant rank* model (M_{rank}), specifically focuses on rank predictions. In particular, M_{rank} assumes that each advertiser seeks to have its ad i displayed at a targeted position j^*, and treats the most frequent observed position for ad i in the training data as its targeted position j^* by assigning a probability value of 90% to j^*. In order to account for variation in agents' positions, the model allows positions other than j^* to appear with equal probabilities that sum up to the remaining $(100\% - 90\%) = 10\%$.

The second simple baseline model, called *historical click* (M_{click}), is tailored for click predictions, assuming that agents expect to receive a constant click through rate for each auction. Given the training data, M_{click} computes $\bar{c}_i$, the average number of clicks per auction that ad i received during the training data's timespan, and uses $\bar{c}_i$ to estimate the number of clicks i will receive in the future by multiplying $\bar{c}_i$ with the number of auctions in which i has a slot.

We further evaluate the estimates produced by M_{logit} against those obtained from the *stochastic* model (M_{stoch}) of Pin and Key [10]. They model an agent called Agent 0 with known value v_0 and weight w_0 submitting bid b_0 against n opponents, who submit random i.i.d bids. Note that in this context, each agent is associated with an ad and a bid value, which diverges from the viewpoint previously used in the other models, M_{rank}, M_{click}, and M_{logit}, that only view each ad as a different agent, who may place multiple bids across time.

M_{stoch} assumes that the agents' weighted bids $w_i b_i / w_0$, from the perspective of Agent 0, are drawn from a known probability distribution, whose cumulative distribution function (c.d.f) is denoted as F. As the number of opponents may vary from one auction to another, M_{stoch} incorporates a discrete probability distribution on the number of opponents, q_n, where $\sum_{n=0}^{N-1} q_n = 1$.

When Agent 0 bids b_0 greater than the reserved price, the probability that it gets the j-th position given that it faces n opponents is:

$$\Pr(j; n) = \binom{n}{j} q(n) F(b_0)^{n-j} (1 - F(b_0))^j. \tag{3}$$

Let $\psi(b_0)$ be the CTR of the slot Agent 0 receives when it bids b_0. The expected number of clicks that Agent 0 receives per auction is computed as:

$$E[\psi(b_0)] = \sum_{n=0}^{N-1} \sum_{j=0}^{n} \Pr(j; n) c_{0j}. \tag{4}$$

Given the training data set, we can construct the distribution of number of opponents q_n and the distribution of weighted bids F for each agent. These distributions allow us to estimate their expected ranks and expected number of clicks per auctions via (3) and (4) respectively.

Recall that an ad may appear in M_{stoch} as different agents, each of which corresponds to a different bid value submitted for the same ad. In order to compare the predictions of M_{stoch} with those of the other models, we apply M_{stoch} to each pair of ad and bid value, and subsequently average over all same-ad pairs to compute the estimates for each ad.

5.2 Estimation Results

We evaluate the models M_{rank}, M_{click}, M_{stoch}, and M_{logit} based on their predictions of ads' ranks and clicks they receive. Note that the M_{rank} baseline only applies to rank prediction, and the M_{click} baseline only applies to click prediction.

Rank Distribution. We measure the predictive power of M_{rank}, M_{stoch}, and M_{logit} with respect to ads' ranks by the likelihood of the testing data induced by each model. In particular, given a model $M \in \{M_{rank}, M_{stoch}, M_{logit}\}$ learned from the training data for an ad i, we compute the log likelihood of the ad's positions in the testing data set D of m auctions in which i won a slot, as follows:

$$L_i(D \mid M) = \sum_{t=1}^{m} \log \Pr_i(\sigma(t) \mid M), \tag{5}$$

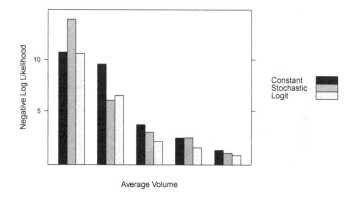

Fig. 3. Rank prediction results, with ads divided into 5 bins according to average volume. Volume increases exponentially towards the right.

where $\text{Pr}_i(\sigma(t) \mid M)$ is the probability that i gets the slot $\sigma(t)$ in auction t, as specified by model M. In order to investigate these models' robustness to data availability, we also varied the training data set's timespan. Due to space constraints we only report in detail on the results from models trained on the whole first three weeks of the data.

We divide the ads into 5 bins according to the volume of auctions in which they are present. For each bin, we compute the average negative log likelihood of ads' ranks per auction and per ad. Figure 3 shows that the logit model M_{logit} consistently outperforms the simple baseline M_{rank} model in every bin, and also provides better rank predictions than the stochastic model M_{stoch} in most bins. As expected, prediction performance improves as the average volume (and hence amount of training data) increases.

Realized Clicks. We next compare the models M_{click}, M_{stoch}, and M_{logit}, by evaluating the estimated number of clicks each ad would receive against the number of realized clicks in the testing data. Given a model M's estimated number of clicks received by ad i per auction, $\hat{c}_i^M$, and the number of realized clicks per auction over the testing data for ad i, c_i^*, we can calculate the relative error[2] for model M as follows:

$$\text{err}_i(D \mid M) = \frac{\mid \hat{c}_i^M - c_i^* \mid}{c_i^*}. \tag{6}$$

We again split the ads into 5 bins, this time based on their average number of clicks, and then compute each bin's relative error as the average of $\text{err}_i(D \mid M)$ over all ads i in the bin for each model $M \in \{M_{\text{click}}, M_{\text{stoch}}, M_{\text{logit}}\}$. Figure 4 demonstrates that the simple baseline M_{click} model performs particularly well for ads that attract fewer hits, beating both M_{stoch} and M_{logit} in the least-clicked ad bin. The logit model predicts clicks noticeably better than M_{click} for ads that

---

[2] We can only compute this relative error for the 51 ads in our dataset that received at least one click.

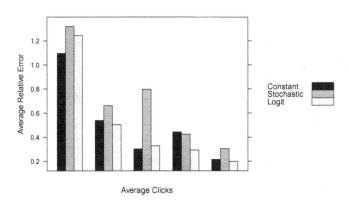

Fig. 4. CTR prediction results, with ads divided into 5 bins according to average number of clicks. Clicks increase exponentially towards the right.

receive more clicks, and moreover, consistently outperforms M_{stoch} in all bins. The predictive power of each model improves as the average number of realized clicks increases, as observed in a different study [10]. That study examined only ads that received at least as much actual clicks as the ads in our two most clicked bins, namely the two right-most bins in Figure 4. They also incorporated results from a baseline model similar to M_{click}, but trained this baseline model on less data than the baseline M_{click} we employed.

Note that in order to make rank and click predictions M_{stoch} has to examine the actual bids placed by agents in the testing data. In contrast, M_{logit} examines an agent's opponents' bids in the testing data to predict realized rank (or more precisely, rank distribution) and clicks; it does not directly draw on the agent's behavior (i.e., bids) to predict. Despite this seeming disadvantage, M_{logit} performs very well against M_{stoch}.

6   Conclusion

We have introduced a novel discrete-choice approach to modeling the bidding behavior of both active and inactive bidders in sponsored search. Our logit model of advertisers' rank decisions produces bidder value estimates that are consistent with basic theory on how values relate to bids, even though these constraints are not incorporated into the regressions. Our empirical evaluation showed that the logit model predicts realized ranks and clicks well, against both simple baselines and a more sophisticated baseline that even draws on agents' actual bidding behavior to make predictions, in contrast to our approach.

The parametric form of utility given in (2) is one of several potential options that we hope to investigate in future work. For instance, we could add position-specific intercepts to the utility specification in order to see whether advertisers value higher slots more than lower slots, all else (i.e., click-through rate) held equal, which would indicate utility for the "branding effect" of slots. We could use a nested logit model [11] that not only relaxes the i.i.d. assumption of the error term, but can also incorporate the variation in the number of bidders competing in an auction. Finally, as our data filtering process left us with a much smaller sample than the original set, we would like to scale up our empirical analysis and include more high-click rather than high-volume ads.

References

1. Athey, S., Nekipelov, D.: A structural model of sponsored search advertising auctions. Tech. rep., Microsoft Research (May 2010)
2. Borgers, T., Cox, I., Pesendorfer, M., Petricek, V.: Equilibrium bids in sponsored search auctions: Theory and evidence (2008) working paper
3. Edelman, B., Ostrovsky, M.: Strategic bidder behavior in sponsored search auctions. Decision Support Systems 43, 192–198 (2070)
4. Edelman, B., Ostrovsky, M., Schwarz, M.: Internet advertising and the Generalized Second Price auction: Selling billions of dollars worth of keywords. American Economic Review 97(1) (March 2007)
5. Goeree, J.K., Holt, C.A., Palfrey, T.R.: Regular quantal response equilibrium. Experimental Economics 8, 347–367 (2005)
6. Hillard, D., Schroedl, S., Manavoglu, E., Raghavan, H., Leggetter, C.: Improving ad relevance in sponsored search. In: Proceedings of the Third ACM International Conference on Web Search and Data Mining, pp. 361–370. ACM, New York (2010)
7. Lahaie, S., Pennock, D.M.: Revenue analysis of a family of ranking rules for keyword auctions. In: Proceedings of the 8th ACM Conference on Electronic Commerce, pp. 50–56 (2007)
8. Lahaie, S., Pennock, D.M., Saberi, A., Vohra, R.V.: Sponsored search auctions. In: Nisan, N., Roughgarden, T., Taros, É., Vazirani, V.V. (eds.) Algorithmic Game Theory, pp. 699–716. Cambridge University Press (2007)
9. Ostrovsky, M., Schwarz, M.: Reserve prices in Internet advertising auctions: A field experiment (2009) working paper
10. Pin, F., Key, P.: Stochastic variability in sponsored search auctions: Observations and models. In: Proceedings of the 12th ACM Conference on Electronic Commerce, pp. 61–70 (2011)
11. Train, K.E.: Discrete Choice Methods with Simulation. Cambridge University Press (2009)
12. Varian, H.R.: Position auctions. International Journal of Industrial Organization 25, 1163–1178 (2007)
13. Zeileis, A., Croissant, Y.: Extended model formulas in R: Multiple parts and multiple responses. Journal of Statistical Software 34(1), 1–13 (2010)

Social Learning in a Changing World

Rafael M. Frongillo[1,*], Grant Schoenebeck[2,**], and Omer Tamuz[3,***]

[1] University of California Berkeley, Computer Science Department
[2] Princeton University, Computer Science Department
[3] Weizmann Institute

Abstract. We study a model of learning on social networks in dynamic environments, describing a group of agents who are each trying to estimate an underlying state that varies over time, given access to weak signals and the estimates of their social network neighbors.

We study three models of agent behavior. In the *fixed response* model, agents use a fixed linear combination to incorporate information from their peers into their own estimate. This can be thought of as an extension of the DeGroot model to a dynamic setting. In the *best response* model, players calculate minimum variance linear estimators of the underlying state.

We show that regardless of the initial configuration, fixed response dynamics converge to a steady state, and that the same holds for best response on the complete graph. We show that best response dynamics can, in the long term, lead to estimators with higher variance than is achievable using well chosen fixed responses.

The *penultimate prediction* model is an elaboration of the best response model. While this model only slightly complicates the computations required of the agents, we show that in some cases it greatly increases the efficiency of learning, and on complete graphs is in fact optimal, in a strong sense.

Keywords: social networks, Bayesian agents, social learning.

1 Introduction

The past three decades have witnessed an immense effort by the computer science and economics communities to model and understand people's behavior on social networks [16]. A particular goal has been the study of how people share information and learn from each other; learning from peers has been repeatedly shown to be a driving force of many economic and social processes (cf. [9,7,19,8]).

* Supported by the National Defense Science & Engineering Graduate Fellowship (NDSEG) Program.
** This work was supported by Simons Foundation Postdoctoral Fellowship and National Science Foundation Graduate Fellowship.
*** Supported by ISF grant 1300/08. Omer Tamuz is a recipient of the Google Europe Fellowship in Social Computing, and this research is supported in part by this Google Fellowship.

N. Chen, E. Elkind, and E. Koutsoupias (Eds.): WINE 2011, LNCS 7090, pp. 146–157, 2011.

1.1   Classical Approaches and Results

Early work by DeGroot [10] considered a set of agents, connected by a social network, that each have a prior belief: a distribution over the possible values of an *underlying state of the world* - say the market value of some company. The agents iteratively observe their neighbors' beliefs and update their own by averaging the distributions of their neighbors. Since DeGroot, a plethora of models for social learning have been proposed and studied.

DeGroot's simple averaging of neighbors' beliefs may seem naive and arbitrary; economists often opt for *rational* models instead. In rational models the agents update their belief not by a fixed rule, but in an attempt to maximize a utility function. It is often assumed that agents are *Bayesian*: they assume some prior distribution on the underlying state and on other agents' behavior, have access to some observations, and maximize the expected value of their utility, using Bayes' Law. Bayesian social learning has a wide literature, with noted work by Aumann [4] and the related *common knowledge* work (cf. [14]), as well as McKelvey and Page [20], Parikh and Krasucki [23], Bala and Goyal [6], Gale and Kariv [13], and many others.

Aumann [4] and Geanakoplos [15] show that a group of Bayesian agents, who each have an initial estimate of an *underlying state*, and repeatedly announce their estimate (in particular, expected value) of this state, will eventually converge to the same estimate. McKelvey and Page [20] extend this result to processes in which "survey results", rather than all the estimates, are repeatedly shared. The social network in these models is the *complete network*; indeed, it seems that non-trivial dynamics and results are achieved already for this (seemingly simple) topology. Aaronson [1] studies the complexity of the computations required of the agents, again with highly non-trivial results.

1.2   Rationality and Bounded Rationality

The term *rational* in economic theory refers to any behavior that maximizes (or even attempts to maximize) some utility function. This is in contrast to, for example, behavior that is heuristic or fixed. Bayesian rationality optimizes in a probabilistic framework that includes a prior and observations, and is, as mentioned above, a commonly used paradigm.

The disadvantage of fully rational, Bayesian models is that the calculations required of the agents can very quickly become intractable, making their applicability to real-world settings questionable; this tension between rationality and tractability is an old recurring theme in behavioral economics models (cf. [24]).

A solution often advocated is *bounded rationality*. Agents still act optimally in bounded rationality models, but only optimize with respect to a restricted set of choices. This usually simplifies the optimization problem that needs to be solved. For example, agents may be required to disregard some of their available information or be restricted in the manner that they calculate their strategy. In addition to serving the goal of more realistically modeling agents, a usual added benefit of bounded rationality is that the analysis of the model becomes easier. We too follow this course.

A standard assumption in this literature is that "actions speak louder than words" (cf. Smith and Sørensen [25]); agents do not participate in a communication protocol intended to optimize the exchange of information, but rather make inferences about each others' private information by observing actions. For example, by observing the price at which a person bids for a stock one may learn her estimate for the future price, but yet not learn all of the information which she used to arrive at this estimate.

1.3   Informal Statement of the Model

We consider a model where the underlying state S is not a constant number - as it is in all of the above mentioned models - but changes with time, as prices and other economic quantities tend to do. In particular we assume that the state $S = S(t)$ performs a random walk; $S(0)$ is picked from some distribution, and at each iteration an i.i.d. random variable is added to it.

The process commences with each agent having some estimator of $S(0)$. We make only very weak assumptions about the joint distribution of these estimators. Then, at each discrete time period t, each agent receives an independent (and identical over time) measurement of $S(t)$, and uses it to update its estimator. Also available to it are the previous *estimates* of its neighbors on a social network. Thus social network neighbors share their beliefs (or rather, observe each others' actions), and information propagates through the network.

While conceivably agents could optimally use all the information available to them to estimate the underlying state, it appears that such calculations are extremely complex. Instead, we explore *bounded rationality* dynamics, assuming that agents are restricted to calculating linear combinations of their observations. We note that if the random walk and the measurements are taken to be Gaussian, then the minimum variance unbiased linear estimator (MVULE) is also the maximum likelihood estimator. A Gaussian random walk is a good first-order approximation for many Economic processes (cf. classical work by Bachelier [5]).

For the first part of the paper, we also require that these linear combinations only involve the agents' neighbors' estimates from the previous time period (and not earlier), as well as their new measurement. In the last model we slightly relax this requirement.

We consider three models. In the *fixed response* model each agent, at each time period, estimates the underlying state by a *fixed* linear combination of its new measurement and the estimates of its neighbors in the previous period. This is a straightforward extension of the DeGroot model to our setting.

In the *best response model*, at each iteration, each agent calculates the MVULE of the underlying state, based on its peers' estimate from the previous round, together with its new measurement. We assume here that at each iteration the agents know the covariance matrix of their estimators. While this may seem like a strong assumption, we note that, under some elaboration of our model, this covariance matrix may be estimated by observing the process for some number of rounds before updating one's estimator. Furthermore, it seems that assumptions in this spirit - and often much stronger assumptions - are necessary in order for agents to perform any kind of optimization. For example, it is not rare in the literature of social Bayesian learning to assume that the agents know the structure of the entire social network graph (e.g., [13,23,2]).

Finally, we introduce the *penultimate prediction* model, which is a simple extension of the best response model, additionally allowing the agents to remember exactly one value from one round to the next. While only slightly increasing the computational requirements on the agents, this model exhibits a sharp increase in learning efficiency.

1.4   Informal Statement of Results

While our long term goal is to understand this process on general social network graphs, we focus in this paper on the *complete network*, which already exhibits mathematical richness.

We consider the system to be in a *steady state* when the covariance matrix of the agents' estimators is constant. On general graphs we show that fixed response dynamics converge to a steady state. On the complete graph we show that best response dynamics also converge to a steady state. Both of these results hold regardless of the initial conditions (i.e., the agents' estimators at time $t = 0$).

We show that the steady state of best response dynamics is not necessarily optimal; there exist fixed response dynamics in which the agents converge to estimators which all have lower variance then the estimators of the steady state of best response dynamics. This shows that every agent can do better than the result of best response by following a socially optimal rule; thus a certain *price of anarchy* is to be paid when agents choose the action that maximizes their short term gain.

Finally, we show that in the penultimate prediction model, for the complete graph, the agents learn estimators which are the optimal (in the minimum variance sense) amongst all linear estimators, and thus outperform those of fixed and best response dynamics.

We define a notion of "socially asymptotic learning": A model has this property when the variance of the agents' steady-state estimators tends towards the information-theoretical optimum with the number of agents. We show that the penultimate prediction model exhibits socially asymptotic learning on the complete graph, while best response and fixed response dynamics fail to do the same.

Due to space constraints, the full proofs have been omitted. For this reason, we direct the interested reader to the full version of the paper [12].

2   Previous Work

Our model is an elaboration of models studied by DeMarzo, Vayanos and Zwiebel [11], as well as Mossel and Tamuz [22,21]. There, the state S is a fixed number picked at time $t = 0$, and each agent receives a single measurement of it. The process thereafter is deterministic, with each agent, at each iteration, recalculating its estimate of S based on its observation of its neighbors' estimates.

In [21] it is shown that if the agents calculate the minimum variance unbiased linear estimator (MVULE) at every turn (remembering all of their observations) then all the agents converge to the optimal estimator of S, i.e. the average of the original measurements. Furthermore, this happens in time that is at most $n \cdot d$, where d is the diameter of the graph.

When agents calculate estimates that are only based on their observations from the previous round, then they do not necessarily converge to the optimal estimator [22]. In fact, it is not known whether they converge at all.

A similar model is studied by Jadbabaie, Sandroni and Tahbaz-Salehi [17]. They explore a bounded rationality setting in which agents receive new signals at each iteration. An agent's private signals may be informative only when combined with those of other agents, and yet their model achieves efficient learning.

Our model is a special case of a model studied by Acemoglu, Nedic, and Ozdaglar [3]. They extend these models by allowing the state to change from period to period. They don't require the change in the state to be i.i.d, but only to have zero mean and be independent in time. Their agents also receive a new, independent measurement of the state at every period, which again need not be identically distributed. They focus on a different regime than the one we study; their main result is a proof of convergence in the case that the variations in the state diminish with time, with variance tending to zero.

In our model the change in the underlying state has constant variance, as does the agents' measurement noise. This allows us to explore steady states, in which the covariance matrix of the agents' estimators does not change from iteration to iteration.

Our model is non-trivial already for a single agent, although here a complete solution is simple, and can be calculated using tools developed for the analysis of *Kalman filters* [18].

3   Notation, Formal Models, and Results

Let $[n] = \{1, 2, \ldots, n\}$ be a set of agents. Let $G = ([n], E)$ be a directed graph representing the agents' social network. We denote by $\partial i = \{j | (i, j) \in E\}$ the neighbors of i, and assume that always $i \in \partial i$.

We consider discrete time periods $t \in \{0, 1, \ldots\}$. The *underlying state of the world* at time t, $S(t)$, is defined as follows. $S(0)$ is a real random variable with arbitrary distribution, and for $t > 0$

$$S(t) = S(t-1) + X(t-1), \tag{1}$$

where $\mathbb{E}[X(t)] = 0$, $\mathrm{Var}[X(t)] = \sigma^2$, and σ is a parameter of the model. The random variables $X(0), X(1), \ldots$ are independent. Hence the underlying state $S(t)$ performs a random walk with zero mean and standard deviation σ.

At time $t = 0$ each agent i receives $Y_i(0)$, an estimator of $S(0)$. The only assumptions we make on their joint distribution is that $\mathbb{E}[Y_i(0)|S(0)] = S(0)$, i.e. the estimators are unbiased, and that $\mathrm{Var}[Y_i(0) - S(0)]$ is finite for all i.

At each subsequent period $t > 0$, each agent i receives $M_i(t)$, an independent measurement of $S(t)$, defined by

$$M_i(t) = S(t) + D_i(t), \tag{2}$$

where $\mathbb{E}\left[D_i(t)\right] = 0$, $\mathrm{Var}\left[D_i(t)\right] = \tau_i^2$, and the τ_i's are parameters of the model. Hence $D_i(t)$ is the measurement error of agent i at time t. Again, the random variables $D_i(t)$ are independent.

At each period $t > 0$, each agent i calculates $Y_i(t)$, agent i's estimate of $S(t)$, using the information available to it. Precisely what information is available varies by the model (and is defined below), but in all cases $Y_i(t)$ is a (deterministic) convex linear combination of agent i's measurements up to and including time t, $\{M_i(t')|t' \leq t\}$, as well as the previous *estimates* of its social network neighbors, $\{Y_j(t')|t' < t, j \in \partial i\}$. Additionally, in the penultimate prediction model, at each round t each agent computes a value $R_i(t)$, and at round $t+1$ uses this value to compute $R_i(t+1)$ and $Y_i(t+1)$. Like $Y_i(t)$, $R_i(t)$ is also a convex linear combination of the same random variables.

In general, we shall assume that the agents are interested in minimizing the expected squared error of their estimators, $\mathbb{E}\left[(Y_i(t) - S(t))^2\right]$; assuming $Y_i(t)$ is unbiased (i.e., $\mathbb{E}\left[Y_i(t)|S(t)\right] = S(t)$) this is equivalent to minimizing $\mathrm{Var}\left[Y_i(t) - S(t)\right]$, which we refer to as the *"variance of the estimator $Y_i(t)$."* We shall assume throughout that the estimators $Y_i(t)$ are indeed unbiased; we elaborate on this in the definitions of the models below.

We shall (mostly) restrict ourselves to the case where the agents use only their neighbors' estimates from the previous iteration, and not from the ones before it. In these cases we write

$$Y_i(t) = A_i(t)M_i(t) + \sum_j P_{ij}(t)Y_j(t - 1). \tag{3}$$

for some $A_i(t)$ and $P_{ij}(t)$ such that $P_{ij} = 0$ whenever $j \notin \partial i$.

We will find it convenient to express such quantities in matrix form. To that end we let $\mathbf{m}(t), \mathbf{y}(t), \mathbf{d}(t) \in \mathbb{R}^n$ be column vectors with entries $M_i(t), Y_i(t), D_i(t)$, and let $\mathbf{P}(t), \mathbf{A}(t), \mathbf{T} \in \mathbb{R}^{n \times n}$ be the weight matrices, with $\mathbf{P} = (P_{ij})_{ij}$, $\mathbf{A}(t) = \mathrm{Diag}(A_1(t), \ldots, A_n(t))$, and $\mathbf{T} = \mathrm{Var}\left[\mathbf{d}(t)\right] = \mathrm{Diag}(\tau_1^2, \ldots, \tau_n^2)$. Using this notation Eq. (3) becomes

$$\mathbf{y}(t) = \mathbf{A}(t)\mathbf{m}(t) + \mathbf{P}(t)\mathbf{y}(t - 1). \tag{4}$$

We will also make use of the *covariance matrix*

$$\mathbf{C}(t) = \mathrm{Var}\left[\mathbf{y}(t) - \mathbf{1}S(t)\right], \tag{5}$$

where $\mathbf{1} \in \mathbb{R}^n$ denotes the column vector of all ones. Hence, we have $C_{ij}(t) = \mathrm{Cov}\left[Y_i(t) - S(t), Y_j(t) - S(t)\right]$, which we refer to as the *"covariance of the estimators $Y_i(t)$ and $Y_j(t)$."*

3.1 Dynamics Models

Best Response. The main model we study is the best response dynamics. Here we assume that at round t, each agent i has access to $M_i(t)$, $\mathbf{y}(t-1)$ and the covariance matrix for these values. At each iteration t, agent i picks $A_i(t)$ and $\{P_{ij}(t)\}_j$ that minimize $C_{ii}(t) = \text{Var}\,[Y_i(t) - S(t)]$, under the constraints that (a) $P_{ij}(t)$ may be non-zero only if $j \in \partial i$, and (b) $\mathbb{E}\,[Y_i(t)|S(t)] = S(t)$, i.e. $Y_i(t)$ is an unbiased estimator of $S(t)$. We show that these minimizing coefficients are a deterministic function of $\mathbf{C}(t-1)$, σ and $\{\tau_i\}$. Hence we assume here that the agents know these values. By this definition $Y_i(t)$ is the *minimum variance unbiased linear estimator* (MVULE) of $S(t)$, given $M_i(t)$ and $\mathbf{y}(t-1)$.

Note that it follows from our definitions that if the estimators $\{Y_i(t-1)\}$ at time $t-1$ are unbiased then, in order for the estimators at time t to be unbiased, it must be the case that

$$A_i(t) + \sum_j P_{ij}(t) = 1. \tag{6}$$

Since at time zero the estimators are unbiased then it follows by induction that Eq. (6) hold for all $t > 0$.

Fixed Response. We shall also consider the case of estimators which are *fixed* linear combinations of the agent's new measurement $M_i(t)$ and its neighbors' estimators at time $t-1$. These we call fixed response estimators. In this case we would have, using our matrix notation:

$$\mathbf{y}(t) = \mathbf{A}\mathbf{m}(t) + \mathbf{P}\mathbf{y}(t-1). \tag{7}$$

The matrices $\mathbf{A}$ and $\mathbf{P}$ are arbitrary matrices that satisfy the following conditions: (a) P_{ij} is positive and non-zero only if $j \in \partial i$, and (b) $\mathbf{y}_i(t)$ is a convex linear combination of $M_i(t)$ and $\{Y_j(t-1)\}_j$. Equivalently, $\mathbf{A}_i + \sum_j P_{ij} = 1$, which is the same condition described in Equation (6).

Penultimate Prediction. Finally, we consider the penultimate prediction model where each agent i can remember one value, which we denote $R_i(t)$, from one round t to the next round $t+1$. We assume that at round t, each agent i has access to $M_i(t)$, $\mathbf{y}(t-1)$, $R_i(t-1)$ and the covariance matrix for these values. We denote $\mathbf{r}(t) = (R_1(t), \dots, R_n(t))$.

We fix $R_i(0) = 0$, and let $R_i(t)$ be agent i's MVULE of $S(t-1)$, given $R_i(t-1)$ and $\mathbf{y}(t-1)$ (note that this is in general *not* equal to $Y_i(t-1)$). $Y_i(t)$ now becomes the MVULE of $S(t)$ given $R_i(t)$ and $M_i(t)$.

3.2 Steady States and Efficient Learning

We say that the system converges to a *steady state* $\mathbf{C}$ when

$$\lim_{t \to \infty} \mathbf{C}(t) = \mathbf{C}.$$

Assuming that agents are constrained to calculating linear combinations of their measurements and neighbors' estimators, the variance of the estimators $Y_i(t)$ of $S(t)$ at time t can be bounded from below by the variance of $Z_i(t)$, where we define $Z_i(t)$ to be the MVULE of $S(t)$ given the initial estimators $\mathbf{y}(0)$, all measurements up to time $t-1$ $\{M_j(s)|j \in [n], s < t\}$ and $M_i(t)$. We therefore define that a process achieves *perfect learning* when $\operatorname{Var}[Y_i(t) - S(t)] = \operatorname{Var}[Z_i(t) - S(t)]$. Note that this definition is a natural one for the complete graph and should be altered for general networks, where a tighter lower bound exists.

If an agent were to know $S(t-1)$ exactly at time t, then, together with $M_i(t)$, its minimum variance unbiased linear estimator for $S(t)$ would be a linear combination of just $S(t-1)$ and $M_i(t)$, because of the Markov property of $S(t)$. In this case it is easy to show that $C_{ii}(t) = \operatorname{Var}[Y_i(t) - S(t)]$ would equal $\sigma^2 \tau_i^2 / (\sigma^2 + \tau_i^2)$. We say that a model achieves *socially asymptotic learning* if for n sufficiently large, as the number of agents tends to infinity, the steady state $\mathbf{C}$ exists and C_{ii} tends to $\sigma^2 \tau_i^2 / (\sigma^2 + \tau_i^2)$ for all i. We stress that this definition only makes sense in models where the number of agents n grows to infinity and therefore is incomparable to perfect learning, which is defined for a particular graph.

4   Statement of the Main Results

The following are our main results. Let $\beta(t) = 1/(\mathbf{1}^\top \mathbf{C}(t)^{-1} \mathbf{1})$.

Theorem 1. *When G is a complete graph, best-response dynamics converge to a unique steady-state, for all starting estimators $\mathbf{y}(0)$ and all choices of parameters $\{\tau_i\}$ and σ. Moreover, the convergence is fast, in the sense that $-\log|\beta(t) - \beta^*| = O(t)$, where $\beta^* = \lim_{t \to \infty} \beta(t)$.*

Theorem 2. *In fixed response dynamics, if $A_i > 0$ for all $i \in [n]$ then system converges to a steady state $\mathbf{C} = \lim_{t \to \infty} \mathbf{C}(t)$ such that*

$$\mathbf{C} = \mathbf{A}^2 \mathbf{T} + \sigma^2 \mathbf{P} \mathbf{1} \mathbf{1}^\top \mathbf{P}^\top + \mathbf{P} \mathbf{C} \mathbf{P}^\top. \tag{8}$$

In particular, $\mathbf{C}$ is independent of the starting estimators $\mathbf{y}(0)$.

Theorem 3. *Let G be a graph with $[n]$ vertices. Fix σ and $\{\tau_i\}_{i \in [n]}$.*

Consider best response dynamics for n agents on G with σ and $\{\tau_i\}_{i \in [n]}$. Let $\mathbf{C}^{br}$ denote the steady state the system converges to.

Consider fixed response dynamics with some $\mathbf{P}$ and $\mathbf{A}$ for n agents on G with σ and $\{\tau_i\}_{i \in [n]}$. Let $\mathbf{C}^{fr}$ denote the steady state the system converges to.

Then there exists a choice of n, G, σ, $\{\tau_i\}$, $\mathbf{A}$ and $\mathbf{P}$ such that $C_{ii}^{br} > C_{ii}^{fr}$ for all $i \in [n]$.

Theorem 4. *If $\sigma, \tau > 0$, no fixed response dynamics can achieve socially asymptotic learning.*

Theorem 5. *Penultimate prediction on the complete graph achieves perfect learning.*

5   Preliminary Analysis

We commence by proving a preliminary proposition on the relation between the coefficients matrices $\mathbf{P}(t)$ and $\mathbf{A}(t)$, and the covariance matrix $\mathbf{C}(t)$ in the best response and fixed response models. This result does not depend on how $\mathbf{P}(t)$ and $\mathbf{A}(t)$ are calculated, and therefore applies to both models.

First, let us calculate the covariance matrix directly. By the definition of $\mathbf{C}(t)$ and by Eq. (4) we have that

$$\mathbf{C}(t) = \mathrm{Var}\left[\mathbf{y}(t) - \mathbf{1}S(t)\right] = \mathrm{Var}\left[\mathbf{A}(t)\mathbf{m}(t) + \mathbf{P}(t)\mathbf{y}(t-1) - \mathbf{1}S(t)\right].$$

Since $S(t) = S(t-1) + X(t-1)$ then we can write

$$\mathbf{C}(t) = \mathrm{Var}\Big[\mathbf{A}(t)\big(\mathbf{m}(t) - \mathbf{1}S(t)\big) + \mathbf{P}(t)\mathbf{y}(t-1)$$
$$- (\mathbf{I} - \mathbf{A}(t))\mathbf{1}\big(S(t-1) + X(t-1)\big)\Big].$$

Since the estimators $\{Y_i(t)\}$ are unbiased then $\mathbf{P}(t)\mathbf{1} = (\mathbf{I} - \mathbf{A}(t))\mathbf{1}$; see the definitions of the models in Section 3.1, and in particular Eq. (6). Hence

$$\mathbf{C}(t) = \mathbf{A}(t)\mathbf{T}\,\mathbf{A}(t)^\top + \mathrm{Var}\left[\mathbf{P}(t)(\mathbf{y}(t-1) - \mathbf{1}S(t-1))\right] + \mathrm{Var}\left[\mathbf{P}(t)\mathbf{1}X(t-1)\right],$$

since $\mathrm{Var}\left[\mathbf{m}(t) - \mathbf{1}S(t)\right] = \mathrm{Var}\left[\mathbf{d}(t)\right] = \mathbf{T}$. Finally, since $\mathrm{Var}\left[\mathbf{y}(t-1)\right] = \mathbf{C}(t-1)$ we can write

$$\mathbf{C}(t) = \mathbf{A}(t)^2\mathbf{T} + \mathbf{P}(t)\mathbf{C}(t-1)\mathbf{P}(t)^\top + \sigma^2\mathbf{P}(t)\mathbf{1}\mathbf{1}^\top\mathbf{P}(t)^\top. \tag{9}$$

5.1   Understanding Best-Response Dynamics

The condition that estimators are unbiased, or $A_i(t) + \sum_j P_{ij}(t) = 1$, means that given $\{P_{ij}(t)\}_j$ one can calculate $A_i(t)$, or alternatively given $\mathbf{P}$ one can calculate $\mathbf{A}$. Hence, fixing σ and $\{\tau_j\}$, $\mathbf{P}(t)$ is a deterministic function of $\mathbf{C}(t-1)$. Since by Eq. (9) $\mathbf{C}(t)$ is a function of $\mathbf{A}(t)$, $\mathbf{P}(t)$ and $\mathbf{C}(t-1)$, then under best response dynamics, $\mathbf{C}(t)$ is in fact a function of $\mathbf{C}(t-1)$. We will denote this function by F, so that $\mathbf{C}(t) = F(\mathbf{C}(t-1))$. Our goal is to understand this map F, and in particular to determine its limiting behavior.

We next analyze in more detail the best response calculation for agent i. This can conceptually be divided into two stages: calculating a best estimator for $S(t)$ from $\mathbf{y}(t-1)$, and then combining that with $M_i(t)$ for a new estimator of $S(t)$.

Let the vector $\mathbf{y}_{\partial i}(t-1) = \{Y_j(t-1) | j \in \partial i\}$ and let $\mathbf{C}_i(t-1) = C_{\partial i, \partial i}(t-1)$ be the covariance matrix of the estimators of the neighbors of agent i.

Denote by $\mathbf{q}_i(t)$ the vector of coefficients for $\mathbf{y}_{\partial i}(t-1)$ that make $Z_i = \mathbf{q}_i(t)^\top \mathbf{y}_{\partial i}(t-1)$ a minimum variance unbiased linear estimator for $S(t)$; note that this is also the estimator for $S(t-1)$. Then we have

$$\mathbf{q}_i(t) = \beta_i(t-1)\mathbf{1}^\top\mathbf{C}_i(t-1)^{-1},$$

where $\beta_i(t-1) = 1/\mathbf{1}^\top\mathbf{C}_i(t-1)^{-1}\mathbf{1}$. It is easy to see that $\mathrm{Var}\left[Z_i - S(t-1)\right] = \beta_i(t-1)$ and thus $\mathrm{Var}\left[Z_i - S(t)\right] = \beta_i(t-1) + \sigma^2$.

$M_i(t)$ is an independent estimator of $S(t)$ with variance τ_i^2. To combine it optimally with Z_i we set

$$A_i(t) = \frac{\beta_i(t-1) + \sigma^2}{\tau_i^2 + \beta_i(t-1) + \sigma^2} \geq \frac{\sigma^2}{\tau_i^2 + \sigma^2}. \tag{10}$$

The optimal weight vector $\mathbf{p}_i(t)$ for agent i (i.e., $\{P_{ij}\}_{j \in \partial i}$) is therefore $\mathbf{p}_i(t) = (1 - A_i(t))\mathbf{q}_i(t)$.

5.2   Complete Graph Case

When G is the complete graph, the agents best-respond similarly, since they all observe the same set of estimators from the previous iteration. We now have $\mathbf{C}_i(t-1) = \mathbf{C}(t-1)$, $\mathbf{q}_i(t) = \mathbf{q}(t)$, and $\beta_i(t-1) = \beta(t-1)$, for all i. For the moment, we will suppress the t. Letting $\boldsymbol{a}$ be the vector with coefficients A_i, we then have $\mathbf{P} = (\mathbf{1} - \boldsymbol{a})\mathbf{q}^\top = \beta(\mathbf{1} - \boldsymbol{a})\mathbf{1}^\top\mathbf{C}^{-1}$. Using this form for $\mathbf{P}$, we can now see that $\mathbf{PCP}^\top = \beta(\mathbf{1} - \boldsymbol{a})(\mathbf{1} - \boldsymbol{a})^\top$. Putting this all together, and adding back the t, we have by Eq. (9) that

$$\mathbf{C}(t) = \mathbf{A}(t)^2\,\mathbf{T} + (\beta(t-1) + \sigma^2)(\mathbf{1} - \boldsymbol{a}(t))(\mathbf{1} - \boldsymbol{a}(t))^\top. \tag{11}$$

Since by equation (10), $A_i(t)$ depends only on $\beta(t-1)$, τ_i, and σ, we see that $\mathbf{C}(t) = F(\mathbf{C}(t-1))$ depends on $\mathbf{C}(t-1)$ only through $\beta(t-1) = 1/\mathbf{1}^\top\mathbf{C}(t-1)^{-1}\mathbf{1}$. Hence we can write $\mathbf{C}(t) = \mathbf{C}(\beta(t-1))$. We now see that we can completely describe the state of the system by a single parameter β, and our map F reduces to the map $f : \beta \mapsto 1/\mathbf{1}^\top\mathbf{C}(\beta)^{-1}\mathbf{1}$. We wish to analyze this function f as a single-parameter discrete dynamical system.

6   Proofs

Due to space restrictions we omit the full proofs of our results and merely provide brief sketches here. The full proofs may be found in the full version of our paper [12] and we encourage the interested reader to look there.

Theorem 1.   The first insight is showing that the best-response dynamics are captured by a single-parameter discrete dynamical system $f(\beta)$, depending only on the variance β of everyone's estimate of $S(t-1)$ after $\mathbf{y}(t-1)$ is revealed. This we outline in Section 5.2. We then bound the derivative $|f'(\beta)| < 1$ and apply the Banach Fixed Point Theorem, giving us both the unique fixed point and the convergence rate. ∎

Theorem 2.   We use equation (9), which simplifies to (8) in the fixed setting. From there we note that $\|\mathbf{P}\|_\infty < 1$ and use properties of matrix norms to show that $\mathbf{C}(t)$ is like a geometric series, and thus converges.

Theorem 3.   We examine two examples: the complete graph with 2 nodes and the complete graph where the number of nodes tends towards infinity. In both examples $\sigma = 1 = \tau_i$ for all i. Calculations confirm that in each of these examples, the optimal symmetric fixed-response dynamics results in lower variance estimators than the best response dynamics for every agent.

Theorem 4. We proceed by showing that, assuming some fixed response dynamics is socially optimal, A_i – the weight that agent i gives itself – must converge to a particular fixed value that depends only on σ and τ_i. However if A_i is fixed, it can be shown that two consecutive rounds cannot be optimal.

Theorem 5. We show that the updates of the agents in the penultimate prediction model simulate the calculations of the Kalman Filter for the penultimate round, which are known to be optimal.

7   Conclusion

This work can be seen as a study of natural extensions of the DeGroot model to the setting where the value to be learned changes over time. The most direct extension is the *fixed response* model. Here we show that while the estimate will keep moving with the true values, its variance will converge to a fixed value. However, in contrast to the DeGroot model, the agents are continually receiving new independent signals, and so have a reference point from which to evaluate the validity of their neighbors' signals. This leads us to propose the *best response* model. We show that in the case of the complete graph, best response dynamics will always converge to a particular fixed response that is (myopically) optimal. However, we also show that it is not necessarily Pareto optimal amongst all fixed responses. Finally, we show that a simple strengthening of the model to allow agents to remember one value can, in certain cases, lead to much improved performance. This can be seen not only as a critique of fixed response dynamics as being too weak to capture natural dynamics, but also as an interesting model to be studied more in its own right.

References

1. Aaronson, S.: The complexity of agreement. In: Proceedings of the Thirty-Seventh Annual ACM Symposium on Theory of Computing, STOC 2005, pp. 634–643. ACM (2005)
2. Acemoglu, D., Dahleh, M., Lobel, I., Ozdaglar, A.: Bayesian learning in social networks (2008)
3. Acemoglu, D., Nedic, A., Ozdaglar, A.: Convergence of rule-of-thumb learning rules in social networks. In: 47th IEEE Conference on Decision and Control, CDC 2008, pp. 1714–1720. IEEE (2008)
4. Aumann, R.J.: Agreeing to disagree. The Annals of Statistics 4(6), 1236–1239 (1976)
5. Bachelier, L.: Théorie de la spéculation. Gauthier-Villars (1900)
6. Bala, V., Goyal, S.: Learning from neighbours. Review of Economic Studies 65(3), 595–621 (1998),
http://ideas.repec.org/a/bla/restud/v65y1998i3p595-621.html

7. Bandiera, O., Rasul, I.: Social networks and technology adoption in northern mozambique*. The Economic Journal 116(514), 869–902 (2006)
8. Besley, T., Case, A.: Diffusion as a learning process: Evidence from hyv cotton (1994) Working Papers
9. Conley, T., Udry, C.: Social learning through networks: The adoption of new agricultural technologies in ghana. American Journal of Agricultural Economics 83(3), 668–673 (2001)
10. DeGroot, M.H.: Reaching a consensus. Journal of the American Statistical Association, 118–121 (1974)
11. DeMarzo, P., Vayanos, D., Zwiebel, J.: Persuasion bias, social influence, and unidimensional opinions. Quarterly Journal of Economics 118, 909–968 (2003)
12. Frongillo, R.M., Schoenebeck, G., Tamuz, O.: Social learning in a changing world. Tech. rep. (September 2011), http://arxiv.org/abs/1109.5482
13. Gale, D., Kariv, S.: Bayesian learning in social networks. Games and Economic Behavior 45(2), 329–346 (2003),
http://ideas.repec.org/a/eee/gamebe/v45y2003i2p329-346.html
14. Geanakoplos, J.: Common knowledge. In: Proceedings of the 4th Conference on Theoretical Aspects of Reasoning about Knowledge, pp. 254–315. Morgan Kaufmann Publishers Inc. (1992)
15. Geanakoplos, J.D., Polemarchakis, H.M.: We can't disagree forever* 1. Journal of Economic Theory 28(1), 192–200 (1982)
16. Jackson, M.O.: The economics of social networks. In: Blundell, R., Newey, W., Persson, T. (eds.) Theory and Applications: Ninth World Congress of the Econometric Society. Advances in Economics and Econometrics, vol. I, pp. 1–56. Cambridge University Press (2006)
17. Jadbabaie, A., Sandroni, A., Tahbaz-Salehi, A.: Non-bayesian social learning, second version. Pier working paper archive, Penn Institute for Economic Research, Department of Economics. University of Pennsylvania (2010)
18. Kalman, R., et al.: A new approach to linear filtering and prediction problems. Journal of basic Engineering 82(1), 35–45 (1960)
19. Kohler, H.P.: Learning in social networks and contraceptive choice. Demography 34(3), 369–383 (1997)
20. McKelvey, R.D., Page, T.: Common knowledge, consensus, and aggregate information. Econometrica: Journal of the Econometric Society, 109–127 (1986)
21. Mossel, E., Tamuz, O.: Efficient bayesian learning in social networks with gaussian estimators. Tech. rep. (September 2010), http://arxiv.org/abs/1002.0747
22. Mossel, E., Tamuz, O.: Iterative maximum likelihood on networks. Advances in Applied Mathematics 45(1), 36–49 (2010)
23. Parikh, R., Krasucki, P.: Communication, consensus, and knowledge* 1. Journal of Economic Theory 52(1), 178–189 (1990)
24. Simon, H.: Reason in Human Affairs. Stanford University Press (1982)
25. Smith, L., Sørensen, P.: Pathological outcomes of observational learning. Econometrica 68(2), 371–398 (2000)

Budget-Balanced and Nearly Efficient Randomized Mechanisms: Public Goods and beyond*

Mingyu Guo[1], Victor Naroditskiy[2], Vincent Conitzer[3], Amy Greenwald[4], and Nicholas R. Jennings[2]

[1] Computer Science Department, University of Liverpool, UK
Mingyu.Guo@liverpool.ac.uk
[2] School of Electronics and Computer Science, University of Southampton, UK
{vn,nrj}@ecs.soton.ac.uk
[3] Department of Computer Science, Duke University, USA
conitzer@cs.duke.edu
[4] Department of Computer Science, Brown University, USA
amy@cs.brown.edu

Abstract. Many scenarios where participants hold private information require payments to encourage truthful revelation. Some of these scenarios have no natural *residual claimant* who would absorb the budget surplus or cover the deficit. Faltings [7] proposed the idea of excluding one agent uniformly at random and making him the residual claimant. Based on this idea, we propose two classes of public good mechanisms and derive optimal ones within each class: Faltings' mechanism is optimal in one of the classes. We then move on to general mechanism design settings, where we prove guarantees on the social welfare achieved by Faltings' mechanism. Finally, we analyze a modification of the mechanism where budget balance is achieved without designating any agent as the residual claimant.

1 Introduction

Many scenarios where participants hold private information require payments to encourage truthful revelation. Some of these scenarios have no natural *residual claimant* who would absorb the budget surplus or cover the deficit (e.g., a group of roommates deciding who gets to use the living room for a weekend party or a company distributing free football tickets among employees). Mechanisms with budget deficit are not very compelling as they require a subsidy. In more compelling surplus-generating (or, weakly budget-balanced) mechanisms, the surplus represents a loss in social welfare (i.e., the sum of the agents' utilities), which can be viewed as the cost of truthfulness. A number of recent papers have investigated what the minimum budget surplus is that still supports truthful reporting and efficient outcomes [14,11,12,4,1,2]. While weak budget

* Part of the work was performed while the first four authors were at Centrum Wiskunde & Informatica (CWI) in Amsterdam. Conitzer gratefully acknowledges NSF IIS-0812113, IIS-0953756, and CCF-1101659, as well as an Alfred P. Sloan fellowship, for support. Naroditskiy and Jennings gratefully acknowledge funding from the UK Research Council for project 'Orchid', grant EP/I011587/1. Greenwald gratefully acknowledges NSF Grant CCF-0905234.

N. Chen, E. Elkind, and E. Koutsoupias (Eds.): WINE 2011, LNCS 7090, pp. 158–169, 2011.
© Springer-Verlag Berlin Heidelberg 2011

balance is a necessary assumption,[1] efficiency is not. In fact, sacrificing efficiency leads to a higher social welfare in certain cases (by having significantly lower net payments than efficient mechanisms) [7,10,5].

The mechanisms we propose here are budget-balanced (i.e., no loss of social welfare is due to the budget surplus) but not efficient. Our work starts with the idea behind Faltings' mechanism [7], which is that we exclude one agent uniformly at random and make him the residual claimant of the payments collected by an efficient mechanism (e.g., the VCG mechanism) in the market with only the remaining agents. Crucially, in order to maintain truthfulness, the outcome must be chosen without considering the private value of the excluded agent. Thus, the chosen outcome may not be the same as the efficient outcome when all agents' values are considered. This results in a social welfare below the value of the efficient outcome. Notice that the loss of social welfare is due only to the non-efficiency of the outcome as mechanisms with a residual claimant are budget-balanced. Since excluding one agent at random results in a randomized outcome function, we speak of expected social welfare. We say that a mechanism is r-competitive if its expected social welfare is at least r of the value of the efficient outcome for all types the agents may have: i.e., we are using a worst-case metric.

We apply the approach of excluding one agent to the public project scenario where a group of agents needs to decide whether or not to build, say, a bridge that comes at a publicly known cost. The public project scenario is fundamental to mechanism design: unlike allocation scenarios, no agent can be excluded from enjoying the benefits of the project if it is undertaken. While maximizing social welfare has been studied extensively in allocation scenarios (see e.g., [14,11,9,10,5]), public good scenarios received relatively less attention. [1] and [2] both studied the problem of designing welfare-maximizing public good mechanisms. [1] studied one dominance relationship between mechanisms, but did not propose any specific mechanisms. [2] studied sequential public good mechanisms with a different notion of truthfulness. Then, there are several general mechanisms that can be applied to public project [3,4,7]. The mechanism described in [3] is not budget-balanced when applied to public project, while the mechanism [4] has a zero competitive ratio. The paper by Faltings [7] is central to many of our results as we discuss next.

First, we derive a competitive ratio for the mechanism proposed by Faltings and prove its optimality within a class of mechanisms. Specifically, we define a class of mechanisms based on the fraction of the cost of the project passed on to the excluded agent. It turns out all mechanisms that assume the excluded agent would cover up to $\frac{1}{n-1}$ of the cost are $\frac{n-1}{n}$-competitive (n is the number of agents). Faltings' mechanism corresponds to the excluded agent covering $\frac{1}{n}$ of the cost and is optimal within the class. Mechanisms that assume the excluded agent covers more than $\frac{1}{n-1}$ of the cost are not $\frac{n-1}{n}$-competitive. A natural question is whether a better mechanism is possible. To this end, we consider a larger mechanism class by taking the mixtures over the above mechanisms. That is, we consider mechanisms that assume the excluded agent would cover a randomized proportion of the cost. We characterize one optimal mechanism within this larger class, which turns out to be $\frac{n}{n+1}$-competitive.

---

[1] An arbitrary social welfare can be achieved when unlimited subsidies are allowed.

The mechanisms above make the excluded agent the recipient of the VCG payments computed without him. The idea of computing VCG payments after excluding one agent has also been used in general quasi-linear domains to design redistribution of VCG payments computed with all agents present. Specifically, in a regular VCG mechanism, the rebate to agent i can be set to $\frac{1}{n}$ of the VCG payments collected in the market without him [3]. The resulting mechanism is efficient but not budget-balanced, and may run a deficit in the public good scenario.

We find that using this redistribution idea together with the inefficient allocation, made after excluding one agent, leads to a budget-balanced mechanism that *does not designate any agent as the residual claimant*. This results in a more fair treatment of all agents, and we call the mechanism *FaltingsFair*. In more detail, we set each agent's payment to be the expected VCG payment he would make after one of the other agents is excluded uniformly at random. This payment is reduced by the rebate described above. The sum of the rebates cancels out the sum of the payments, thus achieving budget balance. Interestingly, this mechanism was already proposed by Faltings in extended versions of his work [6,8], though without the redistribution interpretation. Our analysis sheds new light on this mechanism establishing connections to a standard redistribution function and providing novel proofs.

The rest of this paper is structured as follows. A general model of mechanism design problems is stated in Section 2. Mechanisms with a residual claimant for the public good scenario are studied in Section 3. There we propose two classes of mechanisms and derive optimal ones within each class. In Section 4, we move on to general mechanism design settings. One of the optimal public good mechanisms we derive in Section 3 turns out to be a special case of Faltings' mechanism. We modify this mechanism to remove the residual claimant, which results in the budget-balanced *FaltingsFair* mechanism. Discussion of the results appears in Section 5.

2 Model

The set of agents is denoted by N ($|N| \geq 3$) and the private type of agent $i \in N$ is given by θ_i. The mechanism chooses an outcome $k(\theta')$ from the set of possible outcomes K, based on the profile of reported types θ'. The value of an agent for each outcome depends on his type $v_i(k(\theta'), \theta_i)$, and the utility is quasi-linear. Given an outcome $k \in K$ and a payment $t_i \in \mathbb{R}$, the utility is $u_i(k, t_i, \theta_i) = v_i(k, \theta_i) - t_i$. Let $k^*(\theta)$ denote the *efficient* outcome $k^*(\theta) \in \arg\max_{k' \in K} \sum_i v_i(k', \theta_i)$.

The VCG (also known as Clarke or pivotal) mechanism is defined by the efficient outcome and the following payments *from* the agents: $t_i^{\text{vcg}}(\theta) = \sum_{j \neq i} v_j(k^*(\theta_{-i}), \theta_j) - \sum_{j \neq i} v_j(k^*(\theta), \theta_j)$, where $k^*(\theta_{-i}) \in \arg\max_{k' \in K} \sum_{j \neq i} v_j(k', \theta_j)$.

3 Public Project

In a public project (equivalently, public good) problem, a group of agents needs to decide whether or not to undertake a project such as building a bridge. The two possible outcomes are: do not build the bridge and distribute C among the agents or build the bridge spending C on its construction. Each agent has a private value θ_i for having the bridge built. We define the value of the efficient outcome as $\max(\theta_N, C)$: the sum of agents' values is $\theta_N = \sum_{i \in N} \theta_i$ when the bridge is built and C when it is not built. The valuation function of agent i consistent with this definition of social welfare is

$$v_i(k(\theta), \theta_i) = \begin{cases} \theta_i & \text{if } k(\theta) = 1 \\ \frac{C}{n} & \text{otherwise} \end{cases} \tag{1}$$

Faltings' mechanism [7] is defined as follows (we will call it *Faltings* from now on):

- We exclude one agent uniformly at random.
- The remaining agents use the VCG mechanism to come up with an optimal allocation for themselves.
- The excluded agent acts as the residual claimant. That is, the VCG payments are redistributed to the excluded agent, to achieve budget balance.

Faltings is known to be (dominant-strategy) incentive compatible and budget-balanced.[2] *Faltings* can be generalized to the following class of mechanisms (also incentive compatible and budget-balanced):

- We pick one agent, denoted by a, uniformly at random, and we pretend agent a's reported type is $C - x$ (ignoring what a actually reported).
- All agents, **including** a, participate in a VCG mechanism.
- a acts as the residual claimant. That is, everyone excluding a pays his VCG payment to a. (a does not have to make any payment. Note that incentive compatibility for a is guaranteed because a's report is ignored altogether.)

Mechanisms inside the above class are characterized by the parameter x, where x represents how much the non-excluded agents need to value the project in order for it to be built. When there is no ambiguity, we will simply use mechanism x to refer to the mechanism inside the class that is characterized by x. *Faltings* corresponds to $x = \frac{n-1}{n}C$: For this value of x, the decision is to build if and only if the remaining agents' total valuation is at least $\frac{n-1}{n}C$, which is efficient for the remaining agents.

The parameter x could take any value in $(-\infty, \infty)$, but we only need to consider $x \in [0, C]$ (assuming non-negative types). We recall that x represents how much the non-excluded agents need to value the project in order for it to be built. When $x < 0$, mechanism x is equivalent to mechanism $x = 0$ in terms of social welfare, because both mechanisms always build and they are both budget-balanced. It is never a good idea to set x to be strictly higher than C: if the non-excluded agents' total valuation is at least C, regardless of the excluded agent's type, the optimal decision is to build.

For any $x \in [0, C]$, mechanism x is (ex post) individually rational. Consider an arbitrary agent i. If agent i reports C/n, then he is never pivotal, so he does not pay any VCG payment excluded or not. If the decision is to build, then his utility is θ_i plus the redistribution he received from the others, which is at least 0. If the decision is not to build, then his utility is C/n plus the redistribution he received from the others, which

---

[2] *Faltings* is also (ex post) individually rational in all settings where the VCG mechanism is (ex post) individually rational.

is also at least 0. That is, every agent can guarantee a non-negative utility by reporting C/n. Combining this with the fact that the mechanism is incentive compatible, we can conclude that it is individually rational.

Since the mechanism is always budget-balanced, for the purpose of maximizing social welfare, we can ignore payments when optimizing over x. Thus, for this purpose, we can simplify mechanism x to:

- We exclude one agent uniformly at random.
- If the non-excluded agents' total valuation is at least x, then we build. Otherwise, we do not build.

Theorem 1. *For any $x \in [0, C]$, mechanism x is at most $\frac{n-1}{n}$-competitive.*

Proof. Mechanism 0 always builds. Consider the type profile $(0, 0, \ldots, 0)$. Under mechanism 0, the agents' total utility is 0. The agents' maximum possible total utility $\max\{C, \theta_N\} = \max\{C, 0\} = C$. Hence, mechanism 0 is at most 0-competitive.

Consider $x > 0$. Consider the type profile $(U, 0, \ldots, 0)$, where U is a number larger than C. Under mechanism x, when the agent reporting U is excluded, the decision is not to build (the agents' total utility is C). Otherwise, the decision is to build (the agents' total utility is U). The agents' expected total utility is $\frac{1}{n}C + \frac{n-1}{n}U$. The agents' maximum possible total utility is U. $\lim_{U \to \infty} \frac{\frac{1}{n}C + \frac{n-1}{n}U}{U} = \frac{n-1}{n}$. Hence, mechanism x is at most $\frac{n-1}{n}$-competitive. □

Theorem 2. *Mechanism C is exactly $\frac{n-1}{n}$-competitive.*

Proof. Under mechanism C, if agent i is excluded, then the agents' total utility is at least $\max\{C, \sum_{j \neq i} \theta_j\}$. Averaging over all i, the agents' expected total utility is then at least

$$\frac{1}{n} \sum_{i=1}^{n} \max\{C, \sum_{j \neq i} \theta_j\}.$$

The above expression is no less than

$$\frac{1}{n} \max\{nC, \sum_{i=1}^{n} \sum_{j \neq i} \theta_j\} = \max\{C, \frac{n-1}{n}\theta_N\}.$$

This is always greater than or equal to $\frac{n-1}{n}$ times $\max\{C, \theta_N\}$. That is, mechanism C is exactly $\frac{n-1}{n}$-competitive (Theorem 1 has shown that it is at most $\frac{n-1}{n}$-competitive). □

Theorem 3. *For any $x \in [\frac{n-2}{n-1}C, C)$, mechanism x is also exactly $\frac{n-1}{n}$-competitive.*

Proof. As a result of Theorem 1, we only need to prove that for any $x \in [\frac{n-2}{n-1}C, C)$, mechanism x is at least $\frac{n-1}{n}$-competitive.

For all type profiles with $\theta_N \geq C$, the correct (optimal) decision is to build. That is, for these type profiles, mechanism x is no worse than mechanism C, as mechanism x has a lower threshold for building.

Thus, we only need to prove that mechanism x is $\frac{n-1}{n}$-competitive for all type profiles with $\theta_N < C$. For these type profiles, the correct decision is not to build. That is, if $x_1 \leq x_2$, then for these type profiles, mechanism x_2 is no worse than mechanism x_1, as mechanism x_2 has a higher threshold for building.

Therefore, we only need to prove that mechanism $\frac{n-2}{n-1}C$ is $\frac{n-1}{n}$-competitive for all type profiles with $\theta_N < C$. In other words, we only need to prove that under mechanism $\frac{n-2}{n-1}C$, the agents' expected total utility is at least $\frac{n-1}{n}C$ for all type profiles with $\theta_N < C$.

There are three cases:

1. If under mechanism $\frac{n-2}{n-1}C$, the decision is to build with probability 1, then we have for all i, $\sum_{j \neq i} \theta_j = \theta_N - \theta_i \geq \frac{n-2}{n-1}C$. That is, $\sum_{i=1}^{n}(\theta_N - \theta_i) \geq \sum_{i=1}^{n}(\frac{n-2}{n-1}C)$. Rearranging, $(n-1)\theta_N \geq \frac{n(n-2)}{n-1}C$. Therefore, we have that the agents' total utility θ_N is at least $\frac{n^2-2n}{n^2-2n+1}C \geq \frac{n-1}{n}C$ (recall that $n \geq 3$).

2. If under mechanism $\frac{n-2}{n-1}C$, the decision is to build with probability $\frac{1}{n} \leq p \leq \frac{n-1}{n}$, then the agents' expected total utility is $p\theta_N + (1-p)C$. This expression is decreasing in p, and increasing in θ_N. It is minimized when $p = \frac{n-1}{n}$ and $\theta_N = \frac{n-2}{n-1}C$ ($\theta_N \geq \frac{n-2}{n-1}C$ because there exists i such that $\theta_N - \theta_i \geq \frac{n-2}{n-1}C$). That is, the agents' expected total utility is minimized under type profile $(\frac{n-2}{n-1}C, 0, 0, \ldots, 0)$. For this type profile, the agents' expected total utility is $\frac{1}{n}C + \frac{n-1}{n}\frac{n-2}{n-1}C = \frac{n-1}{n}C$.

3. If under mechanism $\frac{n-2}{n-1}C$, the decision is to build with probability 0, then this mechanism is always making the correct decision. The agents' total utility is C.

□

Theorem 4. *For any $x \in [0, \frac{n-2}{n-1}C)$, mechanism x is strictly less than $\frac{n-1}{n}$-competitive.*

Proof. We have already shown that mechanism 0 is at most 0-competitive in the proof of Theorem 1.

For $x > 0$, consider the type profile $(x, 0, 0, \ldots, 0)$. Under mechanism x, when the agent reporting x is excluded, the decision is not to build, and the agents' total utility is C. When some other agent is excluded, the decision is to build, and the agents' total utility is x. The agents' expected total utility is $\frac{n-1}{n}x + \frac{1}{n}C$. The agents' maximum possible total utility is C. The ratio equals $\frac{(n-1)x}{nC} + \frac{1}{n} < \frac{(n-1)\frac{n-2}{n-1}C}{nC} + \frac{1}{n} = \frac{n-1}{n}$. □

As a summary, we have shown that mechanism x is optimal if and only if $x \in [\frac{n-2}{n-1}C, C]$. Next we consider mixtures of mechanisms with different parameters.

Definition 1. *Mechanism OptMix:*

- *With probability $\frac{1}{n+1}$, we run mechanism 0 (always build);*
- *With probability $\frac{n}{n+1}$, we run mechanism C.*

Theorem 5. *OptMix is exactly $\frac{n}{n+1}$-competitive.*

We note that *OptMix* is more competitive than any individual (non-mixture) mechanism x.

Proof. If $\theta_N < C$, then mechanism C never builds. That is, if $\theta_N < C$, then the agents' expected total utility under *OptMix* is $\frac{1}{n+1}\theta_N + \frac{n}{n+1}C \geq \frac{n}{n+1}C = \frac{n}{n+1}\max\{C, \theta_N\}$.

We have that mechanism C is $\frac{n-1}{n}$-competitive, so if $\theta_N \geq C$, then the agents' expected total utility under mechanism C is at least $\frac{n-1}{n}\theta_N$. Under *OptMix*, the agents' expected total utility is then at least $\frac{1}{n+1}\theta_N + \frac{n}{n+1}\frac{n-1}{n}\theta_N = \frac{n}{n+1}\theta_N = \frac{n}{n+1}\max\{C, \theta_N\}$.

The above shows that *OptMix* is at least $\frac{n}{n+1}$-competitive. Let us consider the type profile $(0, 0, \ldots, 0)$. For this type profile, under *OptMix*, the agents' expected total utility is exactly $\frac{n}{n+1}C = \frac{n}{n+1}\max\{C, \theta_N\}$. Hence, *OptMix* is exactly $\frac{n}{n+1}$-competitive. □

Let *Mix* be an arbitrary mixture of mechanisms with different parameters. Let I be an interval that is a subset of $[0, C]$. We use $P(Mix \in I)$ to denote the probability that a mechanism with parameter $x \in I$ is used. $P(Mix \in [0, C]) = 1$. For *OptMix*, we have $P(OptMix \in [0, 0]) = \frac{1}{n+1}$ and $P(OptMix \in [C, C]) = \frac{n}{n+1}$. We will prove that *Mix* is at most $\frac{n}{n+1}$-competitive. That is, *OptMix* is the most competitive among all mixtures of mechanisms with different parameters.[3]

Theorem 6. *OptMix is the most competitive among all mixtures of mechanisms with different parameters.*

Proof. If $P(Mix \in [0, 0]) < \frac{1}{n+1}$, then let us consider the type profile $(U, 0, \ldots, 0)$, where U is larger than C. When the agent reporting U is excluded (which happens with probability $\frac{1}{n}$), the non-excluded agents' types are all zeros, which means that the probability to build (when the agent reporting U is excluded) is equal to $P(Mix \in [0, 0])$. That is, overall, for this type profile, the probability $\bar{p}$ of not building is at least $\frac{1}{n}(1 - P(Mix \in [0, 0]))$, which is strictly larger than $\frac{1}{n}(1 - \frac{1}{n+1}) = \frac{1}{n+1}$. The agents' expected total utility is $(1 - \bar{p})U + \bar{p}C$. The agents' maximum possible total utility is U. We have $\lim_{U \to \infty} \frac{(1-\bar{p})U + \bar{p}C}{U} = 1 - \bar{p}$. That is, if $P(Mix \in [0, 0]) < \frac{1}{n+1}$, then *Mix* is at most $\frac{n}{n+1}$-competitive. Therefore, if *Mix* is to be no less competitive than *OptMix*, then we must have $P(Mix \in [0, 0]) \geq \frac{1}{n+1}$.

If $P(Mix \in [0, 0]) \geq \frac{1}{n+1}$, then let us consider the type profile $(0, 0, \ldots, 0)$. The probability $\bar{p}$ of not building is most $1 - P(Mix \in [0, 0])$. It follows that $\bar{p} \leq \frac{n}{n+1}$. The agents' expected total utility is $(1 - \bar{p})0 + \bar{p}C$. The agents' maximum possible total utility is C. The ratio equals $\bar{p}$, which is at most $\frac{n}{n+1}$, and it follows that *Mix* is at most $\frac{n}{n+1}$-competitive. □

So far, we have identified many competitive randomized mechanisms. Another natural question to ask is whether there exist competitive deterministic mechanisms. The answer is yes: the VCG mechanism is $\frac{1}{n}$-competitive.

---

[3] *OptMix* is not the unique optimum. Consider a modified version of *OptMix* under which we run mechanism 0 with probability $\frac{1}{n+1}$ and run mechanism $C - \epsilon$ with probability $\frac{n}{n+1}$ (ϵ is a small positive number). When $\theta_N \geq C$, modified *OptMix* is no worse than *OptMix*, since the optimal decision is to build, and modified *OptMix* has a lower threshold for building. When $\theta_N < C - \epsilon$, modified *OptMix* is the same as *OptMix*. When $\theta_N \in [C - \epsilon, C)$, the optimal decision is not to build. The maximum efficiency is C. Under any budget-balanced mechanism (including modified *OptMix*), the agents' expected total utility is between $C - \epsilon$ and C, thus at least $C - \epsilon$. When ϵ is small enough, we have $\frac{C-\epsilon}{C} \geq \frac{n}{n+1}$.

Theorem 7 (Moulin, private communication). *The VCG mechanism is exactly* $\frac{1}{n}$-*competitive for the public project problem.*

To illustrate how poor of a ratio the $1/n$ achieved by VCG is, we now give a very simple mechanism that also obtains this ratio.

Definition 2. *Mechanism* Vote-to-Build: *Let every agent vote whether to build or not. If there is at least one vote toward building, then we build. Otherwise, we do not build.*

If an agent's valuation is at least C/n, then his dominant strategy is to vote toward building. If an agent's valuation is less than C/n, then his dominant strategy is to vote toward not building. *Vote-to-Build* is (ex post) individually rational and budget-balanced (there are no payments involved).

Theorem 8. Vote-to-Build *is exactly* $\frac{1}{n}$-*competitive.*

Proof. If the decision is to build, then there exists i with $\theta_i \geq C/n$. That is, we have $\theta_N \geq C/n$. The ratio $\frac{\theta_N}{\max\{C,\theta_N\}}$ is at least $\frac{1}{n}$, and it reaches $\frac{1}{n}$ when $\theta_N = C/n$ (corresponding to the type profile $(C/n, 0, 0, \ldots, 0)$).

If the decision is not to build, then there is no i with $\theta_i \geq C/n$. It follows that $\theta_N \leq C$. The ratio is then $\frac{C}{\max\{C,\theta_N\}} = \frac{C}{C} = 1$. $\square$

4   General Domains

In Section 3, we showed that *Faltings* is at least $\frac{n-1}{n}$-competitive for the public project problem. Here we show that it remains true for general mechanism design problems, as long as the agents' valuation functions satisfy the following assumption.[4]

Assumption 1. *The valuations of agents are non-negative for all outcomes:* $v_i(k, \theta_i) \geq 0$ *for all* θ_i, k.

Theorem 9. Faltings *is at least* $\frac{n-1}{n}$-*competitive, as long as the agents' valuations satisfy Assumption 1.*

Proof. We begin with observing a few relationships between the value of the efficient outcome when all agents are present and when one agent is excluded. Under Assumption 1, making agent i accept a decision made without him does not decrease the value of that decision.[5] In particular, this applies to the efficient outcome for agents $j \neq i$:

$$\sum_j v_j(k^*(\theta_{-i}), \theta_j) \geq \sum_{j \neq i} v_j(k^*(\theta_{-i}), \theta_j).$$

On the other hand, the total value of agents $j \neq i$ under the outcome efficient for them is at least as high as their total value under the outcome efficient when all agents are present.

---

[4] The assumption places restrictions only on the valuation function and is independent of the mechanism. This is in contrast to the individual rationality property, which requires the *utility* of each agent participating in the mechanism to be above his outside value.

[5] The valuation function in Equation 1 satisfies this property.

$$\sum_{j \neq i} v_j(k^*(\theta_{-i}), \theta_j) \geq \sum_{j \neq i} v_j(k^*(\theta), \theta_j).$$

Combining the two inequalities and summing over all agents, we get

$$\sum_i \sum_j v_j(k^*(\theta_{-i}), \theta_j) \geq \sum_i \sum_{j \neq i} v_j(k^*(\theta_{-i}), \theta_j)$$
$$\geq \sum_i \sum_{j \neq i} v_j(k^*(\theta), \theta_j) = (n-1) \sum_i v_i(k^*(\theta), \theta_i).$$

Dividing by n and focusing on the first and last expressions, we have

$$\frac{1}{n} \sum_i \sum_j v_j(k^*(\theta_{-i}), \theta_j) \geq \frac{n-1}{n} \sum_i v_i(k^*(\theta), \theta_i).$$

The expression on the left-hand side is the expected value of the outcome when the decision is made efficiently after one agent is excluded uniformly at random. The inequality implies that the expected value of the decision under *Faltings* is at least $\frac{n-1}{n}$ of the maximum efficiency. □

Faltings results in a rather unequal treatment of the excluded agent relative to the other agents. In settings where the VCG mechanism collects a lot of revenue, the agents would be envious of the excluded agent.

We next study a more fair payment scheme where each agent pays his expected VCG payment and receives part of his own residual claimant rebate. We call the resulting mechanism *FaltingsFair*. This scheme had been proposed previously by Faltings [6,8]. We derived it independently with formal proofs. The result on the competitive ratio is novel. We discuss this in more detail at the end of this section.

- Exclude an agent a uniformly at random and compute the efficient allocation.
- Collect from each agent i (including a) the payment

$$t_i(\theta) = \frac{1}{n} \sum_{j \neq i} t_i^{\mathrm{vcg}}(\theta_{-j}) - \frac{1}{n} \sum_{j \neq i} t_j^{\mathrm{vcg}}(\theta_{-i}) \tag{2}$$

Expanding each term of the payment, we can rewrite it as follows.

$$t_i^{\mathrm{vcg}}(\theta_{-j}) = \sum_{a \neq i,j} v_a(k^*(\theta_{-i,j}), \theta_a) - \sum_{a \neq i,j} v_a(k^*(\theta_{-j}), \theta_a)$$

$$t_j^{\mathrm{vcg}}(\theta_{-i}) = \sum_{a \neq i,j} v_a(k^*(\theta_{-i,j}), \theta_a) - \sum_{a \neq i,j} v_a(k^*(\theta_{-i}), \theta_a)$$

$$t_i^{\mathrm{vcg}}(\theta_{-j}) - t_j^{\mathrm{vcg}}(\theta_{-i}) = \sum_{a \neq i,j} (v_a(k^*(\theta_{-i}), \theta_a) - v_a(k^*(\theta_{-j}), \theta_a))$$

$$t_i(\theta) = \frac{1}{n} \sum_{j \neq i} \sum_{a \neq i,j} (v_a(k^*(\theta_{-i}), \theta_a) - v_a(k^*(\theta_{-j}), \theta_a))$$

Theorem 10. FaltingsFair *is incentive compatible in expectation, budget-balanced, and for valuations satisfying Assumption 1, $\frac{n-1}{n}$-competitive.*

Proof. First we prove incentive compatibility. Denoting *FaltingsFair*'s allocation function that chooses a residual claimant uniformly at random with k^{rc}, agent i's utility

$u_i(k^{rc}(\theta), t_i, \theta_i) = \left(\frac{1}{n} \sum_j v_i(k^*(\theta_{-j}), \theta_i)\right) - t_i(\theta)$

$= \frac{1}{n} \sum_j v_i(k^*(\theta_{-j}), \theta_i) - \frac{1}{n} \sum_{j \neq i} \sum_{a \neq i,j} (v_a(k^*(\theta_{-i}), \theta_a) - v_a(k^*(\theta_{-j}), \theta_a))$

$= \frac{1}{n} v_i(k^*(\theta_{-i}), \theta_i) + \frac{1}{n} \sum_{j \neq i} \left(\sum_{a \neq j} v_a(k^*(\theta_{-j}, \theta_a)) - \sum_{a \neq i,j} v_a(k^*(\theta_{-i}), \theta_a)\right).$

Removing the terms that agent i does not control with his report, we are left with $\frac{1}{n} \sum_{j \neq i} \sum_{a \neq j} v_a(k^*(\theta_{-j}), \theta_a)$. This expression is maximized when agent i reports the true value θ_i as by the definition of $k^*(\theta_{-j})$

$$\sum_{a \neq j} v_a(k^*(\theta_{-j}), \theta_a) \geq \sum_{a \neq j} v_a(k', \theta_a) \quad \forall j, \ k' \in K.$$

Therefore, incentive compatibility holds.

Next we show budget balance ($\sum_i t_i = 0$).

$$\sum_i \frac{1}{n} \left(\sum_{j \neq i} t_i^{vcg}(\theta_{-j}) - \sum_{j \neq i} t_j^{vcg}(\theta_{-i})\right) = 0$$

$$\sum_i \sum_{j \neq i} t_i^{vcg}(\theta_{-j}) = \sum_i \sum_{j \neq i} t_j^{vcg}(\theta_{-i})$$

The equality follows from the simple identity $\sum_i \sum_{j \neq i} a_{ij} = \sum_i \sum_{j \neq i} a_{ji}$.

Finally, the allocation function is the same as before, thus, *FaltingsFair* has the same competitive ratio as *Faltings*. □

Unlike *Faltings*, *FaltingsFair* is incentive compatible only in expectation with respect to the random outcome function $k(\theta)$. This means that an agent has no incentive to misreport his value before the outcome is chosen,[6] but once the outcome is known, the agent may regret not reporting a different value. Incentive compatibility in expectation is a natural concept for randomized mechanisms as the reporting of values must occur before the outcome is selected.

Our next theorem deals with individual rationality. [7] showed that *Faltings* is (ex post) individually rational in settings where the VCG mechanism is (ex post) individually rational. This is actually the case for valuations satisfying Assumption 1. That is, for valuations that satisfy Assumption 1, *Faltings* is (ex post) individually rational. Similar to the case of incentive compatibility, unlike *Faltings*, *FaltingsFair* is individually rational only in expectation with respect to the random outcome function $k(\theta)$.

Theorem 11. *For valuations satisfying Assumption 1,* FaltingsFair *is individually rational in expectation.*

We now take a closer look at the payment function in Equation 2. The first term is the expected VCG payment in the market with one agent excluded uniformly at random. The second term produces a rebate equal to $\frac{1}{n}$ of the total VCG payments realized without the agent in the market. This rebate has been considered before with the goal of redistributing the VCG surplus in [3,4].

$$h_i^{rc}(\theta_{-i}) = \frac{1}{n} \sum_{j \neq i} t_j^{vcg}(\theta_{-i}) \tag{3}$$

---

[6] Note that unlike the Bayesian incentive compatible "expected externality mechanism" (or dAGVA), our mechanism is dominant-strategy incentive compatible and we have no prior over agents' types.

This rebate, however, may exceed the total VCG revenue resulting in a deficit in some models. In fact, as Cavallo argues in [4], the no-deficit property requires the redistribution to sometimes be smaller than the amount above. Specifically, one can compute the smallest total VCG payment collected from the agents over all values agent i might have. It is this amount that should be used in the redistribution to agent i:

$$h_i^{\min}(\theta_{-i}) = \frac{1}{n} \min_{\theta_i'} \sum_j t_j^{\text{vcg}}(\theta_i', \theta_{-i}) \qquad (4)$$

It is easy to see that in the public good setting, the above rebate (Equation 4) is always zero.[7] Thus, rebates of this form are not helpful in efficient mechanisms in models like those involving public goods. In contrast, in *FaltingsFair*, the rebate (3) results in full budget balance in any model.

The payment rule in Equation 2 was previously proposed by Boi Faltings in a patent [8] and an unpublished paper [6]. There Faltings provides an equivalent definition of the payment rule: instead of considering the rebate function explicitly, the rule directs each agent i to pay $\frac{1}{n}t_i^{\text{vcg}}(\theta_{-j})$ to each agent j. Notice that budget balance follows immediately from this definition. To see that the definition in fact defines the payment rule in Equation 2, notice that the first summation corresponds to the payments agents $j \neq i$ make to agent i and the second summation corresponds to the payments agent i makes to agents $j \neq i$. These definitions provide different interpretations of the mechanism: Faltings views it as the average of the budget-balanced *Faltings* mechanism, while we make explicit the connections to a redistribution function previously considered in the literature.

5   Discussion

We studied randomized mechanisms that are fully budget-balanced and aimed to maximize the expected efficiency, which under budget balance coincides with the expected social welfare. The expected welfare loss of our generally applicable mechanism is only $\frac{1}{n}$ in the worst case leaving little room for improvement. However, whether or not this loss can be reduced with a different randomized mechanism (budget-balanced, or not) remains an open question.

Note that full efficiency is impossible in randomized mechanisms.[8] Thus, the goal of minimizing budget imbalance in an efficient mechanism is not meaningful in this context. However, the question of minimizing budget imbalance in deterministic mechanisms for public good remains open.

Finally, we note that for public project problems, our definition of the value of the efficient outcome adds C to the standard definition of $\max(\theta_N - C, 0) = \max(\theta_N, C) - C$ (see, e.g., [13]). Our results can be interpreted under the standard definition: being

---

[7] If $\theta_N - \theta_i < \frac{n-1}{n}C$, then when $\theta_i = 0$, no agent is pivotal and the total VCG payment is 0; If $\theta_N - \theta_i \geq \frac{n-1}{n}C$, then when $\theta_i = C$, no agent is pivotal and the total VCG payment is also 0.

[8] At least, this is the case for *significantly* randomized mechanisms that do more than just break ties randomly.

r-competitive means we guarantee the welfare of $r(C + \max(\theta_N - C, 0))$. We cannot guarantee the welfare of $r\max(\theta_N - C, 0)$ as Assumption 1 does not hold for the standard valuation function

$$v_i(k(\theta), \theta_i) = \begin{cases} \theta_i - \frac{C}{n} & \text{if } k(\theta) = 1 \\ 0 & \text{otherwise} \end{cases}$$

We are indebted to Hervé Moulin for the idea of using the alternative metric.

The definition of valuations above ensures that the cost C is covered (each agent contributes $\frac{C}{n}$) as long as the sum of the agents' payments is non-negative. The valuation function that we use shifts the standard one by $\frac{C}{n}$. However, in both cases, the cost C is covered: i.e., non-negative total payments result in weak budget balance.

References

1. Apt, K.R., Conitzer, V., Guo, M., Markakis, E.: Welfare Undominated Groves Mechanisms. In: Papadimitriou, C., Zhang, S. (eds.) WINE 2008. LNCS, vol. 5385, pp. 426–437. Springer, Heidelberg (2008)
2. Apt, K.R., Estévez-Fernández, A.: Sequential pivotal mechanisms for public project problems. In: Mavronicolas, M., Papadopoulou, V.G. (eds.) SAGT 2009. LNCS, vol. 5814, pp. 85–96. Springer, Heidelberg (2009)
3. Bailey, M.J.: The demand revealing process: To distribute the surplus. Public Choice 91(2), 107–126 (1997)
4. Cavallo, R.: Optimal decision-making with minimal waste: Strategyproof redistribution of VCG payments. In: AAMAS 2006, Hakodate, Japan (2006)
5. de Clippel, G., Naroditskiy, V., Greenwald, A.: Destroy to save. In: EC 2009 (2009)
6. Faltings, B.: A budget-balanced, incentive-compatible scheme for social choice (2004) (unpublished)
7. Faltings, B.: A Budget-Balanced, Incentive-Compatible Scheme for Social Choice. In: Faratin, P., Rodríguez-Aguilar, J.-A. (eds.) AMEC 2004. LNCS (LNAI), vol. 3435, pp. 30–43. Springer, Heidelberg (2006)
8. Faltings, B.: Social choice determination systems and methods, United States Patent (2011)
9. Gujar, S., Narahari, Y.: Redistribution mechanisms for assignment of heterogeneous objects. Journal of Artificial Intelligence Research 41, 131–154 (2011)
10. Guo, M., Conitzer, V.: Better redistribution with inefficient allocation in multi-unit auctions with unit demand. In: EC 2008, pp. 210–219 (2008)
11. Guo, M., Conitzer, V.: Worst-case optimal redistribution of VCG payments in multi-unit auctions. Games and Economic Behavior 67(1), 69–98 (2009)
12. Guo, M., Conitzer, V.: Optimal-in-expectation redistribution mechanisms. Artif. Intell. 174(5-6), 363–381 (2010)
13. Mas-Colell, A., Whinston, M.D., Green, J.R.: Microeconomic Theory. Oxford University Press, New York (1995)
14. Moulin, H.: Almost budget-balanced VCG mechanisms to assign multiple objects. Journal of Economic Theory 144(1), 96–119 (2009)

Online Stochastic Weighted Matching: Improved Approximation Algorithms

Bernhard Haeupler[1], Vahab S. Mirrokni[2], and Morteza Zadimoghaddam[1]

[1] Massachusetts Institute of Technology, Cambridge MA 02139, USA
{haeupler,morteza}@mit.edu
[2] Google Research, New York NY 10011, USA
mirrokni@gmail.com

Abstract. Motivated by the display ad allocation problem on the Internet, we study the *online stochastic weighted matching* problem. In this problem, given an edge-weighted bipartite graph, nodes of one side arrive online i.i.d. according to a known probability distribution. Recently, a sequence of results by Feldman et. al [14] and Manshadi et. al [20] result in a 0.702-approximation algorithm for the unweighted version of this problem, aka *online stochastic matching*, breaking the $1 - 1/e$ barrier. Those results, however, do no hold for the more general online stochastic weighted matching problem. Moreover, all of these results employ the idea of *power of two choices*.

In this paper, we present the first approximation (0.667-competitive) algorithm for the online stochastic weighted matching problem beating the $1 - 1/e$ barrier. Moreover, we improve the approximation factor of the online stochastic matching by analyzing the more general framework of *power of multiple choices*. In particular, by computing a careful third pseudo-matching along with the two offline solutions, and using it in the online algorithm, we improve the approximation factor of the online stochastic matching for any bipartite graph to 0.7036.

Keywords: online stochastic matching, approximation algorithm, competitive analysis, ad allocation.

1 Introduction

Online bipartite matching is a fundamental problem with many applications in online resource allocation, especially the online allocation of ads on the Internet. In this problem, given a bipartite graph $G(A, I, E)$ with advertisers A and impressions I, and a set E of edges between them. Advertisers in A are fixed and known. Impressions (or requests) in I (along with their incident edges) arrive online. Upon the arrival of an impression $i \in I$, we must assign i to any advertiser $a \in A$ where $(i, a) \in E(G)$. At all times, the set of assigned edges must form a matching. The seminal result of Karp, Vazirani and Vazirani [17] gives an optimal online $1 - 1/e$-competitive algorithm to maximize the size of the matching. This algorithm works in the *adversarial model* where the online algorithm does not know anything about the I or E beforehand.

N. Chen, E. Elkind, and E. Koutsoupias (Eds.): WINE 2011, LNCS 7090, pp. 170–181, 2011.

Motivated by applications in online advertising, a *stochastic online* model has been proposed in which impressions $i \in I$ arrive online according some *known* probability distribution [14]. In this setting, in addition to G, we are given a probability distribution $\mathcal{D}$ over the elements of I. Our goal is then to compute a maximum matching on $\hat{G} = (A, \hat{I}, \hat{E})$, where $\hat{I}$ is drawn from $\mathcal{D}$. This model is particularly well-motivated in the context of online ad allocation in which one can predict the pattern or type of impressions or requests using a vast amount of historical data [14]. A sequence of results give improved approximation algorithms for this problem, beating the approximation factor of $1-1/e$ [14,4,20]. All these results employ the natural technique of the power of two choices [3,21] which was initially used by Feldman, Mehta, Mirrokni, and Muthukrishnan [14] in this context, i.e., they compute *two* offline matchings and using them to guide the online solution. The best known result in this model is a 0.702-approximation in expectation by Manshadi, Oveis Gharan and Saberi [20] which applies the power of two choices paradigm in an adaptive online fashion.

All the above results hold only for *unweighted* bipartite graphs in which the goal is to maximize the size of the matching. However, in many real world scenarios, the value received from matching a node may vary for different nodes. For example, in the context of display online ads, different ads may have different potential values for different impressions (as measured, e.g., by click-throughs). In such settings, advertisers would like their ads to be assigned to well-targeted impressions and the online ad serving algorithm has an important goal of maximizing the overall quality of impressions used for these ads. Moreover, advertisers in some online display ad-campaigns may be willing to pay a different amount every time their ad is shown on a website depending on the type of impression (for example geographic information of users, etc). In such settings, advertisers specify their targeting criteria for the set of impressions they are interested in and may declare how much they are willing to pay for each targeted audience. All such settings can be formalized as a *maximum weighted matching problem* for an edge-weighted bipartite graph G, i.e., given an edge-weighted $G(A, I, E)$ with an edge weight $\omega(e)$ for each edge $e \in E(G)$, the goal is to find a matching of ads in A to impressions in I that maximize the total weight of the matching.

The online version of the maximum weighted matching problem as described above has not been studied in this form. In fact, the only weighted online matching problems studied so far are in the adversarial model, have $1-1/e$-competitive and hold either with the extra assumption of large degrees in the free disposal model [13], or in the ad-weighted special case of this problem in which all edges connected to the same ad $a \in A$ has the same weight [1]. In this paper, we study the unrestricted weighted problem in the stochastic arrival model, and present the first approximation algorithms for it.

Our Results and Techniques. We present two main results under the *i.i.d.* model, generalizing the known techniques in two directions.

1) We first observe a simple $1 - 1/e$-approximation for the online stochastic weighted matching problem, and then improve it to a 0.667-approximation algorithm, breaking the $1 - 1/e$ bottleneck. One significance of this result is in applying the idea of power of two choices in a problem more general than the online stochastic matching problem [14,20]. Previously only $1 - 1/e$-approximation algorithm were known for other similar variants of this problem with small degrees (or small budgets) [10]. In order to beat the $1 - 1/e$ approximation factor, we need to consider a *discounted version* of the offline LP solution, and find the offline solution of this discounted LP. The trick of solving the offline instance for a *discounted instance* may prove useful elsewhere in solving more general stochastic problems. We employ the ideas of power of two choices and sampling from an optimum solution of the matching linear program that have been developed by Feldman et al. [14] and Manshadi et al. [20]. We hope that our application of these ideas paves the way for applying this technique to more general stochastic optimization problems like online stochastic packing and other applications in the OR literature [6,5,8,11].

2) We furthermore apply our discounted LP technique to give an improved competitive algorithm for the online stochastic matching problem. We use the dual of the tightened LP to obtain a new upper bound on the optimal solution. This already leads to a competitive ratio of 0.684. We then even further tighten the LP and sample an additional pseudo-matching from it. Via this pseudo matching we obtain an algorithm with competitive ratio of 0.7036 which improves over the best previously known approximation. Other than the slight improvement in the approximation factor, the significance of this result is in demonstrating the power of multiple choices and providing tightened LPs as a tool to both sample a good subgraph of the expected graph and also obtain better and closely related upper bound on the optimal solution via their dual.

3) Finally, we present simple adaptive online algorithms to solve the online (weighted) stochastic matching problem optimally for a special class of graphs, including the union of two matchings. This algorithm can be used as the final allocation rule for any approximation algorithm that first computes two offline matchings whose edges are then exclusively used in a solution. This improves upon the heuristics used for allocation the algorithms of [14,4], the non-adaptive algorithm of [20] and our algorithm for the weighted case.

Other Related Work. Another stochastic model studied in the context of online stochastic matching is the *random order model* where we assume that I is unknown, but impressions in I arrive in a random order. This has proved to be an important analytical construct for other problems such as secretary-type problems where worst cases are inherently difficult. It is known that in this case even the greedy algorithm has a (tight) competitive ratio of $1 - 1/e$ [15]. Further, no deterministic algorithm can achieve approximation ratio better than 0.75 and no randomized algorithm better than 0.83 [15]. Very recently, improved approximation algorithms have been proposed for this problem [16,18].

Online stochastic weighted matching problem is related to online ad allocation problems, including the *Display Ads Allocation (DA)* problem [13,12,2,22], and the *AdWords (AW)* problem [19,7,9]. In both of these problems, the publisher must assign online impressions to an inventory of ads, optimizing efficiency or revenue of the allocation while respecting pre-specified contracts. In the DA problem, given a set of m advertisers with a set S_j of eligible impressions and demand of at most $n(j)$ impressions, the publisher must allocate a set of n impressions that arrive online. Each impression i has value $w_{ij} \geq 0$ for advertiser j. The goal of the publisher is to assign each impression to one advertiser maximizing the value of all the assigned impressions. The adversarial online DA problem was considered in [13], which showed that the problem is inapproximable without exploiting *free disposal*; using this property (that advertisers are at worst indifferent to receiving more impressions than required by their contract), a simple greedy algorithm is 1/2-competitive, which is optimal. When the demand of each advertiser is large, a $(1 - 1/e)$-competitive algorithm exists [13], and it is tight. The stochastic model of the DA problem is more related to our problem. Following a training-based dual algorithm by Devanur and Hayes [9], training-based $(1 - \epsilon)$-competitive algorithms have been developed for the DA problem and its generalization to various packing linear programs [12,22,2]. These papers develop a $(1 - \epsilon)$-competitive algorithm for online stochastic packing problems in which $\text{OPT}/w_{ij} \geq O(m \log n/\epsilon^3)$ (or $\text{OPT}/w_{ij} \geq O(m \log n/\epsilon^2)$ applying the technique of [2]) and the demand of each advertiser is large, in the random-order and the i.i.d. model. All the above algorithms work only in the presence of extra assumptions, and none of them apply to the online stochastic weighted matching problem as discussed in this paper.

2   Preliminaries

Consider the following online stochastic weighted matching problem in the i.i.d. model: We are given an edge-weighted bipartite graph $G = (A, I, E)$ over advertisers A and impression types I, along with an edge weight $\omega(e)$ for each edge $e \in E(G)$. Let $k = |A|$ and $m = |I|$. We are also given, for each impression type $i \in I$, an integer number e_i of impressions we expect to see. Let $n = \sum_{i \in I} e_i$. We use $\mathcal{D}$ to denote the distribution over I defined by $\Pr[i] = p_i = e_i/n$. Throughout this paper, we assume without loss of generality that each impression occurs with frequency one, i.e., $e_i = 1$ for each $i \in I$. Sometimes, it will be convenient to think of G as the complete graph with "non-available edges" having weight zero.

An instance $(G, \mathcal{D}, n)$ of the *online stochastic matching* problem is as follows: We are given offline access to G and the distribution $\mathcal{D}$. Once online, n i.i.d. draws of impressions $i \sim \mathcal{D}$ arrive, and we must immediately assign an impression i to some advertiser a where $(a, i) \in E$, or not assign i at all. Each advertiser $a \in A$ may only be assigned at most once[1]. Our goal is to assign arriving impressions to advertisers and maximize the total weight of assigned impressions. In the following, we will formally define the objective function of the algorithm.

---

[1] All results in this paper hold for a more general case that each advertiser a has a capacity c_a and advertiser a can be assigned at most c_a times. This more general case can be reduced easily to the degree one case by repeating each node a c_a number of times in the instance.

Let $D(i)$ be the set of draws of impression type i that arrive during the run of the algorithm. We let a scenario $\hat{I} = \cup_{i \in I} D(i)$ be the set of impressions. Let $\hat{G}(\hat{I})$ be the "realization" graph, i.e., with node sets A and $\hat{I}$, and edges $\hat{E} = \{(a, i') : (a, i) \in E, i' \in D(i)\}$. Given an instance $(G, \mathcal{D}, n)$ of the online matching problem, we wish an algorithm ALG for which $E[\text{ALG}(\hat{I})]/E[\text{OPT}(\hat{I})] \geq \alpha$. In this case, we say that the algorithm achieves approximation factor α in expectation. Note that one could also study the stronger notion of approximating with high probability, but it is not hard to see that this notion is not applicable to the weighted matching problem.

3   A 0.667-Competitive Algorithm for Weighted Graphs

In this section, we present a 0.667-competitive algorithm for the online stochastic weighted matching problem. Our algorithm build on the simple $1 - 1/e$-approximation algorithm which we describe next. The algorithm uses the following standard matching LP:

$$\max \boldsymbol{w}^T \boldsymbol{p} \text{ subject to}$$
$$\forall a \in A : \sum_{e \in \Gamma(a)} p_e \leq 1; \quad \forall i \in I : \sum_{e \in \Gamma(i)} p_e \leq 1$$

The algorithm computes an optimal solution vector p^* for this linear program and then whenever an impression i arrives, take an edge $e(i, a)$ (incident to impression i) with probability p_e^* and if a is not taken it matches i to a. We note that this algorithm works also if the arrival rates are non-integral and that it can be made deterministic for integal-arrival rates by selecting an integral solution vector p^*. To analyze the performance of this strategy we see that for each edge $e(i, a)$, the probability that we match i to a in total is at least:

$$\sum_{t=1}^{n} \frac{p_e^*}{n} (1 - 1/n)^{t-1} = p_e^* \frac{1}{n} \frac{(1 - 1/n)^0 - (1 - 1/n)^n}{1 - (1 - 1/n)} \geq p_e^* (1 - 1/e).$$

This is the probability that at some time t, impression i comes, and we pick edge e, and ad a is not taken before time t. We conclude that from this that our expected gain is at least $(1 - 1/e)$ times the optimal solution.

Improved Algorithm. In order to beat the $1 - 1/e$ approximation factor, we consider a *discounted version* of the matching LP and use its solution to precompute matchings that guide our online allocation rule. To develop the LP consider the expected bipartite graph $G = (I, A, E)$ with edge weights $w(e)$ for $e \in E(G)$. For every sequence of impressions, the optimal solution uses a matching strategy to match the incoming impressions to unmatched ads. As every sequence of impressions happen with some probability, every edge e between an impression i, and ad a is taken in the optimal solution with some probability p_e which is the sum of the probabilities of all sequences of impressions in which the optimal solution matches i to a. Note that since every ad a is matched to at most one impression in any situation, for any fixed a the sum $\sum_{e(i,a)} p_e$ is at most one. Furthermore every impression i is coming with rate 1, so for any fixed i, the sum

$\sum_{e(i,a)} p_e$ is at most 1 as well. Since every impression does not come at all with probability $(1 - 1/n)^n \approx 1/e$, we can add the constraint that for each edge e, the probability p_e is at most $1 - 1/e$. In fact, the optimal solution is a feasible solution for the following restricted matching linear program:

$$\max \boldsymbol{w}^T \boldsymbol{p} \text{ subject to}$$
$$\forall a \in A : \sum_{e \in \Gamma(a)} p_e \leq 1; \quad \forall i \in I : \sum_{e \in \Gamma(i)} p_e \leq 1; \quad \forall e \in E(G) : p_e \leq 1 - 1/e$$

Our algorithm is based on a optimal fractional solution p^*: First it computes p^* and then it samples a matching M_s from the graph G such that for any edge e the probability that e is in M_s is equal to p_e^*. Let M_1 be a maximum weighted matching in graph G. Define M' to be the matching $M_1 \setminus M_s$ which is the union of edges in M_1 that are not in M_s. The algorithm uses M_s as the primary matching, and M' as the auxiliary matching. More precisely this translates to the following allocation rule: Upon arrival of an impression i for the first time, assign i to its matched node in M_s if such a matching node exists and it is not already taken. Upon arrival of an impression i for the second time, assign i to its matched node in M' if such a matching node exists and it is not already taken.

Lemma 1. *The expected value of the matching the algorithms outputs is at least in $(1-1/e)\omega(M_s)+0.095(\omega(M'))$ where $\omega(M')$ is the sum of the weights of edges in M'.*

Proof. There are three types of edges the algorithm might take: edges in $M_s \cap M_1$, $M_s \setminus M_1$, and $M' = M_1 \setminus M_s$. We first prove that every edge in $M_s \cap M_1$ gets taken with probability $1 - (1 - 1/n)^n \approx 1 - 1/e$. Consider an edge $e(i,a)$ which is in both M_s and M_1. Note that ad a is not matched to any impression in the auxiliary matching M', so impression i is not competing for ad a with any other impression. Therefore impression i is matched to ad a, if it comes at least once. So with probability $1 - (1 - 1/n)^n \approx 1 - 1/e$, we take this edge in $M_s \cap M_1$.

Now, we prove that every edge in M' gets taken with probability at least 0.148. Consider an edge $e'(i', a') \in M'$, in the worst case a' is matched to some impression i_s in matching M_s (otherwise the probability of taking e' increases). The only case we take e' is when i' comes for the second time, but impression i_s has not arrived yet. Let $1 \leq t_1 < t_2 \leq n$ be the time slots in which i' comes for the first and second time. The probability of taking e' can be formulated and approximated (for large n) as follows:

$$\sum_{t_1=1}^{n-1} (1/n) \sum_{t_2=t_1+1}^{n} (1/n)(1 - 2/n)^{t_2-2} \approx \frac{1 - 1/e^2}{4} - \frac{1}{2e^2} > 0.148.$$

Finally we compute a lower bound for the probability of an edge in $M_s \setminus M_1$ being taken by the algorithm. Consider an edge $e''(i'', a'')$ in $M_s \setminus M_1$. In the worst case, ad a'' is matched to some impression i_1 in matching M_1 (and therefore in matching M_1). We compute the probability that i'' comes before the second i_1 comes. Let t be the time slot in which i'' is arrived for the first time, and i_1 has not come yet, or has come exactly once. This probability can be written and approximated as follows:

$$\sum_{t=1}^{n} \frac{1}{n} \left[(1 - 2/n)^{(t-1)} + \frac{t-1}{n}(1 - 2/n)^{(t-2)} \right] \approx \frac{(1 - 1/e^2)}{2} + 0.148 > 0.58.$$

Putting this together we get that the expected value of the output is at least

$$0.58\omega(M_s \setminus M_1) + (1 - 1/e)\omega(M_s \cap M_1) + 0.148\omega(M')$$
$$\geq 0.58\omega(M_s) + (1 - 1/e - 0.58)w(M_s \cap M_1) + 0.148(M_1 \setminus M_s)$$
$$\geq 0.58\omega(M_s) + (1 - 1/e - 0.58)\omega(M_1) + [0.148 - (1 - 1/e - 0.58)]\omega(M_1 \setminus M_s)$$
$$\geq (1 - 1/e)\omega(M_s) + 0.095\omega(M')$$

where the last step follows since M_1 is maximal and thus at least $\omega(M_s)$.

Theorem 1. *The expected value of the output of the above algorithm is at least 0.667 times the optimal fractional LP solution, and therefore at least 0.667OPT.*

Proof. Using Lemma 1 we know that the expected value of the output of our algorithm is at least $(1 - 1/e)E(\omega(M_s)) + 0.095E(\omega(M'))$. Since every edge is present in M_s with probability p_e^*, the expected value of $\omega(M_s)$ is equal to the value of the LP solution. To upper bound the size of $E(\omega(M')$ we use the additional LP constraint to note that every edge has probability at most $1 - 1/e$ to be sampled into M_s which implies that each edge of M_1 is in M' with probability at least $1/e$. Thus the expected value of $\omega(M')$ is at least $\omega(M_1)/e$. Since both $\omega(M_1)$ and $\omega(M_s)$ are larger than the expected solution of the optimal offline algorithm we obtain that our algorithm has a competitive ratio of at least a $1 - 1/e + 0.095/e = 0.667$.

4   A 0.7036-Approximation for Unweighted Graphs

In this section, we show how to use the power of multiple choices to design an improved 0.7036-approximation for the online stochastic matching problem in unweighted bipartite graphs. To prove this bound we use the dual of a tightened LP to obtain a better upper bound on the optimal offline solution. We first demonstrate this technique on a simpler algorithm using two matchings and then state our new algorithm.

Warm-Up: A 0.684-Approximation Algorithm. We describe a 0.6844-approximation algorithm which is based on computing two matchings and captures the initial idea of using the dual of a tighter linear program to upper bound the optimal value. The algorithm is extremely simple: It computes a maximum weighted matching M_1 in G and a maximum weighted matching M_2 in $G \setminus M_1$ and solely use edges from these two matchings. When an impression comes it first tries to match it along an M_1 edge (if one exists) and then along an M_2 edge. If the impression remains unmatched afterwards it is discarded.

To analyze the performance of this algorithm we classify every edge $e = (i, a) \in M_2$ to belong to one of three different classes:

- class-B: e is adjacent to one edge $(a, i') \in M_1$
- class-C: e is adjacent to one edge $(a', i) \in M_1$
- class-D: e is adjacent to two edges $(a', i), (a, i') \in M_1$

Lemma 2. *The algorithm obtains in expectation a $(1 - 2/e)1/e$ fraction of all class-D edges in M_2, a $(1/e - 1/e^2)$ fraction of all class-B edges and a $(1 - 2/e)$ fraction of all class-C edges.*

Proof. The probability that an impression comes not at all is $(1 - 1/n)^n \approx 1/e$; the probability that an impression comes exactly once is $n(1/n)(1 - 1/n)^{n-1} \approx 1/e$ and therefore the probability that an impression comes at least twice is $1 - (1 - 1/n)^n - (1/n)n(1 - 1/n)^{n-1} \approx 1 - 2/e$. We will pretend that if an impression i comes for the first time but the ad a at the M_1 edge (i, a) of i is already taken by an M_2-edge we will delete the M_2 edge and put in the M_1 edge. This is only for the sake of analysis and since we are in the unweighted case it does not change the value or later decisions of the algorithm. With this, a class-D edge gets picked if impression i comes twice and impression i' does not come at all which happens with probability $(1 - 2/e)1/e$. A class-B edge gets picked if impression i comes at least once and the impression i' does not come at all giving a probability of $(1 - 1/e)1/e$. Lastly a class-C edge gets taken if impression i comes twice which happens with probability at least $1 - 2/e$. ∎

A Simple Upper Bound. We will first argue that $OPT \leq \omega(M_1)$ and $OPT \leq (1 - 1/e)\omega(M_1) + M_2$ are both valid upper bounds on the expected value of the optimal offline solution:

For this we fix an algorithm that given the realization graph computes a maximum weighted matching. This is the optimal offline algorithm we want to compete against. Let p_e be the probability that edge e is used in the solution of the algorithm. It is easy to see that these probabilities form a maximum fractional solution in the standard matching LP. As such it has the same value as our M_1 and the first bound follows. For the second bound we observe that since each impression does not come at all with probability $1/e$ the gain from edges in M_1 for the optimal algorithm can be at most $(1 - 1/e)\omega(M_1)$. In addition to that the value from edges not in M_1 also has to obey the matching LP restrictions in $G \setminus M_1$ which is the graph M_2 is a maximal matching in. The second bound follows.

Using these upper bounds together with Lemma 2 results in a competitive-ratio of 0.667. This competitive ratio is obtained for $\omega(M_2) = \omega(M_1)/e$ and all M_2 edges being of class-D.

Improving the Upper Bound Using the Dual. The following stronger observation is the main tool in our analysis: The gain from edges not in M_1 is not just a solution to the matching LP in $G \setminus M_1$, but also have to obey the fact that any ad can be used at most once in expectation between both M_1 and M_2. Thus we can (significantly) strengthen the constrains in the LP bounding M_2 to:

$$\max \mathbf{1}^T \boldsymbol{p} \text{ subject to}$$
$$\forall a \in A: \sum_{e \in \Gamma_{G \setminus M_1}(a)} p_e \leq 1 - \sum_{e \in \Gamma(a) \cap M_1} p_e$$

In order to find out how much lower the value of this more restricted LP is we switch to the dual. The dual for the weighted matching LP is the minimum vertex cover LP which assigns each vertex a value trying to minimize the sum of all these values subject to the constraint that for each edge e the sum of values on its two endpoints has to be at least one. In the dual of the tightened LP only the weight coefficients in the objective functions have decreased from 1 to $1 - p_e$. This means that the minimum vertex cover in $G \setminus M_1$ (lets call it C) remains a feasible (but not necessarily minimum) solution generating a valid upper bound. While the vertex cover C has a value of $w(M_2)$ using uniform weights, we will prove that (on graphs with many class-D edges on which our algorithm does not perform good) a tighter upper bound on OPT can be obtained by looking at the reweighted value of C:

For this, we analyze the situation of a class-D edge as above. We have two ads a' and a that are matched by M_1 to the impressions i and i' but in M_2 the impression i is matched to a. Let the matching edges be: $e_1 = \{i, a'\}$, $e_2 = \{i', a\}$ and $e_3 = \{i, a\}$ (edges e_1 and e_2 are in M_1, and edge e_3 is in M_2). Let furthermore p_1, and p_2 be the probabilities that edges e_1 and e_2 show up in the offline optimal matching and let c_1 and c_2 be the value of the nodes i and a in the minimum vertex cover on $G \setminus M_1$ corresponding to M_2.

We want to improve upon the upper bound of $OPT \leq (1 - \frac{1}{e})w(M_1) + w(M_2)$ which results from the fact that $p_1, p_2 \leq (1 - 1/e)$ and $c_2 + c_1 = 1$. For this simple situation we can improve over this bound by accounting for the possibility of p_1, p_2 being smaller than (1-1/e) and the new weights of $(1 - p_1)$ and $(1 - p_2)$ in the vertex cover LP for M_2. This improvement is:
$(1 - \frac{1}{e} - p_1) + (1 - \frac{1}{e} - p_2) + p_1 c_1 + p_2 c_2 \geq (1 - \frac{1}{e} - \min\{p_1, p_2\}) + \min\{p_1, p_2\}(c_1 + c_2)$,
which is at least $1 - 1/e$. This implies that we get an improvement of $(1 - 1/e)$ on the previous upper bound for any class-D edge in M_2 that is part of a 3-path in $M_1 \cup M_2$. Unfortunately paths can be longer and the improvements have to be shared between "neighboring" class-D edges. In the general case it is easy to see that one can at least attribute half of this improvement towards any class-D edge e_2. This leads to the improved upper bound of $OPT < (1 - 1/e)\omega(M_1) + \omega(M_2) - D(1 - 1/e)/2$. As the next lemma shows this suffices to prove the 0.6844 competitive ratio:

Lemma 3. *The above algorithm for the online stochastic matching problem has a competitive ratio of at least 0.6844 in expectation.*

Proof. We want to determine the worst ratio of ALG/OPT given the above constraints. For this we first observe that in the worst case $C = 0$. We now fix $r = \omega(M_2)/\omega(M_1)$, $k = D/\omega(M_2)$, scale by $\omega(M_1)^{-1}$ and obtain: $OPT \leq 1$; $OPT \leq (1 - 1/e) + r - rk(1 - 1/e)/2$ and $ALG \geq (1 - 1/e) - rk/e^2 + r(1/e - 1/e^2)$. Observe that both upper bounds on OPT and the lower bound on ALG are linear functions when restricted to r or k. The minimum of ALG/OPT is thus obtained for on a boundary value for r, i.e., $k = 1$ or $k = 0$, or at the point where the tightness of the two upper bounds switches, i.e., when $1/e = r - rk(1 - 1/e)/2$. The case of $r = 0$ results in a competitive ratio of 1 and the case $r = 1$ results in a competitive ratio of at least $(1 - 1/e) - k/e^2 + (1/e - 1/e^2) \geq 1 - 2/e^2 > 0.72$. The last case implies $rk = 2(r - 1/e)/(1 - 1/e)$ and gives a competitive ratio of

$$(1 - 1/e) - (2/e(e - 1))(r - 1/e) + r(1/e - 1/e^2) =$$
$$(1 - 1/e) + 2/e^2(e - 1) - r(2/(e(e - 1))) - 1/e + 1/e^2)$$

We thus have r and k as large as possible, which leads to $k = 1$, $r = 2/(e + 1)$ and a competitive ratio of 0.6844.

A 0.7036-Competitive Algorithm for Online Stochastic Matching. Building on the algorithm above we give a 0.7036-competitive algorithm that employs the power of multiple choices. The analysis crucially exploits the dual LP ideas demonstrated above in a lengthy case analysis. Due to space limitations we omit the analysis here and solely give the algorithm description:

LP Pseudo: max $\Sigma_{e \in E(G)} p_e$ s.t.

$\forall a \in A : \Sigma_{e \in \Gamma(a)} p_e \leq 1$ $\forall i \in I : \Sigma_{e \in \Gamma(i)} p_e \leq 1$

$\forall e \in E(G) : p_e \leq 1 - \dfrac{1}{e}$ $\forall a \in A \ \forall e, e' \in \Gamma(a) : p_e + p_{e'} \leq 1 - \dfrac{1}{e^2}$

Our algorithm uses two matchings M_1 and M_2 as constructed before but employs an additional pseudo-matching that is constructed based on the linear program LP Pseudo. The last inequality of the LP results from the fact that any pair of impressions (in this case, the impressions incident to edges e and e') do not come with probability $(1 - 2/n)^n \approx \frac{1}{e^2}$. The optimum solution, p^* to this LP is thus an upper bound on the expected value of the optimum solution.

For any edge $e' \in M_1 \cup M_2$ we set $p^*_{e'}$ to be zero. We then multiply all probabilities p^* by 2. If the sum of probabilities of some ad $a \in A$ is greater than 1, we reduce the probabilities of some incident edges to a to make it equal to 1. The sum of probabilities of each ad and each impression are at most 1 and 2 respectively. We can make an identical copy of each impression and split the probabilities of the impressions adjacent edges between the edges to the impression and its copy equally to have a fractional matching solution. We then sample a pseudo matching based on this fractional solution. We call the sampled edges a pseudo matching because each impression is adjacent to at most two sampled edges, and each ad is adjacent to at most one. Our algorithm now uses M_1, M_2 and the pseudo matching as follows. When an impression arrives, the priority is to match it based on M_1, if it is not possible we match it based on M_2, and if that is not possible neither, we match it based on the pseudo matching. If the impression is matched to two ads in the pseudo matching, we match it to the ad that has smaller probability of being matched in future (based on its M_1 and M_2 adjacent edges).

Classifying the ads according to what kind of edges they are adjacent to leads to the following result:

Theorem 2. *The expected size of the output of our algorithm is at least 0.7036 times the optimal solution.*

5   Optimal Online Algorithms for Simple Graphs

In this section we briefly discuss our results for solving the online stochastic (weighted) matching problem optimally in a restricted class of graphs.

We define a graph G to have subgraph complexity k if there are at most k subgraphs of G that are connected (and can be obtained by removing ads). Note that, e.g., any cycle or line of length l has a subgraph complexity of at most l^2. Note also that subgraph complexity behaves additively with respect to the vertex-disjoint union of two graphs. From this it follows that the union of two matchings has a subgraph complexity of at most n^2.

The following theorem can be obtained by using dynamic programming to compute the expected matching value $E(C, t)$ for any connected subgraph C when there are exactly t steps left.

Theorem 3. *There is a deterministic algorithm that solves the online stochastic weighted matching problem with T rounds optimally. The algorithm takes $O(kTm)$ time in total for any graph with subgraph complexity k and m edges.*

With this algorithm on hand one can then try to compete against the optimal online or offline algorithm by first reducing the matching graph by throwing away edges to obtain a graph with low subgraph-complexity and then run the optimal online algorithm. This is essentially what is done in the algorithms of [14,4], the non-adaptive algorithm of [20] and our algorithm for the weighted case except that non-optimal heuristics are used to allocate the ads along the edges of two matchings (which form a graph with low subgraph complexity). We leave it as an interesting open question how much the use of this algorithm improves the competitiveness of the above mentioned algorithms and whether it can be useful as a building block for better algorithms.

The last thing to mention is that the above algorithm works in the stated time bound in a much more general setting including edge weights, different non-integral rates for ads and even generalizations to non-bipartite graphs. For the unweighted case with equal frequencies and the union of two matchings (as needed for the algorithms mentioned above) the optimal decision becomes much simpler. Indeed, the following rules suffice: 1.) If the current impression occurs on a path on which at least one side ends on an ad match the impression match it to the ad on the side with the (shorter path to the) ad. 2.) If both sides of the path end on an impression, match it to the side with the longer path. 3.) Otherwise match the impression arbitrarily whenever possible. This strategy can easily be implemented using only an amortized expected $O(\log n)$ time per sampling step.

References

1. Agarwal, G., Goel, G., Karande, C., Mehta, A.: Online vertex-weighted bipartite matching and single-bid budgeted allocation. In: SODA (2011)
2. Agrawal, S., Wang, Z., Ye, Y.: A dynamic near-optimal algorithm for online linear programming. Working paper posted, http://www.stanford.edu/~yyye/

3. Azar, Y., Broder, A.Z., Karlin, A.R., Upfal, E.: Balanced allocations. SIAM J. Comput. 29(1), 180–200 (1999)
4. Bahmani, B., Kapralov, M.: Improved bounds for online stochastic matching. In: de Berg, M., Meyer, U. (eds.) ESA 2010. LNCS, vol. 6346, pp. 170–181. Springer, Heidelberg (2010)
5. Bertsekas, D.: Dynamic programming and optimal control (2007)
6. Bertsekas, D.P., CastanonRollout, D.A.: algorithms for stochastic scheduling problems. Journal of Heuristics 5(1), 89–108 (1999)
7. Buchbinder, N., Jain, K., Naor, J.S.: Online Primal-Dual Algorithms for Maximizing Ad-Auctions Revenue. In: Arge, L., Hoffmann, M., Welzl, E. (eds.) ESA 2007. LNCS, vol. 4698, pp. 253–264. Springer, Heidelberg (2007)
8. de Farias, D.P., Van Roy, B.: On constraint sampling in the linear programming approach to approximate dynamic programming. Math. Oper. Res. 29(3), 462–478 (2004)
9. Devanur, N., Hayes, T.: The adwords problem: Online keyword matching with budgeted bidders under random permutations. In: ACM EC (2009)
10. Devanur, N.R., Jain, K., Sivan, B., Wilkens, C.A.: Near optimal online algorithms and fast approximation algorithms for resource allocation problems. In: Proceedings of the 12th ACM Conference on Electronic Commerce, EC 2011, pp. 29–38. ACM, New York (2011)
11. Farias, V.F., Van Roy, B.: Approximation algorithms for dynamic resource allocation. Oper. Res. Lett. 34(2), 180–190 (2006)
12. Feldman, J., Henzinger, M., Korula, N., Mirrokni, V.S., Stein, C.: Online Stochastic Packing Applied to Display ad Allocation. In: de Berg, M., Meyer, U. (eds.) ESA 2010. LNCS, vol. 6346, pp. 182–194. Springer, Heidelberg (2010)
13. Feldman, J., Korula, N., Mirrokni, V., Muthukrishnan, S., Pál, M.: Online Ad Assignment with Free Disposal. In: Leonardi, S. (ed.) WINE 2009. LNCS, vol. 5929, pp. 374–385. Springer, Heidelberg (2009)
14. Feldman, J., Mehta, A., Mirrokni, V., Muthukrishnan, S.: Online stochastic matching: Beating 1 - 1/e. In: FOCS, p. 1 (2009)
15. Goel, G., Mehta, A.: Online budgeted matching in random input models with applications to adwords. In: SODA, pp. 982–991 (2008)
16. Karande, C., Mehta, A., Tripathi, P.: Online bipartite matching with unknown distributions. In: STOC (2011)
17. Karp, R.M., Vazirani, U.V., Vazirani, V.V.: An optimal algorithm for online bipartite matching. In: Proc. STOC (1990)
18. Mahdian, M., Yan, Q.: Online bipartite matching with random arrivals: A strongly factor revealing lp approach. In: STOC (2011)
19. Mehta, A., Saberi, A., Vazirani, U., Vazirani, V.: Adwords and generalized online matching. In: FOCS (2005)
20. Menshadi, H., OveisGharan, S., Saberi, A.: Offline optimization for online stochastic matching. In: SODA (2011)
21. Mitzenmacher, M.: The power of two choices in randomized load balancing. IEEE Trans. Parallel Distrib. Syst. 12(10), 1094–1104 (2001)
22. Vee, E., Vassilvitskii, S., Shanmugasundaram, J.: Optimal online assignment with forecasts. In: ACM EC (2010)

On Strategy-Proof Allocation
without Payments or Priors

Li Han, Chunzhi Su, Linpeng Tang, and Hongyang Zhang

Shanghai Jiao Tong University
{lihancom.g,suchunzhi,chnttlp,hongyang90}@gmail.com

Abstract. In this paper we study the problem of allocating divisible items to agents without payments. We assume no prior knowledge about the agents. The utility of an agent is additive. The social welfare of a mechanism is defined as the overall utility of all agents. This model is first defined by Guo and Conitzer[7]. Here we are interested in strategy-proof mechanisms that have a good *competitive ratio*, that is, those that are able to achieve social welfare close to the maximal social welfare in all cases. First, for the setting of n agents and m items, we prove that there is no $(1/m + \epsilon)$-competitive strategy-proof mechanism, for any $\epsilon > 0$. And, no mechanism can achieve a *competitive ratio* better than $4/\sqrt{n}$, when $m \geq \sqrt{n}$. Next we study the setting of two agents and m items, which is also the focus of [7]. We prove that the *competitive ratio* of any swap-dictatorial mechanism is no greater than $1/2 + 1/\sqrt{\lceil \log m \rceil}$. Then we give a characterization result: for the case of 2 items, if the mechanism is strategy-proof, symmetric and second order continuously differentiable, then it is always swap-dictatorial. In the end we consider a setting where an agent's valuation of each item is bounded by C/m, where C is an arbitrary constant. We show a mechanism that is $(1/2 + \epsilon(C))$-competitive, where $\epsilon(C) > 0$.

1 Introduction

The agenda of approximate mechanism design without money was first explicitly framed by Procaccia and Tennenholtz in their seminal paper[14], and can be traced back to the work on incentive compatible learning by Deckel et al. [4]. This line of research tries to study how to design truthful mechanisms when payment is not allowed. As noted by Schummer and Vohra[15], "there are many important environments where money cannot be used as a medium of compensation. This constraint can arise from ethical and/or institutional considerations." To this end, Procaccia and Tennenholtz suggests "approximation can be used to obtain strategyproofness without resorting to payments." The *approximation ratio* of a mechanism will be measured on how close it approximates an optimal solution. Following this idea, several models have been studied extensively, for instance, *facility game*[14,9,1], *classification*[10,11,12].

In this paper, we consider the following allocation problem: there are n agents, m heterogeneous, divisible items. Each agent has a private valuation over the

N. Chen, E. Elkind, and E. Koutsoupias (Eds.): WINE 2011, LNCS 7090, pp. 182–193, 2011.

items about which we assume no prior knowledge. Her utility function is linear. A *competitive* allocation mechanism tries to maximize the society's social welfare, that is, the sum of each agent's utility. The competitive ratio is measured by comparing the mechanism's performance to an optimal allocation. We are interested in strategy-proof and at the same time *competitive* mechanisms. While our model is very simple, it also sounds natural. When a central agency tries to allocate various public resources to people efficiently, it faces a similar problem as described in our model. Besides, some resources are divisible in nature, for instance, water, bandwidth, etc. Also, when the items are indivisible and the agents are risk-neutral, the expectation of a randomized mechanism in this case corresponds to a deterministic mechanism in our model.

The problem of resource allocation has been studied in algorithmic game theory on various aspects. R. Johari[8] discusses the problem of allocating an infinitely divisible resource of a fixed capacity to various users, who have their own utility functions and pay money to obtain resources. He gives a proportional mechanism that is quite efficient. There's also work on the allocation of indivisible resources without payments. E. Budish [2] studies a similar combinatorial assignment problem and surveys existing allocation mechanisms. S. Pápai [13] shows that strategy-proof combined with conditions like onto, non-bossiness, etc, can only lead to dictatorship.

Our Results. In the general setting, if we consider an even allocation, that is, allocating each item equally between agents, or a biased plan to allocate all the resources to a single designated agent, the competitive ratios are both $1/m$. In both mechanisms the ratio becomes very small as m grows. Thus the first question arises as:

Question 1. Is there a c-competitive strategy-proof mechanism for any number of agents and items, where $c > 0$?

As it turns out, the answer is negative. We give the following result: there does not exist a $(1/m + \epsilon)$-competitive strategy-proof mechanism, for any $\epsilon > 0$. By a similar technique we also show that the competitive ratio of a strategy-proof mechanism is less than $4/\sqrt{n}$, when $m \geq \sqrt{n}$. This result stands in contrast with the VCG mechanism[16,3,6], which gives an optimal allocation if payments can be used in our model.

Having dealt with the multi-agents setting, next we come to the setting of two agents and any number of items, which is also the focus of Guo and Conitzer[7]. There they used swap-dictatorial(SD) as a basic tool to design strategy-proof mechanisms. The idea of SD is, each agent has some chance to be the dictator, choosing her preferred allocation from a predefined set. The final allocation will be the weighted sum of each agent's choice. We find two interesting results about swap-dictatorial mechanisms. The first is a somewhat surprising link between SD and strategy-proof: In the setting of 2 agents and 2 items, when a mechanism is symmetric and second order continuously differentiable, then strategy-proof coincides with swap-dictatorial. Since items are divisible, the model we are dealing with is inherently a continuous one. The tools from calculus provide us a way to interpret and characterize strategy-proof condition, making the problem much simpler to handle. The second result is that the competitive ratio of an SD mechanism is less than $1/2 + 1/\sqrt{\lceil \log m \rceil}$. In particular this implies when there are too many items, SD is not much better than an even allocation. We remark that when the number of items is small, it is still possible to obtain competitive SD mechanisms. The linear increasing-price[7] mechanism is just swap-dictatorial. In that paper it is also proved that LIP is 0.828-competitive when there are 2 items, nearly matching their established upper bound of 0.841.

Given the negative result on swap-dictatorial mechanisms, it is natural to ask the second question:

Question 2. Is there a c-competitive strategy-proof mechanism for 2 agents, any number of items, where $c > 1/2$?

The question is still *open*. And the only result is that c is smaller than 0.841, as we just mentioned above. Note that our characterization between strategy-proof and SD, if generalized to the any number of items and any mechanism, will give a negative answer to the above question.

Since it appears hard to design a strategy-proof mechanism that beats the 0.5 ratio, and it seems unreasonable to assume that agents' valuations are completely unrestricted, we come to a bounded-valuation setting when an agent's valuation cannot be strongly biased. Here we manage to demonstrate a swap-dictatorial mechanism that is competitive as well, giving a positive answer to Question 2 in a restricted domain.

2   Preliminaries and the Model

We briefly describe our model here, the reader may refer to [7] for more details and discussions.

There are m items, each with capacity 1. These items are allocated to n agents, who keep their valuations on the items in private. The valuation is a *vector* $\boldsymbol{v} = (v_1, \ldots, v_m) \in [0,1]^m$, where $\sum_{i=1}^{m} v_i = 1$. The normalization says when an agent gets all the resources,r she gains one unit of utility. Let $\mathbf{V}$ be the set of valuation vectors. A *valuation matrix* is an $n \times m$ matrix V where each row is a *valuation vector*. We use $\boldsymbol{v}_i$ to denote the i-th row of V, v_{ij} to denote the j-th component of $\boldsymbol{v}_i$. Let $\mathbf{U}$ be the space of valuation matrices. An *allocation vector* $\boldsymbol{o} = (o_1, \ldots, o_m) \in [0,1]^m$. An *allocation matrix* is an $n \times m$ matrix $O = (o_{ij})_{n \times m}$ where $o_{ij} \in [0,1]$ indicates the fraction of item j allocated to agent i, for all $1 \leq i \leq n$, $1 \leq j \leq m$. And $\sum_{i=1}^{n} o_{ij} = 1$, for $1 \leq j \leq m$, i.e., all of the items are allocated. Let $\mathbf{O}$ be the space of allocation matrices.

A *deterministic payment-free* mechanism is a function $M : \mathbf{U} \to \mathbf{O}$. Let $g_i(\boldsymbol{x}, O)$ be the i-th agent's *utility* under allocation O when her valuation is $\boldsymbol{x}$. g_i is additive, that is, $g_i(\boldsymbol{x}, O) = \sum_{j=1}^{m} x_j \cdot o_{ij}$. Let $V(i, \boldsymbol{x})$ be the matrix obtained from substituting the i-th row vector of V by $\boldsymbol{x}$.

M is called *strategy-proof*, if for any *valuation matrix* V, *valuation vector* $\boldsymbol{x}$, $1 \leq i \leq n$, $g_i(\boldsymbol{v}_i, M(V(i, \boldsymbol{v}_i))) \geq g_i(\boldsymbol{v}_i, M(V(i, \boldsymbol{x})))$. In other words, no agent benefits by misreporting his *valuation vector*.

When there are only two agents, for ease of notation, we define the mechanism function as $M : \mathbf{V}^2 \to [0,1]^m$, since it's apparent that the other agent gets $\mathbf{1} - M(\boldsymbol{v}_1, \boldsymbol{v}_2)$, where $\mathbf{1}$ denotes a vector whose components are all 1.[1]

The *social welfare* is defined as $\sum_{i=1}^n g_i(\boldsymbol{v}_i, O)$, that is, the sum of all agents' utilities. The *optimal social welfare* $\gamma(V)$ is the social welfare under an *optimal allocation*, which ideally allocates each item to an agent that values it highest. We measure the competitiveness of a *strategy-proof* mechanism by comparing its achieved *social welfare* to the *optimal social welfare*. More formally, define the *competitive ratio* of a *strategy-proof* mechanism as

$$\min_{V \in \mathbf{U}} \frac{\sum_{i=1}^n g_i(\boldsymbol{v}_i, M(V))}{\gamma(V)}$$

We say that a *strategy-proof* mechanism is α-competitive, if its *competitive ratio* is at least α.

We point out here that randomness does not help provide a more *competitive* mechanism in this model. For if there is an α-competitive strategy-proof randomized mechanism M', then taking the expected outcome of M' gives us a deterministic strategy-proof mechanism that is also α-competitive. However, randomness is still useful for describing a mechanism, as we'll see below.

Definition 1 (Guo, Conitzer[7]). *A mechanism is symmetric if it satisfies:*

1. *Symmetric over the agents: if by swapping the valuations of two agents, their allocations are also swapped correspondingly.*
2. *Symmetric over the items: if by swapping the valuations of two items by each agent, the allocations for these two items are also swapped.*

Let P_{ij} be a permutation matrix that permutes row(or column) i, j. The following proposition is a direct translation of the symmetry condition.

Proposition 1. *A symmetric mechanism M satisfies:*

1. $M(P_{ij}V) = P_{ij} \cdot M(V)$.
2. $M(VP_{ij}) = M(V) \cdot P_{ij}$.

An important property of symmetric mechanisms is the following:

Proposition 2 (Guo, Conitzer[7]). *For any strategy-proof mechanism that is α-competitive, there is a symmetric strategy-proof mechanism that is (at least) α-competitive.*

Next we introduce the family of *swap-dictatorial*[7][2] mechanisms for two agents.

Definition 2. *Let D_1, D_2 be two sets of allocation vectors, $\boldsymbol{v}_1, \boldsymbol{v}_2$ be two valuation vectors. For $i = 1, 2$, let $f_i : \mathbf{V} \to D_i$ be a function such that $f_i(\boldsymbol{v}) \in \mathrm{argmax}_{o \in D_i} \boldsymbol{v} \cdot \boldsymbol{o}$, for any $\boldsymbol{v} \in \mathbf{V}$. A swap-dictatorial mechanism M determined by D_1, D_2 is defined as $M(\boldsymbol{v}_1, \boldsymbol{v}_2) = (f_1(\boldsymbol{v}_1) + \mathbf{1} - f_2(\boldsymbol{v}_2))/2$.*

---

[1] We will always use $\boldsymbol{c}$ to denote a constant vector whose components are all c.

[2] We abbreviate swap-dictatorial by SD sometimes.

There is another intuitive description of SD mechanism: with probability 0.5, agent i becomes the dictator and chooses an allocation from D_i to selfishly maximize her own welfare, leaving the rest to the second agent. The expected outcome will be the resulted allocation. While this description uses randomness, note that SD is actually deterministic.

An SD mechanism is strategy-proof, which can be verified from definition. Intuitively, an agent's utility comes from two parts: one from being the dictator, here there is no incentive to lie; the other from not being the dictator, here she has no influence on the outcome, leaving no benefit from lying either.

A symmetric SD mechanism satisfies two extra conditions:

1. *Symmetric over the items:* if $v = (v_1, \ldots, v_m) \in D_i$, then $(v_{\sigma(1)}, \ldots, v_{\sigma(m)}) \in D_i$, where σ is any permutation.
2. *Symmetric over the agents:* if agent i chooses $u \in D_i$ to maximize her utility at some time, then $u \in D_{-i}$, where $-i$ stands for the other agent. So if we ignore vectors in D_i that are never chosen by agent i, then D_1 and D_2 becomes the same. Since we only care about vectors chosen by an agent for some valuation, in the following we just use the fact that $D_1 = D_2$ for a symmetric SD mechanism.

We have the following characterization for symmetric SD mechanisms.

Theorem 1. *A symmetric strategy-proof mechanism M is SD if and only if for any valuation vector u, v, α,*

$$M(u, v) = M(u, \alpha) + M(\alpha, v) - \frac{1}{2} \tag{1}$$

The proof is omitted here and left to the full version. We also prove:

Corollary 1. *A symmetric strategy-proof mechanism M is SD if and only if for any valuation vectors α, β, u, v,*

$$M(u, \alpha) - M(v, \alpha) = M(u, \beta) - M(v, \beta) \tag{2}$$

3   An Upper Bound for Multiple Agents

Theorem 2. *Fix the number of items m. Let $\epsilon > 0$. There is no $(1/m + \epsilon)$-competitive strategy-proof mechanisms, for some large enough n.*

Proof. We prove by contradiction. Let $n > m$. By Proposition 2, we only need to consider symmetric mechanisms. So assume there is a symmetric mechanism M that is α-competitive on any number of agents, where $\alpha > 1/m$.

Consider the following n by m valuation matrix

$$V = \begin{pmatrix} 1 & 0 & \ldots & 0 & 0 \\ 0 & 1 & \ldots & 0 & 0 \\ & \ldots & & & \\ 0 & \ldots & 1 & 0 & 0 \\ \frac{\epsilon}{n} & \frac{\epsilon}{n} & \ldots & \frac{\epsilon}{n} & 1 - (m-1) * \frac{\epsilon}{n} \\ \frac{\epsilon}{n} & \frac{\epsilon}{n} & \ldots & \frac{\epsilon}{n} & 1 - (m-1) * \frac{\epsilon}{n} \\ & \ldots & & & \end{pmatrix}$$

For row 1 to $m - 2$, each has a 1 in diagonal. Row $m - 1$ to n are the same. ϵ is $2(m - 1)/(\alpha m - 1)$. So it's positive since $\alpha > 1/m$. By Proposition 1, for agent i where $i > m - 2$, their allocation vectors are the same. Thus for each item, an agent gets at most $1/(n - m + 2)$ fraction. And the overall valuations of an agent is 1. This implies the welfare of agent $(m - 1)$ is bounded by $1/(n - m + 2)$.

Now replace agent $(m - 1)$'s type vector by $\boldsymbol{u} = (0, \ldots, 0, 1, 0)$, where the $(m - 1)$-th component is 1. Let V' be the changed valuation matrix.

Let $O = A(V')$. Again by Proposition 1, $o_{11} = \cdots = o_{(m-1)(m-1)}$, denote it x. This is obtained by first exchange row i, j, then column i, j.

Consider the ratio under V'. The observation is, to achieve a good ratio, a mechanism will allocate a relative portion of items to the diagonal 1's. When the amount is large enough then agent $(m - 1)$ has an incentive to lie from $\boldsymbol{v}_{m-1}$ to $\boldsymbol{u}$.

For item 1 to $(m - 1)$, the maximal utility M can achieve is $x + (1 - x) \cdot \epsilon/n$. For item m, the maximal utility M can achieve is 1. And, the optimal allocation gives a utility of $m - (m - 1) \cdot \epsilon/n$. Thus we have:

$$\frac{(x + (1 - x) \cdot \frac{\epsilon}{n}) \cdot (m - 1) + 1}{m - (m - 1) \cdot \frac{\epsilon}{n}} \geq \alpha \Rightarrow x \geq \frac{\frac{\alpha(m - (m-1) \cdot \frac{\epsilon}{n}) - 1}{m - 1} - \frac{\epsilon}{n}}{1 - \frac{\epsilon}{n}} \quad (3)$$

Under allocation O, the welfare of agent $(m - 1)$ in valuation matrix V' is at least $x\epsilon/n$. Meanwhile, in (3), the rightmost formula has limit $(\alpha m - 1)/(m - 1)$, as n grows to infinity. Recall $\epsilon = 2(m - 1)/(\alpha m - 1)$, so formula $x \cdot \epsilon \cdot (n - m + 2)/n$ has limit at least 2 as n becomes infinite. This implies, in particular, for some some large enough n, we have $x\epsilon/n > 1/(n - m + 2)$. So here when agent $(m - 1)$ honestly reports his valuation vector in V, the maximal welfare that can be achieved is $1/(n - m + 2)$. But when she lies as $(0, \ldots, 0, 1, 0)$, the welfare is greater than $x\epsilon/n > 1/(n - m + 2)$, which contradicts with that M is strategy-proof.   □

Note that a $1/m$-competitive mechanism trivially exists: just consider the mechanism that evenly divides each item to each agent. So efficiency really becomes an issue here when there are too many people.

By a refined analysis of the above proof, we can obtain another result quite different in taste.

Theorem 3. *There does not exist a strategy-proof mechanism that achieves a competitive ratio better than $4/\sqrt{n}$, when $m \geq \sqrt{n}$.*

This theorem also implies: as the number of agents and items approaches infinite, the competitive ratio of any strategy-proof mechanism approaches 0.

4   Allocation between Two Agents

4.1   An Upper Bound for Swap-Dictatorial Mechanisms

Now we come to the setting of two agents. As mentioned above, SD is very intuitive, so it becomes very helpful for designing strategy-proof mechanisms.

However, the ratio of SD may not be very good, as shown by the following theorem:

Theorem 4. *The competitive ratio of any swap-dictatorial mechanism is less than* $1/2 + 1/\sqrt{\lceil \log m \rceil}$.

Proof. Again by Proposition 2 it suffices to consider symmetric mechanisms. Let M be a symmetric SD mechanism with a competitive ratio of $1/2 + \delta$. Let O be dictator's choice space. Let $m_1 = 2^{\lceil \log m \rceil}$, $m_{i+1} = m_i/2$.

We define a series of variables for case i. First let the two agent's valuation vectors be:

$$u_i = (x, \ldots, x, y, \ldots, y, 0, \ldots, 0)$$
$$v_i = (y, \ldots, y, x, \ldots, x, 0, \ldots, 0)$$

where there are $m_i/2$ consecutive x, y respectively and $y/x = t = \delta < 1$. And when agent 1 acts as the dictator, she chooses vector $o_i \in O$. Vector o_i is associated with two parameters, a_i, b_i, indicating the average allocation on the portions of x, y respectively. We will show that as i increases, a_i increases by a relative amount in order to keep up the competitive ratio. However a_i cannot be greater than 1, from this seemingly contradiction we derive a bound on the competitive ratio.

By Proposition 1, when agent 2 becomes dictator, it picks $o_i \in O$ with some permutation. So it also takes on average a_i of the x part and b_i of the y part.

Now we compute the ratio for such an allocation, by definition it is greater than $\frac{1}{2} + \delta$:

$$(x \cdot a_i + y \cdot b_i + y \cdot (1 - a_i) + x \cdot (1 - b_i)) \cdot \frac{m_i}{2} \cdot \frac{1}{2} \cdot 2 \geq (\frac{1}{2} + \delta) \cdot x \cdot m_i$$

Note that the optimal utility comes from allocating the first $m_i/2$ items to agent 1 and the next $m_i/2$ items to agent 2.

Rearrange the inequality, we get

$$a_i - b_i \geq \frac{2\delta - t}{1 - t} \tag{4}$$

On the other hand, since agent 1 chooses o_i from the dictator space to maximize utility, it must be greater than that obtained from choosing o_{i-1}, as has been obtained from case $i - 1$. And, by symmetry there is a permutation of o_{i-1} in O such that the average of the first $m_i/2$ components is no less than the average of the second $m_i/2$ components. Denote it o. By comparing agent 1's utility between choosing o_i and o, we obtain:

$$(x \cdot a_i + y \cdot b_i) \cdot m_i/2 \geq (x + y) \cdot a_{i-1} \cdot m_i/2$$
$$\Rightarrow a_i \frac{1}{t+1} + b_i \frac{t}{t+1} \geq a_{i-1}$$

Together with (4) we get $a_i \geq a_{i-1} + t(2\delta - t)/(1 - t^2)$. Since $a_1 \geq 0$, we obtain $a_k \geq (k-1) \cdot t(2\delta - t)/(1 - t^2)$. Let $k = \lceil \log m \rceil$, then:

$$(\lceil \log m \rceil - 1)\frac{t(2\delta - t)}{1 - t^2} \leq a_k \leq 1$$

Substitute t by δ, we have $\delta \leq 1/\sqrt{\lceil \log m \rceil}$. □

4.2 Relation between Swap-Dictatorial and Strategy-Proof Mechanisms

The family of SD mechanism is one kind of strategy-proof mechanisms in our model. Together with symmetry it becomes a useful tool for designing strategy-proof mechanisms. However, it is the only family of strategy-proof mechanism we have found yet, except under some variations like letting the non-dictator choose from a set of allocations that all maximize the utility of the dictator, So, could there be any relation between these concepts? In this subsection we will give a partial result. Before discussing this question, we need to introduce some notations first.

Let $M : \mathbf{V}^2 \to [0,1]^m$ be a mechanism. Let $\boldsymbol{u}$, $\boldsymbol{v}$ be two valuation vectors. Define

$$F : S^2 \to [0,1]^m \qquad (5)$$

where $S = \{(x_1, \ldots, x_{m-1}) : 0 \leq \sum_{i=1}^{m-1} x_i \leq 1 \text{ and } x_i \geq 0, \ \forall \ 1 \leq i \leq m - 1\}$. And $F(u_1, \ldots, u_{m-1}, v_1, \ldots, v_{m-1}) = M(\boldsymbol{u}, \boldsymbol{v})$.

Basically this definition isolates variables upon which a mechanism is defined. It is essential here since we are going to analyze a mechanism mathematically.

Let $int(S)$ be the *interior* of S, i.e., when $0 < \sum_{i=1}^{m-1} x_i < 1$. For a slight abuse of notation, we simply use F to stand for a mechanism and when we say $\boldsymbol{u}$ is a *valuation vector*, it is a $(m-1)$-dimensional vector which can be extended as an agent's valuation. Each component of F can also be viewed as a function on S, we use f_i to denote the i-th component. These notations will be used for the rest of this subsection.

Now we are ready to define continuously differentiable mechanisms.

Definition 3 (Continuously Differentiable Mechanism). *We say a mechanism M is continuously differentiable if and only if f_i is continuously differentiable(or $f_i \in \mathcal{C}$) on T^2, for $i = 1, \ldots, m$, where $T = int(S)$.*[3]

Similarly, M is second order continuously differentiable if and only if $f_i \in \mathcal{C}^2$ when the domain is restricted to T^2, for $i = 1, \ldots, m$.

Now we'll analyze a symmetric strategy-proof mechanism M, which is also second order continuously differentiable. Let F be defined as (5). First we give another condition on whether a mechanism is SD based on differentiable assumption.

---

[3] We take the trouble to distinguish S from the interior of S, since a continuously differentiable function can only be defined on an open set.

Lemma 1. *If for any valuation vectors $\boldsymbol{u}, \boldsymbol{v} \in int(S)$,*

$$\frac{\partial^2 F}{\partial u_i \partial v_j}(\boldsymbol{u}, \boldsymbol{v}) = \boldsymbol{0}, \forall 1 \leq i, j \leq m - 1 \tag{6}$$

Then M is swap-dictatorial.

Proof. For any $\boldsymbol{\alpha}, \boldsymbol{\beta} \in int(S)$, we show that (2) can be inferred from (6). We prove the equality for the first component of F, while others follow similarly. We first do an integration to change the first parameter from $\boldsymbol{u}$ to $\boldsymbol{v}$, followed by another integration on the second parameter to change $\boldsymbol{\alpha}$ to $\boldsymbol{\beta}$.

$$(f_1(\boldsymbol{u}, \boldsymbol{\alpha}) - f_1(\boldsymbol{v}, \boldsymbol{\alpha})) - (f_1(\boldsymbol{u}, \boldsymbol{\beta}) - f_1(\boldsymbol{v}, \boldsymbol{\beta})) \tag{7}$$

$$=(f_1(\boldsymbol{u}, \boldsymbol{\alpha}) - f_1(\boldsymbol{u}, \boldsymbol{\beta})) - (f_1(\boldsymbol{v}, \boldsymbol{\alpha}) - f_1(\boldsymbol{v}, \boldsymbol{\beta})) \tag{8}$$

$$= \int_0^1 dy \, (\boldsymbol{\beta} - \boldsymbol{\alpha}) \cdot \int_0^1 dx \, (\boldsymbol{u} - \boldsymbol{v}) \nabla_x \nabla_y f_1((\boldsymbol{u} - \boldsymbol{v})x + \boldsymbol{v}, (\boldsymbol{\beta} - \boldsymbol{\alpha})y + \boldsymbol{\alpha}) \tag{9}$$

$$=0 \tag{10}$$

Here $\nabla_x \nabla_y f_1$ is a $(m-1) \times (m-1)$ matrix and $(\nabla_x \nabla_y f_1)_{i,j} = \frac{\partial^2 f_1}{\partial x_i \partial y_j}(\boldsymbol{x}, \boldsymbol{y})$. Since $f_1 \in \mathcal{C}^2$, $(\boldsymbol{u} - \boldsymbol{v}) \nabla_x \nabla_y f_1$ is actually $\|\boldsymbol{u} - \boldsymbol{v}\|$ times the directional derivative of $\nabla_y f_1$ along the direction of $\boldsymbol{u} - \boldsymbol{v}$.

So (7) holds in the interior of S. Since f_1 is continuous, by taking a limit it also holds in S. □

Now we are about to give the main result of this subsection. Before that, we first need Clairaut's theorem[5].

Lemma 2 (Clairaut's Theorem). *If $f : \mathbb{R}^n \to \mathbb{R}$ has continuous second partial derivatives at any given point in $\mathbb{R}^n$, say, $(a_1, a_2, \ldots, a_n)$, then for $1 \leq i, j \leq n$,*

$$\frac{\partial^2 f}{\partial x_i \partial x_j}(a_1, \ldots, a_n) = \frac{\partial^2 f}{\partial x_j \partial x_i}(a_1, \ldots, a_n)$$

In words, the partial derivations of this function are commutative at that point.

Lemma 3. *For any valuation vectors $\boldsymbol{u}, \boldsymbol{v} \in int(S)$, we have:*

$$\frac{\partial^2 F}{\partial u_i \partial v_j}(\boldsymbol{u}, \boldsymbol{v}) = \boldsymbol{0}, \forall 1 \leq i, j \leq m - 1 \tag{11}$$

Proof. We merely prove for f_1, the first component of F, without loss of generality. By the strategy-proof condition, the first agent can't make more profits by misreporting his valuation vector, so

$$\frac{\partial f_1}{\partial u_i}(\boldsymbol{u}, \boldsymbol{v}) \cdot \boldsymbol{u}' = 0$$

where $\boldsymbol{u}'$ extends $\boldsymbol{u}$ to an m-dimensional vector, the last component being $1 - \sum_{i=1}^{m-1} u_i$. Taking a partial derivative on v_j,

$$\frac{\partial}{\partial v_j} \frac{\partial f_1}{\partial u_i}(\boldsymbol{u}, \boldsymbol{v}) \cdot \boldsymbol{u}' = 0$$

The derivative can be pushed inside the inner product because $\boldsymbol{u}$ is independent of v_j. Exchanging the role of $\boldsymbol{u}, \boldsymbol{v}$, similarly we get

$$\frac{\partial}{\partial u_i} \frac{\partial f_1}{\partial v_j}(\boldsymbol{u}, \boldsymbol{v}) \cdot \boldsymbol{v}' = 0$$

$\boldsymbol{v}'$ is defined similarly to $\boldsymbol{u}'$.

Since f_1 has continuous second partial derivative at $(\boldsymbol{u}, \boldsymbol{v})$, by Clairaut's theorem

$$\frac{\partial^2 f_1}{\partial u_i \partial v_j}(\boldsymbol{u}, \boldsymbol{v}) = \frac{\partial}{\partial v_j} \frac{\partial f_1}{\partial u_i}(\boldsymbol{u}, \boldsymbol{v}) = \frac{\partial}{\partial u_i} \frac{\partial f_1}{\partial v_j}(\boldsymbol{u}, \boldsymbol{v})$$

In conclusion, $\frac{\partial^2 f_1}{\partial u_i \partial v_j}$ is simultaneously perpendicular to $\boldsymbol{u}', \boldsymbol{v}' \in \mathbb{R}^2$. When $\boldsymbol{u} \neq \boldsymbol{v}$ (i.e. $\boldsymbol{u}' \neq \boldsymbol{v}'$) , it must be the case that $\frac{\partial^2 f_1}{\partial u_i \partial v_j}(u, v) = \boldsymbol{0}$. Since $\frac{\partial^2 f_1}{\partial u_i \partial v_j}(u, v)$ is continuous, it is $\boldsymbol{0}$ when $\boldsymbol{u}$ equals $\boldsymbol{v}$ too.   □

Combining the results of Lemma 1 and Lemma 3, we have

Theorem 5. *In the case of allocating 2 items to 2 agents, if a mechanism M is symmetric and second order continuously differentiable, then M is strategy-proof if and only if M is swap-dictatorial.*

We remark that the assumption of second order continuously differentiable can be extended to the case when the function may have finitely many discontinuous points, since integration can be done in that case too. To find a function beyond this assumption, one may need to think about some quite unnatural functions.

We hope Theorem 5 helps to explain the difficulty we encountered in designing strategy-proof mechanisms that are not SD. It will be interesting to see if Theorem 5 still holds in more general cases, or if there exists other families of strategy-proof mechanisms.

4.3   Bounded Valuation

Section 4.2 gives evidence that symmetric strategy-proof allocation may be, actually, swap-dictatorial. And from Theorem 4 SD can only achieve a competitive ratio of 0.5 when the number of items approaches infinity, which is no better than an even allocation. But can SD do better than even allocation, when we impose some restrictions on the valuation vectors? In this subsection, we see that if an agent's valuation is not too biased, SD can do better than even allocation. To put it formally, we define:

Definition 4 (Bounded Valuation). *Let $\boldsymbol{v} = (v_1, \ldots, v_m)$ be a valid valuation vector, we say $\boldsymbol{v}$ is bounded by T, if $v_i \leq T$, for any $i = 1, \ldots, m$. A valuation space is bounded by T if each vector of the space is bounded by T.*

Let T be C/m. Note that if we allow C to grow arbitrarily large as m grows, then a proof similar to Theorem 4's shows that there is still no SD mechanism that is $(0.5 + \epsilon)$-competitive on a valuation space bounded by C/m, for any $\epsilon > 0$. However, when C is some fixed constant, the proof no longer holds, and we can actually find an SD mechanism that does better than 0.5.

Definition 5 (Sphere Mechanism). *Let* $f(\boldsymbol{u}) = (\frac{u_1 \cdot c}{\|\boldsymbol{u}\|}, \frac{u_2 \cdot c}{\|\boldsymbol{u}\|}, \ldots, \frac{u_m \cdot c}{\|\boldsymbol{u}\|})$, *where* $c = \sqrt{m}/C$ *and* $\|\cdot\|$ *denotes the* L_2-norm. *Given two valuation vectors* $\boldsymbol{u}$, $\boldsymbol{v}$,

$$M(\boldsymbol{u}, \boldsymbol{v}) = \frac{f(\boldsymbol{u}) + 1 - f(\boldsymbol{v})}{2} \tag{12}$$

Here c is chosen such that each component of $f(\boldsymbol{u})$ does not exceed 1.

Our SD has nice mathematical interpretations. The dictator's choice space is:

$$D = \left\{ \frac{c}{\|\boldsymbol{u}\|} \boldsymbol{u} : \boldsymbol{u} \text{ is a valuation vector bounded by } C/m \right\}$$

So all the vectors in the choice space are in a sphere of radius c. To maximize utility, the dictator will choose the vector of the same direction to its valuation vector, i.e., a dictator with valuation vector $\boldsymbol{u}$ will choose $c\boldsymbol{u}/\|\boldsymbol{u}\|$, as M does in (12). Since SD is always strategy-proof, the Sphere mechanism is strategy-proof as well.

Next we analyze the competitive ratio.

Theorem 6. *Let the valuation space* V *be bounded by* C/m. *Then Sphere mechanism is* $(\frac{1}{2} + \epsilon)$-competitive, for some $\epsilon > 0$.

Due to space limitation, we omit the proof and leave it to the full version.

5   Conclusions and Future Research

In this paper we studied allocation problem when there are no payments or priors. While this model is only proposed recently, we hope the results and proof techniques in this paper provide insight into the model. There are still several problem unsettled for this problem. The first is whether there exists a strategy-proof mechanism that beats the 0.5 ratio. There's still a large gap here since the only known result is a 0.841 upper bound. The second is to what extent are strategy-proof equivalent to swap-dictatorial mechanisms in this model. Another direction for future research is to consider other social optimal criterion such as egalitarian criterion, or handle issues like fairness in the model.

Acknowledgments. We are very grateful to Pinyan Lu for introducing this problem to us, as well as his insightful suggestions and warm encouragement throughout our work. And we owe many thanks to the anonymous referees for helpful comments.

References

1. Alon, N., Feldman, M., Procaccia, A.D., Tennenholtz, M.: Strategyproof approximation of the minimax on networks. Mathematics of Operations Research 35(3), 513–526 (2010)
2. Budish, É.: The combinatorial assignment problem: Approximate competitive equilibrium from equal incomes (2009) Working Paper
3. Clarke, E.H.: Multipart pricing of public goods. Public choice 11(1), 17–33 (1971)
4. Dekel, O., Fischer, F., Procaccia, A.D.: Incentive compatible regression learning. In: Proceedings of the Nineteenth Annual ACM-SIAM Symposium on Discrete Algorithms, pp. 884–893 (2008)
5. Partial derivative. In: Hazewinkel, M. (ed.) Encyclopaedia of Mathematics. Springer, Heidelberg (2001)
6. Groves, T.: Incentives in teams. Econometrica: Journal of the Econometric Society, 617–631 (1973)
7. Guo, M., Conitzer, V.: Strategy-proof allocation of multiple items between two agents without payments or priors. In: Proceedings of the 9th International Conference on Autonomous Agents and Multiagent Systems: International Foundation for Autonomous Agents and Multiagent Systems, vol. 1, pp. 881–888 (2010)
8. Johari, R.: The price of anarchy and the design of scalable resource allocation mechanisms. In: Nisan, N., Roughgarden, T., Tardos, É., Vazirani, V. (eds.) Algorithmic Game Theory, ch. 21. Cambridge Univ. Pr. (2007)
9. Lu, P., Sun, X., Wang, Y., Zhu, Z.A.: Asymptotically optimal strategy-proof mechanisms for two-facility games. In: Proceedings of the 11th ACM Conference on Electronic Commerce, pp. 315–324. ACM (2010)
10. Meir, R., Procaccia, A.D., Rosenschein, J.S.: Strategyproof classification under constant hypotheses: A tale of two functions. In: Proceedings of the 23rd AAAI Conference on Artificial Intelligence (AAAI), pp. 126–131 (2008)
11. Meir, R., Procaccia, A.D., Rosenschein, J.S.: On the limits of dictatorial classification. In: Proceedings of the 9th International Conference on Autonomous Agents and Multiagent Systems International Foundation for Autonomous Agents and Multiagent Systems, vol. 1, pp. 609–616 (2010)
12. Meir, R., Procaccia, A.D., Rosenschein, J.S.: Strategyproof classification under constant hypotheses: A tale of two functions. In: Fox, D., Gomes, C.P. (eds.) AAAI, pp. 126–131. AAAI Press (2008)
13. Pápai, S.: Strategyproof and nonbossy multiple assignments. Journal of Public Economic Theory 3(3), 257–271 (2001)
14. Procaccia, A.D., Tennenholtz, M.: Approximate mechanism design without money. In: Proceedings of the tenth ACM Conference on Electronic Commerce, pp. 177–186. ACM (2009)
15. Schummer, J., Vohra, R.V.: Mechanism design without money. In: Nisan, N., Roughgarden, T., Tardos, É., Vazirani, V. (eds.) Algorithmic Game Theory, ch. 10, Cambridge Univ. Pr. (2007)
16. Vickrey, W.: Counterspeculation, auctions, and competitive sealed tenders. The Journal of finance 16(1), 8–37 (1961)

Demand Allocation Games: Integrating Discrete and Continuous Strategy Spaces

Tobias Harks and Max Klimm*

Institut für Mathematik, Technische Universität Berlin
{harks,klimm}@math.tu-berlin.de

Abstract. In this paper, we introduce a class of games which we term *demand allocation games* that combines the characteristics of finite games such as congestion games and continuous games such as Cournot oligopolies. In a strategy profile each player may choose both an action out of a finite set and a non-negative demand out of a convex and compact interval. The utility of each player is assumed to depend solely on the action, the chosen demand, and the aggregated demand on the action chosen. We show that this general class of games possess a pure Nash equilibrium whenever the players' utility functions satisfy the assumptions *negative externality*, *decreasing marginal returns* and *homogeneity*. If one of the assumptions is violated, then a pure Nash equilibrium may fail to exist. We demonstrate the applicability of our results by giving several concrete examples of games that fit into our model.

1 Introduction

The problem of allocating scarce resources to satisfy demands is a central topic in the operations research and optimization literature. While a central planer may compute and implement an optimal allocation, in many applications this may be impossible as the allocation of resources is determined by selfish players. A prominent example for this scenario are congestion games. In a congestion game, there is a set of resources and a pure strategy of a player consists of a subset of resources. The profit of a resource depends only on the number of players choosing the resource, and the utility of a player is the sum of the profits of the chosen resources. Under these assumptions, Rosenthal proved the existence of a pure Nash equilibrium (PNE for short) [25]. Another well-known variant of congestion games arises if players can *fractionally* demand the resources, see Beckmann [2] and Haurie and Marcotte [10] for related models. For this *continuous* variant, the quite general result of Rosen [24] implies the existence of a PNE provided the strategy space is convex and compact and utility functions are concave. In the context of such discrete and continuous classes of games, there are mainly two types of existence theorems for PNE. The first type applies to discrete games (such as classical congestion games and many variants thereof)

---

* This research was supported by the Deutsche Forschungsgemeinschaft within the research training group 'Methods for Discrete Structures' (GRK 1408).

N. Chen, E. Elkind, and E. Koutsoupias (Eds.): WINE 2011, LNCS 7090, pp. 194–205, 2011.

where each player has a *finite* strategy space. For this type, the existence of PNE is proved by either potential function arguments (as in [1,4,5,6,9,22,25]), or by using the combinatorial structure of the finite strategy space (as in [13,16,26]). On the other hand, for continuous games, existence of PNE is usually established via fixed-point theorems of Kakutani (as in [24] for general concave games) and Brouwer (as in [19] for mixed extensions of finite games), or by a monotonicity property of the best reply functions (as in [21,23] for Cournot oligopolies).

While existence of PNE for both extremes is well understood, much less is known for strategic games that exhibit continuous and discrete elements at the same time. To motivate this point we give an example. Consider the classical Cournot oligopoly (cf. [3,30]). In a Cournot oligopoly game, there is a set of firms each producing quantities so as to satisfy an elastic demand. The production cost for every player is modeled by a cost function and the interaction of firms comes from the market price function which is dependent on the total supply on the market. In this form, a Cournot oligopoly game belongs to the class of continuous games and under mild assumptions on the market aggregation function, the existence of a pure Cournot equilibrium follows from Rosen [24]. The situation changes, if there are several (parallel) markets, and each Cournot player can select exactly one market to offer its quantity. The restriction of choosing only one market arises if the market is regulated, e.g., if each firm may only purchase one market license, see for instance Stähler and Upmann [29] for related models. In this case, the strategy of a player is now discrete in the sense that exactly one market can be chosen, and it is continuous in the sense that the production quantity is still continuously variable on the chosen market. Yet, we give another example related to models of population behavior in biology. Suppose there is a set of exhaustible food patches distributed on an area shared by different populations of animals (e.g., sticklebacks as in the experiment of Milinsky [18] or herds of zebras and elephants sharing water locations). Analyzing the equilibrium behavior of such systems belongs to the field of population games, see the book by Sandholm [27] and further references therein. Here, every population is represented by a fixed-sized continuum of infinitesimal small individuals each choosing a food patch. By definition (cf. [27, Chapter 2, condition (v)]) such games are continuous in the sense that the individuals are sufficiently small and may be assigned to different locations even if they belong to the same population. If the populations of animals correspond to swarms or herds the continuity assumption breaks down as swarms or herds move as a whole. Moreover, in reality the size of every population is not fixed but it correlates with the available amount of food supply. For systems having the above described characteristics, a new model is needed that integrates continuous and discrete action spaces.

In this paper, we introduce a class of games which we term *demand allocation games* that comprises the characteristics of the examples above. Suppose we are given a finite set A of actions and a finite set N of players. Each player is associated with a subset $A_i \subseteq A$ of actions allowable to her and a convex and compact interval of non-negative demands. In a strategy profile, a player chooses both a feasible action and a feasible demand for her. We additionally require the following assumptions on the player's utility functions. We assume that the utility of each player is not affected by the strategic choices of players on other actions. This assumption is often referred to as "Independence of Irrelevant Choices", see for instance Konishi et al. [13] and Voorneveld et al. [31] for a similar model with fixed demands. Moreover, we require that the game is anonymous in the sense that the utility of each player depends solely on the aggregate demand of all players playing the same action, which is a common assumption, see e.g. Konishi et al. [13]. It is a useful observation that under these basic assumptions the utility of each player i, when choosing action a_i together with demand d_i can be represented by an indirect utility function $v_i^{a_i} : \mathbb{R}_{\geq 0} \times \mathbb{R}_{\geq 0}$ so that $u_i(a, d) = v_i^{a_i}(d_i, \ell_{-i}^{a_i}(a, d))$, where $\ell_{-i}^{a_i}(a, d) = \sum_{j \in N \setminus \{i\} : a_j = a} d_j$ denotes the aggregated demand (or load) of other players on action a_i. Clearly, in this general form nothing can be said about the existence of pure Nash equilibrium. Therefore, we require more structure about the player's utility functions. We define the following three assumptions on the player's utility functions that capture the properties of the above examples. The first assumption is called "Negative Externality" (EXT for short) and requires that the utility of a player using an action a_i decreases if the aggregate demand of other players playing the same action increases. Informally, the second assumption "Decreasing Marginal Returns" (DMR for short) requires that for every player the marginal return function exists and decreases when both that player's demand and the total demand of the chosen action increase. The last assumption is called "Homogeneity" (HOM for short) and requires that for all $i \in N$, we have $v_i^{a_i} = v_i^{b_i}$ for all $a_i, b_i \in A_i$. This last assumption is clearly the most restrictive and controversial one. We will show, however, that if it is dropped, there are instances without PNE.

Our Results. As our main result, we prove that every demand allocation game satisfying EXT, DMR and HOM possesses a PNE. This result is tight in the sense that if one of the assumptions is dropped, there is a demand allocation game without PNE. For proving this existence result we provide an algorithm that computes a PNE. Our algorithm relies on iteratively computing a (partial) equilibrium on every action separately using Rosen's theorem. Here, a partial equilibrium is a strategy profile that is resilient against unilateral demand deviations. Given a partial equilibrium, the algorithm selects a player that can play a better and best response. After such a best response it recomputes the partial equilibrium and proceeds in the same fashion. We prove that a player-specific load vector of the partial equilibria lexicographically decreases in every iteration and thus, the algorithm terminates. A perhaps surprising property of our proof is that even though we iteratively recompute a partial equilibrium by using Rosen's theorem as a black box, there is enough structure of such a partial equilibrium to prove that the algorithm terminates. We also show that demand allocation games do not have the finite improvement property even if EXT, DMR and HOM are satisfied, thus, they are not potential games. For demand allocation games with only two players, we prove that already EXT and DMR are enough to yield a PNE. In the final section of the paper, we give a series of concrete examples that fit into our model: Cournot games on parallel markets, singleton congestion games with player-specific payoff functions and variable demands, and games in biology.

2   The Model

Let A be a finite set of actions and let N be a finite set of players. For each player $i \in N$ we are given a convex and closed interval $D_i = [\alpha_i, \omega_i] \subseteq \mathbb{R}_{\geq 0}$ of allowable demands and a subset $A_i \subseteq A$ of allowable actions. A *strategy* of player i is a tuple (a_i, d_i) where $a_i \in A_i$ is an allowable action and $d_i \in D_i$ is an allowable demand for player i. A *strategy profile* of the game is a tuple (a, d) where $a = (a_i)_{i \in N}$ is the action vector and $d = (d_i)_{i \in N}$ is the demand vector. We assume that the *utility* of player i under strategy profile (a, d) depends solely on the action a_i and the demand d_i chosen by player i, and the total demand of other players with the same action $\ell_{-i}^{a_i}(a, d) = \sum_{j \in N \setminus \{i\}: a_i = a_j} d_j$. To measure this utility, we introduce for each player i and each of her allowable actions $a_i \in A_i$ an *indirect utility function* $v_i^{a_i} : \mathbb{R}_{\geq 0} \times \mathbb{R}_{\geq 0} \to \mathbb{R}$. The utility of player i under strategy profile (a, d) is then defined as $u_i(a, d) = v_i^{a_i}(d_i, \ell_{-i}^{a_i}(a, d))$. We are interested in establishing conditions on the indirect utility functions that ensure the existence of at least one pure Nash equilibrium. Formally, a strategy profile (a, d) is a *pure Nash equilibrium*, PNE for short, if $u_i(a, d) \geq u_i(a_i', a_{-i}, d_i', d_{-i})$ for all players $i \in N$ and all strategies $(a_i', d_i') \in A_i \times D_i$. We make the following three assumptions on the indirect utility functions $v_i^{a_i}$ of player i and action $a_i \in A_i$. The first assumption is called "Negative Externality" and requires that the utility of every player increases as the total demand of other players with the same action decreases.

Assumption 1 (Negative Externality (EXT)). *For all $i \in N$, $a_i \in A_i$ and $d_i \in D_i$, the indirect utility function $v_i^{a_i}(d_i, \cdot)$ is non-increasing in the second entry, that is $v_i^{a_i}(d_i, \ell_{-i}) \geq v_i^{a_i}(d_i, \ell_{-i}')$ for all $\ell_{-i}, \ell_{-i}' \in \mathbb{R}_{\geq 0}$ with $\ell_{-i} \leq \ell_{-i}'$.*

This assumption is natural when players compete over scare resources to satisfy their demand and has been made explicitly or implicitly in various contexts ranging from traffic and communication networks (i.e. [2,10,12]) to biology (i.e. [18]) and economics (i.e. in Cournot oligopolies [3,30] and Cournot oligopsonies [11,20]).

The second assumption is called "Decreasing marginal returns" and requires that for players with a non-trivial interval of allowable demands, the marginal utility function exists, is continuously differentiable, and decreases if the player's demand and the total demand of the chosen action increase.

Assumption 2 (Decreasing Marginal Returns (DMR)). *For all $i \in N$ with $\alpha_i < \omega_i$, $d_i \in D_i$, $a_i \in A_i$, and $\ell_{-i} \in \mathbb{R}_{\geq 0}$, the marginal return function $\partial v_i^{a_i}(d_i, \ell_{-i}) / \partial d_i$ exists and is continuously differentiable in d_i. Moreover, $\partial v_i^{a_i}(d_i, \ell_{-i}) / \partial d_i > \partial v_i^{a_i}(d_i', \ell_{-i}') / \partial d_i'$ for all $d_i, d_i' \in [\alpha, \omega]$ and $\ell_{-i}, \ell_{-i}' \in \mathbb{R}_{\geq 0}$ with $d_i \leq d_i'$ and $d_i + \ell_{-i} \leq d_i' + \ell_{-i}'$, where at least one of these two inequalities is strict.*

The assumption that the utility of player i is concave in her demand often appears in the literature on Cournot oligopolies (cf. [21,23]) in order to get the existence of an equilibrium. Also many works in telecommunications (cf. [12,28]) justify concavity of the utility function in the demand variable by application-specific characteristics such as the rate-control algorithm used in the TCP protocol. Note that if for some $i \in N$ it holds that $\alpha_i = \omega_i$, then DMR is trivially satisfied.

The next assumption "Homogeneity" imposes that players have no a priori preferences over actions, that is, each player's utility is solely defined by her own demand and the total demand of the chosen action and *not* by the identity of the action itself.

Assumption 3 (Homogeneity (HOM)). *For all $i \in N$, we have $v_i^{a_i} = v_i^{b_i}$ for all $a_i, b_i \in A_i$.*

In games that satisfy HOM, we may write $v_i = v_i^{a_i} = v_i^{b_i}$ for all $a_i, b_i \in A_i$. Note that HOM does not require symmetry among players, i.e., we still allow $v_i \neq v_j$ for $i \neq j$. We only require that every player is indifferent between any two allowable actions as long as their own demand and the total demand on these actions is equal. Clearly, HOM is the most restrictive and controversial assumption. We show, however, that homogeneity is necessary in the sense that if it is dropped, there are games without PNE.

3   Existence of Pure Nash Equilibria

In this section, we will give an existence result for demand allocation games. Specifically, we will show that demand allocation games satisfying the assumptions Negative Externality (EXT), Decreasing Marginal Return (DMR) and Homogeneity (HOM) always possess a PNE. Our results are "tight" in the sense that if any of the three assumptions is dropped, then there are instances without a PNE. To prove our main result, we first introduce the concept of a *partial equilibrium*. Intuitively, a partial equilibrium is a strategy profile that is resilient against unilateral demand deviations. Formally, a strategy profile (a, d) is a partial equilibrium if $u_i(a, d) \geq u_i(a, d_i', d_{-i})$ for all $i \in N$ and $d_i' \in D_i$. Using the result of Rosen [24], we will prove that under assumption DMR for every strategy profile (a, d), there is a partial equilibrium of the form $(a, \tilde{d})$. We say that $(a, \tilde{d})$ is an *associated* partial equilibrium to (a, d).

Proposition 1. *Let G be a demand allocation game. Under assumption DMR, for every strategy profile (a, d), there is an associated partial equilibrium $(a, \tilde{d})$.*

Proof. Pick an arbitrary strategy profile (a, d) of G. Consider the restricted demand allocation game $\tilde{G}$ with $\tilde{A}_i = \{a_i\}$. In $\tilde{G}$ the strategy space of each player reduces to the convex and closed interval $D_i \subseteq \mathbb{R}$. Using DMR, the utility function of each player is continuous and concave in d_i. By Rosen's existence theorem [24, Theorem 1], a pure Nash equilibrium of $\tilde{G}$ exists. Hence, each PNE of $\tilde{G}$ is an associated partial equilibrium to (a, d). □

The following lemma will be important throughout this paper. It expresses the first-order optimality conditions of a partial equilibrium. The proof is straightforward and left to the reader.

Lemma 2. *Let (a,d) be a partial equilibrium. Then, for all $i \in N$ with $\alpha_i < \omega_i$ the following conditions hold: $\partial u_i(a,d) / \partial d_i \leq 0$ if $d_i = \alpha_i$, $\partial u_i(a,d) / \partial d_i = 0$ if $d_i \in (\alpha_i, \omega_i)$, and $\partial u_i(a,d) / \partial d_i \geq 0$ if $d_i = \omega_i$.*

For an action $b \in A$, we define the *active set* on action b under strategy profile (a,d) as $N^b(a,d) = \{i \in N : a_i = b\}$. We need the following lemma.

Lemma 3 (Uniqueness Lemma). *Let (a,d) and (a',d') be two partial equilibria of a demand allocation game satisfying DMR. Then,*

1. *$\ell^b(a,d) = \ell^b(a',d')$ for all $b \in A$ with $N^b(a,d) = N^b(a',d')$,*
2. *$\ell^b(a,d) \leq \ell^b(a',d')$ for all $b \in A$ with $N^b(a,d) \subseteq N^b(a',d')$.*

Proof. Obviously, it suffices to prove 2. Assume by contradiction that there is $b \in A$ with $\ell^b(a,d) > \ell^b(a',d')$ and $N^b(a,d) \subseteq N^b(a',d')$. This implies the existence of a player $i \in N^b(a,d)$ with $d_i > d'_i$. In particular, we have $\omega_i \geq d_i > d'_i \geq \alpha_i$. The conditions of Lemma 2 for a partial equilibrium give $\partial u_i(a,d) / \partial d_i \geq 0$ and $\partial u_i(a',d') / \partial d'_i \leq 0$. We get

$$0 \geq \frac{\partial u_i(a',d')}{\partial d'_i} = \frac{\partial v_i^b\big(d'_i, \ell^b_{-i}(a',d')\big)}{\partial d'_i} \overset{\text{DMR}}{>} \frac{\partial v_i^b\big(d_i, \ell^b_{-i}(a,d)\big)}{\partial d_i} = \frac{\partial u_i(a,d)}{\partial d_i} \geq 0,$$

a contradiction. □

We are now ready to present a procedure for proving the existence of PNE. We claim that the following iterative contraction-switching procedure converges to a PNE.

1. Start with arbitrary strategy profile (a,d)
2. **Contraction phase:** Let $(a,\tilde{d})$ be an associated partial equilibrium
3. **Switching phase:** If there is a player i who can improve unilaterally, pick a best reply $(a'_i, d'_i) \in \arg\max_{(a''_i, d''_i) \in A_i \times D_i} u_i(a''_i, a_{-i}, d''_i, \tilde{d}_{-i})$, set $(a,d) = (a''_i, a_{-i}, d''_i, \tilde{d}_{-i})$ and proceed with 2. Else, return $(a,\tilde{d})$.

Note that in Step 2, we actually call an oracle that gives us an associated partial equilibrium. The oracle takes as input a restricted demand allocation game $\tilde{G}$ and outputs an associated partial equilibrium. By Proposition 1 this is always possible.

In the following, we will show that this procedure ends after finitely many steps (involving finitely many calls of the oracle) and outputs a PNE. The following properties are the key to prove that the contraction-switching procedure terminates.

Lemma 4. *Let G be a demand allocation game satisfying DMR, EXT and HOM, let (a,d) be a partial equilibrium, let (a'_i, d'_i) be a best and better reply of player i and let $(a'_i, a_{-i}, \tilde{d})$ be an associated partial equilibrium. Then, the following properties hold.*

1. $\ell^{a'_i}(a'_i, a_{-i}, d'_i, d_{-i}) < \ell^{a_i}(a, d)$ *(Switching Property)*
2. $\ell^{a'_i}(a'_i, a_{-i}, \tilde{d}) \le \ell^{a'_i}(a'_i, a_{-i}, d'_i, d_{-i})$ *(Contraction Property)*
3. $\ell^{a_i}(a'_i, a_{-i}, \tilde{d}) \le \ell^{a_i}(a, d)$ *(Monotonicity Property)*

Proof. We begin proving the switching property. For the sake of a contradiction, assume $\ell^{a'_i}(a'_i, a_{-i}, d'_i, d_{-i}) \ge \ell^{a_i}(a, d)$. We consider the following three cases:

First case $d'_i > d_i$: As (a, d) is a partial equilibrium and $d_i < d'_i \le w_i$, by Lemma 2 we have $0 \ge \partial u_i(a, d) / \partial d_i$. We calculate

$$0 \ge \frac{\partial u_i(a, d)}{\partial d_i} = \frac{\partial v_i(d_i, \ell^{a_i}_{-i}(a, d))}{\partial d_i}$$
$$\overset{\text{DMR}}{>} \frac{\partial v_i(d'_i, \ell^{a'_i}_{-i}(a'_i, a_{-i}, d'_i, d_{-i}))}{\partial d'_i} = \frac{\partial u_i(a'_i, a_{-i}, d'_i, d'_{-i})}{\partial d'_i} \ge 0,$$

a contradiction. The equalities use the assumption HOM. The last inequality stem from the facts that (a'_i, d'_i) is a best reply of player i and that $d'_i > d_i \ge \alpha_i$.

Second case $d'_i = d_i$: Using $\ell^{a'_i}_{-i}(a'_i, a_{-i}, d'_i, d'_{-i}) \ge \ell^{a_i}_{-i}(a, d)$ and assumptions EXT and HOM, we obtain

$$u_i(a'_i, a_{-i}, d'_i, d_{-i}) = v_i\big(d'_i, \ell^{a'_i}_{-i}(a'_i, a_{-i}, d'_i, d_{-i})\big) \le v_i\big(d_i, \ell^{a_i}_{-i}(a, d)\big) = u_i(a, d).$$

We derive that player i does not improve, a contradiction to the fact that (a'_i, d'_i) is a better reply of player i.

Third case $d'_i < d_i$: Consider the strategy (a_i, d'_i) of player i. Observe that $\ell^{a_i}_{-i}(a, d'_i, d_{-i}) < \ell^{a_i}_{-i}(a, d)$ as $d'_i < d_i$. We obtain

$$u_i(a, d'_i, d_{-i}) = v_i\big(d'_i, \ell^{a_i}_{-i}(a, d'_i, d_{-i})\big)$$
$$\overset{\text{EXT}}{\ge} v_i\big(d'_i, \ell^{a'_i}_{-i}(a'_i, a_{-i}, d'_i, d_{-i})\big) = u_i(a'_i, a_{-i}, d'_i, d_{-i}) > u_i(a, d),$$

where the equalities use the assumption HOM and the first inequality uses the assumption EXT. Thus, (a, d) is not a partial equilibrium, contradiction!

We proceed by proving the contraction property. For a contradiction, suppose that $\ell^{a'_i}(a'_i, a_{-i}, \tilde{d}) > \ell^{a'_i}(a'_i, a_{-i}, d'_i, d'_{-i})$. Then, at least one of the following two cases holds: Either $\tilde{d}_i > d'_i$ or there is a player $j \in N^{a'_i}(a'_i, a_{-i}, \tilde{d}) \setminus \{i\}$ with $\tilde{d}_j > d_j$. If $\tilde{d}_i > d'_i$, we have $\partial u_i(a'_i, a_{-i}, d'_i, d'_{-i}) / \partial d'_i \le 0$ using the fact that (a'_i, d'_i) was a best reply of player i and that $d'_i < \tilde{d}_i \le w_i$. By the assumptions of decreasing marginal values, we obtain

$$0 \ge \frac{\partial u_i(a'_i, a_{-i}, d'_i, d_{-i})}{\partial d'_i} = \frac{\partial v_i(d'_i, \ell^{a'_i}_{-i}(a'_i, a_{-i}, d'_i, d_{-i}))}{\partial d'_i}$$
$$\overset{\text{DMR}}{>} \frac{\partial v_i\big(\tilde{d}_i, \ell^{a'_i}_{-i}(a'_i, a_{-i}, \tilde{d})\big)}{\partial \tilde{d}_i} = \frac{\partial u_i(a'_i, a_{-i}, \tilde{d})}{\partial \tilde{d}_i} \ge 0,$$

a contradiction.

If there is on the other hand $j \in N^{a'_i}(a'_i, a_{-i}, \tilde{d}) \setminus \{i\}$ with $\tilde{d}_j > d_j$, then we have $\partial u_j(a, d) / \partial d_j \leq 0$ as (a, d) was a partial equilibrium and $d_j < \tilde{d}_j \leq \omega_j$. We then get the same contradiction as for player i.

The monotonicity property follows directly from Lemma 3. $\square$

We are now ready to state and prove our main result.

Theorem 5. *For demand allocation games, assumptions DMR, EXT, HOM yield the existence of a PNE.*

Proof. By using the previous lemmas, we show that the contraction-switching procedure terminates for any given starting profile (a, d). First notice that there are only finitely many action vectors $a = (a_i)_{i \in N}$ as both the number of players and the number of actions is finite. We will show that each possible action vector is visited at most once in the contraction-switching procedure.

To this end, we consider for a strategy profile (a, d), the vector $L(a, d) = (\ell^{a_i}(a, d))_{i \in N}$. We shall prove that $L(a, d)$ strictly decreases with respect to the sorted lexicographical order $\prec_{\text{lex}}$ that is defined as follows. For two vectors $u, v \in \mathbb{R}^n_{\geq 0}$ we say that u is *sorted lexicographically* smaller than v, written $u \prec_{\text{lex}} v$, if there is an index $k \in \{1, \dots, n\}$ such that $u_{\pi(i)} = v_{\psi(i)}$ for all $i < k$ and $u_{\pi(k)} < v_{\psi(k)}$ where π and ψ are permutations that sort the vectors u and v non-increasingly, that is, $u_{\pi(1)} \geq u_{\pi(2)} \geq \cdots \geq u_{\pi(n)}$ and $v_{\psi(1)} \geq v_{\psi(2)} \geq \cdots \geq v_{\psi(n)}$. To see that $L(a, d)$ lexicographically decreases, let (a, d) be a partial equilibrium and let (a'_i, d'_i) be a best and better reply of player i. Denote by $(a'_i, a_{-i}, \tilde{d})$ the partial equilibrium associated with strategy profile $(a'_i, a_{-i}, d'_i, d_{-i})$. Clearly, for every player $j \in N \setminus (N^{a_i}(a, d) \cup N^{a'_i}(a, d))$ we have $L_j(a, d) = L_j(a'_i, a_{-i}, d'_i, d_{-i})$. The switching property proven in Lemma 4 ensures that the load on the new action a'_i stays strictly below that of the old action a_i, that is, $\ell^{a'_i}(a'_i, a_{-i}, d'_i, d_{-i}) < \ell^{a_i}(a, d)$. The contraction property ensures that, after the new set of players on the new action a'_i settles to an associated partial equilibrium, the total demand will not increase, that is, $\ell^{a'_i}(a'_i, a_{-i}, \tilde{d}) \leq \ell^{a'_i}(a'_i, a_{-i}, d'_i, d_{-i})$. It follows that $\ell^{a'_i}(a'_i, r_{-i}, \tilde{d}) < \ell^{a_i}(a, d)$. Also, by the monotonicity property we have $\ell^{a_i}(a'_i, a_{-i}, \tilde{d}) \leq \ell^{a_i}(a, d)$. Thus, we have shown that the entry $L_i(\cdot)$ of player i strictly decreases and that none of the changed entries becomes larger than $L_i(a, d)$, hence, the vector $L(\cdot)$ lexicographically decreases after one iteration of the contraction-switching procedure. This fact, together with the uniqueness of the load vector proven in Lemma 3, implies that the algorithm never visits the same action vector twice and, thus, terminates after finitely many steps. $\square$

Note that the existence result of Theorem 5 is tight; if one of the assumption three assumptions DMR, EXT, and HOM is dropped then one can construct a game satisfying the other two assumptions that does not have a PNE. We can also provide an example of a game satisfying DMR, EXT, and HOM that has an improvement cycle. Thus, demand allocation games are not potential games, in general. Formal proofs of the above results appear in the full version of this paper.

3.1   Two Player Demand Allocation Games

In this section, we turn to the case of two players. We will show that any two-player demand allocation game that satisfies the assumptions EXT and DMR possesses a PNE.

Theorem 6. *For two-player demand allocation games, assumptions EXT and DMR yield the existence of a PNE.*

Proof. We shall prove that the following procedure computes a PNE. Start with the empty strategy profile and let player 1 choose a best reply (a_1, d_1). Then, let player 2 choose a best reply (a_2, d_2) to (a_1, d_1). If $a_1 \neq a_2$, we have reached a PNE as EXT implies that player 1 has no interest in switching to action a_2. The only interesting case is $a_1 = a_2$. Let $\tilde{x} = (a_1, a_2, \tilde{d}_1, \tilde{d}_2)$ be an associated partial equilibrium to $x = (a_1, a_2, d_1, d_2)$. We first show that $\tilde{d}_1 \leq d_1$. For a contradiction, suppose $\tilde{d}_1 > d_1$. Because $d_1 < \tilde{d}_1 \leq \omega_1$, we have $\partial v_1^{a_1}(d_1, 0) / \partial d_1 \leq 0$ as (a_1, d_1) was a best reply. On the other hand, we have $\partial v_1^{a_1}(\tilde{d}_1, \tilde{d}_2) / \partial \tilde{d}_1 \geq 0$ as $\tilde{d}_1 > d_1 \geq \alpha_1$ and $\tilde{x}$ is a partial equilibrium. We obtain $0 \leq \partial v_1^{a_1}(\tilde{d}_1, \tilde{d}_2) / \partial \tilde{d}_1 < \partial v_1^{a_1}(d_1, 0) / \partial d_1 \leq 0$, by the assumption DMR, a contradiction.

Next, we show $u_2(\tilde{x}) \geq u_2(x)$. To see this, note that $u_2(\tilde{x}) = v_2^{a_1}(\tilde{d}_2, \tilde{d}_1) \geq v_2^{a_1}(d_2, \tilde{d}_1) \geq v_2^{a_1}(d_2, d_1) = u_2(x)$, where the first inequality uses the fact that $\tilde{x}$ is a partial equilibrium and the second inequality stems from the assumption EXT and the fact that $\tilde{d}_1 \leq d_1$.

Because $u_2(\tilde{x}) \geq u_2(x)$ and (a_2, d_2) was a best reply of player 2, there is no improvement move of player 2 from $\tilde{x}$. If player 1 does not want to deviate as well, $\tilde{x}$ is a PNE and we are done. If on the other hand (a_1', d_1') is a best reply of player 1, we let player 1 deviate and let player 2 play a best reply (a_2', d_2'). Note that player 2 will only adapt her demand, that is $a_2' = a_2 = a_1$. It is shown in the proof of Lemma 3 that the equilibrium demand of a player does not increase as the load increases, thus, $d_2' \geq \tilde{d}_2$. Then, player 1 will not want to switch again to action a_1. Also player 2 will not deviate as her payoff may only decrease when adapting her demand. Hence, we have reached a PNE.   □

Note that the above result is tight in the sense that if either DMR or EXT are dropped, then there exist two-player games without a PNE.

4   Examples

We now give several examples of games that fall into the class of demand allocation games.

Cournot Competition on Parallel Markets. Cournot games (cf. Cournot [3], Mas-Colell et al. [15] and Tirole [30]) are among the most fundamental models of strategic interaction between firms. In a Cournot game, players correspond to firms that produce a homogeneous product. In each strategy, each firm chooses its production quantity d_i out of a compact and convex interval $[\alpha_i, \omega_i]$ of allowable production quantities. The price for which these quantities are sold is given by a non-increasing market reaction function $P : \mathbb{R}_{\geq 0} \to \mathbb{R}_{\geq 0}$ that maps the total supply of the market $\ell = \sum_{i \in N} d_i$ to the market price for selling the produced quantity. Given a strategy profile $d = (d_i)_{i \in N}$, the utility of firm i is given as $u_i(d) = P(\ell)\, d_i - C_i(d_i)$, where $C_i : [\alpha_i, \omega_i] \to \mathbb{R}$ is a non-decreasing production cost function of player i.

Demand allocation games contain a natural generalization of Cournot games that we term *Cournot games on parallel markets*. In such games, there is a set A of markets each endowed with a non-increasing market reaction function $P_a, a \in A$. The markets are called *identical* if $P_a = P_b$ for all $a, b \in A$. In each strategy profile, each player chooses both a market a_i out of a player-specific set $A_i \subseteq A$ of allowable markets and a production quantity $d_i \in [\alpha_i, \omega_i]$. Given a strategy profile (a, d), the utility of player i is then defined as $u_i(a, d) = P_{a_i}(\ell^{a_i}(a, d))\, d_i - C_i(d_i)$. Cournot games on identical parallel markets with continuously differentiable and strictly concave market reaction function and continuously differentiable and convex production cost functions are demand allocation games satisfying assumptions EXT, DMR, and HOM and thus possess a PNE. For games with two players (originally studied by Cournot), a PNE exists even if HOM is violated.

Singleton Congestion Games. The class of congestion games is a well-studied class of games introduced by Rosenthal [25]. As congestion games with weighted players and/or player-specific costs may fail to have a PNE (see the counterexamples given in [6,7,14] for weighted congestion games and [16,17] for games with player-specific costs) many authors focused on singleton strategies. Here, a PNE is guaranteed to exists, even when players are weighted (see [1,4,5,9,26]) or costs are player-specific (see [13,16]). However, games with weighted players *and* player-specific costs need not possess a PNE [16].

In many situations, however, the assumption that the demand of each player is *fixed* is unrealistic. In a previous work [8], we studied congestion games with elastic demands. In that work, we show that affine or certain exponential cost functions yield the existence of a PNE. We did not study, however, the case of player-specific costs. Demand allocation games include *singleton congestion games with variable demands and player-specific costs* as a special case. In such games, the incentive of each player i to use higher demands is stimulated by a reward function $U_i : \mathbb{R}_{\geq 0} \to \mathbb{R}$ that defines the reward received from the chosen demand. Given a strategy profile (a, d), the utility of player i is defined as $u_i(a, d) = U_i(d_i) - c_i^{a_i}(\ell^{a_i}(a, d))$, where $\ell^{a_i}(a, d) = \sum_{j \in N : a_j = a_i} d_j$ is the load of resource a_i under strategy profile (a, d). Singleton congestion games with variable demands and player-specific costs are demand allocation games. If reward functions are continuously differentiable and strictly concave functions and for each player all costs functions are equal, continuously differentiable, non-decreasing and convex they satisfy assumptions EXT, DMR, and HOM and thus possess a PNE. For two-player games, we can drop assumption HOM and still get the existence of a PNE.

Games in Biology. Consider population behavior in biology as described in the introduction. The food patches correspond to the actions and the population-specific costs $c_i^{a_i}(\ell^{a_i}(a,d))$ capture the rivalry for food supply. The size of population i is given by an inverse demand function, say $f_i : \mathbb{R}_{\geq 0} \to \mathbb{R}_{\geq 0}$ that is decreasing in the population specific costs. Thus, defining $v_i(d_i, \ell^{a_i}(a,d)) = \int_0^{d_i} f_i(z) - c_i^{a_i}(\ell^{a_i}(a,d) - d_i + z)\, dz$ models the tradeoff between food supply and population size, see also Milchtaich (cf. [16]) for a detailed discussion of congestion games used in biology. His actual model, however, involves fixed demands only.

References

1. Andelman, N., Feldman, M., Mansour, Y.: Strong price of anarchy. Games Econom. Behav. 65(2), 289–317 (2009)
2. Beckmann, M., McGuire, C., Winsten, C.: Studies in the Economics and Transportation. Yale University Press (1956)
3. Cournot, A.: Recherches Sur Les Principes Mathematiques De La Theorie De La Richesse, Hachette, Paris (1838)
4. Even-Dar, E., Kesselman, A., Mansour, Y.: Convergence time to Nash equilibrium in load balancing. ACM Trans. Algorithms 3(3), 1–21 (2007)
5. Fotakis, D.A., Kontogiannis, S.C., Koutsoupias, E., Mavronicolas, M., Spirakis, P.G.: The Structure and Complexity of Nash Equilibria for a Selfish Routing Game. In: Widmayer, P., Triguero, F., Morales, R., Hennessy, M., Eidenbenz, S., Conejo, R. (eds.) ICALP 2002. LNCS, vol. 2380, pp. 123–134. Springer, Heidelberg (2002)
6. Fotakis, D., Kontogiannis, S., Spirakis, P.: Selfish unsplittable flows. Theor. Comput. Sci. 348(2-3), 226–239 (2005)
7. Goemans, M., Mirrokni, V., Vetta, A.: Sink equilibria and convergence. In: Proc. 46th Annual IEEE Sympos. Foundations Comput. Sci., pp. 142–154 (2005)
8. Harks, T., Klimm, M.: Congestion games with variable demands. In: Krzysztof, R. (ed.) Proc. 13th Biannual Conf. Theoretical Aspects of Rationality and Knowledge, pp. 111–120 (2011)
9. Harks, T., Klimm, M., Möhring, R.H.: Strong Nash Equilibria in Games with the Lexicographical Improvement Property. In: Leonardi, S. (ed.) WINE 2009. LNCS, vol. 5929, pp. 463–470. Springer, Heidelberg (2009)
10. Haurie, A., Marcotte, P.: On the relationship between Nash-Cournot and Wardrop equilibria. Networks 15, 295–308 (1985)
11. Johari, R., Tsitsiklis, J.: Efficiency loss in cournot games. Technical report, LIDS-P-2639, Laboratory for Information and Decision Systems. MIT (2005)
12. Kelly, F., Maulloo, A., Tan, D.: Rate Control in Communication Networks: Shadow Prices, Proportional Fairness, and Stability. J. Oper. Res. Soc. 49, 237–252 (1998)
13. Konishi, H., Le Breton, M., Weber, S.: Equilibria in a model with partial rivalry. J. Econom. Theory 72(1), 225–237 (1997)
14. Libman, L., Orda, A.: Atomic resource sharing in noncooperative networks. Telecommun. Syst. 17(4), 385–409 (2001)
15. Mas-Colell, A., Whinston, M., Green, J.: Microeconomic Theory. Oxford University Press (1995)
16. Milchtaich, I.: Congestion games with player-specific payoff functions. Games Econom. Behav. 13(1), 111–124 (1996)

17. Milchtaich, I.: The Equilibrium Existence Problem in Finite Network Congestion Games. In: Spirakis, P.G., Mavronicolas, M., Kontogiannis, S.C. (eds.) WINE 2006. LNCS, vol. 4286, pp. 87–98. Springer, Heidelberg (2006)
18. Milinsky, M.: An evolutionary stable feeding strategy in sticklebacks. Z. Tierpsychol. (51), 36–40 (1979)
19. Nash, J.: Non-cooperative games. PhD thesis. Princteon (1950)
20. Naylor, R.: Pay discrimination and imperfect competition in the labor market. J. Econ. 60(2), 177–188 (1994)
21. Novshek, W.: On the existence of Cournot equilibrium. Rev. Econom. Stud. 52(1), 85–98 (1985)
22. Panagopoulou, P., Spirakis, P.: Algorithms for pure Nash equilibria in weighted congestion games. ACM J. Exp. Algorithmics 11(2.7), 1–19 (2006)
23. Roberts, J., Sonnenschein, H.: On the existence of Cournot equilibrium without concave profit functions. J. Econom. Theory 22, 112–117 (1976)
24. Rosen, J.: Existence and uniqueness of equilibrium points in concave n-player games. Econometrica 33(3), 520–534 (1965)
25. Rosenthal, R.: A class of games possessing pure-strategy Nash equilibria. Internat. J. Game Theory 2(1), 65–67 (1973)
26. Rozenfeld, O., Tennenholtz, M.: Strong and Correlated Strong Equilibria in Monotone Congestion Games. In: Spirakis, P.G., Mavronicolas, M., Kontogiannis, S.C. (eds.) WINE 2006. LNCS, vol. 4286, pp. 74–86. Springer, Heidelberg (2006)
27. Sandholm, W.H.: Population Games and Evolutionary Dynamics. MIT Press (2010)
28. Shenker, S.: Fundamental design issues for the future internet. IEEE J. Sel. Area Comm. 13, 1176–1188 (1995)
29. Stähler, F., Upmann, T.: Market entry regulation and international competition. Rev. Internat. Econ. 16(4), 611–626 (2008)
30. Tirole, J.: The Theory of Industrial Organization. MIT Press (1988)
31. Voorneveld, M., Borm, P., van Megen, F., Tijs, S., Facchini, G.: Congestion games and potentials reconsidered. Internat. Game Theory Rev. 1(3-4), 283–299 (1999)

Controlling Infection by Blocking Nodes and Links Simultaneously*

Jing He[1], Hongyu Liang[1], and Hao Yuan[2]

[1] Institute for Interdisciplinary Information Sciences, Tsinghua University
{he-j08,lianghy08}@mails.tsinghua.edu.cn
[2] Department of Computer Science, City University of Hong Kong
haoyuan@cityu.edu.hk

Abstract. In this paper we study the problem of controlling the spread of undesirable things (viruses, epidemics, rumors, etc.) in a network. We present a model called the *mixed generalized network security model*, denoted by $\mathrm{MGNS}(d)$, which unifies and generalizes several well-studied infection control model in the literature. Intuitively speaking, our goal under this model is to secure a subset of nodes and links in a network so as to minimize the expected total loss caused by a possible infection (with a spreading limit of d-hops) plus the cost spent on the preventive actions. Our model has wide applications since it incorporates both node-deletion and edge-removal operations. Our main results are as follows:

1. For all $1 \le d < \infty$, we present a polynomial time $(d+1)$-approximation algorithm for computing the optimal solution of $\mathrm{MGNS}(d)$. This improves the approximation factor of $2d$ obtained in [19] for a special case of our model. We derive an $O(\log n)$-approximation for the case $d = \infty$. Moreover, we give a polynomial time $\frac{3}{2}$-approximation for $\mathrm{MGNS}(1)$ on bipartite graphs.

2. We prove that for all $d \in \mathbb{N} \cup \{\infty\}$, it is $\mathcal{APX}$-hard to compute the optimum cost of $\mathrm{MGNS}(d)$ even on 3-regular graphs. We also show that, assuming the Unique Games Conjecture [13], we cannot obtain a $(\frac{3}{2} - \epsilon)$-approximation for $\mathrm{MGNS}(d)$ in polynomial time. Our hardness results hold for the special case $\mathrm{GNS}(d)$ in [19] as well.

3. We show that an optimal solution of $\mathrm{MGNS}(d)$ can be found in polynomial time for every fixed $d \in \mathbb{N} \cup \{\infty\}$ if the underlying graph is a tree, and the infection cost and attack probability are both uniform. Our algorithm also works for the case where there are budget constraints on the number of secured nodes and edges in a solution. This in particular settles an open question from [21] that asks whether there exists an efficient algorithm for the minimum average contamination problem on trees.

---

* The first two authors were supported in part by the National Basic Research Program of China Grant 2007CB807900, 2007CB807901, and the National Natural Science Foundation of China Grant 61033001, 61061130540, 61073174. Portions of this work were supported by a grant from City University of Hong Kong (Project No. 7200218).

N. Chen, E. Elkind, and E. Koutsoupias (Eds.): WINE 2011, LNCS 7090, pp. 206–217, 2011.

1   Introduction

During the recent years, much effort has been devoted to the study on the structure of various types of networks such as social networks, wireless sensor networks, computer networks, transportation networks, and the World Wide Web. An important and active subject is to study the *information diffusion process* in the situations where we want some news, topics, thoughts or products to spread quickly in the network, such as viral marketing [8]. This idea is formalized by Kempe, Kleinberg and Tardos [12] as a combinatorial problem called the *influence maximization problem*, which has since then been extensively studied under various settings (see, e.g., [6,10,15,20]).

In contrast, another important line of research is to study how to prevent or limit the spread of undesirable things through the network, such as the progation of computer viruses and worms over computer networks, the fast spreading of malicious rumors through social networks, and the spread of infections or epidemics (such as Swine Flu and H1N1) among groups of people. In all these circumstances we need to eliminate or at least control the evolution of the bad things over the whole network, which is usually achieved by taking some preventive measures before the emergence of these undesirable things, and isolating or restricting the behaviors of some individuals if the infection has already been spread through the network. An important issue in real-world applications is the balance between the cost spent on prevention and the expected loss caused by infection. For example, installing anti-virus softwares on the computers is a natural response to the possible virus attack, but it may cost a lot of money and bring inefficiency to the protected computers due to high maintenance cost or memory requirement.

An elegant model that integrates both the security and infection costs has been formalized by Aspnes, Chang and Yampolskiy [3]. In their model, we seek for a subset of nodes on which we shall install the anti-virus softwares (call such nodes *secure*). A virus-attack is initiated by choosing one node from the network uniformly at random, and this node, if not secure, will infect all other nodes that are reachable from it in the network with all secure nodes removed. The goal is to minimize the cost for installing softwares (*security cost*) plus the expected total loss caused by the virus (*infection cost*). They consider both centralized (optimization) and game-theoretic settings. The model is substantially generalized by Kumar et al. [19] by allowing individual security and infection costs and arbitrary distribution of the virus-attack probability, and by introducing a parameter d into the model that represents the distance within the network that an infection can spread. This new model is called the *generalized network security model*, denoted GNS(d). Thus, GNS(d) is able to capture networks with less infection power or limited local information, such as ad hoc wireless networks. An issue with GNS(d) is that it lacks the power of modeling the action of restricting the interconnections between individuals in the network (instead of simply removing them from the network), which, in the graph language, corresponds to blocking edges in the graph instead of deleting nodes. In spirit of such consideration, the contamination minimization model where edges are supposed to be blocked is raised by [16] and has been further studied in, e.g., [17,18,21].

In this paper, we present a model for minimizing the spread of infection that unifies and further generalizes the two aforementioned approaches, which we call the *mixed generalized network security model*, denoted by MGNS(d). In our model, each node has its own *security cost* and *infection cost* as in GNS(d), and each edge has its own *link-blocking cost* that represents the lost caused by the removal of the edge. The attack probability distribution can be arbitrary as in GNS(d). The insecure node that is attacked initially will infect exactly those nodes that are within distance at most d from it in the *attack graph* obtained by removing all secure nodes and blocked edges from the original network. The cost of a solution is equal to the total expected infection cost of the nodes plus the cost for securing nodes and blocking edges in this solution. The goal is then to find a solution with minimum cost. Our main results in this paper, some of which improve on the previously best known results achieved for special cases of our model, are given in the following.

1. For all $1 \leq d < \infty$, we present a polynomial time $(d+1)$-approximation algorithm for computing the optimal solution of MGNS(d) based on the primal-dual method. This improves the approximation factor of $2d$ obtained in [19] for GNS(d), which is a special case of MGNS(d). (We note that it is possible to design a reduction from MGNS(d) to GNS($2d$), which will give us a $4d$-approximation for MGNS(d) using the algorithm in [19]. However, the reduction loses a lot of information about the topology of the underlying network.) For the case $d = \infty$, we derive an $O(\log n)$-approximation for MGNS(∞) that matches the result of [19] for GNS(∞). Moreover, we give a polynomial time $\frac{3}{2}$-approximation for MGNS(1) on bipartite graphs.

2. We prove that for all $d \in \mathbb{N} \cup \{\infty\}$, it is $\mathcal{APX}$-hard to compute the optimum cost of GNS(d) even if the graph is 3-regular and all costs and probability are uniform, thus ruling out the possibility of designing PTAS for the problem. We also show that, assuming the Unique Games Conjecture [13], we cannot obtain a $(\frac{3}{2} - \epsilon)$-approximation for GNS(d) in polynomial time. To our knowledge these are the first inapproximability results for GNS(d). Since GNS(d) is a special case of MGNS(d), all the hardness results trivially apply to MGNS(d).

3. We show that an optimal solution of MGNS(d) can be found in polynomial time for every fixed $d \geq 1$ or $d = \infty$ if the underlying graph is a tree, and the infection cost and attack probability are both uniform. Our algorithm can handle all $d \leq O(\sqrt{\log n})$ in polynomial time on bounded-degree trees. Our algorithm also works for the case where budget constraints are put on the number of nodes and edges that can be secured and blocked respectively in a wanted solution. In particular, this settles an open question of [21] that asks whether there exists an efficient algorithm for the *minimum average contamination problem* on trees (which will be mentioned later in more detail). We remark that the tree structure, despite being special, has applications in hierarchically-organized networks such as company relationships.

Paper Organization. In the rest part of this section, we rigorously define our model and compare it with some previous work. In Section 2 we present approximation algorithms for MGNS(d). Hardness of approximation results for MGNS(d) are given in Section 3. Section 4 copes with tree instances of MGNS(d). Finally, in Section 5 we conclude the whole paper and propose some open problems and future research directions.

1.1   Our Model for Infection Control

In this subsection we explain the mixed generalized network security model MGNS(d) in more detail, where $d \in \mathbb{N}^+ \cup \{\infty\}$ is a parameter that, intuitively, reflects the "degree of infectivity" within the network. Although we will describe our model in terms of preventing virus-spreading in computer networks, one should keep in mind that the model is capable of many other situations where we wish to minimize the propagation of undesirable things. Specifically, our model MGNS(d) comprises the following ingredients:

Contact Graph, Costs and Strategy. The contact graph is an undirected graph $G = (V, E)$, where $V = \{1, 2, \ldots, n\}$ denotes the set of computers in a connected network, and $E \subseteq V^2$ specifies the underlying topology of the network. Thus, an edge $\{u, v\} \in E$ indicates that nodes (computers) u and v are directly connected, so that u can potentially affect v if it is infected by a computer virus or worm, and vice versa. For each $v \in V$, let C_v denote the *security cost* of v (for installing an anti-virus software on v), and L_v the *infection cost* of v (for recovering it from a virus attack). For each $e \in E$, let C'_e denote the *link-blocking cost* of e (for the lost caused by the removal of e). All the costs are non-negative. In a *strategy (solution)*, we need to decide on which nodes to install anti-virus softwares and which edges to block. A node with anti-virus software installed on it is called *secure*, and otherwise is called *insecure*. Similarly we have *blocked* and *unblocked* edges. A solution S is also identified with $V_S \cup E_S$, where $V_S \subseteq V$ is the set of secure nodes in S and $E_S \in E$ is the set of blocked edges in S. The *attack graph* of a solution is the graph obtained from G by removing all secure nodes and blocked edges.

Infection Model and Social Cost. We assume that the virus is initiated at *exactly one* node chosen from V according to the *attack probability distribution* $\{w_v \mid v \in V\}$, where $\sum_{v \in V} w_v = 1$. Write $w(S) := \sum_{v \in S} w_v$ for $S \subseteq V$. A secure node will neither suffer from the virus nor transmit the virus to other nodes (although it can be chosen as the attacked node), whereas an insecure node, if chosen as the attacked node, will infect exactly those nodes at distance at most d from it in the attack graph (including itself). For a strategy S, let $V_S^{\leq d}(v)$ denote the set of nodes at distance at most d from v in the attack graph of S. Then the *social cost* of S (denoted by $cost(S)$) is defined as:

$$cost(S) = \underbrace{\sum_{v \in V_S} C_v}_{\substack{\text{cost for} \\ \text{installing softwares}}} + \underbrace{\sum_{e \in E_S} C'_e}_{\substack{\text{cost for} \\ \text{blocking links}}} + \sum_{v \in V \setminus V_S} \underbrace{L_v \cdot w(V_S^{\leq d}(v))}_{\substack{\text{expected cost for} \\ \text{recovering } v \text{ from infection}}} .$$

Goal. In the *centralized* setting of MGNS(d), the goal is to find a strategy with minimum social cost, or *social optimum*. We can also define the *decentralized* (game-theoretic) model, in which the user needs to decide whether to install the anti-virus software on his/her computer and whether to disconnect some of the links with other users in the network. In this paper we concentrate on the centralized setting of MGNS(d), while leaving explorations of the decentralized model to future work.

1.2   Related Work

As stated before, our model MGNS(d) incorporates and generalizes several infection prevention models that have been studied recently. We list some problems considered in the literature that are either special cases of or related to the problem of computing the social optimum of MGNS(d).

- Consider the instances of MGNS(d) where $d = \infty$, $C'_e = \infty$ for all $e \in$
  . E, all nodes have the same security cost C and infection cost L, and the attack probability distribution is uniform over nodes. When restricted on such instances, MGNS(d) coincides with the model proposed by Aspnes, Chang and Yampolskiy [3], who gave an $O(\log^{1.5} n)$-approximation for computing the social optimum, based on the sparsest cut algorithm of Arora, Rao and Vazirani [1]. The approximation ratio is subsequently improved to $O(\log n)$ independently by [5] and [19], which is also the currently best known result for this problem.

- Restricted on the instances where $C'_e = \infty$ for all $e \in E$ (i.e., all the edges should remain unblocked in any reasonable solution), our model is equivalent to the *generalized network security* model GNS(d) introduced by Kumar et al. [19]. They present a $2d$-approximation for computing the social optimum of GNS(d) for all $d < \infty$ by rounding a natural linear program for the problem. This result is subsumed by our $(d + 1)$-approximation for MGNS(d). They also give an $O(\log n)$-approximation for GNS(∞) based on a reduction to the *minimum weighted vertex multicut problem* [9], improving the $O(\log^{1.5} n)$ factor of [3] and matching the result independently obtained in [5].

- Under the case where $d = \infty$, $C_v = \infty$ for all $v \in V$, $w_v = 1/n$ for all $v \in V$, and both the infection costs and link-blocking costs are uniform, the problem of computing the social optimum of MGNS(d) is similar to the *minimum average contamination problem* studied by Li and Tang [21], which originates from a (stochastic) link-blocking model initiated by Kimura, Saito and Motoda [16]. The difference between our setting and theirs is that they put a budget constraint K on the number of edges that can be removed from the network. In [21], a $(1 + \epsilon, O(\frac{\log n}{\epsilon}))$-bicriteria approximation algorithm and a $(\frac{5}{3} - \epsilon)$-inapproximability result are given for the minimum average contamination problem. Note that their problem is harder than ours (with an additional budge constraint) and thus their hardness factor is stronger than ours. However, they only consider the case $d = \infty$, while our hardness result applies to all d. Also, our polynomial-time algorithm for tree instances of MGNS(d) holds for the budgeted case as well.

- Another related problem that has mainly been studied in the operations research forum is the *critical node problem* [2,4,7] defined as follows: given a node-weighted graph $G = (V, E)$, a connection cost $c(u, v)$ for each pair of nodes $\{u, v\} \in V^2$, and a parameter K, the goal is to find a subset of nodes whose total weight does not exceed K such that the total connection cost (counted for all connected pairs of nodes) is minimized. This problem is similar to MGNS(∞) with $C'_e = \infty$ for all $e \in E$ and $w_v = 1/n$ for all $v \in V$, but with additional budget constraints and more general cost functions. The problem is NP-complete on general graphs with unit costs and unit weights [2], and on trees with unit weights [7]. For the unit-cost case (which makes the problem fit in our model with $d = \infty$) in a tree of size n, Di Summa et al. [7] show that the problem is solvable in $O(n^7)$ time. Our polynomial-time algorithm for (budgeted) MGNS(d) on trees substantially generalizes their result to all fixed d.

2   Approximation Algorithm for MGNS(d)

In this section we concern with the computation of the social optimum of MGNS(d). As the problem is NP-hard, we focus on the perspective of approximation, and obtain the following results.

Theorem 1. *For any $d \geq 1$, there is a polynomial time $(d + 1)$-approximation algorithm for computing the social optimum of MGNS(d). (Here d need not be a constant.)*

Theorem 2. *There is a polynomial time $O(\log n)$-approximation for the social optimum of MGNS(∞).*

Theorem 3. *There is a polynomial time $\frac{3}{2}$-approximation algorithm for computing the social optimum of MGNS(1) with bipartite contact graphs.*

We only prove Theorems 1 and 2 here. The proof of Theorem 3 will appear in the full version of this paper.

First consider the case $1 \leq d < \infty$. Let $\mathcal{I}$ be an instance of MGNS(d) with contact graph $G = (V, E)$ where $V = \{1, 2, \ldots, n\}$. If $C_i < w_i L_i$ for some $i \in V$, then clearly i should be secured in any optimum solution. Thus, we assume in what follows that $C_i \geq w_i L_i$ for all $i \in V$. We write an integer program to formulate the social optimum of $\mathcal{I}$. For each $k \in V \cup E$, let x_k be a binary variable that is 1 if and only if k is secure (or blocked, depending on whether k is a node or an edge). For a path p, let V_p and E_p denote the sets of nodes and edges on p, respectively. For all $1 \leq i < j \leq n$, let $P^d_{i,j}$ denote the collection of all simple paths from i to j of length at most d (note that $P^d_{i,j}$ can be empty and can also be of exponential size), and $y_{i,j}$ be a binary variable that is 1 if and only if there exists at least one path $p \in P^d_{i,j}$ on which all nodes are insecure and all edges are unblocked. Thus, $y_{i,j} = 1$ iff i and j can infect each other in the attack graph. Then the following integer program IP1 characterizes precisely the social optimum of $\mathcal{I}$:

IP1: Min $\sum_{i \in V} C_i x_i + \sum_{\{i,j\} \in E} C'_{\{i,j\}} x_{\{i,j\}} + \sum_{i \in V} L_i \left(w_i (1 - x_i) + \sum_{j \in V \setminus \{i\}} w_j y_{i,j} \right)$

subject to: $y_{i,j} + \sum_{k \in V_p \cup E_p} x_k \geq 1 \qquad \forall 1 \leq i < j \leq n$ and $p \in P_{i,j}^d$

$$y_{i,j} = y_{j,i} \qquad \forall 1 \leq i < j \leq n$$

$$x_k \in \{0,1\} \qquad \forall k \in V \cup E$$

$$y_{i,j} \in \{0,1\} \qquad \forall 1 \leq i,j \leq n, i \neq j.$$

We write $C'_i = C_i - w_i L_i$ for each $i \in V$ (with a little abuse of notation since C' is originally defined for edge costs), $L_{i,j} = w_i L_j + w_j L_i$ for all $1 \leq i < j \leq n$, and $C = \sum_{1 \leq i \leq n} w_i L_i$. Note that $C'_i \geq 0$ for all $i \in V$ by our assumption before. Rearranging terms, unifying the first two summations, and combining the occurrences of $y_{i,j}$ and $y_{j,i}$ in the objective function of IP1, we get a simpler yet equivalent formulation IP2 as follows:

IP2:   Min $\sum_{k \in V \cup E} C'_k x_k + \sum_{1 \leq i < j \leq n} L_{i,j} y_{i,j} + C$ subject to:

$$y_{i,j} + \sum_{k \in V_p \cup E_p} x_k \geq 1 \qquad \forall 1 \leq i < j \leq n \text{ and } p \in P_{i,j}^d$$

$$x_k \in \{0,1\} \qquad \forall k \in V \cup E$$

$$y_{i,j} \in \{0,1\} \qquad \forall 1 \leq i < j \leq n.$$

Observe that IP2, with the constant part C discarded, can be regarded as an instance of the weighted set cover problem when treating the length-at-most-d paths as the elements to be covered. When d is fixed, the instance of set cover is constructible in polynomial time. Also, in this set cover instance, every element appears in at most $2d + 2$ sets, because each constraint in IP2 involves at most $2d + 2$ variables (note that each $p \in P_{i,j}^d$ consists of at most $d+1$ vertices and d edges). Therefore, a polynomial time $(2d + 2)$-approximation exists for IP2 (see, e.g., [11]) and thus also for MGNS(d). Notice that, by reducing the problem to set cover, we can only handle constant d, and cannot hope for a poly-time $(2d + 2 - \epsilon)$-approximation due to the $(k - \epsilon)$-hardness of k-uniform hypergraph vertex cover [14], assuming the Unique Games Conjecture [13].

We next show that we can obtain an approximation factor of $d+1$ for all d (not necessarily fixed) by utilizing the special structure of IP2, thus saving a factor of 2 from the set cover approach. To achieve this, we relax the last two constraints of IP2 to $x_k \geq 0$ and $y_{i,j} \geq 0$ respectively, and ignore the constant part C in the objective function. This gives us a linear programming relaxation (which might still have super-polynomial size) of the original instance, which we call LP. (We do not state LP explicitly since it is very similar to IP2.) Obviously, $OPT(LP) + C \leq OPT(IP2) = OPT(IP1)$, where $OPT(P)$ is the optimum objective value of the mathematical program P.

We now write the dual formulation of LP. Let $P^d = \cup_{1 \leq i < j \leq n} P_{i,j}^d$. For each $p \in P^d$, introduce a dual variable z_p, which corresponds to the constraint

$y_{i,j} + \sum_{k \in V_p \cup E_p} x_k \geq 1$ in LP (where i and j are the endpoints of p). The dual program DU can be written as follows:

$$\text{DU:} \quad \text{Max} \sum_{p \in P^d} z_p \quad \text{subject to:}$$

$$\sum_{p \in P_{i,j}^d} z_p \leq L_{i,j} \quad \forall 1 \leq i < j \leq n$$

$$\sum_{\substack{p \in P^d \\ k \in V_p \cup E_p}} z_p \leq C_k' \quad \forall k \in V \cup E$$

$$z_p \geq 0 \quad \forall p \in P^d.$$

By the strong duality theorem, $OPT(DU) = OPT(LP)$. We now find a solution to IP2 by Algorithm 1, which basically consists of a primal-dual procedure and a "pruning" phase. Since the number of variables in DU can be super-polynomial in n for non-constant d, the naïve implementation of Algorithm 1 may not run in polynomial time. Nevertheless, we will show later that the running time can be reduced to $n^{O(1)}$ regardless of d; stating the algorithm in its current form is just to simplify the analysis of its performance guarantee. Let S denote the solution to IP2 returned by Algorithm 1, and $Z = \{z_p \mid p \in P^d\}$ be the solution to DU obtained in Algorithm 1 (which is not explicitly returned). Let $value(S)$ denote the objective value of the solution S.

Algorithm 1. Constructing a feasible solution for IP2

1: $x_k \leftarrow 0$, $\forall k \in V \cup E$; $y_{i,j} \leftarrow 0$, $\forall 1 \leq i < j \leq n$.
2: $z_p \leftarrow 0$, $\forall p \in P^d$; also, set all z_p to be "unfrozen."
3: **while** there are still unfrozen variables **do**
4:     Choose any unfrozen variable, say z_p, that appears in some constraint of DU. Raise the value of z_p until some constraint in DU, say c, becomes tight. (Pick an arbitrary one if there are more than one tight constraints.)
5:     **if** c is "$\sum_{p \in P^d : k \in V_p \cup E_p} z_p \leq C_k'$" for some $k \in V \cup E$ **then**
6:         $x_k \leftarrow 1$
7:     **else if** c is "$\sum_{p \in P_{i,j}^d} z_p \leq L_{i,j}$" for some $1 \leq i < j \leq n$ **then**
8:         $y_{i,j} \leftarrow 1$
9:     **end if**
10:    Freeze all variables that occur in some (newly appeared) tight constraint.
11: **end while**
12: **for all** $1 \leq i < j \leq n$ **do**
13:    **if** $x_i = 1$ or $x_j = 1$ **then**
14:       $y_{i,j} \leftarrow 0$; $x_{\{i,j\}} \leftarrow 0$ if $\{i,j\} \in E$.
15:    **end if**
16: **end for**
17: **return** $\{x_k \mid k \in V \cup E\} \cup \{y_{i,j} \mid 1 \leq i < j \leq n\}$.

Lemma 1. *Z is a feasible solution to DU, and S is a feasible solution to IP2.*

The proof of Lemma 1 is easy and thus omitted.

Lemma 2. $value(S) \leq (d+1)OPT(IP2)$.

Proof. For each variable v of IP2, let $c(v)$ denote the constraint in DU that corresponds to v. Call a constraint $c(v)$ *active* if $v = 1$ in the solution S. By Line 4 of Algorithm 1, every active constraint $c(v)$ (say) is tight, and hence the contribution of this v to $value(S)$ (which is the coefficient of v in the objective function of IP2) equals to the sum of z_p's contained in $c(v)$. Therefore, $value(S) = \sum_{p \in P^d} t_p z_p$, where t_p is the number of active constraints containing z_p.

Now fix an arbitrary $p = (i_0, i_1, \ldots, i_t) \in P^d, t \leq d$. The set of constraints in which z_p appears is $\{c(y_{i_0,i_t})\} \cup \{c(x_{i_j}) \mid 0 \leq j \leq t\} \cup \{c(x_{\{i_j,i_{j+1}\}}) \mid 0 \leq j \leq t-1\}$, which can be partitioned into the following $t+1$ subsets:

$$\{c(x_{i_0}), c(x_{\{i_0,i_1\}})\}, \{c(x_{i_1}), c(x_{\{i_1,i_2\}})\}, \ldots, \{c(x_{i_{t-1}}), c(x_{\{i_{t-1},i_t\}})\}, \{c(x_{i_t}), c(y_{i_0,i_t})\}.$$

Due to the function of the FOR loop, at most one constraint from each subset is active. Thus z_p appears in at most $t + 1 \leq d + 1$ active constraints. Recalling that the objective function of IP2 embraces an additional part C, we have

$$value(S) \leq C + (d+1) \sum_{p \in P^d} z_p \leq C + (d+1)OPT(DU)$$
$$= C + (d+1)OPT(LP) \leq C + (d+1)(OPT(IP2) - C)$$
$$\leq (d+1)OPT(IP2),$$

completing the proof of Lemma 2.   □

Lemmas 1 and 2 ensure that S is a $(d+1)$-approximate solution to IP2. We next explain how to make Algorithm 1 run in poly-time for all d. Consider the following two operations:

(1) Find an unfrozen variable of DU if there exists at least one.
(2) Given a variable z_p, find all the constraints in DU that contain z_p.

Lemma 3. *If operations (1) and (2) can be done in polynomial time, then Algorithm 1 can be implemented to run in polynomial time.*

Proof. Suppose (1) and (2) can be done in polynomial time. Since DU has at most $\binom{n}{2} + n \leq n^2$ constraints and each time only one variable raises its value, we can keep the current LHS and RHS values of each constraint, and are thus able to know which constraints are tight. Hence Line 10 can be realized implicitly since a variable is frozen iff it appears in some tight constraint. To implement Line 4, we first apply (1) to find an unfrozen variable (say z_p) if there exists one, and then use (2) to find a constraint containing z_p that has the smallest difference between RHS and LHS values; this difference is exactly the amount that z_p can be raised. The other steps in Algorithm 1 can clearly be implemented to run in poly-time. The lemma is thus proved.   □

Lemma 4. *We can accomplish (1) and (2) in polynomial time.*

Proof. We use $c(v)$ to denote the constraint in DU that corresponds to the variable v of IP2. First note that (2) is easy to implement: For each variable z_p where p has endpoints i and j, z_p appears exactly in the constraints corresponding to $y_{i,j}$ or x_k for some $k \in V_p \cup E_p$. Thus we focus on (1). As shown in the proof of Lemma 3, we know the set of tight constraints in DU, and a variable is unfrozen if and only if it does not appear in any tight constraint. For $p \in P^d$, the variable z_p does not appear in $c(x_k)$ (where $k \in V \cup E$) iff $k \notin V_p \cup E_p$, and z_p does not appear in $c(y_{i,j})$ (where $1 \le i < j \le n$) iff p is not a path between i and j. We do the following: Construct a graph G' from G by deleting all $k \in V \cup E$ from G for which $c(x_k)$ is tight. Then, for every $1 \le i < j \le n$ such that $c(y_{i,j})$ is not tight, check whether there exists a path p from i to j in G' of length at most d; if so, then the corresponding variable z_p must be unfrozen due to our previous analysis. Also, by this procedure we will find an unfrozen variable if there exists at least one. Clearly this process can be finished in polynomial time.   □

Now Theorem 1 follows directly from Lemmas 1, 2, 3 and 4.

We next turn to the case $d = \infty$ and prove Theorem 2. We reduce MGNS(∞) to GNS(∞) as follows: Construct a graph G' by subdividing each edge $e \in E$ with a new vertex v_e. Let $w(v_e) = 0, C_{v_e} = C'_e$ and $L_{v_e} = 0$ for all $e \in E$. It is easy to argue that the problem of finding the social optimum of GNS(∞) on this new instance is equivalent to that of MGNS(∞) on the original one. Now, applying the poly-time approximation algorithm for GNS(∞) given in [19], we get a solution for MGNS(∞) with approximation ratio $O(\log |V(G')|) = O(\log n)$. This finishes the proof of Theorem 2.

We remark that a similar reduction can reduce an instance of MGNS(d) to that of GNS($2d$). Using the approximation algorithm in [19], we obtain a solution for MGNS(d) with approximation factor $4d$, which is nearly four times larger than the ratio guaranteed by Theorem 1. This is in part due to the fact that such a reduction loses some information of the graph topology, which is important to our algorithm.

3   Hardness of Approximation for GNS(d)

In this section we present inapproximability results for GNS(d), a special case of our model MGNS(d). Thus, all the hardness results trivially apply to MGNS(d). The proof of the following two theorems will appear in the full version of this paper.

Theorem 4. *For every $d \in \mathbb{N} \cup \{\infty\}$, computing the social optimum of GNS(d) is $\mathcal{APX}$-hard, even if the contact graph is 3-regular and all types of costs as well as the attack probability distribution are uniform.*

Theorem 5. *Assuming Unique Games Conjecture, for any $d \in \mathbb{N} \cup \{\infty\}$ and any fixed $\epsilon > 0$, we cannot approximate the social optimum of GNS(d) to a factor of $\frac{3}{2} - \epsilon$ in polynomial time.*

4   Polynomial Algorithm for MGNS(d) on Trees

In this section we consider a special class of instances of MGNS(d), in which the underlying contact graph of the instance is a tree, and the infection cost and attack probability are both uniform. Our main results are as follows, whose rigorous proofs will appear in the full version of this paper.

Theorem 6. *For every fixed $d \geq 1$ or $d = \infty$, we can find in polynomial time an optimal solution of a tree-instance of MGNS(d) with uniform infection cost and attack probability, even if there are budget constraints, i.e., given two integers K and K', a solution can secure at most K nodes and block at most K' edges.*

Theorem 7. *For all $d \leq O(\sqrt{\log n})$, we can find in polynomial time an optimal solution to (budgeted) MGNS(d) if the instance has uniform infection cost and attack probability, and its contact graph is a tree of bounded degree.*

Theorem 6 in particular settles an open problem from [21] that asks if there is a polynomial time algorithm for the minimum average contamination problem, which corresponds to the special case of budgeted MGNS(d) on trees where every node has security cost ∞ and all other costs as well as the attack probability distribution are uniform.

5   Conclusions and Future Research

We propose in this paper the mixed generalized network security model MGNS(d), which generalizes several other models for infection control. We present approximation and inapproximability results for the problem of computing the optimum solution of MGNS(d), and exact polynomial-time algorithms for tree instances with uniform infection cost and attack probability distribution. Some of our results lead immediately to improvements upon the previously best known results achieved for some special cases of our model.

There are many interesting questions left that deserve further explorations. Regarding the optimization of social cost, a big open question is whether we can break the $O(\log n)$ factor for MGNS(∞) or GNS(∞), or there is a matching hardness of approximation result. Also for MGNS(d) where $d < \infty$, there remains a large gap between the upper bound of $d+1$ and the lower bound of $\frac{3}{2} - \epsilon$ on the approximation ratio. Another research issue is the formulation and investigation of the decentralized or game-theoretic counterpart of our model, where a user can decide whether to install an anti-virus software, and might also be able to block some of the links to other users. Finally, incorporating other propagation models (e.g., the independent cascade model, or the linear threshold model) into MGNS(d) may lead to more accurate modeling of some applications.

References

1. Arora, S., Rao, S., Vazirani, U.: Expander flows, geomeric embeddings and graph partitioning. In: Proceedings of the 35th ACM STOC (2004)

2. Arulselvan, A., Commander, C.W., Elefteriadou, L., Pardalos, P.M.: Detecting critical nodes in sparse graphs. Comput. Oper. Res. 36(7), 2193–2200 (2009)
3. Aspnes, J., Chang, K.L., Yampolskiy, A.: Inoculation strategies for victims of viruses and the sum-of-squares partition problem. J. Comput. Syst. Sci. 72(6), 1077–1093 (2005); Preliminary version in SODA 2005
4. Borgatti, S.: Identifying sets of key players in a social network. Comput. Math. Org. Theory 12, 21–34 (2006)
5. Chen, P.-A., David, M., Kempe, D.: Better vaccination strategies for better people. In: Proceedings of the 11th ACM EC (2010)
6. Chen, W., Wang, Y., Yang, S.: Efficient influence maximization in social networks. In: Proceedings of the 15th ACM SIGKDD International Conference on Knowledge Discovery and Data Mining (2009)
7. Di Summa, M., Grosso, A., Locatelli, M.: Complexity of the critical node problem over trees. Comput. Oper. Res. 38(12), 1766–1774 (2011)
8. Domingos, P., Richardson, M.: Mining the network value of customers. In: Proceedings of the 7th ACM SIGKDD International Conference on Knowledge Discovery and Data Mining (2001)
9. Garg, N., Vazirani, V.V., Yannakakis, M.: Approximate max-flow min-(multi)cut theorems and their applications. SIAM J. Comput. 25(2), 235–251 (1993); Preliminary version in STOC 1993
10. Goyal, A., Bonchi, F., Lakshmanan, L.V.S.: Learning influence probabilities in social networks. In: Proceedings of the 3rd ACM WSDM (2010)
11. Halperin, E.: Improved approximation algorithms for the vertex cover problem in graphs and hypergraph. SIAM J. Comput. 31(5), 1608–1623 (2002)
12. Kempe, D., Kleinberg, J., Tardos, É.: Maximizing the spread of influence through a social network. In: Proceedings of the 9th ACM SIGKDD International Conference on Knowledge Discovery and Data Mining (2003)
13. Khot, S.: On the power of unique 2-prover 1-round games. In: Proceedings of the 34th ACM STOC (2002)
14. Khot, S., Regev, O.: Vertex cover might be hard to approximate to within $2 - \epsilon$. J. Comput. Syst. Sci. 74(3), 335–349 (2003); Preliminary version in CCC 2003
15. Kimura, M., Saito, K.: Tractable Models for Information Diffusion in Social Networks. In: Fürnkranz, J., Scheffer, T., Spiliopoulou, M. (eds.) PKDD 2006. LNCS (LNAI), vol. 4213, pp. 259–271. Springer, Heidelberg (2006)
16. Kimura, M., Saito, K., Motoda, H.: Minimizing the spread of contamination by blocking links in a network. In: Proceedings of the 23rd AAAI Conference on Artificial Intelligence (2008)
17. Kimura, M., Saito, K., Motoda, H.: Solving the Contamination Minimization Problem on Networks for the Linear Threshold Model. In: Ho, T.-B., Zhou, Z.-H. (eds.) PRICAI 2008. LNCS (LNAI), vol. 5351, pp. 977–984. Springer, Heidelberg (2008)
18. Kimura, M., Saito, K., Motoda, H.: Blocking links to minimize contamination spread in a social network. ACM Trans. Knowl. Discov. Data. 3(2) (2009)
19. Anil Kumar, V.S., Rajaraman, R., Sun, Z., Sundaram, R.: Existence theorems and approximation algorithms for generalized network security games. In: Proceedings of the 30th ICDCS (2010)
20. Leskovec, J., Krause, A., Guestrin, C., Faloutsos, C., VanBriesen, J., Glance, N.S.: Cost-effective outbreak detection in networks. In: Proceedings of the 13th ACM SIGKDD Conference on Knowledge Discovery and Data Mining (2007)
21. Li, A., Tang, L.: The Complexity and Approximability of Minimum Contamination Problems. In: Ogihara, M., Tarui, J. (eds.) TAMC 2011. LNCS, vol. 6648, pp. 298–307. Springer, Heidelberg (2011)

A General Framework for Computing Optimal Correlated Equilibria in Compact Games

(Extended Abstract)*

Albert Xin Jiang and Kevin Leyton-Brown

Department of Computer Science,
University of British Columbia, Vancouver, Canada
{jiang,kevinlb}@cs.ubc.ca

Abstract. We analyze the problem of computing a correlated equilibrium that optimizes some objective (e.g., social welfare). Papadimitriou and Roughgarden [2008] gave a sufficient condition for the tractability of this problem; however, this condition only applies to a subset of existing representations. We propose a different algorithmic approach for the optimal CE problem that applies to *all* compact representations, and give a sufficient condition that generalizes that of Papadimitriou and Roughgarden [2008]. In particular, we reduce the optimal CE problem to the *deviation-adjusted social welfare problem*, a combinatorial optimization problem closely related to the optimal social welfare problem. This framework allows us to identify new classes of games for which the optimal CE problem is tractable; we show that graphical polymatrix games on tree graphs are one example. We also study the problem of computing the optimal *coarse correlated equilibrium*, a solution concept closely related to CE. Using a similar approach we derive a sufficient condition for this problem, and use it to prove that the problem is tractable for singleton congestion games.

1 Introduction

A fundamental class of computational problems in game theory is the computation of *solution concepts* of finite games. Much recent effort in the literature has concerned the problem of computing a sample Nash equilibrium [Chen & Deng, 2006; Daskalakis *et al.*, 2006; Daskalakis & Papadimitriou, 2005; Goldberg & Papadimitriou, 2006]. First proposed by Aumann [1974; 1987], correlated equilibrium (CE) is another important solution concept. Whereas in a mixed strategy Nash equilibrium players randomize independently, in a correlated equilibrium the players can coordinate their behavior based on signals from an intermediary.

Correlated equilibria of a game can be formulated as probability distributions over pure strategy profiles satisfying certain linear constraints. The resulting linear feasibility program has size polynomial in the size of the normal form representation of the game. However, the size of the normal form representation grows exponentially in the number

---

* All proofs are omitted in this extended abstract. A full version is available at
http://arxiv.org/abs/1109.6064

N. Chen, E. Elkind, and E. Koutsoupias (Eds.): WINE 2011, LNCS 7090, pp. 218–229, 2011.
© Springer-Verlag Berlin Heidelberg 2011

of players. This is problematic when games involve large numbers of players. Fortunately, most large games of practical interest have highly-structured payoff functions, and thus it is possible to represent them compactly. A line of research thus exists to look for *compact game representations* that are able to succinctly describe structured games, including work on graphical games [Kearns *et al.*, 2001] and action-graph games [Bhat & Leyton-Brown, 2004; Jiang *et al.*, 2011]. But now the size of the linear feasibility program for CE can be exponential in the size of compact representation; furthermore a CE can require exponential space to specify.

The problem of computing a sample CE was recently shown to be in polynomial time for most existing compact representations [Papadimitriou & Roughgarden, 2008; Jiang & Leyton-Brown, 2011]. However, since in general there can be an infinite number of CE in a game, finding an arbitrary one is of limited value. Instead, here we focus on the problem of computing a correlated equilibrium that optimizes some objective. In particular we consider optimizing linear functions of players' expected utilities. For example, computing the best (or worst) social welfare corresponds to maximizing (or minimizing) the sum of players' utilities, respectively. We are also interested in computing optimal coarse correlated equilibrium (CCE) [Hannan, 1957]. It is known that the empirical distribution of any no-external-regret learning dynamic converges to the set of CCE, while the empirical distribution of no-internal-regret learning dynamics converges to the set of CE (see e.g. [Nisan *et al.*, 2007]). Thus, optimal CE / CCE provide useful bounds on the social welfare of the empirical distributions of these dynamics.

We are particularly interested in the relationship between the optimal CE / CCE problems and the problem of computing the optimal social welfare outcome (i.e. strategy profile) of the game, which is exactly the optimal social welfare CE problem without the incentive constraints. This is an instance of a line of questions that has received much interest from the algorithmic game theory community: "How does adding incentive constraints to an optimization problem affect its complexity?" This question in the mechanism design setting is perhaps one of the central questions of algorithmic mechanism design [Nisan & Ronen, 2001]. Of course, a more constrained problem can in general be computationally easier than the relaxed version of the problem. Nevertheless, results from complexity of Nash equilibria and algorithmic mechanism design suggest that adding *incentive constraints* to a problem is unlikely to decrease its computational difficulty. That is, when the optimal social welfare problem is hard, we tend also to expect that the optimal CE problem will be hard as well. On the other hand, we are interested in the other direction: when it is the case for a class of games that the optimal social welfare problem can be efficiently computed, can the same structure be exploited to efficiently compute the optimal CE?

The seminal work on the computation of optimal CE is [Papadimitriou & Roughgarden, 2008]. This paper considered the optimal linear objective CE problem and proved that the problem is NP-hard for many representations including graphical games, polymatrix games, and congestion games. On the tractability side, Papadimitriou and Roughgarden [2008] focused on so-called "reduced form" representations, meaning representations for which there exist player-specific partitions of the strategy profile space into payoff-equivalent outcomes. They showed that if a particular *separation problem* is polynomial-time solvable, the optimal CE problem is polynomial-time solvable as well. Finally, they showed that this separation problem is polynomial-time solvable for bounded-treewidth graphical games, symmetric games and anonymous games.

Perhaps most surprising and interesting is the *form* of Papadimitriou and Roughgarden's sufficient condition for tractability: their separation problem for an instance of a reduced-form-based representation is essentially equivalent to solving the optimal social welfare problem for an instance of that representation with the same reduced form but possibly different payoffs. In other words, if we have a polynomial-time algorithm for the optimal social welfare problem for a reduced-form-based representation, we can turn that into a polynomial-time algorithm for the optimal social welfare CE problem. However, Papadimitriou and Roughgarden's sufficient condition for tractability only applies to reduced-form-based representations. Their definition of reduced forms is unable to handle representations that exploit linearity of utility, and in which the structure of player p's utility function may depend on the action she chose. As a result, many representations do not fall into this characterization, such as polymatrix games, congestion games, and action-graph games. Although the optimal CE problems for these representations are NP-hard in general, we are interested in identifying tractable subclasses of games, and a sufficient condition that applies to all representations would be helpful.

In this article, we propose a different algorithmic approach for the optimal CE problem that applies to *all* compact representations. By applying the ellipsoid method to the dual of the LP for optimal CE, we show that the polynomial-time solvability of what we call the *deviation-adjusted social welfare problem* is a sufficient condition for the tractability of the optimal CE problem. We also give a sufficient condition for tractability of the optimal CCE problem: the polynomial-time solvability of the *coarse deviation-adjusted social welfare problem*. We show that for reduced-form-based representations, the deviation-adjusted social welfare problem can be reduced to the separation problem of Papadimitriou and Roughgarden [2008]. Thus the class of reduced forms for which our problem is polynomial-time solvable contains the class for which the separation problem is polynomial-time solvable. More generally, we show that if a representation can be characterized by "linear reduced forms", i.e. player-specific linear functions over partitions, then for that representation, the deviation-adjusted social welfare problem can be reduced to the optimal social welfare problem. As an example, we show that for graphical polymatrix games on trees, optimal CE can be computed in polynomial time. Such games are not captured by the reduced-form framework.[1]

On the other hand, representations like action-graph games and congestion games have *action-specific* structure, and as a result the deviation-adjusted social welfare problems and coarse deviation-adjusted social welfare problems on these representations are structured differently from the corresponding optimal social welfare problems. Nevertheless, we are able to show a polynomial-time algorithm for the optimal CCE problem on *singleton congestion games* [Ieong et al., 2005], a subclass of congestion games. We use a symmetrization argument to reduce the optimal CCE problem to the coarse

---

[1] In a recent paper Kamisetty et al. [2011] has independently proposed an algorithm for optimal CE in graphical polymatrix games on trees. They used a different approach that is specific to graphical games and graphical polymatrix games, and it is not obvious whether their approach can be extended to other classes of games.

deviation-adjusted social welfare problem with player-symmetric deviations, which can be solved using a dynamic-programming algorithm. This is an example where the optimal CCE problem is tractable while the complexity of the optimal CE problem is not yet known.

2 Problem Formulation

Consider a simultaneous-move game $G = (\mathcal{N}, \{S_p\}_{p \in \mathcal{N}}, \{u^p\}_{p \in \mathcal{N}})$, where $\mathcal{N} = \{1, \ldots, n\}$ is the set of players. Denote a player p, and player p's set of pure strategies (i.e., actions) S_p. Let $m = \max_p |S_p|$. Denote a pure strategy profile $s = (s_1, \ldots, s_n) \in S$, with s_p being player p's pure strategy. Denote by S_{-p} the set of partial pure strategy profiles of the players other than p. Let u^p be the vector of player p's utilities for each pure profile, denoting player p's utility under pure strategy profile s as u_s^p. Let w be the vector of social welfare for each pure profile, that is $w = \sum_{p \in \mathcal{N}} u^p$, with w_s denoting the social welfare for pure profile s.

Throughout the paper we assume that the game is given in a representation with *polynomial type* [Papadimitriou, 2005; Papadimitriou & Roughgarden, 2008], i.e., that the number of players and the number of actions for each player are bounded by polynomials of the size of the representation.

2.1 Correlated Equilibrium

A *correlated distribution* is a probability distribution over pure strategy profiles, represented by a vector $x \in \mathbb{R}^M$, where $M = \prod_p |S_p|$. Then x_s is the probability of pure strategy profile s under the distribution x.

Definition 1. *A correlated distribution x is a* correlated equilibrium *(CE) if it satisfies the following* incentive constraints: *for each player p and each pair of her actions $i, j \in S_p$, we have $\sum_{s_{-p} \in S_{-p}} [u_{is_{-p}}^p - u_{js_{-p}}^p] x_{is_{-p}} \geq 0$, where the subscript "$is_{-p}$" (respectively "$js_{-p}$") denotes the pure strategy profile in which player p plays i (respectively j) and the other players play according to the partial profile $s_{-p} \in S_{-p}$.*

Intuitively, when a trusted intermediary draws a strategy profile s from this distribution, privately announcing to each player p her own component s_p, p will have no incentive to choose another strategy, assuming others follow the suggestions. We write these incentive constraints in matrix form as $Ux \geq 0$. Thus U is an $N \times M$ matrix, where $N = \sum_p |S_p|^2$. The rows of U are indexed by (p, i, j), where p is a player and $i, j \in S_p$ are a pair of p's actions. Denote by U_s the column of U corresponding to pure strategy profile s. These incentive constraints, together with the constraints $x \geq 0$, $\sum_{s \in S} x_s = 1$, which ensure that x is a probability distribution, form a linear feasibility program that defines the set of CE. The problem of computing a maximum social welfare CE can be formulated as the LP

$$\max w^T x \qquad (P)$$

$$Ux \geq 0, \ x \geq 0, \ \sum_{s \in S} x_s = 1$$

Another solution concept of interest is *coarse correlated equilibrium* (CCE). Whereas CE requires that each player has no profitable deviation even if she takes into account the signal she receives from the intermediary, CCE only requires that each player has no profitable *unconditional deviation*.

Definition 2. *A correlated distribution x is a* coarse correlated equilibrium *(CCE) if it satisfies the following incentive constraints: for each player p and each of his actions $j \in S_p$, we have $\sum_{(i,s_{-p}) \in S} [u^p_{is_{-p}} - u^p_{js_{-p}}] x_{is_{-p}} \geq 0$.*

We write these incentive constraints in matrix form as $Cx \geq 0$. Thus C is an $(\sum_p |S_p|) \times M$ matrix. By definition, a CE is also a CCE.

The problem of computing a maximum social welfare CCE can be formulated as the LP

$$\max w^T x \qquad\qquad (CP)$$

$$Cx \geq 0, \ x \geq 0, \ \sum_{s \in S} x_s = 1.$$

3   The Deviation-Adjusted Social Welfare Problem

Consider the dual of (P),

$$\min t \qquad\qquad (D)$$

$$U^T y + w \leq t\mathbf{1}$$

$$y \geq 0.$$

We label the (p, i, j)-th element of $y \in \mathbb{R}^N$ (corresponding to row (p, i, j) of U) as $y^p_{i,j}$. This is an LP with a polynomial number of variables and an exponential number of constraints. Given a separation oracle, we can solve it in polynomial time using the ellipsoid method. A separation oracle needs to determine whether a given (y, t) is feasible, and if not output a hyperplane that separates (y, t) from the feasible set. We focus on a restricted form of separation oracles, which outputs a violated constraint for infeasible points.[2] Such a separation oracle needs to solve the following problem:

Problem 1. Given (y, t) with $y \geq 0$, determine if there exists an s such that $(U_s)^T y + w_s > t$; if so output such an s.

The left-hand-side expression $(U_s)^T y + w_s$ is the social welfare at s plus the term $(U_s)^T y$. Observe that the (p, i, j)-th entry of U_s is $u^p_s - u^p_{js_{-p}}$ if $s_p = i$ and is zero otherwise. Thus $(U_s)^T y = \sum_p \sum_{j \in S_p} y^p_{s_p, j} \left(u^p_s - u^p_{js_{-p}} \right)$. We now reexpress $(U_s)^T y + w_s$ in terms of *deviation-adjusted utilities* and *deviation-adjusted social welfare*.

---

[2] This is a restriction because in general there exist separating hyperplanes other than the violated constraints. For example Papadimitriou and Roughgarden [2008]'s algorithm for computing a sample CE uses a separation oracle that outputs a convex combination of the constraints as a separating hyperplane.

Definition 3. *Given a game, and a vector $y \in \mathbb{R}^N$ such that $y \geq 0$, the deviation-adjusted utility for player p under pure profile s is*

$$\hat{u}_s^p(y) = u_s^p + \sum_{j \in S_p} y_{s_p,j}^p \left(u_s^p - u_{js_{-p}}^p \right).$$

The deviation-adjusted social welfare is $\hat{w}_s(y) = \sum_p \hat{u}_s^p(y)$.

By construction, the deviation-adjusted social welfare $\hat{w}_s(y) = \sum_p u_s^p + \sum_p \sum_{j \in S_p} y_{s_p,j}^p \left(u_s^p - u_{js_{-p}}^p \right) = (U_s)^T y + w_s$. Therefore, Problem 1 is equivalent to the following *deviation-adjusted social welfare problem*.

Definition 4. *For a game representation, the* deviation-adjusted social welfare problem *is the following: given an instance of the representation and rational vector $(y, t) \in \mathbb{Q}^{N+1}$ such that $y \geq 0$, determine if there exists an s such that the deviation-adjusted social welfare $\hat{w}_s(y) > t$; if so output such an s.*

Proposition 1. *If the deviation-adjusted social welfare problem can be solved in polynomial time for a game representation, then so can the problem of computing the maximum social welfare CE.*

Let us consider interpretations of the dual variables y and the deviation-adjusted social welfare of a game. The dual (D) can be rewritten as $\min_{y \geq 0} \max_s \tilde{w}_s(y)$. By weak duality, for a given $y \geq 0$ the maximum deviation-adjusted social welfare $\max_s \tilde{w}_s(y)$ is an upper bound on the maximum social welfare CE. So the task of the dual (D) is to find y such that the resulting maximum deviation-adjusted social welfare gives the tightest bound.[3] At optimum, y corresponds to the concept of "shadow prices" from optimization theory; that is, y_{ij}^p equals the rate of change in the social welfare objective when the constraint (p, i, j) is relaxed infinitesimally. Compared to the maximum social welfare CE problem, the maximum deviation-adjusted social welfare problem replaces the incentive constraints with a set of additional penalties or rewards. Specifically, we can interpret y as a set of nonnegative prices, one for each incentive constraint (p, i, j) of (P). At strategy profile s, for each incentive constraint (p, i, j) we impose a penalty equal to y_{ij}^p times the amount the constraint (p, i, j) is violated by s. Note that the penalty can be negative, and is zero if $s_p \neq i$. Then $\tilde{w}_s(y)$ is equal to the social welfare of the modified game.

Practical Computation. The problem of computing the expected utility (EU) given a mixed strategy profile has been established as an important subproblem for both the sample NASH problem and the sample CE problem, both in theory [Daskalakis *et al.*, 2006; Papadimitriou & Roughgarden, 2008] and in practice [Blum *et al.*, 2006; Jiang *et al.*, 2011]. Our results suggest that the deviation-adjusted social welfare problem is of similar importance to the optimal CE problem. This connection is more than theoretical: our algorithmic approach can be turned into a practical method for computing optimal CE. In particular, although it makes use of the ellipsoid method, we can easily

---

[3] An equivalent perspective is to view y as Lagrange multipliers, and the optimal deviation-adjusted SW problem as the Lagrangian relaxation of (P) given the multipliers y.

substitute a more practical method, such as simplex with column generation. In contrast, Papadimitriou and Roughgarden [2008]'s algorithmic approach for reduced forms makes two nested applications of the ellipsoid method, and is less likely to be practical.

3.1 The Coarse Deviation-Adjusted Social Welfare Problem

For the optimal social welfare CCE problem, we can form the dual of (CP)

$$\min t \tag{1}$$
$$C^T y + w \leq t\mathbf{1}$$
$$y \geq 0$$

Definition 5. *We label the (p, j)-th element of y as y_j^p. Given a game, and a vector $y \in \mathbb{R}^{\sum_p |S_p|}$ such that $y \geq 0$, the coarse deviation-adjusted utility for player p under pure profile s is $\tilde{u}_s^p(y) = u_s^p + \sum_{j \in S_p} y_j^p(u_s^p - u_{js_{-p}}^p)$. The coarse deviation-adjusted social welfare is $\tilde{w}_s(y) = \sum_p \tilde{u}_s^p(y)$.*

Proposition 2. *If the coarse deviation-adjusted social welfare problem can be solved in polynomial time for a game representation, then the problem of computing the maximum social welfare CCE is in polynomial time for this representation.*

The coarse deviation-adjusted social welfare problem reduces to the deviation-adjusted social welfare problem. To see this, given an input vector y for the coarse deviation-adjusted social welfare problem, we can construct an input vector $y' \in \mathbb{Q}^N$ for the deviation-adjusted social welfare problem with $y_{ij}'^p = y_j^p$ for all $p \in \mathcal{N}$ and $i, j \in S_p$.

4 The Deviation-Adjusted Social Welfare Problem for Specific Representations

In this section we study the deviation-adjusted social welfare problem and its variants on specific representations. Depending on the representation, the deviation-adjusted social welfare problem is not always solvable in polynomial time. Indeed, Papadimitriou and Roughgarden [2008] showed that for many representations the problem of optimal CE is NP-hard. Nevertheless, for such representations we can often identify tractable subclasses of games. We will argue that the deviation-adjusted social welfare problem is a more useful formulation for identifying tractable classes of games than the separation problem formulation of Papadimitriou and Roughgarden [2008], as the latter only applies to reduced-form-based representations.

4.1 Reduced Forms

Papadimitriou and Roughgarden [2008] gave the following reduced form characterization of representations.

Definition 6 ([Papadimitriou & Roughgarden, 2008]). *Consider a game* $G = (\mathcal{N}, \{S_p\}_{p \in \mathcal{N}}, \{u^p\}_{p \in \mathcal{N}})$. *For* $p = 1, \dots, n$, *let* $P_p = \{C_p^1 \dots C_p^{r_p}\}$ *be a partition of* S_{-p} *into* r_p *classes. The set* $\mathcal{P} = \{P_1, \dots, P_n\}$ *of partitions is a* reduced form *of* G *if* $u_s^p = u_{s'}^p$ *whenever (1)* $s_p = s'_p$ *and (2) both* s_{-p} *and* s'_{-p} *belong to the same class in* P_p. *The* size *of a reduced form is the number of classes in the partitions plus the bits required to specify a payoff value for each tuple* (p, k, ℓ) *where* $1 \le p \le n, 1 \le k \le r_p$ *and* $\ell \in S_p$.

Intuitively, the reduced form imposes the condition that p's utility for choosing an action s_p depends only on which *class* in the partition P_p the profile of the others' actions belongs to. Papadimitriou and Roughgarden [2008] showed that several compact representations such as graphical games and anonymous games have natural reduced forms whose sizes are (roughly) equal to the sizes of the representation. We say such a compact representation has a *concise reduced form*. Intuitively, such a reduced form describes the structure of the game's utility functions.

Let $\mathcal{S}_p(k, \ell)$ denote the set of pure strategy profiles s such that $s_p = \ell$ and s_{-p} is in the k-th class C_p^k of P_p, and let $u_{(k,\ell)}^p$ denote the utility of p for that set of strategy profiles. Papadimitriou and Roughgarden [2008] defined the following *Separation Problem* for a reduced form.

Definition 7 ([Papadimitriou & Roughgarden, 2008]). *Let* $\mathcal{P}$ *be a reduced form for game* G. *The* Separation Problem *for* $\mathcal{P}$ *is the following: Given rational numbers* $\gamma_p(k, \ell)$ *for all* $p \in \{1, \dots, n\}, k \in \{1, \dots, r_p\}$, *and* $\ell \in S_p$, *is there a pure strategy profile* s *such that* $\sum_{p,k,\ell : s \in \mathcal{S}_p(k,\ell)} \gamma_p(k, \ell) < 0$? *If so, find such* s.

Since $s \in \mathcal{S}_p(k, \ell)$ implies $s_p = \ell$, the left-hand side of the above expression is equivalent to $\sum_p \sum_{k : s \in \mathcal{S}_p(k, s_p)} \gamma_p(k, s_p)$. Furthermore, since s belongs to exactly one class in P_p, the expression is a sum of exactly n summands.

Papadimitriou and Roughgarden [2008] proved that if the separation problem can be solved in polynomial time, then a CE that maximizes a given linear objective in the players' utilities can be computed in time polynomial in the size of the reduced form. How does Papadimitriou and Roughgarden [2008]'s sufficient condition relate to ours, provided that the game has a concise reduced form? We show that the class of reduced form games for which our deviation-adjusted social welfare problem is polynomial-time solvable contains the class for which the separation problem is polynomial-time solvable.

Proposition 3. *Let* $\mathcal{P}$ *be a reduced form for game* G. *Suppose the separation problem can be solved in polynomial time. Then the deviation-adjusted social welfare problem can be solved in time polynomial in the size of the reduced form.*

We now compare the the deviation-adjusted social welfare problem with the optimal social welfare problem for these representations. We observe that the deviation-adjusted social welfare problem can be formulated as an instance of the optimal social welfare problem on another game with the same reduced form but different payoffs. Can we claim that the existence of a polynomial-time algorithm for the optimal social welfare problem for a representation implies the existence of a polynomial-time algorithm for the social welfare problem (and thus the optimal CE problem)? This is not necessarily the case, because the representation might impose certain structure on the utility functions that are not captured by the reduced forms, and the polynomial-time algorithm for the optimal social welfare problem could depend on the existence of such structure. The deviation-adjusted social welfare problem might no longer exhibit such structure and thus might not be solvable using the given algorithm.

Nevertheless, if we consider a game representation that is "completely characterized" by its reduced forms, the deviation-adjusted social welfare problem is equivalent to the decision version of the optimal social welfare outcome problem for that representation. To make this more precise, we say a game representation is a *reduced-form-based representation* if there exists a mapping from instances of the representation to reduced forms such that it maps each instance to a concise reduced form of that instance, and if we take such a reduced form and change its payoff values arbitrarily, the resulting reduced form is a concise reduced form of another instance of the representation.

Corollary 1. *For a reduced-form-based representation, if there exists a polynomial-time algorithm for the optimal social welfare problem, then the optimal social welfare CE problem and the max-min welfare CE problem can be solved in polynomial time.*

Of course, this can be derived using the separation problem for reduced forms without the deviation-adjusted social welfare formulation. On the other hand, the deviation-adjusted social welfare formulation can be applied to representations without concise reduced forms. In fact, we will use it to show below that the connection between the optimal social welfare problem and the optimal CE problem applies to a wider classes of representations than just reduced-form-based representations.

4.2 Linear Reduced Forms

One class of representations that does not have concise reduced forms are those that represent utility functions as sums of other functions, such as polymatrix games and the hypergraph games of Papadimitriou and Roughgarden [2008]. In this section we characterize these representations using linear reduced forms, showing that linear-reduced-form-based representations satisfy a property similar to Corollary 1.

Roughly speaking, a linear reduced form has multiple partitions for each agent, rather than just one; an agent's overall utility is a sum over utility functions defined on each of that agent's partitions.

Definition 8. *Consider a game* $G = (\mathcal{N}, \{S_p\}_{p \in \mathcal{N}}, \{u^p\}_{p \in \mathcal{N}})$. *For* $p = 1, \ldots, n$, *let* $P_p = \{P_{p,1}, \ldots, P_{p,t_p}\}$, *where* $P_{p,q} = \{C_{p,q}^1 \ldots C_{p,q}^{r_{pq}}\}$ *is a partition of* S_{-p} *into* r_{pq} *classes. The set* $\mathcal{P} = \{P_1, \ldots, P_n\}$ *is a* linear reduced form *of* G *if for each* p *there exist* $u^{p,1}, \ldots, u^{p,t_p} \in \mathbb{R}^M$ *such that for all* s, $u_s^p = \sum_q u_s^{p,q}$, *and for each* $q \leq t_p$, $u_s^{p,q} = u_{s'}^{p,q}$ *whenever (1)* $s_p = s'_p$ *and (2) both* s_{-p} *and* s'_{-p} *belong to the same class in* $P_{p,q}$. *The* size *of a reduced form is the number of classes in the partitions plus the bits required to specify a number for each tuple* (p, q, k, ℓ) *where* $1 \leq p \leq n$, $1 \leq q \leq t_p$, $1 \leq k \leq r_{pq}$ *and* $\ell \in S_p$.

We write $u_{(k,\ell)}^{p,q}$ for the value corresponding to tuple (p, q, k, ℓ), and for $\mathbf{k} = (k_1, \ldots, k_{t_p})$ we write $u_{(\mathbf{k},\ell)}^p \equiv \sum_q u_{(k_q,\ell)}^{p,q}$.

Example 1 (polymatrix games). In a polymatrix game, each player's utility is the sum of utilities resulting from her bilateral interactions with each of the $n - 1$ other players: $u_s^p = \sum_{p' \neq p} e_{s_p}^T A^{pp'} e_{s_{p'}}$, where $A^{pp'} \in \mathbb{R}^{|S_p| \times |S_{p'}|}$ and $e_{s_p} \in \mathbb{R}^{|S_p|}$ is the unit vector corresponding to s_p. The utility functions of such a representation require only $\sum_{p,p' \in \mathcal{N}} |S_p| \times |S_{p'}|$ values to specify. Polymatrix games do not have a concise reduced-form encoding, but can easily be written as linear-reduced-form games. Essentially, we create one partition for every matrix game that an agent plays, with each class differing in the action played by the other agent who participates in that matrix game, and containing all the strategy profiles that can be adopted by all of the other players. Formally, given a polymatrix game, we construct its linear reduced form with $P_p = \{P_{p,q}\}_{q \in \mathcal{N} \setminus \{p\}}$, and $P_{p,q} = \{C_{p,q}^\ell\}_{\ell \in S_q}$ with $C_{p,q}^\ell = \{s_{-p} | s_q = \ell\}$. □

Most of the results in Section 4.1 straightforwardly translate to linear reduced forms.

Corollary 2. *For a linear-reduced-form-based representation, if there exists a polynomial-time algorithm for the optimal social welfare problem, then the optimal social welfare CE problem and the max-min welfare CE problem can be solved in polynomial time.*

Graphical Polymatrix Games. A polymatrix game may have graphical-game-like structure: player p's utility may depend only on a subset of the other player's actions. In terms of utility functions, this corresponds to $A^{pp'} = 0$ for certain pairs of players p, p'. As with graphical games, we can construct the (undirected) graph $G = (\mathcal{N}, E)$ where there is an edge $\{p, p'\} \in E$ if $A^{pp'} \neq 0$ or $A^{p'p} \neq 0$. We call such a game a graphical polymatrix game. This can also be understood as a graphical game where each player p's utility is the sum of bilateral interactions with her neighbors.

A tree polymatrix game is a graphical polymatrix game whose corresponding graph is a tree. Consider the optimal CE problem on tree polymatrix games. Since such a game is also a tree graphical game, Papadimitriou and Roughgarden [2008]'s optimal CE algorithm for tree graphical games can be applied. However, this algorithm does not run in polynomial time, because the representation size of tree polymatrix games can be exponentially smaller than that of the corresponding graphical game (which grows exponentially in the degree of the graph). Nevertheless, we give an polynomial-time algorithm for the deviation-adjusted social welfare problem for such games, which then implies the following theorem.

Theorem 1. *Optimal CE in tree polymatrix games can be computed in polynomial time.*

4.3 Representations with Action-Specific Structure

The above results for reduced forms and linear reduced forms crucially depend on the fact that the partitions (i.e., the structure of the utility functions) depend on p but do not depend on the action chosen by player p. There are representations whose utility functions have action-dependent structure, including congestion games [Rosenthal, 1973], local effect games [Leyton-Brown & Tennenholtz, 2003], and action-graph games [Jiang *et al.*, 2011]. For such representations, we can define a variant of the reduced form that has action-dependent partitions. However, unlike both the reduced form and linear reduced form, the deviation-adjusted utilities no longer satisfy the same partition structure as the utilities. Intuitively, the deviation-adjusted utility at s has contributions from the utilities of the strategy profiles when player p deviates to different actions. Whereas for linear reduced forms these deviated strategy profiles correspond to the same class as s in the partition, we now consider different partitions for each action to which p deviates. As a result the deviation-adjusted social welfare problem has a more complex form that the optimal social welfare problem.

Singleton Congestion Games. Ieong *et al.* [2005] studies a class of games called singleton congestion games and showed that the optimal PSNE can be computed in polynomial time. Such a game can be formulated as an instance of congestion games where each action contains a single resource, or an instance of symmetric AGGs where the only edges are self edges.

Formally, a singleton congestion game is specified by $(\mathcal{N}, \mathcal{A}, \{f^\alpha\}_{\alpha \in \mathcal{A}})$ where $\mathcal{N} = 1, \ldots, n$ is the set of players, $\mathcal{A}$ the set of actions, and for each action $\alpha \in \mathcal{A}$, $f^\alpha : [n] \to \mathbb{R}$. The game is symmetric; each player's set of actions $S_p \equiv \mathcal{A}$. Each strategy profile s induces an action count $c(\alpha) = |\{p|s_p = \alpha\}|$ on each α: the number of players playing action α. Then the utility of a player that chose α is $f^\alpha(c(\alpha))$. The representation requires $O(|\mathcal{A}|n)$ numbers to specify.

Before attacking the optimal social welfare CCE problem, we first note that the optimal social welfare problem can be solved in polynomial time by a relatively straightforward dynamic-programming algorithm which is a simplified version of Ieong *et al.* [2005]'s algorithm for optimal PSNE in singleton congestion games. Can we leverage the algorithm for the optimal social welfare problem to solve the coarse deviation-adjusted social welfare problem? Our task here is slightly more complicated: in general the coarse deviation-adjusted social welfare problem no longer has the same symmetric structure due to the fact that y can be asymmetric. However, when y is player-symmetric (that is, $y_j^p = y_j^{p'}$ for all pairs of players (p, p')), then we recover symmetric structure.

Lemma 1. *Given a singleton congestion game and player-symmetric input y, the coarse deviation-adjusted social welfare problem can be solved in polynomial time.*

Therefore if we can guarantee that during a run of ellipsoid method for (1) all input queries y to the separation oracle are symmetric, then we can apply Lemma 1 to solve the problem in polynomial time. We observe that for any symmetric game, there must exist a *symmetric* CE that optimizes the social welfare. This is because given an optimal CE we can create a mixture of permuted versions of this CE, which must itself be a CE by convexity, and must also achieve the same social welfare by symmetry. However, this argument in itself does not guarantee that the y we obtain by the method above will be symmetric. Instead, we observe that if we solve (1) using a ellipsoid method with a player-symmetric initial ball, and use a separation oracle that returns a player-symmetric cutting plane, then the query points y will be player-symmetric. We are able to construct such a separation oracle using a symmetrization argument.

Theorem 2. *Given a singleton congestion game, the optimal social welfare CCE can be computed in polynomial time.*

References

Aumann, R.: Subjectivity and correlation in randomized strategies. Journal of Mathematical Economics 1(1), 67–96 (1974)

Aumann, R.: Correlated equilibrium as an expression of Bayesian rationality. Econometrica: Journal of the Econometric Society, 1–18 (1987)

Bhat, N., Leyton-Brown, K.: Computing Nash equilibria of action-graph games. In: UAI: Proceedings of the Conference on Uncertainty in Artificial Intelligence, pp. 35–42 (2004)

Blum, B., Shelton, C., Koller, D.: A continuation method for Nash equilibria in structured games. JAIR: Journal of Artificial Intelligence Research 25, 457–502 (2006)

Chen, X., Deng, X.: Settling the complexity of 2-player Nash-equilibrium. In: FOCS: Proceedings of the Annual IEEE Symposium on Foundations of Computer Science, pp. 261–272 (2006)

Daskalakis, C., Fabrikant, A., Papadimitriou, C.: The Game World is Flat: The Complexity of Nash Equilibria in Succinct Games. In: Bugliesi, M., Preneel, B., Sassone, V., Wegener, I. (eds.) ICALP 2006. LNCS, vol. 4051, pp. 513–524. Springer, Heidelberg (2006)

Daskalakis, C., Papadimitriou, C.: Three-player games are hard. In: ECCC, TR05-139 (2005)

Goldberg, P.W., Papadimitriou, C.H.: Reducibility among equilibrium problems. In: STOC: Proceedings of the Annual ACM Symposium on Theory of Computing, pp. 61–70 (2006)

Hannan, J.: Approximation to Bayes risk in repeated plays. In: Dresher, M., Tucker, A., Wolfe, P. (eds.) Contributions to the Theory of Games, vol. 3, pp. 97–139. Princeton University Press (1957)

Ieong, S., McGrew, R., Nudelman, E., Shoham, Y., Sun, Q.: Fast and compact: A simple class of congestion games. In: AAAI: Proceedings of the AAAI Conference on Artificial Intelligence, pp. 489–494 (2005)

Jiang, A., Leyton-Brown, K.: Polynomial computation of exact correlated equilibrium in compact games. In: EC: Proceedings of the ACM Conference on Electronic Commerce (2011), http://arxiv.org/abs/1011.0253

Jiang, A.X., Leyton-Brown, K., Bhat, N.: Action-graph games. Games and Economic Behavior 71(1), 141–173 (2011)

Kamisetty, H., Xing, E.P., Langmead, C.J.: Approximating correlated equilibria using relaxations on the marginal polytope. In: ICML (2011)

Kearns, M., Littman, M., Singh, S.: Graphical models for game theory. In: UAI: Proceedings of the Conference on Uncertainty in Artificial Intelligence, pp. 253–260 (2001)

Leyton-Brown, K., Tennenholtz, M.: Local-effect games. In: IJCAI: Proceedings of the International Joint Conference on Artificial Intelligence, pp. 772–780 (2003)

Nisan, N., Ronen, A.: Algorithmic mechanism design. Games and Economic Behavior 35, 166–196 (2001)

Nisan, N., Roughgarden, T., Tardos, E., Vazirani, V. (eds.): Algorithmic game theory. Cambridge University Press, Cambridge (2007)

Papadimitriou, C.: Computing correlated equilibria in multiplayer games. In: STOC: Proceedings of the Annual ACM Symposium on Theory of Computing, pp. 49–56 (2005)

Papadimitriou, C., Roughgarden, T.: Computing correlated equilibria in multi-player games. Journal of the ACM 55(3), 14 (2008)

Rosenthal, R.: A class of games possessing pure-strategy Nash equilibria. International Journal of Game Theory 2, 65–67 (1973)

Buy-Sell Auction Mechanisms in Market Equilibrium

Sanjiv Kapoor

Illinois Institute of Technology, Chicago, USA
kapoor@iit.edu

Abstract. In this paper we consider the problem of computing market equilibrium when utilties are homothetic concave functions. We use the Fisher market model. The problem of finding a tâtonnement process for equilibrium in this case has been the subject of recent papers and determining an approximation is of considerable interest. Our buy-sell algorithm starts with an arbitrary price vector and converges to an ϵ-equilibrium price vector in time proportional to $O(1/\epsilon^2)$. This process attempts to closely mimic the convergence process of real-life markets.

1 Introduction

This paper addresses the computation of market equilibrium, a topic that has been considered with vigor in the recent past. Given a market with n traders and m goods, where the traders are endowed with money or/and goods and wish to optimize their utilities, the problem aims to determine a price vector and an allocation such that no trader has any incentive to trade and there is no excess demand of any good. The problem now has a long history even in the computer science community. Historically, the problem was first proposed in 1891 by Fisher. Independently, Leon Walras (1894) proposed the notion of general equilibrium. Walras proposed that a general equilibrium could be achieved by a price-adjustment process called *Tâtonnement*. The existence of equilibrium prices, under some conditions, was established by Arrow and Debreu [1] using a non-constructive proof. In Fisher's model, buyers (traders) initially have endowment of money while the sellers initially have items. Moreover, the buyers do not have any value for money and the sellers do not have any value for the items. In a more general model, proposed by Walras, the traders have an initial endowment of goods (instead of money). The amount of money available for use by the trader is dependent on the final price of the goods. Starting with the work of [4], there has been considerable recent progress in understanding the complexity of computing equilibrium prices in a market. Polynomial time algorithms have been proposed for several special cases using primal-dual auction algorithms and convex programming techniques [5, 8, 7, 2, 10]. One of the main recent complexity results is that the problem is PPAD-Hard [3].

Of particular historical interest are algorithms that are tâtonnement process. Pioneering research by Scarf on iterative methods for computing market equilibria via the computation of fixed points can be found in [11]. The typical real-life market mechanisms are composed of buying and selling microsteps which can be modeled as a version of combination of English and Dutch auctions. A good example is the stock market where depending on the liquidity the price of individual stocks keeps rising or falling in small steps. A stock sale is made when the asking and bidding price converges. The asks and

N. Chen, E. Elkind, and E. Koutsoupias (Eds.): WINE 2011, LNCS 7090, pp. 230–241, 2011.

bids change depending on the valuation of the good. It is intriguing to model the market processes and verify that the algorithm converges to a market equilibrium in polynomial time. Given that in current trading environment, millions of trades are executed in a short span of time it is conceivable that the markets actually achieve equilibrium in a short time and the fluctuations that arise in the market is a result of changing valuations based on a variety of factors, including general economic conditions.

Towards this end we consider the buy-sell bidding model, an extension of the natural increasing bids auction mechanism. In this paper we describe a mechanism which consists of buying and selling phases. A sell-phase occurs when a good is in excess supply and a buy-phase occurs when traders have money and have demand for a good. This approach appears to more closely model the tâtonnement process of real life markets and conceivably the process envisaged by Walras. Contrast this approach with functional models of price change that dominates the tâtonnement literature and attempt to model the rate of change of the price as a function of the excess demand.

We show that this bidding model can be interpreted as a primal-dual mechanism which attempts to achieve the optimality conditions (KKT conditions) via a sequence of price and allocation changes. Of further interest is the application of the method to problems not easily addressed by the current polynomial time methods. In particular, we consider the market equilibrium problem when the utilities are homothetic concave functions. Results for equilibria in the case of homothetic utility functions can be found in the work of Jain et al. [9] where they generalize the standard Eisenberg-Gale program that characterizes the equilibrium for homogeneous concave functions. The ellipsoid method is used for solving the mathematical program. For homogeneous functions an approach using indirect utility funtions is also presented in [6]. We believe that devising a tâtonnement process based on excess demand or supply will elucidate the workings of real market processes.

In this paper we propose to analyse a new mechanism termed the *Buy-Sell Auction* mechanism, for the market equilibrium problem in the Fisher model. This mechanism corresponds to a primal-dual method, Previous techniques which were largely applicable to utilities satisfying *gross-substitutibility* used the montonicity of price. The buy-sell bidding model allows the price to both increase and decrease. The main result of the paper is to show that the buy-sell bidding mechanism provides an approximation scheme for concave homothetic utilities. The result also applies for an extended class of utiltiies which satisfy a generalized version of Euler's homogeneous function theorem. We show that the complexity is proportional to $O(1/\epsilon^2)$ where ϵ is an error parameter.

2 Market Model

We consider a generalized market model with non-differentiable, continuous concave utilities (our algorithm will be specific to differentiable functions). The market consists of a set of m goods (S) and a set of n traders (T). Trader i has an initial endowment e_i of money. The total amount of good j available in the market is given by a_j (Assume w.l.o.g, via scaling, that $\min_j a_j \geq 1$). The utilities of the traders on these goods are assumed to be defined by a function $F_i(X_i)$, where $F_i(X_i)$ is the utility function of the ith trader and X_i is the vector of allocation of goods to the ith trader.

Given the prices $p_1, p_2, \ldots, p_m$ of the m goods, a trader would like to buy goods with high utility per unit money and sell goods with low utility per unit money. Thus, in equilibrium, trader i will keep only those goods that maximize $F_i(X_i)$. Let x_{ij} represent the amount of good j obtained by trader i. Let $\underline{P}$ represent the $m \times 1$ vector of prices and $\underline{X}$ represent the $n \times m$ matrix of allocations, where the $(i, j)^{th}$ entry is x_{ij}. The pair $(\underline{X}, \underline{P})$ forms a market equilibrium iff (a) there is neither a surplus nor a deficiency of any good (including money); (b) all the traders get goods that maximize their utility. The prices $\underline{P}$ are called market clearing prices and $\underline{X}$ is called an equilibrium allocation.

As in [7], the conditions for market equilibrium can be mathematically represented as:

$$\forall j : \sum_{i=1}^{n} x_{ij} = a_j \tag{1}$$

$$\forall i : \sum_{j=1}^{m} x_{ij} p_j = e_i \tag{2}$$

$$\forall i, \exists \alpha_i, \forall j : x_{ij} > 0 \Rightarrow \alpha_i p_j \in \partial_j (F_i(X_i)) \tag{3}$$

where $x_{ij} \geq 0, p_j \geq 0$. Equation (1) implies that there is no deficiency or surplus of any good. Equation (2) implies that there is no deficiency or surplus of money. Equation (3) implies that every trader is allocated only those goods that maximize the trader's utility, $F(X_i)$, subject to the budget constraint (2) of the trader. Our algorithm will show that the following approximate optimality conditions are satisfied :

$$\text{Goods} - \text{sold} - \text{out} : \forall j : \sum_{i=1}^{n} x_{ij} = a_j \tag{4}$$

$$\text{Trader} - \text{endowment} : \forall i : e_i(1 - \epsilon) \leq \sum_{j=1}^{m} x_{ij} p_j \leq (1 + \epsilon) e_i \tag{5}$$

$$\text{Trader} - \text{Optimality} : \forall i, \exists \alpha_i', \forall j \exists \alpha_{ij} : x_{ij} > 0 \Rightarrow \alpha_{ij} p_j \in \partial_j (F_i(X_i)) \tag{6}$$

$$\text{where } \forall i, \forall j \ \alpha_{ij}(1 - \epsilon) \leq \alpha_i' \leq \alpha_{ij}(1 + \epsilon).$$

3   A Primal-Dual Framework

The Model and Properties Consider the exchange model. We look at equilibrium solutions that maximize the social utiltity function $\sum_i U_i(X_i)$. In this section we assume that $U_i(X_i)$ is a differentiable function that is concave and homothetic. Recall that homogeneous functions are characterized by the property that $f(\alpha x) = \alpha f(x)$ and homothetic functions are monotone transforms of homogeneous functions. Furthermore, $V_{ij}(0)$ is bounded by a large number K. We define a primal-dual framework for discovering the equilibrium. In fact we show that this mechanism is equivalent to optimizing the social benefit. Previous research has shown a similar result for linear utility functions [7]. Consider the welfare program:

$$\max \sum_i U_i(X_i) \tag{7}$$

$$\forall j : \sum_{i=1}^{n} x_{ij} = a_j \tag{8}$$

$$\forall i : \sum_{j=1}^{m} x_{ij} p_j \le \sum_j a_{ij} p_j \tag{9}$$

where $x_{ij} \ge 0, p_j \ge 0$. Further, consider the langrangean function:

$$\min_{\alpha, \beta} \max_{X, p} \sum_i U_i(X) + \sum_i \alpha_i \left(\sum_j a_{ij} p_j - \sum_j x_{ij} p_j \right) + \sum_j \beta_j \left(a_j - \sum_i x_{ij} \right)$$

where $i \in [1 \ldots n]$ and $j \in [1 \ldots m]$. At fixed p, for achieving maximum over X, the following conditions must hold:

$$\forall i, j : x_{ij} > 0 \rightarrow \alpha_i p_j + \beta_j = V_{ij}(X_i^*)$$

where X_i^* is the optimum solution, resulting in the function

$$\min \sum_i U_i(X^*) + \sum_i \alpha_i \sum_j a_{ij} p_j + \sum_j \beta_j a_j - \sum_i \sum_j \beta_j x_{ij}^* - \sum_i \alpha_i \sum_j x_{ij}^* p_j$$

or

$$\min \sum_i U_i(X^*) + \sum_j \beta_j a_j + \sum_j \alpha_i a_{ij} p_j - \sum_i \sum_j V_{ij} x_{ij}^*$$

Since, by Euler's homegeneous function theorem, $U_i(X^*) = \sum_{ij} V_{ij} x_{ij}^*$ for homogeneous functions, the following dual program can be derived for homogeneous functions:

$$\min \sum_{ij} \alpha_i a_{ij} p_j + \sum_j \beta_j a_j \tag{10}$$

$$\forall i, j : \alpha_i p_j + \beta_j \ge V_{ij} \tag{11}$$

$$\alpha_i \ge 0; \ \beta_j \ge 0 \tag{12}$$

where $V_{ij} = \frac{\partial U_i(X_i)}{\partial x_{ij}}$. We can show that our approach is also applicable when an affine version of the above condition holds, i.e: $U_i(X^*) = \sum_{ij} V_{ij} x_{ij}^* + Y$ where Y is a function independent of X_i. Finding a p such that there is a dual solution with $\beta_j = 0, \forall j$, provides the market equilibrium. We next provide a primal-dual framework (Algorithm 1) to solve the problem in the Fisher model, when the dual problem corresponding to the optimization of the i consumer with homogeneous utilities becomes

$$\min \alpha_i e_i \tag{13}$$

$$\forall i, j : \alpha_i p_j \ge V_{ij} \tag{14}$$

$$\alpha_i \ge 0; \tag{15}$$

Algorithm 1. Primal-Dual Mechanism

---

while Equilibrium conditions (4),(5) and (6) not met **do**
    while $\exists$ Good j such that $p_j > 0$ and $\sum_i x_{ij} < a_j$ **do**
        Reduce price and reallocate goods /*Comment: SELL-PHASE */
    end while
    while $\exists$ Trader i such that $r_i > 0$ $(r_i = \sum_j a_{ij} p_j - \sum_j x_{ij} p_j)$ **do**
        Acquire good j s.t. $j = \arg\max V_{ij}/p_j$ /* Comment: BUY-PHASE */
        Raise price p_j if necessary
        $\forall$ traders k: Balance_α for trader k,
    end while
end while

Note that for homothetic functions, represented by $u_i = g_i(f_i(X_i))$, where f_i is homogeneous and g_i a monotone transformation, the optimum of f_i is at X_i^* iff the optimum of u_i is at at X_i^*. The optimality conditions for the above program (for fixed p) are

$$p_j \cdot \{\sum_{i=1}^{n} x_{ij} - a_j\} = 0 \qquad (16)$$

$$x_{ij} \cdot \{\alpha_i p_j - V_{ij}\} = 0 \qquad (17)$$

The first condition follows from Walras's identity and the second from the optimality of the trader's demands. In this frame-work we consider algorithms where both these conditions may be relaxed: we start with the condition that all goods are allocated initially. This is easily ensured for the Fisher Model by starting with very low prices. During the algorithm it is ensured that for all goods j, $x_{kj} \cdot \{\alpha_k p_j - V_{kj}\} = 0$. In the remainder of the paper we detail this frame-work to provide efficient algorithms. This frame-work differs from previous auction algorithms in that prices are not assumed to be monotone.

4 Algorithmic Framework

Overview: In this section we design an auction based method for markets where utilities are defined by functions that are homothetic concave functions. We note that unlike previous previous auction methods, the price of goods does not monotonically increase. The algorithm is iterative, with each iteration having two phases, the *Sell-phase* and the *Buy-phase*. In the Sell-phase the price of goods that are over-supplied is reduced thus allowing more goods to be purchased by traders. In the Buy-phase excess revenue with traders is spent on purchasing goods, increasing their price , if necessary. At the end of this phase, for every trader i, the bang-per-buck, or α_i will be checked and it will be ensured that all goods provide the maximum bang-per-buck. This is done in a process that allows the trader to rebid on goods. Before we define the algorithm, we need some definitions that would be useful to characterize the state of the system at any stage.

Definition 1. *A trader i is said to have* balanced bang/buck *if* $\forall j, k$ *s.t.* $x_{ij}, x_{ik} > 0$, $(1 - \epsilon) \leq \frac{V_{ij}/p_j}{V_{ik}/p_k} \leq (1 + \epsilon)$.

At various stages within the algorithm we may have goods that are not completely sold off. We need the following definition:

Definition 2. *The set of goods that are oversupplied, i.e. goods which are not completely sold off, constitute the set $\mathcal{O}$, defined as:* $\mathcal{O} = \{j | \sum_i x_{ij} < a_j\}$.

We let D_i be the demand set for every consumer, i..e $D_i = \{j | j = \arg\max_j \{\alpha_{ij}\}\}$ where $\alpha_{ij} = \frac{V_{ij}}{p_j}$ and $\alpha_i = \max_j \alpha_{ij}$. Then we can define an approximate demand set:

Definition 3. *The set of goods $D_i^\epsilon = \{j | \alpha_{ij} \geq \alpha_i(1 - \epsilon)\}$ is a set of goods that are approximate maximum bang-per-buck goods for trader i.*

For simplicity we will use D_i instead of D_i^ϵ.

We now describe the algorithm, details of which are in the accompanying psedo-code. The main algorithm (**Algorithm Main**) is in two phases, the *Sell-phase* and the *Buy-phase*. The goal of the algorithm is to spend the endowments of the traders while maintaining the following invariants

(i) Condition 6 (*Trader-Optimality*)
(ii) Condition 4 (*Goods-Sold-Out*)

Algorithm Main is as follows:

1. Assume an initial allocation of goods, i.e. $x_{ij} = \epsilon$.
2. Initialize:Compute $\alpha_i, D_i \forall i$;
3. Compute $\mathcal{O}$;
4. **While** (($\exists$ a trader i s.t. $r_i \geq \epsilon e_i$) OR ($\exists$ a good j s.t. $j \in \mathcal{O}$)
    (a) *Sell-Phase*;
    (b) *Buy-Phase*
    End While

During the course of the algorithm, the endowment remaining unspent with each traders will decrease and the algorithm ends when condition 5 (Trader-endowment) is also satisfied. The *Sell-phase* considers goods that are not completely sold out, i.e. goods in the set $\mathcal{O}$. Each such good j is sold to a buyer i such that $j \in D_i$ and trader i has residual money. If the good is still not sold out, then its price is reduced using **Procedure** *Reduce_price*. Note that, as in previous version of the auction algorithm [7], a good at price p_j has actually been sold to traders either at price, p_j or at price $p_j/(1 + \epsilon)$. We let h_{ij} and y_{ij} be the amount of good j acquired by trader i at level p_j and $p_j/(1 + \epsilon)$, respectively. The availability of goods at the lower price level is indicated by $y_{kj} > 0$, for some k, i.e. at least one trader has acquired the good at price p_j.

In the *Buy-phase*, the pre-condition established is that all goods are sold out and if there is a trader, i, with remaining money, she acquires a good, say j, in the demand set D_i. The acquisition of good j by a trader i is done in procedure *OutbidN*. The Procedure *OutbidN* is a complicated form of an outbid procedure which allows a trader i to acquire a good j from another trader, say k, who has acquired the good at price $p_j/(1 + \epsilon)$. There are two cases (i) Suppose such a trader k exists. Then when trader k is outbid, j may still be in D_k, the demand set of trader k. The trader k then bids back for good j withdrawing, possibly, money from another good, say j'. This process is termed

Recover. The other good j' is then added to the set of goods that are oversupplied. (ii) If there is no trader k who has acquired good j at $p_j/(1 + \epsilon)$ (i.e. such that $y_{kj} > 0$), then the price of good j is raised.

We will assume that there exists a polynomial time oracle $\mathcal{G}$, with time complexity denoted by $\mathcal{G}$ itself, that determines the optimal allocation, given concave utilities, a set of prices and endowments. We now give a description of procedure *Recover*. Other procedures are clear from the description in the algorithm details.

Procedure Recover: Suppose trader k gains M units of money since another trader, say i, has outbid trader k on good j. If trader k prefers good j to other goods, we term that condition as *Change-Benefit* and denote it by $\mathcal{CB}$. The trader who is now unsatisfied, i.e for whom optimality conditions are not met, belongs to a set $\mathcal{U}(j)$ defined as follows:

- $\mathcal{U}(j) = \{k|\mathcal{CB}(k, j)$ is true $\}$ where
  $\mathcal{CB}(k, j) = True$ iff there exists a good j', currently acquired by trader k, such that $V_{kj}/p_j > V_{kj'}/p_{j'}(1 + \epsilon)$.
  In this case it benefits k to acquire j instead of j'.
- The good j' is termed a *witness* for $\mathcal{CB}(k, j)$.

Note that when a trader k tries to retain good j by acquiring good j at a higher price, she funds this using available money and may withdraw money from other goods which may have less utility. There are two cases:

1. The good j is not available at the lower price: then the price of the good may have to be raised.
2. Good j is available at the lower price and is currently with another trader i'.

In the second case, the good is acquired from trader i' who is then added to the set $\mathcal{U}(j)$ if the condition $\mathcal{CB}(i', j)$ is satisfied for trader i'. This process continues until condition 6 (Trader-Optimality) is true for all the traders.

4.1 Analysis

We will use the following properties about the value of the optimal solution to the traders optimization problem:

$$U_i^*(p) = \max\{U_i(X_i)|X_i^T \cdot p \le e_i, e^T \cdot p = 1\}$$

where $U_i(X_i)$ is strictly concave and monotonically increasing: Let p_1 and p_2 be two price vectors. We define $p_1 >_d p_2$ iff p_1 dominates p_2 component-wise, with strict dominance in at least one component.

Lemma 1. *Let p^a and p^b be two price vectors in R^m, such that $p^a >_d p^b$. Then $U_i^*(p^a) < U_i^*(p^b)$*

Proof. Consider the optimum solution, $X_i^*(a)$ at price p^a, giving utility $U_i^*(p^a)$ to trader i. Let p^b differ from p^a in the kth component, i.e. $p_k^b < p_k^a$. Let $X_i(b)$ be a solution defined as follows:

$$x_{ij}(b) = x_{ij}^*(a), \forall j \ne k, x_{ik}(b) = x_{ik}^*(a) + \delta$$

where $\delta = x_{ik}^*(a)(p_k^a - p_k^b)/p_k^b > 0$. The solution $X_i(b)$ is feasible since the money spent by trader i on good k is the same as the money spent in the solution $X_i(a)$. Furthermore $x_{ij}(b) > x_{ij}^*(a)$. Thus $U_i^*(p^b) > U_i^*(p^a)$. Note that for homogeneous function $U_i^*(p) = \alpha_i e_i$ (equation 13) where α_i is the langrange variable representing the bang-per-buck achieved by the allocation for trader i. Thus for these functions, α_i shows an increase. $\qquad\square$

Sell-Phase

1: **while** ($\exists$ a good j s.t. $j \in \mathcal{O}$) **do**
2:    **if** $\exists$ a trader i s.t. $r_i \geq \epsilon e_i$ and $j \in D_i$ **then**
3:       Allocate good j to trader i s.t. $r_i \geq 0$ and Condition 6 (*Trader-Optimality*) is
         satisfied, i.e. $\alpha_{ij}p_j \geq V_{ij}(1 - \epsilon)$. The optimal strategy for trader i is evaluated
         using oracle $\mathcal{G}$ (the process is termed $Allocate(i, j)$);
4:    **else**
5:       $Reduce\_price(j)$;
6:    **end if**
7: **end while**

The following holds for the Sell-Phase:

Lemma 2. *(i) Condition 6 (* Trader-Optimality *) is satisfied for every trader at the end of the Sell-Phase.*
(ii) Condition 4 (Goods-Sold-Out) is satisfied for every good at the end of the Sell-Phase.
(iii)Residual Money Property: $\forall i, r_i$ *residual money does not increase at the end of the Sell phase.*

Proof. The *Sell-Phase* is invoked when there is a good that is *oversupplied*. Let $j \in \mathcal{O}$ be such a good. There are two cases. In the first case, if there is a trader i such that good j is demanded by trader i, i.e. $j \in D_i$, and trader i has residual money, then good j is allocated to trader i. Thus (i) and (iii) are clearly ensured. In the second case, there is no such trader and *Reduce_price* is invoked.

When *Reduce_price* is invoked, first the price of good j is reduced by re-assigning good j at price $p_j/(1 + \epsilon)$ to traders that have acquired the good at price p_j. A trader i is considered for this operation if it satisfies the condition: $h_{ij} > 0$. Let e_{ij} be the endowment spent on good j by trader i We proceed by considering each trader one by one. Let the amount of goods acquired after this step by trader i be denoted by y_{ij}'. Note that $y_{ij}' = y_{ij} + h_{ij}*(1+\epsilon)$. We term as z_j the amount of good j that is oversupplied. In the first case, if $\delta_{ij} = y_{ij}' - x_{ij} \leq z_j$ then this re-assignment is complete and increases the amount of good j bought by trader i.

Alternately $y_{ij}' - x_{ij} > z_j$, where z_j is the current surplus of good j. In this case the amount re-assigned to trader i must satisfy (i) $y_{ij}'p_j/(1 + \epsilon) + h_{ij}'p_j = e_{ij}$ and (ii) $y_{ij}' + h_{ij}' = x_{ij} + z_j$. After this assignment, the price of the good remains unchanged, the good is completely allocated (i.e. $j \notin \mathcal{O}$) and the procedure terminates.

Since the procedure considers traders one by one (while loop at line 1), at the end of the while loop either the good is left oversupplied or is completely sold out.

Suppose the good is still oversupplied. Note that at this step all traders have acquired good j at price $p_j/(1+\epsilon)$. The price of good j is then reduced, the demand sets $D_i, \forall i$ re-computed and the procedure invokes $Release\_Acquire()$ to ensure that the optimality conditions are satisfied, since now it is possible that there exist a trader i such that good j is preferred over some other good k, where $x_{ik} > 0$.

$Release\_Acquire()$ considers all traders that may have the *Trader-Optimality* condition violated and transfers endowment from the allocated good k to j using a re-computation of the optimum allocation for the trader i.

Claim. Procedure $Release\_Acquire$ ensures that for all traders, condition *Trader-Optimality* is satisfied and $\sum_i r_i$ has not increased. (proof omitted)

Note that at the end of this procedure there may exist a good in the set $\mathcal{O}$. The Sell-Phase repeats if $\mathcal{O}$ is non-empty. It suffices to show that the phase ends. We note that at each step of the phase, the goods decrease in price only. When the price of a good j is reduced, consider the amount of good x_{ij} sold to trader i. x_{ij} increases and so does $\sum_i x_{ij}$ since in the procedure, traders acquire more of good j. A sequence of price reductions ends since when the price is reduced to a sufficiently small amount, the value of which will be detailed below, the traders have enough money to acquire all the goods.   □

Buy-Phase

```
 1: for ( i=1 to i=n) do
 2:     if trader i is s.t. r_i ≥ εe_i then
 3:         while there is no reduction in r_i by a factor of (1 + ε) do
 4:             if ∃ good j ∈ 𝒪 ∩ D_i then
 5:                 Allocate good j to trader i without violating approximate optimality via
                    Allocate(i, j);
 6:                 Update 𝒪;
 7:             else
 8:                 if ∃k s.t. y_kj > 0 then
 9:                     OutbidN(i, k, j);
10:                 else
11:                     Raise_price(j);
12:                     ∀i, Update D_i.
13:                 end if
14:             end if
15:         end while
16:     end if
17: end for
```

We consider the complexity of the *Sell-Phase*. No price increase happens during the *Sell-Phase*. We show that the total number of price reductions are bounded. We let P_{maxI} be the maximum price of any good in any equilibrium solution and let $E = \sum_i e_i$. P_{maxI} is bounded by E/a_{min}. Furthermore let $a_{min} = \min_j a_j$ and $a_{max} = \max_j a_j$. W.l.o.g. assume $a_{min} \geq 1$. Note that $\mathcal{G}$ is the complexity of the Oracle that computes the optimum allocation.

Lemma 3. *The   Sell-Phase   terminates   in $O(nm\mathcal{G} \log_{1+\epsilon} P)$ steps   where*
$P = \max\{\frac{P_{maxI}}{P'}, (P_{maxI} \cdot (\frac{ma_{max}V}{E}))\}$ *where $V = \max_{i,j,k} \sup_{x_{ij}, x_{ik}}$*
$\{V_{ij}(x_{ij})/V_{ik}(x_{ik})\}$ *and* $P' = \min_{ij} e_i/a_j$.

To analyze the *Buy-Phase* we consider the key procedure, *OutbidN*. We show below
that this procedure maintains the required invariance, Condition 6. We first consider the
sub-procedure, *Recover*, in *OutbidN*. The procedure *Recover* ensures that approximate
optimality is maintained. Suppose $x_{kj} > 0$. The procedure first checks if the residual
money with trader k will ensure approximate optimality. If not, then let $j' \in J'$ be a
good such that $x_{kj'} > 0$ and $\alpha_{kj'} < \alpha_{kj}(1 - \epsilon)$. The procedure reduces allocation of
goods in J' and increases allocation of good j until trader k has balanced bang/buck.
The optimal allocation is determined by the oracle $\mathcal{G}$.

Procedure *Reduce_price(j)*;
    while ($\exists$ a trader i s.t. $h_{ij} > 0$ and $\Sigma_l x_{lj} < a_j$) **do**
        Let $z_j = a_j - \sum_k x_{kj}$.
        Let e_{ij} be endowment spent by trader i on good j where $e_{ij} = h_{ij} p_j + y_{ij} p_j/(1 + \epsilon)$.
        Reassign good j to trader i at price $p_j/(1 + \epsilon)$: term the assignment y'_{ij}
        If $\sum_i x_{ij} = a_j$ and endowment of trader i is not completely spent, then change the
        assignment to acquire good j in amounts y'_{ij} and h'_{ij}, at prices p_j and $p_j/(1 + \epsilon)$,
        respectively, to satisfy:
        (i) endowment of trader i is spent ($y'_{ij} p_j/(1 + \epsilon) + h'_{ij} p_j = e_{ij}$)
        (ii) the good j is sold out, i.e $y'_{ij} + h'_{ij} = x_{ij} + z_j$.
    end while
    if ($\Sigma_i h_{ij} = 0$) **then**
        $p_j \leftarrow p_j/(1 + \epsilon); \forall i : h_{ij} = y_{ij}$;
        Recompute D_i, $\forall i$;
        Release_Acquire(j);
    end if

Lemma 4. *Recover terminates with balanced bang/buck balanced for trader k.*

We next analyze the *OutbidN* procedure.

Lemma 5. *Assume that at the beginning of OutbidN, Condition 6 (Trader-Optimality)
is satisfied. Then OutbidN terminates with Condition 6 satisfied for all traders. Further
OutbidN requires $O(nm(\mathcal{G} + m))$ steps and the total residual money $\sum_i r_i$ does not
increase during the procedure.*

The invariances maintained by each phase have been discussed above. We concentrate
on the complexity. We call a *Buy-Sell Phase* of the algorithm, one iteration of the *while*
loop (line 4) in *Algorithm Main2*. In that phase goods that are not sold out completely
are first considered in the *Sell-phase* and then every trader is considered once for reduction of her residual endowment in the *Buy-Phase* .

Procedure $Release - Acquire(j)$;

1: **while** ($\exists$ a trader i s.t. $r_i \geq \epsilon e_i$ and good $j \in D_i, j \in \mathcal{O}$) OR ($\exists$ a trader i s.t. good $j \in D_i, j \in \mathcal{O}$ and $\exists$ a good k s.t. $x_{ik} \geq 0$ and $k \notin D_i$) **do**
2:    Compute optimum allocation for trader i;
3:    Increase current allocation of good j to trader i by a factor of $(1 + \epsilon)$ so as to ensure approximate optimal allocation.
4:    Update the set $\mathcal{O}$.
5: **end while**

Procedure $OutbidN(i, k, j)$;

1: outbid(i,j,k); (The outbid procedure acquires good j from trader k and assigns to trader i at a higher price level limited by r_i or y_{kj}.)
2: **if** $(\mathcal{CB}(k, j))$ **then**
3:    Add k to $\mathcal{U}(j)$;
4: **end if**
5: **while** $\mathcal{U}(j) \neq \phi$ **do**
6:    Pick $k \in \mathcal{U}(j)$;
7:    $Recover$;
8:    If $\mathcal{CB}(k, j)$ then add k to $\mathcal{U}(j)$
9: **end while**
10: $\forall j'$ such that $\sum_i x_{ij'} < a_{j'}$ add j' to $\mathcal{O}$;

Procedure $Recover$

1: **if** $r_k > 0$ **then**
2:    $outbid(k, j, k')$ such that trader k acquires j to at most the level it had before (say x^p_{kj} units), where k' is a trader that has acquired good at price $p_j/(1 + \epsilon)$; If not possible, raise price of good j and acquire good j (x^p_{kj} units) from the original trader i at price $p_j(1 + \epsilon)$.
3:    If $\mathcal{CB}(k', j)$ then add k' to $\mathcal{U}(j)$
4: **else**
5:    Let J' be the set of goods that are witness to $\mathcal{CB}(k, j)$;
6:    Compute optimal allocation for trader k using his current endowment.
7:    Allocate good j reducing allocation of goods in J'
    by transferring money from j' to bid for good j and acquire good j back (at most the amount, say x^p_{kj}, that was reduced during the outbid procedure at price p_j) from trader i' at price p_j if possible.
    If $\mathcal{CB}(i', j)$ then add i' to $\mathcal{U}(j)$
8:    If not possible raise price of good j and acquire good j (x_j units) from the original trader i at price $p_j(1 + \epsilon)$.
9: **end if**

Lemma 6. *The* Buy Phase *ends with a reduction in* $\sum_i r_i$ *by a factor of at least* $(1+\epsilon)$.

Finally, let $P = \max\{\frac{P_{max}I}{P'}, (P_{max}I \cdot (\frac{ma_{max}V}{E}))\}$ where $V = \max_{i,j,k} \sup_{x_{ij},x_{ik}} \{V_{ij}(x_{ij})/V_{ik}(x_{ik})\}$.

Theorem 1. *The Buy-Sell algorithm finds the market equilibrium for homothetic concave functions in* $O(f(n, m, \log P))$ *steps where* $f(n, m, p)$ *is a polynomial function.*

5   Conclusions

In this paper we show how a realistic buy-sell mechanism converges to market equilibrium for homothetic functions. Helpful discussions with Rahul Garg and Vijay Vazirani are acknowledged. Research supported in part by NSF CNS-0916743.

References

1. Arrow, K., Debreu, G.: Existence of an Equilibrium for a Competitive Economy. Econometrica 22, 265–290 (1954)
2. Codenotti, B., McCune, B., Varadarajan, K.: Market equilibrium via the excess demand function. In: STOC 2005: Proceedings of the Thirty-Seventh Annual ACM Symposium on Theory of Computing, pp. 74–83. ACM Press, New York (2005)
3. Codenotti, B., Saberi, A., Varadarajan, K., Ye, Y.: Leontief economies encode nonzero sum two-player games. In: SODA 2006: Proceedings of the Seventeenth Annual ACM-SIAM Symposium on Discrete Algorithm, pp. 659–667. ACM, New York (2006)
4. Deng, X., Papadimitriou, C., Safra, S.: On the Complexity of Equilibria. In: 34th ACM Symposium on Theory of Computing (STOC 2002), Montreal, Quebec, Canada (May 2002)
5. Devanur, N., Papadimitriou, C., Saberi, A., Vazirani, V.: Market Equilibrium via a Primal-Dual-Type Algorithm. In: 43rd Symposium on Foundations of Computer Science (FOCS 2002), pp. 389–395 (2002); Journal version to appear in the Journal of the ACM
6. Fleischer, L., Garg, R., Kapoor, S., Khandekar, R., Saberi, A.: Market equilibrium using indirect utility function. In: WINE (2008)
7. Garg, R., Kapoor, S.: Auction algorithms for market equilibrium. Math. Oper. Res. 31(4), 714–729 (2006)
8. Jain, K.: A Polynomial Time Algorithm for Computing the Arrow-Debreau Market equilibrium for Linear Utilities. In: FOCS (2004)
9. Jain, K., Vazirani, V.V., Ye, Y.: Market equilibria for homothetic, quasi-concave utilities and economies of scale in production. In: Proceedings of the Sixteenth Annual ACM-SIAM Symposium on Discrete Algorithms, SODA 2005, pp. 63–71. Society for Industrial and Applied Mathematics, Philadelphia (2005)
10. Nisan, N., Roughgarden, T., Tardos, E., Vazirani, V.V. (eds.): Algorithmic Game Theory (2007)
11. Scarf, H.: The approximation of fixed points of a continuous mapping. Siam Journal on Applied Mathematics 15 (1967)

Behavioral Conflict and Fairness in Social Networks

Stephen Judd[1], Michael Kearns[1], and Yevgeniy Vorobeychik[2]

[1] Computer and Information Science, University of Pennsylvania, Philadelphia, PA
[2] Sandia National Laboratories, Livermore, CA

Abstract. We report on a series of behavioral experiments in social networks in which human subjects continuously choose to play either a dominant role (called a *King*) or a submissive one (called a *Pawn*). Kings receive a higher payoff rate, but only if all their network neighbors are Pawns, and thus the maximum social welfare states correspond to *maximum independent sets*. We document that *fairness* is of vital importance in driving interactions between players. First, we find that payoff disparities between network neighbors gives rise to conflict, and the specifics depend on the network topology. However, allowing Kings to offer "tips" or side payments to their neighbors substantially reduces conflict, and consistently increases social welfare. Finally, we observe that tip reductions lead to increased conflict. We describe these and a broad set of related findings.

1 Introduction

Reporting on a series of behavioral experiments involving a particular class of coordination tasks on social networks, we demonstrate the central importance of *fairness* and *conflict* in interactions between players which entail exclusively financial consequences. The experiments were held in a single session with 36 human subjects, each controlling the state of a single node in an exogenously imposed social network. In our first set of experiments, each subject could choose to be a *King* or a *Pawn*. A King is paid at a higher rate (twice as much as a Pawn), but only if no network neighbor is in *conflict* with him or her by also having chosen to be a King; a King in conflict receives no payments. Players can asynchronously change their state at any time. Since only one of any two neighbors can be a King for either to be paid, such a configuration is inherently "unfair", giving rise to considerable tensions between pure self-interest and fairness considerations. Our second set of experiments thus involved an additional element: Kings that had no conflicts were able to designate a *tip* (or side payment) which was equally divided among their Pawn neighbors.

Networked King-Pawn games may broadly be seen as modeling economic interactions in which each local neighborhood can support only one dominant player. For example, in organized crime it is often the case that only one clan or faction is permitted to rule a locality, and incursions against the incumbent often result in violent clashes that are damaging to both sides. We may also

N. Chen, E. Elkind, and E. Koutsoupias (Eds.): WINE 2011, LNCS 7090, pp. 242–253, 2011.

consider geographic sovereignty as an example — governments oversee property, and attempts by neighboring nations to overtake that property may result in costly wars.

A game theoretic understanding of dynamic coordination games such as these offers several approaches. One considers a stylized one-shot game modeling a long-run outcome, which in our setting exhibits no conflict in a pure strategy Nash equilibrium, and in which any positive tip level is strictly dominated. Another is a repeated game model, with a concomitant explosion of equilibria. Our behavioral results contradict the predictions of a one-shot model: tipping is marked when allowed, and persists even at the end of games. A repeated game model, on the other hand, has too many equilibria to offer a meaningful prediction, and does not suggest any fundamental difference between our two settings. In our experiments, on the other hand, we document numerous qualitative and quantitative differences between the first (no tips) and the second (with tips) settings.

One of our most notable observations is that social welfare is uniformly higher when tips are allowed. Even if we allow that such outcomes are consistent with *some* equilibrium of a repeated-game model, we must still appreciate why this, and not another, equilibrium is ultimately chosen by human subjects. We argue that (the lack of) perceived fairness is primarily responsible for the observed differences between the two experimental settings: inequality in payoffs creates considerable conflict in the first setting, and tips ameliorate conflict by bridging the payoff gaps in the second setting. This finding is robust to the network topology and has broad implications, including the suggestion that reducing income inequality may actually raise social welfare. While there has been well-documented evidence of the importance of fairness considerations in ultimatum games [7,4], we note that our setting involves global coordination on networks, rather than bargaining, and it is not a priori clear that equity plays a role in facilitating or hampering coordination.

A well-known theory in macroeconomics suggests that wages are resistant to reduction, since people view reductions in their wages as inherently unfair even if their real value is preserved [1]. Roughly mapping tips in our setting to wages, we find behavioral evidence for this "downward rigidity". Specifically, we observe that for similar average tip levels, a tip reduction resulted in considerably more conflict. Furthermore, we find that the amount of conflict in response to tip reductions actually rises with average tip pay rate—higher earners appear to respond more strongly to pay cuts.

The experiments described here are part of a broader and ongoing program of behavioral experiments in strategic and economic interaction on social networks conducted at Penn [14,11,13], and are an effort to apply the methods of behavioral game theory [6] to the study of social networks.

1.1   Related Literature

The games we study are networked generalizations of repeated or continuous versions of the game of *Chicken* or *Hawk-Dove* [9], 2-player instances and certain generalizations of which have been studied extensively in the lab [15,17,3]. The subject of fairness in human interactions has a very long history as well. Sociologists and social psychologists view it as central to many social phenomena, and have well-developed theories of fair exchange and reciprocity (exchange/equity theory) [5]. The economic experiments of Fehr and Gächter [8] show that people frequently punish non-altruistic behavior and derive pleasure from doing so. Akerlof and Yellen develop a hypothesis of wage effort based on fairness considerations [2] which allows them to offer an explanation of unemployment and supports the general observation that wages tend to be downwardly rigid [1]. Rabin [18], Fehr and Schmidt [7], and Bolton and Ockenfels [4] offer alternative theories that incorporates fairness into more traditional game theoretic models.

The term "social welfare" will be used here to mean the total payoff to all players in a game. It is worth noting that maximizing the social welfare of our game is isomorphic to the Maximum Independent Set problem, which is a canonical NP-Complete problem [10]. In this study, we construct games in such a way that a Pareto optimal pure strategy Nash equilibrium of its one-shot version solves the maximum independent set problem. In that regard, this work is similar to the experiments in which subjects were placed as nodes in a graph and tasked with coordinating on a *proper coloring*—another canonical NP-Complete problem [14,12].

2   Experimental Design

In our experiments players were mapped to nodes on exogenously specified networks. Each player was given one of two action (role) choices: to be a King or a Pawn. As a King, the player would enjoy a high pay rate ($1/minute), but payments only accrue if *none of his neighbors are also Kings*. A *conflict* is a situation in which there are two neighbors both selecting King — and both earn zero. Being a Pawn, in contrast, is risk-free: no matter what their neighbors choose, Pawns earn a steady income, albeit only half of a King's ($0.50/minute). Payments accrued continuously for each player, pro-rated by the time spent in each of the three possible local states (King without conflicts, King with conflicts, and Pawn). Players could asynchronously update their choices at any time.

A conflict-free configuration of Kings forms an *independent set*. Since Kings are paid only when all their neighbors are Pawns, social welfare is maximized when Kings form a *maximum independent set*, though computing such a maximum is NP-Hard in general.

We ran two variants of this basic King-Pawn scenario. The first was precisely as described above. In the second, we allowed players to offer tips to each other. Tips are payable only while a King is *non-conflicting* (i.e., he is a King and all of his neighbors are Pawns), and when payable they are divided equally among neighbors. Tip offer values were an amount between 0 and 100% of a King's pay rate, but were restricted to quantum steps of 10% (i.e., 10 cents/min). We call this second scenario the "tips" setting, in contrast to the former, which we call the "no-tips" setting.

A natural question to ask is whether allowing players to exchange tips is at all consequential according to traditional game theory. Let us thus observe that *in the tips setting, non-conflicting Kings should never offer tips at Nash equilibrium of a one-shot game corresponding to our setup.* This observation also holds in the last stage of finite repeated games. Since the experiments involve a known time limit and our clock has a finite granularity, we can view them as finite-period repeated games; in such a repeated game, positive tipping could indeed occur even in a subgame perfect equilibrium (except in the last stage), since a mixed strategy equilibrium of a stage game can offer a credible threat.

All experiments were held in a single session lasting multiple hours with 36 University of Pennsylvania students as participants. We ran two sets of 19 experiments, one set for the no-tips and another for the tips setting. Each experiment had a fixed network topology, and subjects were randomly assigned to nodes. All experiments lasted exactly two minutes.

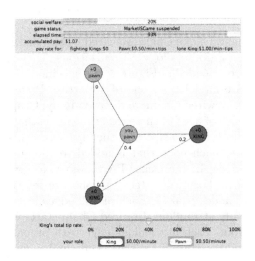

Fig. 1. A screenshot of a player's GUI for the tips scenario. The central node represents the player using the GUI. The numbers displayed near the circles indicate tip offers. The slider designates a choice for the tip offer. The buttons at the bottom of the screen allow a player to choose to be a king or a pawn. In the no-tips setting all allusions to tips (including the slide bar and tip amounts near the nodes) are removed.

A screenshot of the tips GUI is shown in Figure 1; for the no-tips setting, the tip rate bar was simply absent. Each player could see his neighbors and relationships between them, as well as their role and tip choices, but could not see relationships or actions of anyone else. All actions were asynchronous. Role changes or tip adjustments could be made at any time during the game. The session was closely proctored and physical partitions were erected to ensure no communication between subjects.

In both the no-tips and the tips settings we ran three experiments each on six network topologies (Bipartite, Preferential Attachment Tree, Dense

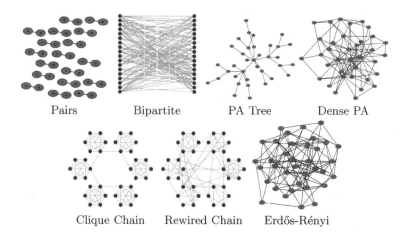

Fig. 2. Sample network topologies used in experiments

Preferential Attachment, Clique Chain, Rewired Chain, and Erdős-Rényi) and a single experiment on a Pairs topology. Visualizations of typical candidates from each topology class are provided in Figure 2. If a specific topology is a class with a stochastic generative model (i.e., one of Bipartite, Preferential Attachment Tree or Dense Graph, Rewired Chain, and Erdős-Rényi Graph), we generated a different network in each of a set of three experiments on that topology, but used the same graphs in both the no-tips and the tips settings. In the Preferential Attachment (PA) Tree, each node that is added to the graph is connected to exactly one existing node. In the Dense PA topology, a new node is connected to three existing nodes. For Erdős-Rényi graphs, we set the probability p of an edge between two nodes to 0.15. Each of 102 edges in the Bipartite graphs paired players uniformly at random. Rewired Chain starts with a Clique Chain as a baseline and rerouts each intra-clique edge with probability 0.2. More detailed descriptions and motivation for these and similar generative models can be found in [16].

The *clustering coefficient* of networks is relevant to our results. It is defined as the number of closed triplets divided by the number of connected triplets of vertices. Figure 3 (left) and later bar plots organize the networks in increasing value of their clustering coefficient. In figures throughout, we mark a network by *** if the reported result or difference is significant with $P < 0.01$, while ** indicates $P < 0.05$, and * corresponds to $P < 0.1$ significance level. When such a result is attributed to both the no-tips and tips settings, we marked the pair with the lowest significance level observed.

3   Results

3.1   Collective Wealth and Tipping

Game-theoretic solutions (applied most directly) do not predict a fundamental difference arising from allowing players to exchange tips. Figures 3 and 4,

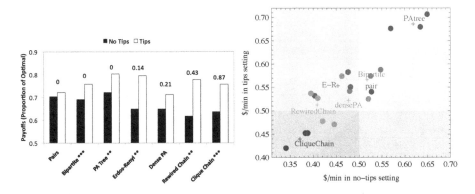

Fig. 3. Left: Achieved welfare (average payoff per player per minute) in the no-tips (black) and tips (white) settings; networks ordered from left to right by clustering coefficient, which is also displayed above each set of bars. Right: Average rate of earnings in each game. Each of the 19 networks is shown without averaging into its replication group; all 19 of them fall above the gray triangle, indicating uniform improvement in the tips setting. The + marks are located at the averages of the replication groups. The shaded zones are where performance is below Pawn rate.

however, demonstrate a systematic improvement in welfare under the tips setting. The impact of tips on welfare varies greatly, and is substantial for some networks.

In Figure 3 (left) we report the *relative* social efficiencies (behaviorally realized social welfare as a proportion of the theoretically optimal social welfare) for the different network topologies (averaged over trials), under both the no-tips and tips settings. Clique Chain, Rewired Chain, and Erdős-Rényi networks exhibit the greatest payoff improvements under tips (around 15% of optimal welfare). Note that these are also three of the four most clustered networks; more on that later. The Pairs network — where there is no "network" per se but rather 18 separate one-on-one games — shows the least improvement, suggesting that the social welfare benefits of tips increase with network complexity. Payoff improvements were significant ($P < 0.05$) in 5 of 7 network architectures (shown in Figure 3, left), and overall improvement in welfare was significant with $P < 0.01$. One may suggest that the reason for the improved outcomes in the tips setting can be attributed to learning effects. To rule out this explanation, we correlated the experiment sequence index with corresponding welfare outcome separately in the no-tips and tips settings. The correlation coefficient was small in the no-tips setting, somewhat larger when tips were allowed, but not statistically significant in either case; it seems clear in any case that subjects had not learned to play the game any better during the no-tips sequence.

Figure 3 (right) illustrates the *absolute* average rates of income for all 19 networks in each of the two settings. The PA trees stand out as being particularly wealthy in both settings; the CliqueChains performed below Pawn level in both settings. The ER graphs are all in the upper left quadrant; they all move from sub-Pawn losers to relative winners when tips are allowed. This figure also

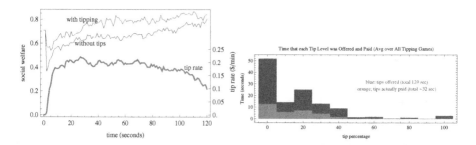

Fig. 4. Left: The top two lines are the average pay rate (as a proportion of optimum) over time in the two settings, averaged over all experiments. Their scale is on the left. The bottom (thick) line is the tip rate, averaged over all tips experiments. Its scale is on the right. Right: Distribution of tips, both offered and accepted.

demonstrates that not only the averages, but all 19 individual network topologies yielded higher social welfare under tips.

Some of the aggregate dynamics (Figure 4, left) reveal the effects of tipping. Average welfare improvement due to tips is persistent throughout the span of experiment. Furthermore, we find that the use of tips, when allowed, is rather substantial: tips accounted for 13% of all income in the tips setting. The lowest curve in the figure shows the average tip rate offered by players over time (average taken over all tips experiments). The tip rate is initialized to zero, but jumps almost immediately after an experiment starts, and persists at around 20% for the bulk of the experiment. It falls off gradually between 70 and 100 seconds and then faster after the 100 second mark, but, even at the end of the experiments, average tip rate persists at around 10%, well above equilibrium level.

We observe that welfare rises over the span of a game in both settings. Furthermore, social welfare *increases* over the last 90 seconds, even as tipping *decreases*. This observation suggests an alternative hypothesis that tips serve as a coordination device, similar to cheap talk, to help select an equilibrium. While such an explanation seems difficult since it would require players to first coordinate on a global meaning of tips in order to use it as a communication device, and is further undermined by the observed persistence of non-negligible tipping at the ends of games, we cannot fully rule it out given our experimental design. We found no significant correlation between tip rates and experiment index within the session, suggesting no long-term adaptation of tipping behavior.

Figure 4 (right) shows the amount of time tip sliders spent in each of the 11 possible states (averaged over all players and games). In the case of one King adjacent to a single Pawn, the tip amount that divides the income equally is 25%. One of the modes in this histogram is slightly below that, at 20%. In the vaguely similar setting of the Ultimatum Game [6] there is a mode of 30 or 40%, also slightly below equitable.

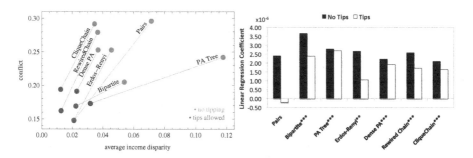

Fig. 5. Left: The connection between income disparity and conflict. Both quantities are reduced when tips are allowed. Right: The connection (linear regression coefficient) between income disparity and conflict (after normalizing both quantities) by network class for the no-tips and tips settings.

3.2   Conflict and Fairness

Thus far we have established the substantial use of tips when available, and their consistent improvement of social welfare. But what behavioral processes underly these phenomena? Here we propose and support the following hypothesis: *Subjects used conflict — which reduces the wealth of all players involved — to express perceived unfairness or inequality. Tipping reduces unfairness and consequently reduces conflict, thereby raising the average payoffs of all players and facilitating coordination.*

We begin this analysis by contrasting quantitative measures of income inequality between the no-tips and tips scenarios. Consider first just the horizontal axis of Figure 5 (left), which measures average income disparity (defined as the average squared difference in payoffs between network neighbors). Since tip levels persist well above zero, and that money is being routed to other players, it is significant and unsurprising that income disparity falls when tipping is allowed. What is more interesting is that tipping appears to roughly equalize payoff asymmetries *across networks*, which were substantially more variable in the no-tips case. For example, PA Tree networks that had shown large income inequality in the no-tips setting are now much closer to other network types. We found a significant correlation (0.49, $P < 0.04$) between income disparity under the no-tips setting and tips exchanged when they are allowed. Our interpretation is that the more a game is perceived as unfair, the greater the role that tips must play in bridging income gaps between players.

The role of tipping in reducing income inequality is only one part of our hypothesis. Additionally, we posit that conflict expresses a *perception* of unfairness. Since tips reduce inequity, we propose that they alleviate the tension that leads to conflict; thus, tips effectively replace or substitute for conflict when they bridge inequality gaps. To support the idea that tips substitute for conflict, we expect to see substantial reduction in conflict between players when tips are allowed. Figure 5 (left) shows this on its vertical axis: the amount of conflict between players (specifically, the average proportion of the game that a player spent in conflict, with average taken over players and games) is systematically lower in the tips setting. Nevertheless, it is difficult to establish a clear relationship between income disparity and conflict. We conjecture that what matters is *perceived*, rather than *observed* (or measured) unfairness, as suggested by *equity theory* [5]. For example, it may seem fair that low degree nodes receive higher income due to the natural advantage their network position offers. We can test this conjecture by considering the correlation between income inequality or conflict with average disparity of degrees between network neighbors; however, we did not find such correlations to be significant in our setting. Instead, we found that the *clustering coefficient* exhibited significant correlation with time players spent in conflict in the no-tips setting (0.62, P-value < 0.01); correlation between the same quantities is considerably smaller and not significant in the tips setting.

As more direct support that conflict communicates perceived unfairness, we looked at *individual* level correlations between the time that a player spends in conflict that he initiates and ultimately terminates, and that player's perceived income disparity, defined as zero when his income is higher than a neighbor's and the squared payoff difference otherwise, and averaged over all of his neighbors.[1] The correlation between these quantities is 0.345 ($P < 0.001$) in the no-tips setting and 0.25 ($P < 0.001$) in the tips setting. These correlations suggest that when players perceive unfairness in their predicament, they are much more likely to engage in conflict with neighbors. On the other hand, the correlation is markedly weaker in the tips setting, providing further evidence for substitution between conflict and tips. One may hypothesize that conflict serves the purpose of punishment to motivate coordinated, better outcomes, similar to Prisoner's Dilemma; below we refute this by showing that conflict decidedly does not pay.

We next consider again the correlation between perceived income inequality and conflict, separated by individual network. However, rather than simply looking at correlations between the two quantities, we regress time a player spends in conflict on his perceived income disparity. In Figure 5 (right) we report the regression coefficient. The figure does not appear to exhibit much systematic difference in the linear relationship between conflict and perceived income disparity across networks. While there does appear to be a slight negative trend as the clustering coefficient increases in the no-tips setting, we did not find it to be statistically significant. Nevertheless, the relationship is clearly positive— significantly so in all graph classes except "Pairs".

Conflict appears to also serve as a means of tip bargaining. Let C be the time (in seconds) that a player spends in conflict that he both initiates and terminates. Define T as *tip income rate*, that is, average tip income per minute that a player spends as a Pawn. Let W be the wealth of a player for the entire game. The correlation between C and T is 0.19 ($P < 0.001$), while the correlation between C and W is -0.51 ($P < 0.001$). The positive correlation between C and T generalizes across 5 of 7 network architectures (significant in all 5); the only exceptions are Clique Chain and Rewired Chain. Thus, while conflict may show

---

[1] This definition of fairness closely mirrors the notion introduced by Fehr and Schmidt [7].

some success in negotiating a higher tip income rate, it yields an unambiguous loss in the long run.

To quantify the tradeoff between time spent in conflict and tip income rate, as well as conflict and wealth, we fit linear regression models to both sets of data pairs. We find (with coefficients having $P < 0.001$) that every second that a player engages his neighbors in conflict earns him (on average) an additional 0.2 cents in tips. Regressing wealth against conflict, on the other hand, tells us (with even higher significance for both regression coefficients) that every second in conflict *costs* a player 1.2 cents on average. The struggle for a bigger tip yields meager rewards and ultimately costs a player more than it is worth.

3.3   Downward Rigidity of Tips

One explanation of high unemployment offered in macroeconomic theory posits that wages are *downwardly rigid*, as people view wage decreases as unfair, even if these decreases maintain the real value of wages (e.g., when there is deflation) [2,1]. As a result, employers prefer to offer above-market wages to ensure that worker productivity remains high; what results is a shortage of jobs relative to the number of people seeking work.

There is a suggestion in Figure 4 (left) that tip changes are downwardly rigid. After being quickly established at the 20% level, they are very slow to head toward equilibrium level and never fall even half way back to zero.

Figure 6 (left) supports the hypothesis that downward changes are viewed as unfair more directly. The comparisons in the figure are between players who made at least one tip reduction and those who made none. The players who did make a tip reduction suffered more conflict than those who did not, even as average tip income rates were roughly equal between the groups. Additionally, *as tip rates increase, tip reductions actually entail more, not less, conflict.* To test the significance of this, we looked at finer discretized tip income rate intervals and correlated midpoints of these with average increases in conflict time. The resulting correlation was 0.99 and highly significant ($P < 0.001$). This result cannot be explained by suggesting that higher tippers also made greater tip reductions: we found no significant correlation between tip pay rate and average size of a tip cut.

3.4   Individual Nodes

The previous discussion pertains to the communal patterns of behavior, but there were also interesting variations at the level of individual nodes.

One natural question to ask is whether a node's degree had an impact on its wealth and role choices. We found significant negative correlation between a node's degree and wealth in both settings (correlation of -0.33 in the no-tips setting, -0.26 in the tips setting, both with $P < 0.001$). Thus, having a high degree was, overall, a handicap. However, breaking this down by network class (Figure 6, right) we find that the negative relationship between degree and income is only significant in three networks (PA Tree, Erdos-Renyi, and Dense PA) in the no-tips setting and in only the first two in the tips setting. This is

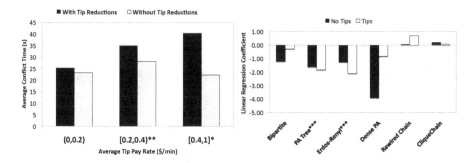

Fig. 6. Left: Average time in conflict at similar average tip pay rates with (black) and without (white) a negative tip change. Right: The connection (linear regression coefficient) between degree and income by network class for the no-tips and tips settings.

not too surprising: all the preferential attachment networks exhibit relatively low degree variation, and in rewired and clique chain graphs it is even less.

Nodes with a high degree spent considerably less time as King (correlation is -0.37 in the no-tips and -0.4 in the tips settings, both with $P < 0.001$). This finding is consistent across network classes. In contrast, the total time spent as King had significant *positive* correlation with wealth overall, 0.18 in the no-tips setting and 0.31 in the tips experiments ($P < 0.001$ in both). However, this conclusion is somewhat nuanced when dissected by network class: in 4 of the 7 network classes, the relationship between time spent as King and wealth is clearly positive in at least one of the game settings (no-tips or tips), but it is highly significant and *negative* in the two most highly clustered networks, rewired and clique chain. Thus, while generally being a King carries an advantage, it is more trouble than it's worth in highly clustered networks (presumably, because Kings face far too much conflict there from other neighbors vying for power).

While high degree nodes had a disadvantage, they were partially compensated for their handicap when tipping was allowed: the correlation between degree and tip income was 0.27 ($P < 0.001$); they naturally also dished out significantly less in tips to their neighbors (correlation between degree and tips paid was -0.23 with $P < 0.01$). Both these findings are consistent across network topologies.

4   Conclusion

One of our key observations is that allowing players to exchange tips substantially increases social welfare. Furthermore, we note that although conflict is clearly damaging to all parties, players systematically engage in it, although substantially less when tipping is allowed. We explain the impact of tipping on the amount of conflict between players by noting that tips equalize incomes between network neighbors. When players view their neighbors' income as unfairly higher than theirs, they engage in conflict, perhaps to punish the high earners. Greater equality in wealth therefore reduces the propensity to engage neighbors in conflict.

Since tip exchanges are pure transfers of wealth in our setting, classical economic theory would not anticipate any impact of tips on average profits. It is thus rather remarkable that tipping raises social welfare in our experiments. The positive welfare impact of tipping (and greater equality of wealth distribution) has considerable implications for policy, as it suggests that bridging income inequality may raise social welfare. Alternatively, our findings suggest that when compensation, resources, or tasks are distributed unequally, transfers of money or gifts may go a long way in alleviating interpersonal conflict.

References

1. Akerlof, G., Shiller, R.J.: Animal Spirits: How Human Psychology Drives the Economy and Why It Matters for Global Capitalism. Princeton University Press (2009)
2. Akerlof, G.A., Yellen, J.L.: The fair wage-effort hypothesis and unemployment. The Quarterly Journal of Economics 105(2), 255–283 (1990)
3. Berninghaus, S.K., Keser, C., Vogt, B.: Strategy choice and network effects. In: Ockenfels, A., Sadrieh, A. (eds.) The Selten School of Behavioral Economics, ch. 8, pp. 89–109. Springer, Heidelberg (2010)
4. Bolton, G.E., Ockenfels, A.: ERC: A theory of equity, reciprocity, and competition. American Economic Review 90(1), 166–193 (2000)
5. Brown, R.: Social Psychology, 2nd edn. The Free Press (1986)
6. Camerer, C.F.: Behavioral Game Theory. Princeton University Press (2003)
7. Fehr, E., Schmidt, K.: Fairness, competition, and inequlity. Quarterly Journal of Economics 114(3), 817–868 (1999)
8. Fehr, E., Gächter, S.: Cooperation and punishment in public goods experiments. American Economic Review 90(4), 980–994 (2000)
9. Fudenberg, D., Levine, D.K.: The Theory of Learning in Games. The MIT Press (1999)
10. Garey, M.R., Johson, D.S.: Computers and Intractability: A Guide to the Theory of NP-Completeness. W.H. Freeman (1979)
11. Judd, S., Kearns, M.: Behavioral experiments in networked trade. In: Proceedings of the ACM Conference on Electronic Commerce (2008)
12. Judd, S., Kearns, M., Vorobeychik, Y.: Behavioral dynamics and influence in networked coloring and consensus. Proceedings of the National Academy of Sciences 107(34), 14978–14982 (2010)
13. Kearns, M., Judd, S., Tan, J., Wortman, J.: Behavioral experiments on biased voting in networks. Proceedings of the National Academy of Sciences 106(5), 1347–1352 (2009)
14. Kearns, M., Suri, S., Montfort, N.: An experimental study of the coloring problem on human subject networks. Science 313(5788), 824–827 (2006)
15. Neugebauer, T., Poulsen, A., Schram, A.: Fairness and reciprocity in the Hawk Dove game. Journal of Economic Behavior and Organization 66(2), 243–250 (2008)
16. Newman, M.: Networks: An Introduction. Oxford University Press (2010)
17. Oprea, R., Henwood, K., Friedman, D.: Separating the Hawks from the Doves: Evidence from continuous time laboratory games. Working paper (2010)
18. Rabin, M.: Incorporating fairness into game theory and economics. American Economic Review 83(5), 1281–1302 (1993)

Efficient Ranking in Sponsored Search

Sébastien Lahaie and R. Preston McAfee

Yahoo! Research
{lahaies,mcafee}@yahoo-inc.com

Abstract. In the standard model of sponsored search auctions, an ad is ranked according to the product of its bid and its estimated click-through rate (known as the quality score), where the estimates are taken as exact. This paper re-examines the form of the efficient ranking rule when uncertainty in click-through rates is taken into account. We provide a sufficient condition under which applying an exponent—strictly less than one—to the quality score improves expected efficiency. The condition holds for a large class of distributions known as natural exponential families, and for the lognormal distribution. An empirical analysis of Yahoo's sponsored search logs reveals that exponent settings substantially smaller than one can be efficient for both high and low volume keywords, implying substantial deviations from the traditional ranking rule.

1 Introduction

Sponsored search is today considered one of the most effective marketing vehicles available online. As the stakes have grown, the auction mechanism has seen several revisions over the years to improve efficiency and revenue. When first introduced by GoTo in 1998, ads were ranked purely by bid; later, in 2002, Google adopted the mechanism and introduced a quality score to weigh bids in proportion to clicks received [5], a practice now shared by every major search engine. In the basic model of sponsored search auctions [10], the quality score corresponds to an ad's position-normalized click-through rate (CTR). Under the assumption that CTRs are measured *exactly*, it is simple to verify that ranking ads in order of quality score times bid is economically efficient.

In this paper we re-examine the form of the efficient ranking rule, taking into account the inherent uncertainty in CTR estimates. Even for high-volume keywords, CTRs are notoriously difficult to estimate because clicks are rare events and new ads constantly enter the system. We consider a parametrized family of ranking rules that order ads according to scores of the form $c^\gamma b$, where c is the estimated position-normalized CTR, b is the bid, and $\gamma \in [0,1]$. This family was proposed by Lahaie and Pennock [9] in the context of sponsored search; they showed that settings of γ strictly less than 1 can improve *revenue*. Their model assumes that CTR estimates are exact. In this work we show that, in the presence of CTR uncertainty, using γ less than 1 can be justified on *efficiency* grounds.

Our main result identifies a sufficient condition under which setting γ strictly less than 1 improves efficiency. The condition relates quality scores based on

N. Chen, E. Elkind, and E. Koutsoupias (Eds.): WINE 2011, LNCS 7090, pp. 254–265, 2011.
© Springer-Verlag Berlin Heidelberg 2011

historical click data (e.g., taking c to be the empirical CTR, normalized for position) to a Bayes estimator of the CTR. We show that the condition holds for a wide class of distributions known as natural exponential families, which includes the normal, Poisson, gamma, and binomial distributions among others. We further show that it holds for the lognormal distribution, which we found to be the best model of Yahoo's CTR estimates. We observe that γ is linked to the concept of *shrinkage* in Bayesian inference [4], and draw on this connection to empirically estimate the efficient γ for several keywords in Yahoo's sponsored search market. Our empirical analysis reveals that settings of γ substantially smaller than 1 can be efficient for both high and low volume keywords.

The remainder of the paper is organized as follows. Section 2 introduces the model, including the manner in which we incorporate uncertainty in CTR estimates. Section 3 identifies a condition under which using γ less than 1 improves efficiency. Section 4 shows that the result holds for natural exponential families as well as the lognormal distribution; it also provides concrete examples of the efficient ranking rules for the beta and lognormal distributions. Section 5 reports on our data analysis of Yahoo's sponsored search logs to uncover the efficient settings of γ in practice. Section 6 concludes.

2   The Model

In this paper we restrict our attention to a single keyword, with a fixed set of agents competing for ad placement whenever a query on the keyword is performed. There are K slots on the page to be allocated among N agents, where $N > K$. In a sponsored search auction each agent i places a bid b_i, and the ads are ranked in decreasing order of $w_i b_i$ where w_i is a weight, or *quality score*, assigned by the search engine. When an ad is clicked, the corresponding agent pays the lowest bid it could have placed while maintaining its position; this is known as the *second-price* payment rule.

While the second-price rule amounts to the Vickrey payment with a single slot, this is no longer the case with multiple slots, and it is well-known that for $K > 1$ sponsored search auctions are not truthful [1]. In general an agent has an incentive to shade its bid b_i below its true value per click (i.e., willingness to pay) v_i. Nonetheless, under the widely accepted solution concept of *envy-free equilibrium* [3,14], it is the case that agents bid in such a way that they are ranked according to $w_i v_i$, because $w_i b_i$ is an increasing function of $w_i v_i$. Furthermore, our adaptation of the click-through rate model in this paper does not affect the agents' incentives, because their ranking only depends on the weights w_i and not their own click-through estimates. Therefore, in what follows, our results and statements in terms of bids will continue to hold if these are replaced with values, assuming envy-free equilibrium, and we can set aside incentive concerns to focus on the problem of efficient ranking.

The determine an efficient ranking the search engine develops an estimate of the *click-through rate* (CTR) ρ_{ij} that ad i would obtain if placed in slot j. We assume that CTRs are *separable*, meaning they factor according to $\rho_{ij} = c_i x_j$

into an advertiser effect c_i and a position effect x_j. Because clicks are stochastic, the advertiser effect is treated as a random variable that follows a probability model $c_i \sim p(\cdot|\theta_i)$, parametrized by θ_i, with mean $\mu_i = \mathbf{E}[c_i \,|\, \theta_i]$. Position effects could also be modeled as random variables in principle, but in this work we treat them as known constants.

While separability is only an approximation to actual CTR patterns [2], it is still relevant for the search engine to estimate position-normalized advertiser effects because $w_i = \mu_i$ is a natural choice for the quality score. If $s : K \to N$ is an allocation of slots, where slot j goes to agent $s(j)$, then under separability the efficiency of the allocation is:

$$\mathbf{E}\left[\sum_{j=1}^{K} x_j c_{s(j)} b_{s(j)} \,\middle|\, \theta_1, \ldots, \theta_N \right] = \sum_{j=1}^{K} x_j \mu_{s(j)} b_{s(j)}.$$

As it is (typically) the case that $x_1 > x_2 > \cdots > x_K$, it is then efficient to take $w_i = \mu_i$ and rank agents in decreasing order of $\mu_i b_i$ [8]. In this work, we relax the assumption that the probability model for each c_i is known exactly and consider how this uncertainty can affect the form of the efficient ranking rule. When discussing CTR modeling, we will often suppress the subscript i when not referring to a specific advertiser, as we do until the end of this section.

To incorporate uncertainty in the probability model due to limited data, we introduce a prior $\theta \sim q(\cdot)$ on the model parameter. Given a vector of m observations $\mathbf{c} = (c^1, \ldots, c^m)$ for the advertiser effect, a generic approach to ranking is to compute a statistic $t(\mathbf{c})$ of the data, and set the weight w to be a function of the statistic. For instance, one could compute the maximum likelihood estimate $\hat{\theta}(\mathbf{c})$ given the data and use the corresponding statistic

$$t_M(\mathbf{c}) = \mathbf{E}[c \,|\, \hat{\theta}(\mathbf{c})] \tag{1}$$

as a weight in order to rank the agents. We will refer to (1) as the *maximum likelihood statistic*. This is often straightforward to compute (e.g., for distributions such as the Bernoulli, normal, and Poisson it is the empirical mean). The maximum likelihood approach is unbiased as the amount of data grows, but in practice click observations are limited. To properly incorporate uncertainty in the presence of limited data, we can instead use a Bayesian approach. In this case the parameter distribution is updated via Bayes rule which sets $q(\theta|\mathbf{c}) \propto p(\mathbf{c}|\theta)q(\theta)$, where $p(\mathbf{c}|\theta) = \prod_{i=1}^{m} p(c^i|\theta)$, and the posterior mean is then

$$t_B(\mathbf{c}) = \mathbf{E}[c \,|\, \mathbf{c}] = \int_{\Theta} \mathbf{E}[c \,|\, \theta] \, q(\theta \,|\, \mathbf{c}) \, d\theta, \tag{2}$$

where Θ is the domain of the parameter θ. We will refer to (2) as the *Bayes statistic*. While this statistic leads to efficient ranking incorporating all uncertainty, it can be more challenging to compute depending on the probability model for advertiser effects and the prior used because of the integration. There is also the issue of setting the initial prior.

In the remainder of the paper we will focus our attention on ranking rules that set $w = t(\mathbf{c})^\gamma$ for $\gamma \in [0, 1]$. With $\gamma = 1$, using statistic (2) is efficient, and using statistic (1) is efficient in the limit as the amount of data grows. This is the usual form of ranking rule used in sponsored search, taking the statistic as a quality score. With $\gamma = 0$, on the other hand, we rank purely by bid, a rule that was used in the very first sponsored search auctions [5]. As we will see, the virtue of this class of ranking rules is that it allows one to use γ to incorporate uncertainty into the ranking, increasing efficiency, while using simpler statistics such as (1) rather than (2) and obviating the need to choose an initial prior.

Formally, assuming bids have been fixed, a *ranking rule* σ defines an allocation of slots to agents for every set of observations $\underline{\mathbf{c}} = (\mathbf{c}_1, \ldots, \mathbf{c}_N)$ of advertiser effects, so that $\sigma(\cdot\,;\underline{\mathbf{c}}) : K \to N$. The expected efficiency of a ranking rule is defined as

$$\mathbf{E}\left[\sum_{j=1}^{K} x_j t_B(\mathbf{c}_{\sigma(j;\underline{\mathbf{c}})}) b_{\sigma(j;\underline{\mathbf{c}})} \right],$$

where the expectation is with respect to the distribution over sampled observations. In what follows, we use $V(\gamma)$ to denote the expected efficiency of the ranking rule that uses $w = t(\mathbf{c})^\gamma$ to weigh bids, for a given statistic t. We are interested in the settings of γ that are most efficient.

3   Main Condition

Our main result[1] provides a sufficient condition for the use of a $\gamma < 1$ exponent on the chosen ranking statistic $t(\mathbf{c})$ on efficiency grounds, rather than revenue grounds as in Lahaie and Pennock [9]. In their approach, the purpose of the exponent is to handicap stronger bidders (with higher advertiser effects), leading to higher competition and increased revenue. In this work, the exponent reflects the contribution of the prior in the Bayes statistic (2).

Theorem 1. *Assume that agents are ranked according to the weights $t(\mathbf{c}_i)$ for $i = 1, \ldots, N$. Then we have $V'(1) < 0$ if the quantity*

$$\frac{\mathbf{E}\,[t_B \mid t]}{t} \tag{3}$$

is decreasing *in the statistic $t \equiv t(\mathbf{c})$, where $t_B \equiv t_B(\mathbf{c})$.*

The conditions given in the theorem imply that efficiency is improved by using $\gamma = 1 - \epsilon$ rather than $\gamma = 1$, for some $\epsilon > 0$. The theorem does not claim that using $t(\mathbf{c})^\gamma$ as a weight, with a properly chosen $\gamma < 1$, is exactly efficiency. When using a statistic such as the empirical advertiser effect for ranking, the condition that (3) be decreasing should hold, intuitively, because t_B is a mixture of the empirical effect and the prior. Therefore the expectation t_B should not respond strongly to a change in the observation t. This intuition is corroborated for a large class of distributions in the next section.

---

[1] Proofs are available from the authors as an appendix.

4   Exponential Families

To usefully apply our main theorem, one needs the ability to evaluate the expectation of the Bayes statistic given the value of the ranking statistic used in practice. As suggested in Section 2, a convenient choice for the latter is the maximum likelihood statistic, which often evaluates to the empirical mean of the observed advertiser effects. In this section we consider a rich collection of distributions, known as *exponential families*, to which the theorem applies and which cover most of the standard distributions one might use for CTR modeling. Exponential families have closed forms for the maximum likelihood statistic, and have convenient conjugate priors which make the Bayes statistic tractable to analyze. The properties of exponential families that we introduce here are standard and can be found in [12,15].

An exponential family is a parametrized distribution with density that takes the form

$$p(c|\theta) = f(c) \exp\left[\theta \cdot \phi(c) - g(\theta)\right]. \tag{4}$$

Here f is a base density over advertiser effects, and θ is known as the *natural parameter*. The term $\phi(c)$ is the *sufficient statistic*. We will restrict our attention to families with scalar-valued sufficient statistics; this implies that the natural parameter θ is also a scalar. The term $g(\theta)$ is a normalization constant, and the domain of the natural parameter is those θ for which the normalizer is finite: $\Theta = \{\theta : g(\theta) < +\infty\}$. It is known to be convex—for the case of a scalar natural parameter, the domain is a (possibly unbounded) interval.

In general, the maximum likelihood estimate $\hat{\theta}(\mathbf{c})$ for the natural parameter, given a vector of m observations $\mathbf{c} = (c^1, \ldots, c^m)$, cannot be evaluated analytically. However, the expectation of the sufficient statistic under this estimate is simply

$$\mathbf{E}[\phi(c) \mid \hat{\theta}(\mathbf{c})] = \frac{1}{m} \sum_{i=1}^{m} \phi(c^i), \tag{5}$$

namely the empirical mean of the sufficient statistic. An exponential family has a conjugate prior of the form

$$p(\theta|\nu, n) = \exp\left[\nu \cdot \theta - n \cdot g(\theta) - h(\nu, n)\right].$$

This is again an exponential family, but with a two-dimensional natural parameter (ν, n), and here $h(\nu, n)$ is the normalizing constant. Given the m observations $(c^1, \ldots, c^m)$, the parameters of the conjugate distribution are updated according to the rule:

$$n \leftarrow n + m$$
$$\nu \leftarrow \nu + \sum_{i=1}^{m} \phi(c^i)$$

Note that the latter parameter is essentially updated according to the maximum likelihood statistic (5). Therefore, exponential families provide a tractable form for the maximum likelihood statistic, and define a clear relationship between this statistic and the posterior distribution. This makes them amenable to the application of Theorem 1.

4.1   Natural Exponential Families

A *natural* exponential family is one where the sufficient statistic is simply $\phi(c) = c$. In this case, the maximum likelihood statistic coincides with the empirical mean, because according to (5) we have

$$t_M(\mathbf{c}) = \mathbf{E}[c \mid \hat{\theta}(\mathbf{c})] = \frac{1}{m}\sum_{i=1}^{m} c^i.$$

Many of the most prominent univariate distributions are natural exponential families, such as the normal, Poisson, gamma, exponential, Weibull, binomial, and Bernoulli distributions [12]. For all of these distributions, the condition (3) in our main theorem applies when using the maximum likelihood statistic for ranking, as the next result shows.

Proposition 1. *Assume advertiser effects are distributed according to a natural exponential family, and that advertisers are ranked according to weights $t_M(\mathbf{c})^\gamma$. Then there is an $\epsilon > 0$ such that using $\gamma = 1 - \epsilon$ improves expected efficiency over $\gamma = 1$.*

To gain some intuition for the result, it is helpful to consider a concrete instance of a natural exponential family. In one interpretation of the separable CTR model, the position effect is the probability that the user will look at a slot, and the advertiser effect is the probability the ad is clicked given that it is viewed [8]. As clicks are binary events, the Bernoulli distribution—a natural exponential family—is then a straightforward choice of model for advertiser effects. Assume that $c \sim \text{Bernoulli}(p)$ and that $p \sim \text{Beta}(n\mu, n(1-\mu))$—the beta distribution is the conjugate prior for the Bernoulli. The mean of the latter is μ, while the empirical mean $\bar{c}$ is both the maximum likelihood statistic and a sufficient statistic for the Bayes update. After the update we have

$$p \mid \bar{c} \sim \text{Beta}\left(n\mu + m\bar{c}, n(1-\mu) + m(1-\bar{c})\right),$$

which has a mean of $\gamma\bar{c} + (1-\gamma)\mu$ where $\gamma = \frac{m}{n+m}$. Because the parameter p for the Bernoulli is its mean, the posterior mean of p is also the posterior mean of e. The term (3) in our main theorem therefore evaluates to

$$\gamma + (1-\gamma)\frac{\mu}{\bar{c}},$$

which is decreasing in $\bar{c}$, as expected. However, Theorem 1 only states that using some $\gamma < 1$ as an exponent on $\bar{c}$ improves efficiency here—it does *not* state that ranking according to $\bar{c}^\gamma b$ is efficient. The closed form solution to the update implies that to rank two bidders efficiently, we should make the comparison

$$b_1 \cdot [\gamma\bar{c}_1 + (1-\gamma)\mu] \overset{?}{>} b_2 \cdot [\gamma\bar{c}_2 + (1-\gamma)\mu], \tag{6}$$

which takes a linear rather than exponential form. We see that when the prior is uninformative ($n = 0$) or there is ample data ($m \to \infty$), then $\gamma \to 1$ and we rank by $\bar{c}b$. When there is no data, $\gamma = 0$ and we rank purely by bid. Note that to rank efficiently according to (6), one needs an estimate of the prior mean μ.

4.2 Lognormal Distribution

While the probability interpretation of the advertiser and position effects is intuitively appealing, in practice the search engine may use a different factorization of CTRs that does not lead to effects in $[0, 1]$. However, it is clear that the effects should be non-negative. The lognormal distribution has support on the positive reals and so could prove a convenient choice to model advertiser effects—this turned out to be the case in our empirical analysis, as we report in Section 5 later on. We will show in this section that Theorem 1 applies to this distribution as well; in fact, using a certain $\gamma \in (0,1)$ exponent is *exactly* efficient for this distribution.

The lognormal is an exponential family, but not a *natural* exponential family, because it has sufficient statistic $\phi(c) = \log c$. Recall that an effect c is lognormal if $\log c \sim \mathcal{N}(\mu, \sigma_c^2)$. We assume the variance is known, and that $\mu \sim \mathcal{N}(\nu, \sigma_\mu^2)$—the normal distribution is the conjugate prior for the normal. Given m observations, let $\bar{\ell} = \frac{1}{m} \sum_{i=1}^{n} \log c_i$ denote the empirical mean of the sufficient statistic. Let $\hat{c} = \left(\prod_{i=1}^{m} c_i \right)^{1/m}$ denote the geometric mean of the observations, and observe that we have $\hat{c} = \exp(\bar{\ell})$. It is known that the expected value of $\exp(y)$ for $y \sim \mathcal{N}(\mu, \sigma^2)$ is $\exp(\mu + \sigma^2/2)$, so we have

$$t_M(\mathbf{c}) = \exp(\bar{\ell} + \sigma_c^2/2) = \hat{c} \cdot \exp(\sigma_c^2/2). \tag{7}$$

That is, the maximum likelihood statistic is proportional to the geometric mean, so the latter is a natural ranking statistic in this context. On the other hand, letting $\tau_c = \sigma_c^{-1}$ and $\tau_\mu = \sigma_\mu^{-1}$, the Bayes update leads to the posterior

$$\mu \,|\, \bar{\ell} \sim \mathcal{N}\left((1 - \gamma)\nu + \gamma\bar{\ell}, \ \left(\tau_\mu^2 + \tau_c^2\right)^{-1} \right), \tag{8}$$

where $\gamma = m\tau_c^2/(\tau_\mu^2 + m\tau_c^2)$. A straightforward evaluation of (2) therefore gives

$$\begin{aligned} t_B(\mathbf{c}) &= \exp[(1 - \gamma)\nu + \gamma\bar{\ell} + \sigma_\mu^2/2 + \sigma_c^2/2] \\ &= \hat{c}^\gamma \cdot \exp[(1 - \gamma)\nu + \sigma_\mu^2/2 + \sigma_c^2/2] \end{aligned} \tag{9}$$

The next result is now immediate, but because of its relevance in practice we record it as a proposition.

Proposition 2. *Assume advertiser effects follow a lognormal distribution. Then ranking according to $\hat{c}^\gamma$, with $\gamma = m\tau_c^2/(\tau_\mu^2 + m\tau_c^2) \in (0,1)$, maximizes expected efficiency.*

When there is ample data ($m \to +\infty$) or the prior is uninformative ($\tau_\mu \to 0$), it is efficient to rank according to $\hat{c}b$. When there is no data ($m = 0$), we rank purely by bid. Note that under the lognormal distribution the prior mean cancels out when comparing weighted bids. This compares favorably to the linear form of the efficient ranking rule we derived for the beta distribution in (6), where it is necessary to estimate the prior mean; however, the prior variance is still needed to determine the efficient γ.

5   Empirical Data Analysis

In this section we report on an empirical analysis of Yahoo's sponsored search logs to get a sense of the settings of γ that are efficient in practice. The theory so far has established that, under reasonable modeling assumptions, using an exponent of $\gamma = 1 - \epsilon$ on the empirical advertiser effect would improve efficiency, for some $\epsilon > 0$. However, if the ϵ need only be very small according to the data, these results would have little bearing on real sponsored search auctions.

5.1   Data Description

We collected data by considering all the keywords in the month of June 2010 that had at least one advertisement. From these keywords we retained those where, over the month, the total number of clicks on ads was at least 2, and the average depth was at least 2. The depth of a query is the number of ads shown, which can range from 0 to 12 on Yahoo. The keywords were stratified into 10 deciles by search volume, and we randomly selected 20 from each decile for a total of 200 keywords. While the sampling is not proportional, we are not interested in aggregating statistics across deciles; proportional sampling would lead to a dataset overwhelmed by tail keywords with sparse click data.

For each ad shown on a keyword, and every position the ad was placed in, we have the total number of searches and clicks as well as the position effect. A position here is defined not just by the rank of the ad, but also where it was placed on the page (top, bottom, side), and how its competitors were laid out. For instance, showing an ad at the third rank when there are two ads at the top (i.e., first on the side) is not the same as showing the ad at that same rank when no ads are at the top (i.e., third on the side): the different positioning leads to a different position effect. There are a total of 60 distinct positions in our dataset. For each position we have a position effect hard-coded by Yahoo; while these were occasionally revised over the month, the changes were typically minimal. The relative standard deviations of the position effects over the month had a median of 0% and mean of 2% over the keywords and advertisers. We therefore take these effects as constants, consistent with our earlier assumptions.

Our dataset has 117K records, one for each keyword-ad-position triplet, and contains information on 19K distinct ads, for an average of 95 ads per keyword over the month and 587 records per keyword (naturally the distribution is heavily skewed). We define the observed advertiser effect for an ad at a certain position on a given keyword as the position-normalized empirical click-through rate:

$$\frac{\text{clicks}}{\text{searches} \cdot \text{position effect}}$$

The observed effects do not all lie in $[0, 1]$: they have a median of 0.002 and mean of 8.12 in our data. Figure 1 indicates that the observed ad effects are well modeled by a lognormal distribution, restricting our attention to ads that received at least one click. For this probability model, the results of Section 4.2 show that there is a setting of γ for each keyword that is exactly efficient.

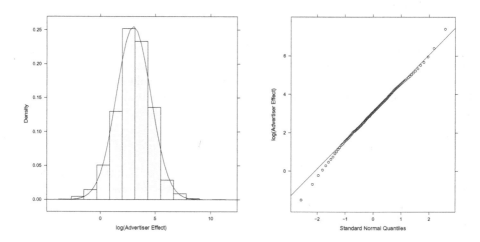

Fig. 1. Lognormality of the observed advertiser effects (position-normalized CTRs). The left panel shows the empirical distribution for ads that have at least one click over the month, together with the best-fit normal distribution. The right panel gives the theoretical quantile-quantile plot.

5.2   Hierarchical Model

To empirically estimate the optimal γ for different keywords we develop a hierarchical Bayesian model of advertiser effects. We have seen through (8) that with the lognormal distribution (among others), γ can be viewed as the weight on the empirical advertiser effect in a convex combination between it and the prior mean. In Bayesian inference this is known as the *shrinkage* factor [4,11], and we can obtain shrinkage estimates as a by-product of a hierarchical model.

We fit a model to each individual keyword. Given a keyword, the units are ad-position pairs i, and we denote the position-normalized empirical CTR for this pair by y_i. Let $j[i]$ denote the ad in unit i. We fit the following basic one-way hierarchical model [6]:

$$\log y_i \sim \mathcal{N}(\alpha_{j[i]}, \sigma_y^2) \qquad (10)$$

$$\alpha_j \sim \mathcal{N}(\mu_\alpha, \sigma_\alpha^2) \qquad (11)$$

where i ranges over all the units and j over all the ads. (To avoid taking the log of 0, we recoded empirical effects of 0 to 10^{-5}, which is an order of magnitude smaller than the smallest positive observed effect in our dataset.) We assign uninformative uniform priors to σ_y, μ_α, and σ_α. The posterior distribution was evaluated using the Gibbs sampler provided by the JAGS program [13], and 1000 draws from the posterior were taken to estimate model statistics, in particular γ. For each draw γ was estimated using the following approach proposed by

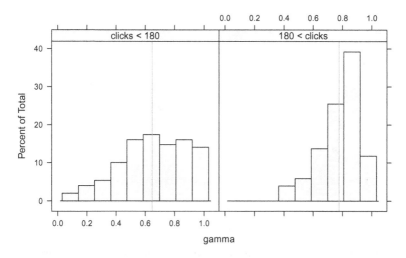

Fig. 2. Empirical distribution of estimated γ's for keywords with small and large numbers of clicks over the month. The reference lines indicate the means. For keywords with small numbers of clicks, the distribution is more uniform, whereas for keywords that attract many clicks γ skews towards 1.

Gelman and Pardoe [7]. Letting $\epsilon_j = \alpha_j - \mu_\alpha$ for each advertiser j, we set

$$\gamma = \frac{\mathbf{V}_j \mathbf{E}[\epsilon_j]}{\mathbf{E}[\mathbf{V}_j \epsilon_j]}, \tag{12}$$

where $\mathbf{V}$ represents the finite-sample variance operator, $\mathbf{V}_j \epsilon_j = \frac{1}{n-1} \sum_j (\epsilon_j - \bar{\epsilon}_j)^2$, and $\mathbf{E}$ in this context is the finite-sample mean. The denominator in (12) is the unexplained component of the variance in the α_j's, while the numerator is the variance among the point estimates of the ϵ_j's. We will have γ close to 1 if the latter is large relative to the former, meaning that α_j's usually lie closer to the empirical mean of the advertiser's effect. On the other hand, if the latter is small relative to the former, then the estimated α_j cluster more closely to μ_α and so the prior mean is given higher weight. Gelman and Pardoe [7] demonstrate that (12) can be viewed as a Bayesian analog to the definition of γ we saw earlier: $\gamma = m\tau_c^2/(\tau_\mu^2 + m\tau_c^2)$. We report on the γ evaluated according to (12) with the expectations taken over the 1000 draws.

Figure 2 shows the distribution of the resulting γ's over the 200 keywords. We identified different patterns in the distribution depending on whether we consider low or high click keywords; here high means greater than 180 clicks per month, or 6 clicks per day on average. For low click keywords the distribution of γ is more uniform, with mean and median both at 0.64. High click keywords see γ more skewed towards 1, as one would intuitively expect, with a mean of 0.78 and a median of 0.82. Note that under both regimes the mean is substantially below 1, which shows that using a rule of the form $c^\gamma b$ could improve efficiency for many keywords.

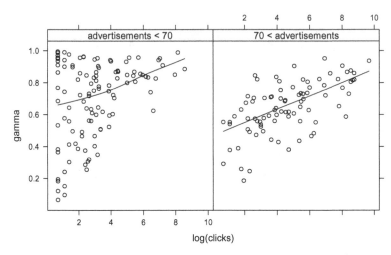

Fig. 3. Estimated γ for keywords with small and large numbers of advertisers over the month. The Loess curves show that under both regimes γ increases on average as the keyword receives more clicks, but for keywords with small numbers of advertisers and clicks there is substantial variability.

Figure 3 shows the empirical results from a different perspective. We again have two different regimes: keywords with few and many ads. Here a keyword has many ads if more than 70 distinct ads were shown over the month. For keywords with many ads there is a clear relationship between the volume of clicks and γ. This is intuitive since more clicks means more accurate CTR estimates. For keywords with few ads there is still a general upward trend, but there is substantial variability in the γ estimates, attributable to the dearth of data. In both cases the most relevant range for tuning γ seems to be $[0.6, 1]$.

6   Discussion

To conclude let us discuss a few limitations and extensions of this analysis. A key assumption implicit in the use of (12), and throughout the paper, is that each ad sees the same amount of observations m. In practice this is of course not the case, especially as ads are constantly added to the system. With uneven amounts of data among ads on a keyword, the estimate (12) amounts to a weighted combination of the different shrinkage factors for the individual ads. To rank efficiently, one would have to use ad-specific γ's. This is not very appealing because the contribution of the prior mean in (9) no longer cancels out when comparing weighted means, leading to a more complicated ranking rule. A better understanding of the efficiency trade-offs between keyword- and ad-specific γ's is in order.

In our analysis, we base our estimate of the shrinkage factor γ on the empirical advertiser effects, but in practice the search engine uses machine-learned effects to rank. While these correlate well with realized advertiser effects, it would be informative to understand exactly how γ should be set given the search engine's estimates. One possibility is to introduce them into (10) as a linear predictor for realized effects. However, the resulting γ from such a model would not be the recommended exponent for the machine-learned effects. In fact, because the predictor would reduce the errors in the numerator of (10), this would misleadingly pull (12) towards 0. Developing sound ways to estimate γ with machine-learned effects is an important next step in this line of research.

References

1. Aggarwal, G., Goel, A., Motwani, R.: Truthful auctions for pricing search keywords. In: Proceedings of the 7th ACM Conference on Electronic Commerce, pp. 1–7 (2006)
2. Athey, S., Nekipelov, D.: A structural model of sponsored search advertising auctions. Tech. rep., Microsoft Research (May 2010)
3. Edelman, B., Ostrovsky, M., Schwarz, M.: Internet advertising and the Generalized Second Price auction: Selling billions of dollars worth of keywords. American Economic Review 97(1) (March 2007)
4. Efron, B., Morris, C.: Data analysis using Stein's estimator and its generalizations. Journal of the American Statistical Association 70(350), 311–319 (1975)
5. Fain, D.C., Pedersen, J.O.: Sponsored search: A brief history. In: Second Workshop on Sponsored Search (2006)
6. Gelman, A., Carlin, J.B., Stern, H.S., Rubin, D.B.: Bayesian Data Analysis. Chapman and Hall/CRC (2003)
7. Gelman, A., Pardoe, I.: Bayesian measures of explained variance and pooling in multilevel (hierarchical) models. Technometrics 48(2), 241–251 (2006)
8. Lahaie, S.: An analysis of alternative slot auction designs for sponsored search. In: Proceedings of the 7th ACM Conference on Electronic Commerce, pp. 218–227 (2006)
9. Lahaie, S., Pennock, D.M.: Revenue analysis of a family of ranking rules for keyword auctions. In: Proceedings of the 8th ACM Conference on Electronic Commerce, pp. 50–56 (2007)
10. Lahaie, S., Pennock, D.M., Saberi, A., Vohra, R.V.: Sponsored search auctions. In: Nisan, N., Roughgarden, T., Taros, É., Vazirani, V.V. (eds.) Algorithmic Game Theory, pp. 699–716. Cambridge University Press (2007)
11. Louis, T.A.: Estimating a population of parameter values using Bayes and empirical Bayes methods. Journal of the American Statistical Association 79(386), 393–398 (1984)
12. Morris, C.N.: Natural exponential families with quadratic variance functions. The Annals of Statistics 10(1), 65–80 (1982)
13. Plummer, M.: JAGS: A program for analysis of Bayesian graphical models using Gibbs sampling, www-ice.iarc.fr/~{}martyn/software/jags/
14. Varian, H.R.: Position auctions. International Journal of Industrial Organization 25, 1163–1178 (2007)
15. Wainwright, M.J., Jordan, M.I.: Graphical Models, Exponential Families, and Variational Inference. Now Publishers Inc. (2008)

The Complexity of Approximate Nash Equilibrium in Congestion Games with Negative Delays[*]

Frédéric Magniez[1], Michel de Rougemont[1,2],
Miklos Santha[1,3], and Xavier Zeitoun[1,4]

[1] LIAFA, Univ. Paris Diderot, CNRS, Paris, France
frederic.magniez@univ-paris-diderot.fr
[2] Univ. Paris 2, Paris, France
mdr@liafa.jussieu.fr
[3] Centre for Quantum Technologies, National University of Singapore, Singapore
santha@liafa.jussieu.fr
[4] Univ. Paris Sud, Orsay, France
xavier.zeitoun@lri.fr

Abstract. We extend the study of the complexity of computing an ε-approximate Nash equilibrium in symmetric congestion games from the case of positive delay functions to delays of arbitrary sign. Our results show that with this extension the complexity has a richer structure, and it depends on the exact nature of the signs allowed. We first prove that in symmetric games with increasing delay functions and with α-bounded jump the ε-Nash dynamic converges in polynomial time when all delays are negative, similarly to the case of positive delays. We are able to extend this result to monotone delay functions. We then establish a hardness result for symmetric games with increasing delay functions and with α-bounded jump when the delays can be both positive and negative: in that case computing an ε-approximate Nash equilibrium becomes PLS-complete, even if each delay function is of constant sign or of constant absolute value.

1 Introduction

Congestion games were introduced by Rosenthal [19] to model shared resources by selfish players. In these games the strategies of each player correspond to some collection of subsets of a given set of common resources. The cost of a strategy is the sum of the costs of the selected resources, where the cost of a particular resource depends on the number of players having chosen this resource. This dependence is described in the specification of the game by some integer valued delay function for each resource.

---

[*] Partially supported by the French ANR Defis program under contract ANR-08-EMER-012 (QRAC project). Research at CQT is funded by the Singapore Ministry of Education and the National Research Foundation.

Congestion games can describe several interesting routing and resource allocation scenarios in networks. More importantly from a game theoretic perspective, they have some particularly attractive properties. Rosenthal has proven that they belong to the class of potential games where, for each player, an improvement (decrease) in his cost is reflected by an improvement in a global function, the potential function. This implies, in particular, that congestion games always have a pure Nash equilibrium. More precisely, a Nash equilibrium can be reached by the so called Nash dynamics, in which an unsatisfied player switches his strategy to a better one, which decreases his cost function. Since the same improvement is mirrored in the potential function, which can not be decreased infinitely, this process indeed has to converge to an equilibrium in a finite number of steps. In an exact potential game the changes in the individual cost functions and the potential function are not only identical in sign, but also in the exact value. Monderer and Shapley [17] have proved that congestion games and exact potential games are equivalent.

The existence of a potential function for congestion games allows us to cast searching for a Nash equilibrium as a local search problem. The states, that is the strategy profiles of the players, are the feasible solutions, and the neighborhood of a state consists of all authorized changes in the strategy of a single player. Then local optima correspond to states where no player can improve individually his cost, that is exactly to Nash equilibria. The potential of a state can be evaluated in polynomial time, and similarly a neighboring state of lower potential can be exhibited, provided that there exists one. This means that the problem of computing a Nash equilibrium in a congestion game belongs to the complexity class PLS, Polynomial Local Search, defined in [13,18]. The class PLS is a subclass of TFNP [16], the family of NP search problems for which a solution is guaranteed to exist. While PLS is not harder than NP ∩ coNP, it is widely believed to be computationally intractable. Fabrikant et al. [10] have shown that computing a Nash equilibrium in congestion games is PLS-complete. In addition, they have explicitly constructed games in which the Nash dynamics takes exponential time to converge. It is worth to note that it is also highly unlikely that computing a mixed Nash equilibrium in general games is feasible in polynomial time, even when the number of players is restricted to two [9,4].

It is therefore natural to look for relaxed versions, and in particular approximations, of Nash equilibria which might be computed in polynomial time. Approximate Nash equilibria of various games have been defined and studied both in the additive [14,15,5,7,8,11,21] and in the multiplicative models of approximation [6,2]. Here we consider multiplicative ε-approximate Nash equilibria, for $0 < \varepsilon < 1$, that is states where no single player can improve his cost by more than a factor of ε by unilaterally changing his strategy. In this context, the analogous concept of the Nash dynamics is the ε-Nash dynamics, where only ε-moves are permitted, which improve the respective player's cost at least by a factor of ε. Rosenthal's potential function arguments imply again that the ε-Nash dynamics converges to an ε-approximate Nash equilibrium.

In a very interesting positive result, Chien and Sinclair [6] proved that in congestion games with four specific constraints the ε-Nash dynamics indeed does converge fast, in polynomial time. The four constraints require the game to be increasing, positive, symmetric, and with α-bounded jump. The first three constraints are rather standard. A congestion game is increasing (respectively positive) if of all delay functions are non-decreasing (respectively non-negative). It is symmetric if all players have the same strategies. The last constraint puts a limit on the speed of growth of the delay functions. They define an increasing and positive congestion game to be with α-bounded jump, for some $\alpha \geq 1$, if the delay functions can not grow more than a factor α when their argument is increased by one. Their result states that in increasing, positive and symmetric congestion games with α-bounded jump, the ε-Nash dynamics converges in polynomial time in the input length, α and $1/\varepsilon$.

Could it be that the ε-Nash dynamic converges fast in every congestion game? Skopalik and Vöcking have found a very strong evidence for the contrary. In a negative result [20], they proved that for every polynomial time computable $0 < \varepsilon < 1$, computing an ε-approximate Nash equilibrium is PLS-complete, that is as hard as computing a Nash equilibrium. In fact, they result is even stronger, it shows the PLS-completeness of the problem for increasing positive games.

In this paper we extend these studies to the case when the delays can be also negative, that is some resources might have the special status of improving the cost of the players when they are chosen. We consider negative games where the delay functions may be either increasing or decreasing. We first prove that in negative symmetric games with α-bounded jump, when all delay functions are increasing, the ε-Nash dynamics converges in polynomial time, just as in the case of positive increasing games. We then extend this result to games where all delay functions are monotone, that is either increasing or decreasing. We then prove a hardness result: computing an ε-approximate Nash equilibrium in symmetric and increasing games with α-bounded jump becomes PLS-complete when delay functions of arbitrary sign are allowed. In fact, our result is somewhat stronger: the PLS-completeness holds even when all delay functions are of constant sign or when all the delays are of constant absolute value.

2   Preliminaries and Results

We recall the notions of congestion games, local search problems and approximate Nash equilibrium. We also give motivations and applications of this work.

Congestion games. For a natural number n, we denote by $[n]$ the set $\{1, \ldots, n\}$. For an integer $n \geq 2$, an *n-player game in normal form* is specified by a set of *(pure) strategies* S_i, and a *cost* function $c_i : S \to \mathbb{Z}$, for each player $i \in [n]$, where $S = S_1 \times \cdots \times S_n$ is the set of *states*. For $s \in S$, the value $c_i(s)$ is the cost of player i for state s. A game is *symmetric* if $S_1 = \ldots = S_n$.

For a state $s = (s_1, \ldots, s_n) \in S$, and for a pure strategy $t \in S_i$, we let (s_{-i}, t) to be the state $(s_1, \ldots, s_{i-1}, t, s_{i+1}, \ldots, s_n) \in S$. A *pure Nash equilibrium* is a state s such that for all i, and for all pure strategies $t \in S_i$, we have $c_i(s) \le c_i(s_{-i}, t)$. In general games do not necessarily have a pure Nash-equilibrium.

A specific class of games which always have a pure Nash equilibrium are *congestion games*, where the cost functions are determined by the shared use of resources. More precisely, an n-player congestion game is a 4-tuple $G = (n, E, (d_e)_{e \in E}, (S_i)_{i \in [n]})$, where E is a finite set of *edges* (the common resources), $d_e : [n] \to \mathbb{Z}$ is a *delay function*, for every $e \in E$, and $S_i \subseteq 2^E$ is the set of pure strategies of player i, for $i \in [n]$. Given a state $s = (s_1, \ldots, s_i, \ldots, s_n)$, let the *congestion* of e in s be $f_e(s) = |\{i \in [n] : e \in s_i\}|$. The cost function of user i is defined then as $c_i(s) = \sum_{e \in s_i} d_e(f_e(s))$. Intuitively, each player uses some set of resources, and the cost of each resource e depends on the number of players using it, as described by the delay function. To simplify the notation, we will specify a symmetric congestion game by a 4-tuple $G = (n, E, (d_e)_{e \in E}, Z)$, where by definition the set of pure strategies of every player is $Z \subseteq 2^E$. We will refer to Z as the set of *available* strategies.

A delay function d_e is increasing if $d_e(t) \le d_e(t+1)$, for all $t \in [n-1]$, and it is decreasing if $-d_e$ is increasing. We say that d_e is monotone if it is increasing or decreasing. A congestion game is *increasing* (respectively *decreasing*, *monotone*) if all delay functions are increasing (respectively decreasing, monotone).

That congestion games have indeed a Nash equilibrium can be easily shown by a potential function argument, due to Rosenthal [19], as follows. Let us define the potential function ϕ on the set of states as $\phi(s) = \sum_{e \in E} \sum_{t=1}^{f_e(s)} d_e(t)$. If $s = (s_1, \ldots, s_i, \ldots, s_n)$ and $s' = (s_{-i}, s_i')$ are two states differing only for player i then $\phi(s) - \phi(s') = c_i(s) - c_i(s')$ since both of these quantities are in fact equal to $\sum_{e \in s_i \setminus s_i'} d_e(f_e(s)) - \sum_{e \in s_i' \setminus s_i} d_e(f_e(s'))$. Therefore, in any state which is not a pure Nash equilibrium, there is always a player that can change unilaterally his strategy so that the induced new state has a smaller potential. In fact the decrease in the cost function and in the potential are identical. This means that a finite sequence of such individual changes, the so-called *Nash dynamics*, necessarily results in a pure Nash equilibrium since the integer valued potential function can not decrease forever. Therefore congestion games can be casted as local search problems, and the computing of a Nash equilibrium can be interpreted as the search of a local optimum.

Local search problems. A local search problem is defined by a 4-tuple $\Pi = (\mathcal{I}, F, (v_I)_{I \in \mathcal{I}}, (N_I)_{I \in \mathcal{I}})$, where $\mathcal{I}$ the set of instances, F maps every instance $I \in \mathcal{I}$ to a finite set of feasible solutions $F(I)$, the objective function $v_I : F(I) \to \mathbb{Z}$ gives the value $v_I(S)$ of a feasible solution, and $N_I(S) \subseteq F(I)$ is the neighborhood of $S \in F(I)$. Given an instance I, the goal is to find a feasible solution $S \in F(I)$ such that is also local minimum, that is for all $S' \in N_I(S)$, it satisfies $v_I(S) \le v_I(S')$. A local search problem is in the class PLS [13,18] if there exist polynomial algorithms in the instance length to compute: an initial solution S_0; the membership in $F(I)$; the objective value $v_I(S)$; and a feasible solution $S' \in N_I(S)$ such that $v_I(S') < v_I(S)$ whenever S is not a local minimum. Computing a Nash equilibrium of congestion games is then indeed in PLS: Given an instance G, the feasible solutions $F(G)$ are the states S, the value $v_G(s)$ of a state s is its potential $\phi(s)$, and the neighborhood $N_G(s)$ consists of those states which differ in one coordinate from s.

The notion of PLS-reducibility was introduced in [13]. A problem $\Pi = (\mathcal{I}, F, (v_I)_{I \in \mathcal{I}}, (N_I)_{I \in \mathcal{I}})$ is PLS-reducible to $\Pi' = (\mathcal{I}', F', (v'_I)_{I \in \mathcal{I}'}, (N'_I)_{I \in \mathcal{I}'})$ if there exist polynomial time computable functions $f : \mathcal{I} \to \mathcal{I}'$ and $g_I : F(f(I)) \to F(I)$, for $I \in \mathcal{I}$, such that if S' is a local optimum of $f(I)$ then $g_I(S')$ is local optimum of I. Complete problems in PLS are not believed to be solvable by efficient procedures. Therefore, it is highly unlikely that there exists at all a polynomial time algorithm for computing a pure equilibrium in congestion games. Indeed, Fabrikant, Papadimitriou and Talwar [10] have shown that this problem is PLS-complete, even for symmetric games.

Approximate Nash equilibrium. Several relaxations of the notion of equilibrium have been considered in the form of approximations. Let $0 < \varepsilon < 1$. In our context ε will be a constant or some polynomial time computable function in the input length. A *ε-approximate Nash equilibrium* is a state s such that for all $i \in [n]$, and for all strategies $t \in S_i$, we have
$$c_i(s) - c_i(s_{-i}, t) \leq \varepsilon |c_i(s)|.$$
Otherwise, we say that that (s_{-i}, t) is an *ε-move* for player i if
$$c_i(s) - c_i(s_{-i}, t) > \varepsilon |c_i(s)|.$$
Clearly s is an ε-approximate Nash equilibrium if no player has an ε-move.

The *ε-Nash dynamics* is defined as a sequence of ε-moves, where a player with the *largest absolute gain* makes the change in his strategy, when several players with ε-move are available. Analogously to the exact case, the ε-Nash dynamics converges to an ε-approximate Nash equilibrium. computing an ε-approximate Nash equilibrium is also a problem in PLS. When casting this as a local search, the only difference with the exact equilibrium case is that the neighborhoods are restricted to states which are reachable by an ε-move.

Related results. In [6] Chien and Sinclair have considered the rate of convergence of the ε-Nash dynamics in symmetric congestion games with three additional restrictions on the delay functions. We say that a delay function d_e is positive if the delays $d_e(t)$ are non-negative integers for all $1 \leq t \leq n$. A congestion game is *positive* if all delay functions are positive. Let $\alpha \geq 1$. A positive and increasing delay function is with α-bounded jump if the delays satisfy $d_e(t+1) \leq \alpha d_e(t)$, for all $t \geq 1$. We can think of α as being a constant, or a polynomial time computable function in the input length of the game. Obviously, a positive delay function with α-bounded jump can never take the value 0. A positive game is *with α-bounded jump* if all delay functions are with α-bounded jump. Chien and Sinclair have shown that in symmetric, positive, increasing games with bounded jump the ε-Nash dynamics converges in polynomial time.

Theorem 1 (Chien and Sinclair [6]). *For every $\alpha \geq 1$ and $0 < \varepsilon < 1$, in n-player symmetric, positive and increasing congestion games with α-bounded jump the ε-Nash dynamics converges from any initial state in $O(n\alpha\varepsilon^{-1} \log(nmD))$ steps, where $m = |E|$, and $D = \max\{d_e(n) : e \in E\}$ is an upper bound on the delay functions.*

The hope that the ε-Nash dynamics converges fast in generic congestion games was crushed by Skopalik and Vöcking [20], even for positive increasing games.

Theorem 2 (Skopalik and Vöcking [20]). *For every polynomial time computable $0 < \varepsilon < 1$, computing an ε-approximate Nash equilibrium in a positive and increasing congestion game is* PLS-*complete.*

Motivations. In this paper we mainly study the complexity of computing an ε-approximate Nash equilibrium in congestion games where the delay functions can also have negative values. Negative delays are motivated by real scenarios worth of investigations. *Profit maximizing games* are defined exactly as congestion games, except that each player tries to maximize its cost. These games are easily seen to be equivalent to congestion games when the delay functions are multiplied by a -1 factor. *Market sharing games* [3,12], also studied in the context of content distribution in service networks, are specific profit maximizing games, where the delay functions are positive and decreasing as the value of a resource is shared. They are equivalent to congestion games with negative and increasing delay functions. *Market social games*, introduced in section 3.2, generalize market sharing games where the value of some resources, such as Web pages, may increase with the number of players who selected them, whereas some other resources are shared as in market sharing games. They are equivalent to congestion games with negative increasing and decreasing delay functions, that is negative monotone delay functions.

3 Negative Games

We start now the study of computing ε-approximate Nash equilibria in congestion games where the delay functions can take negative values. In this section we impose the restriction that the delay functions have only negative values. We further suppose that the games are symmetric, monotone and α-bounded. We show in a result analogous to Theorem 1 that for any polynomial time computable α and ε, the ε-Nash dynamics converges in polynomial time. We then point out that this result applies to symmetric market sharing and social games.

We say that a delay function d_e is negative if the delays $d_e(t)$ are negative integers for all $1 \leq t \leq n$. A congestion game is *negative* if all delay functions are negative. Let $\alpha \geq 1$. A negative and increasing delay function is with α-bounded jump if the delays satisfy $d_e(t+1) \leq d_e(t)/\alpha$, for all $t \geq 1$. A negative and decreasing delay function d_e is with α-bounded jump if $-d_e$ is with α-bounded jump. A negative and monotone game is *with α-bounded jump* if all delay functions are with α-bounded jump.

We show our positive result first for increasing games, then we generalize it to monotone games.

3.1 Increasing Delay Functions

Theorem 3. *For every $\alpha \geq 1$ and every $\varepsilon > 0$, in an n-player symmetric, negative, increasing congestion game with α-bounded jump the ε-Nash dynamics* *converges from any initial state in $O((\alpha n^2 + nm)\varepsilon^{-1}\log(nmD))$ steps where $m = |E|$, and $D = \max\{-d_e(1) : e \in E\}$ is an upper bound on magnitude of the delay functions.*

Proof. We will suppose without loss of generality that every edge appears in some strategy, since otherwise the edge can be discarded from E. We first define a positive potential function which will be appropriate to measure the progress of the ε-Nash dynamics. Let ψ be defined over the states as $\psi(s) = -\sum_{e \in E}\sum_{t=f_e(s)+1}^{n} d_e(t)$. The function ψ is clearly positive, and we claim that it is a potential function, that is $\psi(s) - \psi(s') = c_i(s) - c_i(s')$ if the states s and s' differ only in their ith coordinate. This follows immediately from the fact that for every state s, we have $\psi(s) = \phi(s) - k$, where $\phi(s) = \sum_{e \in E}\sum_{t=1}^{f_e(s)} d_e(t)$ is the Rosenthal potential function, and k is the constant $\sum_{e \in E}\sum_{t=1}^{n} d_e(t)$. Observe that $\psi(s)$ is bounded from above by nmD, for every state s.

For an arbitrary initial state $s^{(0)}$, let $s^{(k)}$ be the kth state of the ε-Nash dynamics process. We claim that $\psi(s^{(k+1)}) \leq \psi(s^{(k)})(1 - \varepsilon/4(\alpha n^2 + nm))$, for every k, which clearly implies the theorem. Suppose that $s^{(k)} = s = (s_1, \ldots, s_n)$ is not an ε-equilibrium, and let i be the player which can make the largest gain ε-move. To prove our claim, we will show that there exists a strategy s_i' for player i such that $c_i(s) - c_i(s_{-i}, s_i') \geq \varepsilon\psi(s)/4(\alpha n^2 + nm)$, and we observe that an ε-move can only be better for player i than playing strategy s_i'.

The first idea is to try to prove, analogously to the case of positive games, that for some player j, the opposite of its cost $-c_j(s)$ is a polynomial fraction of $\psi(s)$. Unfortunately this is not necessarily true. The sum $\sum_{j=1}^{n} c_j(s)$ is not necessarily a polynomial fraction of $\psi(s)$ because edges whose congestion is 0 in s do not contribute to the former, but do contribute the latter. Therefore we introduce the function ψ' as ψ restricted to the edges with nontrivial congestion, that is by definition $\psi'(s) = -\sum_{e \in E | f_e(s) \neq 0}\sum_{t=f_e(s)+1}^{n} d_e(t)$. The following Lemma shows that some of the $-c_j(s)$ is at least a polynomial fraction of $\psi'(s)$.

Lemma 1. *There exists a player j such that $-c_j(s) \geq \psi'(s)/n^2$.*

Proof. We claim that $-n\sum_{j=1}^{n} c_j(s) \geq \psi'(s)$, from which the statement clearly follows. To prove the claim we proceed by the following series of (in)equalities:

$$-n\sum_{j=1}^{n} c_j(s) = -n\sum_{e \in E | f_e(s) \neq 0} f_e(s)\, d_e(f_e(s)) \geq -n\sum_{e \in E | f_e(s) \neq 0} d_e(f_e(s))$$

$$\geq -\sum_{e \in E | f_e(s) \neq 0}\sum_{t=f_e(s)+1}^{n} d_e(t) = \psi'(s),$$

where the second inequality holds because the delay functions are non-decreasing. $\square$

We fix a value j which satisfies Lemma 1 for the rest of the proof. To upper bound $\psi(s)$, we also have to consider the edges of congestion 0, besides the edges which are accounted for in $\psi'(s)$. We have

$$\psi'(s) - n \sum_{E \in E | f_e(s)=0} d_e(1) \geq \psi(s),$$

again because the delays are non-decreasing. This implies that either $\psi'(s) \geq \psi(s)/2$ or $-n \sum_{e \in E | f_e(s)=0} d_e(1) \geq \psi(s)/2$, and the proof proceeds by distinguishing these two cases.

Case 1: $\psi'(s) \geq \psi(s)/2$. We then reason in two sub-cases by comparing the value of $c_i(s)$ to $\psi'(s)/2\alpha n^2$. If $-c_i(s) \geq \psi'(s)/2\alpha n^2$, then let s'_i be the strategy which makes the biggest gain for player i. Then we have
$$c_i(s) - c_i(s_{-i}, s'_i)) \geq -\varepsilon c_i(s) \geq \varepsilon\psi(s)/4\alpha n^2,$$
where first inequality holds since the move of player i is an ε-move, and the second inequality is true because of the hypotheses. If $-c_i(s) < \psi'(s)/2\alpha n^2$, then let $s'_i = s_j$, the strategy of player j in state s. Observe that s_j is an available strategy for player i since the game is symmetric. Then
$$c_i(s) - c_i(s_{-i}, s'_i)) \geq c_i(s) - c_j(s)/\alpha \geq \psi'(s)/\alpha n^2 - \psi'(s)/2\alpha n^2 \geq \psi(s)/4\alpha n^2.$$
Here the first inequality is true because the game is with α-bounded jump. The second inequality follows from the hypothesis and because $-c_j(s) \geq \psi'(s)/n^2$. Finally, the third inequality holds because $\psi'(s) \geq \psi(s)/2$.

Case 2: $-n \sum_{e \in E | f_e(s)=0} d_e(1) \geq \psi(s)/2$. Then for some edge with $f_e(s) = 0$, we have $-d_e(1) \geq \psi(s)/2nm$. Let's fix such an edge e. We distinguish two sub-cases now by comparing the value of $c_i(s)$ to $d_e(1)/2$. If $c_i(s) \leq d_e(1)/2$ then let s'_i be the strategy which makes the biggest gain for player i. Then, similarly to the first sub-case of Case 1, using the hypotheses and that player i's move is an ε-move, we have
$$c_i(s) - c_i(s_{-i}, s'_i)) \geq -\varepsilon c_i(s) \geq \varepsilon\psi(s)/4nm.$$
If $c_i(s) > d_e(1)/2$ then let s'_i be some strategy that contains the edge e. There exists such a strategy since useless edges were discarded from E. Then $f_e(s_{-i}, s'_i)) = 1$ since $f_e(s) = 0$ and s and (s_{-i}, s'_i) differ only for the ith player. This, in turn, implies that $c_i(s_{-i}, s'_i)) \leq d_e(1)$, since the delays are negative. Therefore
$$c_i(s) - c_i(s_{-i}, s'_i)) \geq c_i(s) - d_e(1) \geq -d_e(1)/2 \geq \psi(s)/4nm,$$
where the last two inequalities follow from the hypotheses.   □

Market Sharing Games. In market sharing games [3,12] n players sell their goods on subsets of m markets $E = \{e_1, \ldots, e_m\}$, and they try to maximize their gains. Each market e has a value $v(e) > 0$. If t sellers choose a market e, they share its value and each earn $v(e)/t$. The gain of player i on a strategy profile $s = (s_1, \ldots, s_n)$, with $s_i \subseteq E$, is $\sum_{e \in s_i} v(e)/f_e(s)$, where $f_e(s)$ is the number of sellers on the market e. A symmetric market sharing game with markets strategies $Z \subseteq 2^E$ is a congestion game $(n, E, (d_e)_{e \in E}, Z)$ with delay functions $d_e(t) = -v(e)/t$, which are increasing, negative and with 2-bounded jump.

Corollary 1. *In symmetric market sharing games the ε-Nash dynamics converges in polynomial time.*

3.2   Monotone Delay Functions

We extend Theorem 3 to monotone congestion games where the resources can be partitioned into two sets: $E^\uparrow$ with increasing delay functions and $E^\downarrow$ with decreasing delay functions. Notice that if $E^\uparrow$ is empty, then the task of finding a Nash equilibrium becomes trivial. Indeed, if the strategy s^* minimizes $\sum_{e \in s} d_e(n)$ over all available strategies, then the state where all players select s^* is an equilibrium.

Theorem 4. *For every $\alpha \geq 1$ and every $\varepsilon > 0$, in an n-player symmetric, negative, monotone congestion game with α-bounded jump the ε-Nash dynamics converges from any initial state in $O((\alpha n^2 + nm)\varepsilon^{-1} \log(nmD))$ steps where $m = |E|$, and $D = \max\{-d_e(t) : e \in E, t \in [n]\}$ is an upper bound on the magnitude of the delay functions.*

The proof is similar to the proof of theorem 3 using the potential function $\psi(s) = -\sum_{e \in E^\uparrow} \sum_{t=f_e(s)+1}^{n} d_e(t) + \sum_{e \in E^\downarrow} \sum_{t=1}^{f_e(s)} d_e(t)$.

Market Social Games. Let us call a symmetric *market social game* a congestion game $(n, E, (d_e)_{e \in E}, Z)$ where the market E is partitioned into $E^\uparrow, E^\downarrow$. Each market $e \in E$ has a value $v(e) > 0$. The delay functions are defined as $d_e(t) = -v(e)/t$ when $e \in E^\uparrow$, and $d_e(t) = -t.v(e)$ when $e \in E^\downarrow$. The delays are clearly negative increasing on $E^\uparrow$ and negative decreasing on $E^\downarrow$. They are also with 2-bounded jump. We interpret $f_e(s)$ as the number of sellers on the market e. These games generalize the market sharing games as some resources are shared between the players, whereas some other resources have a value which increases with the number of players.

Corollary 2. *In symmetric market social games the ε-Nash dynamics converges in polynomial time.*

4   Games without Sign Restriction

In this section we deal with congestion games with no restriction on the sign of the delay functions. Our overall result is that in that case computing an ε-approximate Nash equilibrium is PLS-hard, even when the remaining restrictions of Chien and Sinclair are kept, that is when the game is symmetric, increasing and with α-bounded jump, for $\alpha \geq 1$. Observe that the smaller α the stronger is the hardness result, therefore we deal only with constant α. Our first step is to observe that a simple consequence of Theorem 2 is that computing an ε-approximate Nash equilibrium in positive and increasing games remains PLS-complete even if we additionally suppose that the game is symmetric. Our reductions will use the hardness of this latter problem. The proof of this statement is a PLS-reduction of the search of an ε-approximate Nash equilibrium in positive and increasing games to the same problem in symmetric, positive, increasing games. This reduction is basically identical to the analogous reduction for pure Nash equilibria, due to Fabrikant, Papadimitriou and Talwar [10].

Theorem 5. *For every polynomial time computable $0 < \varepsilon < 1$, computing an ε-approximate Nash equilibrium in a symmetric, positive, increasing congestion game is* PLS-*complete.*

We need to discuss now the right notion of α-bounded jump when the jump occurs from a negative to a positive value in the delay function. One possibility could be to require $d_e(t+1) \leq -\alpha d_e(t)$ when $d_e(t) < 0$ and $d_e(t+1) \geq 0$, but there are also other plausible definitions. In fact, we will avoid to give a general definition because it turns out that this is not necessary for our hardness results. Indeed, we will be able to establish a hardness result for congestion games where there is no jump at all around 0, that is for delay functions of constant sign (still some of the delay functions can be negative while some others positive). We say that a congestion game is *non-alternating*, if every delay function is positive or negative. Let $\alpha > 1$ be a constant. A non-alternating congestion game is *with α-bounded jump* if all delay functions are with α-bounded jump.

What happens when $\alpha = 1$? If the delays are constant functions, a pure Nash equilibrium can be determined trivially. Indeed, the cost functions of the individual players are independent from the strategies of the other players, and therefore any choice of a least expensive strategy, for each player, forms a Nash equilibrium.

Nonetheless, if we authorize a jump around 0, then even if the jump changes only the sign without changing the absolute value (which corresponds intuitively to the case $\alpha = 1$ in that situation), the game becomes already hard. We say that a delay function d_e is a *flip function*, if there exists a positive integer c such that for some $1 \leq k \leq n$, the function satisfies:

$$d_e(t) = \begin{cases} -c & \text{if } t < k, \\ c & \text{if } t \geq k. \end{cases}$$

Flip functions are either constant positive functions, or they are simple step functions, which are constant negative up to some point, where an alternation occurs which keeps the absolute value. After the alternation the function remains constant positive. A congestion game is a *flip* game if all delay functions are flip functions. The next two theorems state our hardness results respectively for non-alternating games with α-bounded jump and for flip games.

Theorem 6. *For every constant $\alpha > 1$, and for every polynomial time computable $0 < \varepsilon < 1$, computing an ε-approximate Nash equilibrium in n-player symmetric, non-alternating, increasing congestion games with α-bounded jump is* PLS-*hard.*

Proof. As stated in Theorem 5 computing an ε-approximate Nash equilibrium in a symmetric, positive, increasing congestion game is PLS-complete [20]. We present a PLS-reduction from this problem to the problem of computing an ε-approximate Nash equilibrium in a symmetric, non-alternating, positive game with α-bounded jump.

Let $G = (n, E, (d_e)_{e \in E}, Z)$ a symmetric, positive, increasing congestion game, and let $\alpha > 1$ be a constant. In our reduction we map G to the symmetric game $G' = (n, E', (d_{e'})_{e' \in E'}, Z')$ that we define now. For each $e \in E$, we set $E_e = \{e_1, e_2^+, e_2^-, \ldots, e_n^+, e_n^-\}$, and for every $z \subseteq E$, we define $z' = \bigcup_{e \in z} E_e$ (and therefore $E' = \bigcup_{e \in E} E_e$). The set of available strategies is defined as $Z' = \{z' : z \in Z\}$. Finally the delay functions are defined as follows. The delay d_{e_1} is simply the constant function $d_e(1)$. For $k \geq 2$, we set

$$d_{e_k^+}(t) = \begin{cases} (d_e(k) - d_e(k-1))\frac{\alpha}{\alpha^2 - 1} & \text{if } t < k, \\ (d_e(k) - d_e(k-1))\frac{\alpha^2}{\alpha^2 - 1} & \text{if } t \geq k, \end{cases}$$

and

$$d_{e_k^-}(t) = \begin{cases} -(d_e(k) - d_e(k-1))\frac{\alpha}{\alpha^2 - 1} & \text{if } t < k, \\ -(d_e(k) - d_e(k-1))\frac{1}{\alpha^2 - 1} & \text{if } t \geq k. \end{cases}$$

The game G' is clearly non-alternating, increasing and with α-bounded jump.

Observe that there is a bijection between the states of G and G'. Indeed, the states of G' are of the form $s' = (s_1', \ldots, s_n')$, where $s = (s_1, \ldots, s_n) \in Z^n$ is a state of G. For the reduction we will simply show that if s' is an ε-approximate Nash equilibrium in G' then s is an ε-approximate Nash equilibrium in G (our construction satisfies also the reverse implication). In fact, we show a stronger statement about cost functions: for every state s, and for every player i, the cost of player i for s in G is the same as the cost of player i for s' in G'.

The edges e_k^+ and e_k^- are such that the sum of their delay functions emulates the jump $d_e(k) - d_e(k-1)$ when $t \geq k$. Therefore the sum of the delays corresponding to edges in E_e is just d_e which is expressed in the following lemma.

Lemma 2. *For every edge $e \in E$, and $1 \leq t \leq n$, $\sum_{e' \in E_e} d_{e'}(t) = d_e(t)$.*

We now claim the following strong relationship between the cost functions in the two games.

Lemma 3. *For all state $s = (s_1, \ldots, s_n)$ in G, and for every player i, we have $c_i(s') = c_i(s)$, where $s' = (s_1', \ldots, s_n')$.*

By Lemma 3 we can deduce an ε-approximate Nash equilibrium for G, given an ε-approximate Nash equilibrium for G'. This concludes the proof.   □

Theorem 7. *For every polynomial time computable $0 < \varepsilon < 1$, computing an ε-approximate Nash equilibrium in n-player symmetric, flip congestion games is PLS-hard.*

Proof. The proof is very similar to the proof of the previous theorem. In the reduction the delay functions, for $2 \leq k \leq n$, are defined as

$$d_{e_k^+}(t) = (d_e(k) - d_e(k-1))/2 \quad \text{for every } t,$$

and

$$d_{e_k^-}(t) = \begin{cases} -(d_e(k) - d_e(k-1))/2 & \text{if } t < k, \\ (d_e(k) - d_e(k-1))/2 & \text{if } t \geq k. \end{cases}$$

□

References

1. Ackermann, H., Röglin, H., Vöcking, B.: On the impact of combinatorial structure on congestion games. Journal of the ACM 55(6) (2008)
2. Albers, S., Lenzner, P.: On approximate Nash equilibria in network design. In: Proc. of International Workshop on Internet and Network Economics, pp. 14–25 (2010)
3. Awerbuch, B., Azar, Y., Epstein, A., Mirrokni, V., Skopalik, A.: Fast convergence to nearly optimal solutions in potential games. In: Proc. of ACM Conference on Electronic Commerce, pp. 264–273 (2008)
4. Chen, X., Deng, X.: Settling the complexity of two-player Nash equilibrium. In: Proc. of IEEE Symposium on Foundations of Computer Science, pp. 261–272 (2006)
5. Chen, X., Deng, X.: Computing Nash equilibria: approximation and smoothed complexity. In: Proc. of IEEE Symposium on Foundations of Computer Science, pp. 603–612 (2006)
6. Chien, S., Sinclair, A.: Convergence to approximate Nash equilibria in Congestion Games. In: Proc. of the ACM-SIAM Symposium on Discrete Algorithms, pp. 169–178 (2007)
7. Daskalakis, C., Mehta, A., Papadimitriou, C.: A note on approximate Nash equilibria. In: Proc. of Workshop on Internet and Network Economics, pp. 297–306 (2006)
8. Daskalakis, C., Mehta, A., Papadimitriou, C.: Progress in approximate Nash equilibria. In: Proc. of ACM Conference on Electronic Commerce, pp. 355–358 (2007)
9. Daskalakis, C., Goldberg, P., Papadimitriou, C.: The complexity of computing a Nash equilibrium. In: Proc. of ACM Symp. on Theory of Computing, pp. 71–78 (2006)
10. Fabrikant, A., Papadimitriou, C., Talwar, K.: The complexity of pure Nash equilibria. In: Proc. of ACM Symposium on Theory of Computing, pp. 604–612 (2004)
11. Feder, T., Nazerzadeh, H., Saberi, A.: Approximating Nash equilibria using small-support strategies. In: Proc. of ACM Conference on Electronic Commerce, pp. 352–354 (2007)
12. Goemans, M., Li, L., Mirrokni, V., Thottan, M.: Market sharing games applied to content distribution in ad-hoc networks. In: Proc. of ACM Symposium on Mobile Ad Hoc Networking and Computing, pp. 55–66 (2004)
13. Johnson, D., Papadimitriou, C., Yannakakis, M.: How easy is local search? Journal of Computer and System Sciences 37(1), 79–100 (1988)
14. Kearns, M., Mansour, Y.: Efficient Nash computation in large population games with bounded influence. In: Proc. of Conference on Uncertainty in Artificial Intelligence, pp. 259–266 (2002)
15. Lipton, R., Markakis, E., Mehta, A.: Playing large games using simple strategies. In: Proc. of ACM Conference on Electronic Commerce, pp. 36–41 (2003)
16. Megiddo, N., Papadimitriou, C.: On total functions, existence theorems, and computational complexity. Theoretical Computer Science 81, 317–324 (1991)
17. Monderer, D., Shapley, L.: Potential Games. Games and Economic Behavior 14, 124–143 (1996)
18. Papadimitriou, C., Yannakakis, M.: Optimization, approximation, and complexity classes. In: Proc. of ACM Symp. on Theory of Computing, pp. 229–234 (1988)
19. Rosenthal, R.: A class of games possessing pure-strategy Nash equilibria. International Journal of Game Theory 2, 65–67 (1973)
20. Skopalik, E., Vöcking, B.: Inapproximability of pure Nash equilibria. In: Proc. of ACM Symposium on Theory of Computing, pp. 355–364 (2008)
21. Tsaknakis, H., Spirakis, P.: An optimization approach for approximate Nash equilibria. In: Proc. of Workshop on Internet and Network Economics, pp. 42–56 (2007)

On Worst-Case Allocations in the Presence of Indivisible Goods[*]

Evangelos Markakis and Christos-Alexandros Psomas

Athens University of Economics and Business
Department of Informatics
{markakis,alexpsomi}@gmail.com

Abstract. We study a fair division problem, where a set of indivisible goods is to be allocated to a set of n agents. In the continuous case, where goods are infinitely divisible, it is well known that proportional allocations always exist, i.e., allocations where every agent receives a bundle of goods worth to him at least $\frac{1}{n}$. With indivisible goods however, this is not the case and one would like to find worst case guarantees on the value that every agent can have. We focus on algorithmic and mechanism design aspects of this problem.

An explicit lower bound was identified by Hill [5], depending on n and the maximum value of any agent for a single good, such that for any instance, there exists an allocation that provides at least this guarantee to every agent. The proof however did not imply an efficient algorithm for finding such allocations. Following upon the work of [5], we first provide a slight strengthening of the guarantee we can make for every agent, as well as a polynomial time algorithm for computing such allocations. We then move to the design of truthful mechanisms. For deterministic mechanisms, we obtain a negative result showing that a truthful $\frac{2}{3}$-approximation of these guarantees is impossible. We complement this by exhibiting a simple truthful algorithm that can achieve a constant approximation when the number of goods is bounded. Regarding randomized mechanisms, we also provide a negative result, under the restrictions that they are Pareto-efficient and satisfy certain symmetry requirements.

1 Introduction

Fair division problems have attracted the attention of various scientific disciplines, including among others, mathematics, economics, and political science. Ever since the first attempt for a formal treatment by Steinhaus, Banach, and Knaster [9], many challenging questions have emerged and a vast literature has developed, see e.g., [2,8]. In the recent years, this area has also gained popularity in computer science, as most of the questions that have been posed are algorithmic in nature.

The objective in fair division problems is to allocate a set of goods to a set of n agents in a way that leaves every agent satisfied. In the continuous case, the available resources are typically represented by the interval [0, 1], whereas in the discrete case, we have a set of distinct, indivisible goods. Each agent has a valuation function, which

---

[*] Research supported by the Basic Research Funding Program of the Athens University of Economics and Business. A version with all missing proofs is available at the authors' webpages.

N. Chen, E. Elkind, and E. Koutsoupias (Eds.): WINE 2011, LNCS 7090, pp. 278–289, 2011.

is usually normalized to be a probability distribution on the set of goods. Given such a setup, many solution concepts have been proposed as to what constitutes a fair solution, including *proportionality, envy-freeness, equitability* and many variants of them. The most related concept to our work is proportionality, meaning that every agent receives a bundle of the goods that is worth at least $\frac{1}{n}$, according to his valuation function.

In the continuous case, it has long been shown that proportional allocations always exist [9]. In the presence of indivisible goods however, this is not the case. If, for example, we have two agents who have a very high value for one of the goods, then any allocation will leave one of the two people unhappy. In this work we are interested in finding allocations with worst case guarantees in instances with indivisible goods. In particular, we focus on algorithmic and mechanism design aspects of this problem.

1.1   Related Work and Contribution

Consider a set of indivisible items, and a set of n agents with additive valuations. Given that proportional allocations do not always exist for indivisible goods, a natural question is whether we can provide any lower bound as to the value that we can ensure to every agent. This question was studied by Hill in [5], where an explicit such guarantee was given. In particular, a certain function was identified, denoted by $V_n(\cdot)$ (defined in Section 2), such that, when the maximum value of a good is at most α, then there always exists an allocation where every agent can receive a bundle that is worth to him at least $V_n(\alpha)$. Clearly when α is large, $V_n(\alpha)$ may be 0. For smaller values of α however this is a positive result, showing that we can ensure a relatively fair solution.

Regarding the complexity of finding such allocations, the result of [5] does not yield an efficient algorithm. The proof is based on certain combinatorial arguments, which however result in an exponential algorithm. Motivated by this fact, we start with studying the question of efficiently producing allocations that respect the bound of $V_n(\alpha)$. Our main result in Section 3 is that (i) we can have a slight strengthening of the guarantee of [5] so that every agent can have a bundle worth at least $V_n(\alpha_i)$, where α_i is the maximum value in the valuation of agent i (hence an agent is not penalized if someone else has a much higher maximum value than him) (ii) we provide a simple polynomial time algorithm for computing such an allocation.

In Section 4, we move to mechanism design aspects. We show that no deterministic truthful mechanism can guarantee an allocation that is worth at least $\frac{2}{3} \cdot V_n(\alpha_i)$ for every agent i, even for two agents. The proof of this statement is achieved in two steps: first we argue about *permutation-respecting* mechanisms, i.e., mechanisms that when faced with a permutation of a given input, return a permutation of the initial output. We later use this to argue about general mechanisms. We then complement this negative result by a simple algorithm showing that for two agents and a small number of goods, we can have a truthful, constant approximation to $V_n(\alpha_i)$. In Subsection 4.2, we turn to randomized algorithms. The picture is far less clear there and in general, there have been very few attempts for randomized algorithms in cake-cutting, such as [3,7]. We focus on *truthful in expectation* mechanisms and present an impossibility result for Pareto-efficient mechanisms under certain symmetry requirements.

Finally, in Section 5, we study a slightly different question. Since proportional allocations do not always exist, can we at least decide when this is the case? In [4] it has already been proved that deciding the existence of proportional allocations, is NP-hard. We strengthen this result by providing a different reduction showing that it is NP-hard to decide even if there exists an allocation where every person gets a bundle worth at least $1/cn$ for any constant $c \geq 1$.

We should note here that a related problem is to compute a max-min fair allocation, where the objective is to maximize the value of the least happy person. This is an NP-hard problem and several approximation algorithms have been proposed [1]. Our problem is less demanding, since we do not want to compute a max-min fair allocation but simply an allocation that reaches the threshold of $V_n(\alpha_i)$ for every agent i. An approximate solution to max-min fairness problem does not necessarily achieve this.

2   Definitions and Preliminaries

Let $N = \{1, ..., n\}$ be a set of n agents and $M = \{1, ..., m\}$ be a set of m indivisible goods. The input to our problem is a valuation matrix V so that v_{ij} is the utility derived by agent i for obtaining good j. We assume the usual normalization in fair division that $\sum_j v_{ij} = 1$. Let $S_{n,m}$ be the set of $n \times m$ matrices satisfying this requirement.

An allocation of the goods is denoted by a tuple $(S_1, ..., S_n)$, where S_i is the set allocated to player i, such that $\bigcup_i S_i = M$ and $S_i \cap S_j = \emptyset$, (implying that the algorithm has to allocate the whole set M). The total value of player i for an allocation $(S_1, ..., S_n)$, is $v_i(S_i) = \sum_{j \in S_i} v_{ij}$. In [5], Hill defined the following function.

Definition 1. *Given any integer $n \geq 2$, let $V_n : [0,1] \rightarrow [0, n^{-1}]$ be the unique nonincreasing function satisfying $V_n(\alpha) = \frac{1}{n}$ for $\alpha = 0$, and for $\alpha > 0$:*

$$V_n(\alpha) = \begin{cases} 1 - k(n-1)\alpha & \text{if } \alpha \in I(n,k) \\ 1 - \frac{(k+1)(n-1)}{(k+1)n-1} & \text{if } \alpha \in NI(n,k) \end{cases}$$

where for any integer $k \geq 1$,

$$I(n,k) = \left[\frac{k+1}{k((k+1)n-1)}, \frac{1}{kn-1} \right], \quad NI(n,k) = \left(\frac{1}{(k+1)n-1}, \frac{k+1}{k((k+1)n-1)} \right)$$

Definition 2. *For integers $n, k \geq 1$, let $r(n,k) = \frac{1}{kn-1}$, i.e., the right endpoint of the interval $I(n,k)$. Similarly let $l(n,k) = \frac{k+1}{k((k+1)n-1)}$, the left endpoint of $I(n,k)$.*

Example 1. In Figure 1, one can see the function $V_n(\cdot)$ for $n = 2$ and $n = 3$. For larger n, it has a similar form. The function alternates between decreasing and constant segments. The intervals $I(n,k)$ in Definition 1 correspond to the decreasing segments, whereas the intervals $NI(n,k)$ correspond to the constant segments. For example, for $n = 2$, looking at the function from right to left, we can see that the rightmost decreasing segment is $I(2,1) = [\frac{2}{3}, 1]$, which is followed by $NI(2,1) = (\frac{1}{3}, \frac{2}{3})$, followed by $I(2,2) = [0.3, \frac{1}{3}]$, then followed by $NI(2,2) = (\frac{1}{5}, 0.3)$, and so on.

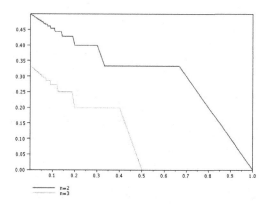

Fig. 1. The function $V_n(\cdot)$ for $n = 2$ and $n = 3$

The function $V_n(\cdot)$ has a number of useful properties that will be needed in the following sections:

Fact 1. *The function $V_n(\cdot)$ satisfies:*

1. $V_n(\frac{1}{n-1}) = V_n(r(n, 1)) = 0$, *and for any* $\alpha \geq \frac{1}{n-1}$, $V_n(\alpha) = 0$.
2. *The equation* $\alpha = V_n(\alpha)$ *has a unique solution at* $\alpha = \frac{1}{2n-1} = r(n, 2)$. *For* $\alpha < r(n, 2)$, $V_n(\alpha) > \alpha$ *and for* $\alpha > r(n, 2)$, $V_n(\alpha) < \alpha$.

Theorem 2. *[5] For any instance, with $\max_{i,j} v_{ij} \leq \alpha$, there is an allocation where the total value of every player is at least $V_n(\alpha)$.*

The result of [5] actually holds in a more general model, where the valuations of the agents are probability distributions on $[0, 1]$ which are allowed to have atoms of size at most α. This includes our setting.

3   The Algorithm

The proof of Theorem 2 does not imply an efficient algorithm for producing the desired allocation, as part of the proof relies on existential arguments. Furthermore, in the tight example provided in [5], all agents have the same maximum valuation, α. It could still be feasible to provide better guarantees to agents whose maximum value is lower than the maximum value over all agents, i.e., the fact that some agents have very high values should penalize only themselves.

The main result of this Section is the following theorem, which provides (i) a slight strengthening of [5] in terms of the guarantee that each agent can have, and (ii) a simple, efficient algorithm for finding such allocations.

Theorem 3. *There exists a polynomial time algorithm that for any instance, it produces an allocation such that, each player i receives a bundle with total value at least $V_n(\alpha_i)$, where $\alpha_i = \max_j v_{ij}$.*

The algorithm is achieved by obtaining a better understanding of the behavior of $V_n(\cdot)$. Although the algorithm itself turns out to be quite simple, the analysis is more involved and is based on a series of Lemmas. We start our analysis with proving some useful properties of the function $V_n(\cdot)$. The next fact simply says that for $\alpha \in NI(n, k)$, $V_n(\alpha)$ is the same as the value at the right endpoint of the decreasing segment to the left of α, or as the value at the left endpoint of the segment to the right of α.

Fact 4. *If $\alpha \in NI(n, k)$, for some $k \geq 1$, then $V_n(\alpha) = V_n(r(n, k+1)) = V_n(l(n, k))$.*

The main ingredients that make our algorithm work are the properties stated in Lemma 1 and Lemma 2. Lemma 1 below provides a relation between $V_n(\cdot)$ and $V_{n-1}(\cdot)$, needed for the inductive analysis of the algorithm later on. We omit its proof from this version.

Lemma 1. *For fixed $n \geq 3$ and $k \geq 1$, and for $\alpha \in I(n, k) \cup NI(n, k)$,*

$$(1 - k\alpha)V_{n-1}\left(\frac{\alpha}{1 - k\alpha}\right) \geq V_n(\alpha) \tag{1}$$

The next Lemma is crucial as it provides an upper bound on the left-over value of the remaining items, according to the preferences of a given agent i, once we satisfy the agent. Clearly the upper bound holds for any other agent that cannot be satisfied by the bundle allocated to agent i.

Lemma 2. *For $n \geq 2$, let i be an agent with $\alpha_i < V_n(\alpha_i)$. Suppose we start allocating items to player i, starting from his most desirable item and proceeding in decreasing order, as induced by his valuation, until the total value for i, say s, equals or exceeds $V_n(\alpha_i)$. Then,*

$$s \leq \begin{cases} k\alpha_i \left(= \frac{1 - V_n(\alpha_i)}{n-1}\right), & \text{if } \alpha_i \in I(n, k), \text{ for some } k \geq 1 \\ k \cdot l(n, k), & \text{if } \alpha_i \in NI(n, k), \text{ for some } k \geq 1 \end{cases}$$

Proof. **Case 1:** $\alpha_i \in I(n, k)$ for some $k \geq 1$.

Suppose the statement of the lemma is not true, i.e., once we surpass $V_n(\alpha_i)$ for agent i, we have $s > k\alpha_i$. Then, before the final item is given to player i, the t first items had total value c, where

$$V_n(\alpha_i) > c > (k - 1)\alpha_i \tag{2}$$

Otherwise it would not be possible to exceed $k\alpha_i$ with the last item. Note that we can assume that $t \geq 1$, since $\alpha_i < V_n(\alpha_i)$. Let S be the set of these t items, and let w be their average value, $w = \sum_{j \in S} v_{ij}/t$. We can write their total value as $c = tw$, and we can easily see that $\alpha_i \geq w \geq k\alpha_i - V_n(\alpha_i) = kn\alpha_i - 1$. The leftmost inequality is trivial. As for the rightmost inequality, since we examine the goods in the decreasing order of agent i's valuation, the next good allocated to i after the first t goods would have value at most w. But then if $w < k\alpha_i - V_n(\alpha_i)$, it would be impossible to exceed $k\alpha_i$. Given this range for the value of w, we now consider the possible values that t can take. Even if all t items had value exactly α_i (i.e., $w = \alpha_i$), we would still need more than $k - 1$ items to get to c, by (2). On the other hand, if all items had value exactly $kn\alpha_i - 1$, we would need strictly less than $\frac{V_n(\alpha_i)}{kn\alpha_i - 1}$ items, again by (2). Hence,

$$\frac{1 - k(n-1)\alpha_i}{kn\alpha_i - 1} > t > k - 1 \tag{3}$$

Let x be the difference between the upper and the lower bounds of t:

$$x = \frac{1 - k(n-1)\alpha_i}{kn\alpha_i - 1} - k + 1 = \frac{k\alpha_i}{kn\alpha_i - 1} - k$$

We show that $x \leq 1$. If $x > 1$, then we would get that: $\frac{k\alpha_i}{kn\alpha_i - 1} > k + 1 \Rightarrow k\alpha_i > k^2 n\alpha_i + kn\alpha_i - k - 1 \Rightarrow \alpha_i < \frac{k+1}{k((k+1)n-1)} = l(n, k)$, which is impossible, since $\alpha_i \in I(n, k)$.

Since $x \leq 1$, the right hand side of (3) can be at most k. But this means that t, which is an integer, is greater than $k - 1$ and strictly less than k, a contradiction.

Case 2: $\alpha_i \in NI(n, k)$ for some $k \geq 1$. The proof for this case is based on similar arguments and is omitted.

Algorithm 1. ALLOCATE(V, N, M)

Input: $V \in S_{n,m}$, N (set of agents), M (set of goods)
Output: Allocation of items such that agent i receives a bundle worth at least $V_n(\alpha_i)$, $\forall i$.

1: **for** $i = 1$ to n **do**
2:     Set $S_i = \emptyset$
3: **end for**
4: **while** $\nexists\, i$ with $v_i(S_i) \geq V_n(\alpha_i)$ **do**
5:     **for** every player i **do**
6:         $S_i = S_i \cup \{$next highest item in agent i's order$\}$
7:     **end for**
8: **end while**
9: Pick an agent i with $v_i(S_i) \geq V_n(\alpha_i)$ //pick arbitrarily in case of ties
10: Allocate S_i to agent i
11: **if** $|N| = 2$ **then**
12:     Allocate all other items to remaining agent
13: **else**
14:     **for** every row $k \neq i$ **do**
15:         row k = (row k) * $1/(1 - v_k(S_i))$ //normalization before going to next round
16:     **end for**
17:     V' = new normalized matrix after also removing row i and columns corresponding to S_i
18:     run $ALLOCATE(V', N \setminus \{i\}, M \setminus S_i)$
19: **end if**

Proof of Theorem 3

We are now ready to prove the main result of this Section. The previous lemmas give rise to a simple algorithm seen above. In short, the algorithm satisfies first the player i that needs the least number of items to achieve $V_n(\alpha_i)$. This is done by maintaining for each agent, a decreasing ordering of the goods, as induced by the agent's valuation. Once the algorithm finds such an agent i, the corresponding items are given to i, who is then removed from this process. We then perform a normalization so that the remaining items add to 1 for all the remaining players, and we start again, trying to find an agent

j that needs the least number of items to achieve $V_{n-1}(\tilde{\alpha}_j)$, where $\tilde{\alpha}_j$ is the highest value of j, after the normalization. The algorithm continues in the same fashion until everybody is satisfied. The algorithm clearly terminates in a polynomial number of steps. Regarding the implementation, one also needs to ensure that the value $V_n(\alpha)$ is efficiently computable for rational inputs. This can be easily established and we omit its proof.

Claim. For any rational number $\alpha \in [0, 1]$, $V_n(\alpha)$ can be computed in polynomial time.

The proof of correctness of the algorithm is by induction on the number of players.
• *Induction Basis:* $n = 2$. Without loss of generality we can assume that the first player receives the first bundle allocated by the algorithm, and that it consists of t items. Then we know that this bundle is worth at least $V_2(\alpha_1)$ for agent 1 and he is settled. We now argue about the second agent.

Suppose $\alpha_1 < V_2(\alpha_1)$, which implies that $t > 1$ and $\alpha_2 < V_2(\alpha_2)$, otherwise the algorithm would have allocated first to agent 2 his best item. Let s_2 be the value of the t items of agent 1, according to the valuation of agent 2. We know that even with his $t-1$ most desirable items, agent 2 cannot exceed $V_2(\alpha_2)$. Hence Lemma 2 can be applied.
Case 1: $s_2 \geq V_2(\alpha_2)$. s_2 is at most as much as the value of the t most desirable items of agent 2. If $\alpha_2 \in I(n, k)$, for some $k \geq 1$, then Lemma 2 gives $s_2 \leq 1 - V_2(\alpha_2)$, i.e., $1 - s_2 \geq V_2(\alpha_2)$. If $\alpha_2 \in NI(n, k)$, for some $k \geq 1$, then $s_2 \leq k \cdot l(n, k)$. Hence

$$1 - s_2 \geq 1 - k \cdot l(n, k) = \frac{(k+1)(n-1)-1}{(k+1)n-1}$$

It is now an easy calculation to show that $1 - s_2 \geq V_2(\alpha_2)$. Thus, for all possible values of α_2 under this case, we have $1 - s_2 \geq V_2(\alpha_2)$. Since the bundle allocated to agent 1 is worth s_2 to agent 2, the remaining items are worth to him at least $V_2(\alpha_2)$.
Case 2: If $s_2 < V_2(\alpha_2) \leq \frac{1}{2}$, the remaining items will surely be worth at least $V_2(\alpha_2)$.

The above take care of the case that $\alpha_1 < V_2(\alpha_1)$. Suppose now that $\alpha_1 \geq V_2(\alpha_1)$. Then $t = 1$, and it suffices to show that we can satisfy agent 2 with the remaining items. If $\alpha_2 \geq V_2(\alpha_2)$, this implies that $\alpha_2 \geq \frac{1}{3}$, and $V_2(\alpha_2) \leq \frac{1}{3}$, since the equation $\alpha_2 = V_2(\alpha_2)$ has a solution at $\frac{1}{3}$, by Fact 1. If $\alpha_2 \in [\frac{1}{3}, \frac{2}{3}]$, all the remaining items are worth to him at least $1 - \alpha_2 \geq \frac{1}{3} \geq V_2(\alpha_2)$. If $\alpha_2 > \frac{2}{3}$, then $V_2(\alpha_2) = 1 - \alpha_2$, because $\alpha_2 \in I(n, 1)$. Hence again all the remaining items are worth to him at least $V_2(\alpha_2)$. Finally, if $\alpha_2 < V_2(\alpha_2) \leq \frac{1}{n} = \frac{1}{2}$, the remaining items will surely be worth more than $V_2(\alpha_2)$.
• *Induction Step:* Suppose the algorithm is correct for $n - 1$ agents and consider an instance with n agents. Without loss of generality we can assume that player 1 gets the first bundle of t items, which is worth to him at least $V_n(\alpha_1)$.
Case 1: $\alpha_1 < V_n(\alpha_1)$

Since player 1 gets the items first, $\alpha_i < V_n(\alpha_i)$ holds for all i. For the remaining $n - 1$ players, the induction hypothesis guarantees that for $i = 2, ..., n$, agent i will end up with an allocation worth at least $V_{n-1}(\tilde{\alpha}_i)$, where $\tilde{\alpha}_i$ is the new highest value of agent i after the normalization (Line 15). Let s_i be the total value of agent i for the t items that agent 1 received. It suffices to show that $(1 - s_i)V_{n-1}(\tilde{\alpha}_i) \geq V_n(\alpha_i)$. But $\tilde{\alpha}_i \leq \frac{\alpha_i}{1-s_i}$, and since V_n is nonincreasing, we need to show that

$$(1 - s_i)V_{n-1}\left(\frac{\alpha_i}{1 - s_i}\right) \geq V_n(\alpha_i) \tag{4}$$

The agents $2, ..., n$ may belong to one of the following two groups:

(i) Consider an agent i with $\alpha_i \in I(n, k)$, for some $k \geq 1$. Since at the end of the $(t-1)$-th round, no player was selected by the algorithm, the $t-1$ most desirable goods for agent i do not exceed $V_n(\alpha_i)$. Hence Lemma 2 can be applied for the t highest items of i, which in turn implies that $s_i \leq k\alpha_i$. Substituting and by using Lemma 1, we see that (4) holds.

(ii) Consider an agent i with $\alpha_i \in NI(n, k)$, for some $k \geq 1$. Then by Lemma 2, we have $s_i \leq k \cdot l(n, k)$. To show (4), it suffices to show

$$(1 - k \cdot l(n, k))V_{n-1}\left(\frac{l(n, k)}{1 - k \cdot l(n, k)}\right) \geq V_n(\alpha_i)$$

Since $l(n, k) \in I(n, k)$, by Lemma 1, and by Fact 4 we have

$$(1 - k \cdot l(n, k))V_{n-1}\left(\frac{l(n, k)}{1 - k \cdot l(n, k)}\right) \geq V_n(l(n, k)) = V_n(\alpha_i)$$

Case 2: $\alpha_1 \geq V_n(\alpha_1)$. Then agent 1 receives only one good, his most desirable item. Consider an agent $i \in \{2, ..., n\}$. Suppose $\alpha_i \in I(n, k) \cup NI(n, k)$, for some $k \geq 1$.

Since agent 1 receives only one good, $s_i \leq a_i$, where s_i is, as in Case 1, the value of i for the item allocated to agent 1. Hence $s_i \leq k\alpha_i$. The recursive call for the $n - 1$ players guarantees to i, $V_{n-1}(\tilde{\alpha}_i)$, and $\tilde{\alpha}_i \leq \frac{\alpha_i}{1-s_i} \leq \frac{\alpha_i}{1-k\alpha_i}$. By (4), all we need to show is

$$(1 - k\alpha_i)V_{n-1}\left(\frac{\alpha_i}{1 - k\alpha_i}\right) \geq V_n(\alpha_i)$$

But this holds because of Lemma 1.

Finally, the examples provided in [5] show that Theorem 3 is tight.

4 Mechanism Design

4.1 Deterministic Algorithms

Given an instance V, let $A(V)$ be the outcome of a deterministic algorithm A. Let also V_i denote the i-th row of V, and V_{-i} denote the remaining matrix, excluding V_i. Let (V_i', V_{-i}) be the matrix that is produced from V, when we replace V_i with V_i'.

Definition 3. *A mechanism is truthful if for any instance V, and any other possible declaration V_i' of i:*

$$v_i(A(V)) \geq v_i(A(V_i', V_{-i}))$$

It is relatively easy to show that Algorithm 1 is not truthful (by constructing appropriate examples). In fact, as we show below, even if we ask for a large enough approximation to the value of $V_n(\alpha_i)$, no truthful algorithm can satisfy this.

Theorem 5. *Even for two agents, there is no deterministic truthful algorithm that guarantees a total value that is strictly better than $\frac{2}{3} \cdot V_n(\alpha_i)$, for every player i.*

We will first argue about the impossibility of truthfulness for a restricted class of algorithms. Given an instance I, and a permutation $\pi \in S^m$, we denote by $\pi(I)$ the instance where every row of the matrix is permuted by π. We call a deterministic algorithm *permutation-respecting with regard to instance I*, if, for every permutation π, the allocation produced by the algorithm at $\pi(I)$ is $(\pi(S_1), ..., \pi(S_n))$, where $(S_1, ..., S_n)$ is the allocation produced by the algorithm at I. Thus, such an algorithm does not depend on the identities of the goods for the specific instance but only on their values. We call an algorithm simply *permutation-respecting* if this holds for any instance I.

Lemma 3 below and more generally the approach of proving first the impossibility result for permutation-respecting mechanisms, may be of independent interest in proving other impossibility results in fair division.

Lemma 3. *If a permutation-respecting algorithm A guarantees a value that is strictly better than $\frac{2}{3} \cdot V_n(\alpha_i)$ to each player i, then A cannot be truthful.*

Proof. We will argue about instances with 2 players and 4 goods. Suppose there is a permutation-respecting algorithm A that is truthful and always guarantees a value better than $\frac{2}{3} \cdot V_n(\alpha_i)$, for every player i. Consider the following instances:

$$V' = \begin{bmatrix} \frac{2}{3} & \frac{1}{9} & \frac{1}{9} & \frac{1}{9} \\ \frac{2}{3} & \frac{1}{9} & \frac{1}{9} & \frac{1}{9} \end{bmatrix}, V = \begin{bmatrix} \frac{1}{4} & \frac{1}{4} & \frac{1}{4} & \frac{1}{4} \\ \frac{1}{4} & \frac{1}{4} & \frac{1}{4} & \frac{1}{4} \end{bmatrix}$$

In V', since $V_2(\frac{2}{3}) = \frac{1}{3}$, and $\frac{2}{3} \cdot V_2(\frac{2}{3}) = \frac{2}{9}$, the only choices for algorithm A is to give the first item to one player and all other items to the other player. Without loss of generality, we can assume that A gives the set $\{1\}$ to player 1 and $\{2, 3, 4\}$ to player 2.

In V, $V_2(\frac{1}{4}) = 0.4$, and $\frac{2}{3} \cdot V_2(\frac{1}{4}) > \frac{1}{4}$. Hence each player should receive exactly two items, therefore the choices for A is to give player 1 one of the sets $\{1, 2\}$, $\{1, 3\}$, $\{1, 4\}$, $\{2, 3\}$, $\{2, 4\}$, $\{3, 4\}$. We will show that at least one player has an incentive to lie in every case. We analyze all cases below. We will make use of the instances

$$W = \begin{bmatrix} \frac{2}{3} & \frac{1}{9} & \frac{1}{9} & \frac{1}{9} \\ \frac{1}{4} & \frac{1}{4} & \frac{1}{4} & \frac{1}{4} \end{bmatrix}, W' = \begin{bmatrix} \frac{1}{9} & \frac{2}{3} & \frac{1}{9} & \frac{1}{9} \\ \frac{1}{4} & \frac{1}{4} & \frac{1}{4} & \frac{1}{4} \end{bmatrix}$$

- A gives one of $\{1, 2\}, \{1, 3\}, \{1, 4\}$ to player 1 in instance V.
  Then in W, since A must guarantee more than $\frac{2}{3} \cdot V_2(\frac{1}{4}) > 0.25$ to player 2, it has to give at most two items to player 1. If the first item is not given to player 1, then player 2 would have an incentive to lie when his valuation is as in V', and report $(\frac{1}{4}, \frac{1}{4}, \frac{1}{4}, \frac{1}{4})$. Hence A has to give player 1 at least the first item. If A gives to player 1 only the first item, he can get more if he reports $(\frac{1}{4}, \frac{1}{4}, \frac{1}{4}, \frac{1}{4})$ and switches to V. If it gives him one of $\{1, 2\}, \{1, 3\}, \{1, 4\}$, then player 2 can raise his utility by reporting $(\frac{2}{3}, \frac{1}{9}, \frac{1}{9}, \frac{1}{9})$, and switching to V'. Hence there is always an incentive for lying by one of the players.
- A gives $\{2, 3\}$ or $\{2, 4\}$ to player 1 in instance V. Consider instance W'. In analogy to the previous case, A can give to player 1 at most two items. Furthermore, player

1 has to receive the second item, as otherwise he will not achieve more than $\frac{2}{3} \cdot$ $V_n(\frac{2}{3}) = \frac{2}{9}$. If A gives to player 1 only the second item, he has an incentive to lie, and report $(\frac{1}{4}, \frac{1}{4}, \frac{1}{4}, \frac{1}{4})$ so as to switch to instance V. If A gives one of $\{2, 1\}, \{2, 3\}$, or $\{2, 4\}$, to player 1, then player 2 has an incentive to lie and report $(\frac{1}{9}, \frac{2}{3}, \frac{1}{9}, \frac{1}{9})$. In this case, since A is permutation-respecting, player 2 would receive the set $\{1, 3, 4\}$ and would be better off.

- A gives $\{3, 4\}$ to player 1. Similar arguments show that one of the two players has an incentive to lie.

This shows that in all cases, regarding the possible allocations of A in instance V, someone has an incentive to lie. Hence algorithm A cannot be truthful.

The proof of Theorem 5 can now be completed by arguing about the general case. Suppose the statement of the theorem is not true and let A be such a truthful algorithm. We can then prove the following (proof omitted):

Lemma 4. *A is permutation-respecting with regard to instance V'.*

The proof of Lemma 3 only required that an algorithm is permutation-respecting with regard to instance V'. Hence by Lemma 4 the proof of Theorem 5 is complete.

Below we state a very simple algorithm that shows that the bound of Theorem 5 is almost tight when we have two agents and four goods. The same algorithm shows that we cannot have the same lower bound for the case of two or three goods.

Algorithm 2

Given the reported input V, allocate to player 1 his most desirable good and allocate the remaining goods to player 2.

Theorem 6. *For 2 agents and m goods, Algorithm 2 is truthful and always returns an allocation worth at least $\rho V_2(\alpha_i)$ for $i = 1, 2$, where $\rho = \min\{1, \frac{1}{m V_2(m)}\}$.*

In Table 1, we see the ratio ρ of Algorithm 2, as the number of goods increases. The ratio equals 0.625 when $m = 4$, which is very close to the lower bound of $\frac{2}{3}$. The worst case of Algorithm 2 when $m = 4$, is achieved when the valuation of player 1 is $(\frac{1}{4}, \frac{1}{4}, \frac{1}{4}, \frac{1}{4})$. In general, the worst case scenario for this mechanism happens when player 1 values all items equally.

Table 1. The guarantee provided by Algorithm 2, for various values of m

Number of goods	2	3	4	5	...	m
ratio ρ	1	1	0.625	$\frac{1}{2}$	...	$\frac{1}{m V_2(1/m)}$

It still remains an open problem to determine whether a constant factor approximation to $V_n(\alpha_i)$ can be achieved when the number of goods is large, even for two agents. The difficulties arise from the fact that we have a multi-parameter domain (each player submits m numbers), which makes it more challenging to argue about truthfulness.

4.2 Randomized Algorithms

Given the negative results of the previous subsection, we initiate the study of randomized algorithms. Once we allow randomization, we need to decide on the quality of the solution that we want the algorithm to return as well as on the notion of truthfulness.

Given an instance V reported by the players, let $\mathcal{F}(V)$ be the set of all allocations in which every player receives a bundle worth at least $V_n(\alpha_i)$, according to the reported valuations. An allocation that belongs to $\mathcal{F}(V)$ will be called *feasible*. We say that a randomized algorithm A is *universally feasible* if it always outputs a feasible allocation. Similarly, we say that an algorithm is *universally truthful* if an agent never has an incentive to lie, regardless of the randomness used by the algorithm. An algorithm that is universally feasible and universally truthful cannot exist given the negative results of Section 4.1. Furthermore, an algorithm that is universally truthful and produces allocations that give $V_n(\alpha_i)$ only in expectation is trivial to construct: simply allocate all the goods to a player i uniformly at random.

Given the above, the most appropriate setting is to insist on universally feasible mechanisms but relax the notion of truthfulness. An algorithm is *truthful in expectation* if no agent can improve his expected payoff by misreporting. Hence, we want to investigate the possibility of truthful in expectation algorithms, that are probability distributions on $\mathcal{F}(V)$.

We exhibit that certain classes of algorithms cannot achieve the goals we want. The family of algorithms we focus on, satisfy Pareto-efficiency and some symmetry requirements. In particular, we consider the following properties:

(P1) The algorithm outputs only Pareto-efficient outcomes.
(P2) Two feasible outcomes that give exactly the same utilities to all players have the same probability.
(P3) Two feasible outcomes, where the utility vector of the first is a permutation of the utility vector of the second, have the same probability.

An example of an algorithm with such properties is to find all feasible allocations, eliminate those that are Pareto-inefficient and select one of those left uniformly at random. In general, any uniform distribution on a subset of Pareto-efficient allocations satisfies the above properties. It is however easy to construct examples showing that such algorithms are not truthful in expectation. In fact, we have the following:

Theorem 7. *Even for two agents, there is no truthful in expectation algorithm that satisfies (P1)-(P3) and guarantees, for every instance, a value of $V_n(\alpha_i)$, for every player i.*

5   On the Complexity of Finding Better Allocations

In this Section, we study a slightly different question. In instances where there is not much conflict, one might be able to produce allocations that exceed the worst case guarantee of $V_n(\alpha_i)$. The question that arises is whether we can compute such improved allocations on instances that admit them or even decide when do they exist. A particularly interesting case is to determine which instances admit proportional allocations. With this in mind, Demko and Hill [4] proved the following:

Theorem 8. *[4] It is NP-hard to decide if there exists a proportional allocation.*

This is done via a reduction from the PARTITION problem, as for two identical agents, any such allocation would imply a partitioning of the goods in two sets of equal value. Clearly there must be some value $\beta \in [V_n(\alpha), \frac{1}{n}]$ ($\alpha = \max \alpha_i$), for which we can decide if an allocation where everybody receives at least β exists. So far, we only know that $\beta \geq V_n(\alpha)$. Below we exhibit that this value cannot be close to $\frac{1}{n}$.

Theorem 9. *For any constant $c \geq 1$, it is NP-hard to decide if there exists an allocation where every player receives a bundle worth at least $1/cn$.*

The proof is based on adjusting appropriately a reduction from 3-dimensional matching, used in [6] for the inapproximability of makespan in job scheduling.

6 Future Work

There are many interesting open questions. The most important one is to obtain a better understanding of truthful mechanisms. Surprisingly, even for two agents we are not yet aware if there exists a deterministic truthful algorithm that provides a constant factor approximation to $V_n(\alpha_i)$, when the number of goods is not $O(1)$. One of the main challenges for resolving this, is the fact that we have a multi-parameter domain. As is the case in other contexts, arguing about truthfulness is harder in multi-parameter domains, as players have plenty of flexibility in finding possible ways of lying.

Acknowledgements. We would like to thank Ted Hill for many valuable discussions on this topic. We would also like to thank Jarek Byrka for suggesting to us the question of strengthening the guarantee in Section 3 from $V_n(\alpha)$ to $V_n(\alpha_i)$, for every agent i.

References

1. Asadpour, A., Saberi, A.: An approximation algorithm for max-min fair allocation of indivisible goods. In: ACM Symposium on Theory of Computing (STOC), pp. 114–121 (2007)
2. Brams, S.J., Taylor, A.D.: Fair Division: from Cake Cutting to Dispute Resolution. Cambrige University press (1986)
3. Chen, Y., Lai, J., Parkes, D., Procaccia, A.: Truth, justice, and cake cutting. In: AAAI, pp. 756–761 (2010)
4. Demko, S., Hill, T.: Equitable distribution of indivisible items. Mathematical Social Sciences 16, 145–158 (1988)
5. Hill, T.: Partitioning general probability measures. The Annals of Probability 15(2), 804–813 (1987)
6. Lenstra, J.K., Shmoys, D.B., Tardos, E.: Approximation algorithms for scheduling unrelated parallel machines. Mathematical Programming 46, 259–271 (1990)
7. Mossel, E., Tamuz, O.: Truthful Fair Division. In: Kontogiannis, S., Koutsoupias, E., Spirakis, P.G. (eds.) SAGT 2010. LNCS, vol. 6386, pp. 288–299. Springer, Heidelberg (2010)
8. Robertson, J.M., Webb, W.A.: Cake Cutting Algorithms: be fair if you can. AK Peters (1998)
9. Steinhaus, H.: The problem of fair division. Econometrica 16, 101–104 (1948)

Natural Models for Evolution on Networks

George B. Mertzios[1], Sotiris Nikoletseas[2],
Christoforos Raptopoulos[2], and Paul G. Spirakis[2]

[1] School of Engineering and Computing Sciences, Durham University, UK
george.mertzios@durham.ac.uk
[2] Computer Technology Institute and University of Patras, Greece
nikole@cti.gr, raptopox@ceid.upatras.gr, spirakis@cti.gr

Abstract. Evolutionary dynamics have been traditionally studied in
the context of homogeneous populations, mainly described by the Moran
process [15]. Recently, this approach has been generalized in [13] by
arranging individuals on the nodes of a network (in general, directed).
In this setting, the existence of directed arcs enables the simulation
of extreme phenomena, where the fixation probability of a randomly
placed mutant (i.e. the probability that the offsprings of the mutant
eventually spread over the whole population) is arbitrarily small or
large. On the other hand, undirected networks (i.e. undirected graphs)
seem to have a smoother behavior, and thus it is more challenging to find
suppressors/amplifiers of selection, that is, graphs with smaller/greater
fixation probability than the complete graph (i.e. the homogeneous
population). In this paper we focus on undirected graphs. We present
the first class of undirected graphs which act as suppressors of selection,
by achieving a fixation probability that is at most one half of that of
the complete graph, as the number of vertices increases. Moreover,
we provide some generic upper and lower bounds for the fixation
probability of general undirected graphs. As our main contribution,
we introduce the natural alternative of the model proposed in [13].
In our new evolutionary model, all individuals interact *simultaneously*
and the result is a compromise between aggressive and non-aggressive
individuals. That is, the behavior of the individuals in our new model
and in the model of [13] can be interpreted as an *"aggregation"* vs. an
"all-or-nothing" strategy, respectively. We prove that our new model
of mutual influences admits a *potential function*, which guarantees the
convergence of the system for any graph topology and any initial fitness
vector of the individuals. Furthermore, we prove fast convergence to the
stable state for the case of the complete graph, as well as we provide
almost tight bounds on the limit fitness of the individuals. Apart from
being important on its own, this new evolutionary model appears to
be useful also in the abstract modeling of control mechanisms over
invading populations in networks. We demonstrate this by introducing
and analyzing two alternative control approaches, for which we bound
the time needed to stabilize to the "healthy" state of the system.

Keywords: Evolutionary dynamics, undirected graphs, fixation probability, potential function, Markov chain, fitness, population structure.

N. Chen, E. Elkind, and E. Koutsoupias (Eds.): WINE 2011, LNCS 7090, pp. 290–301, 2011.
© Springer-Verlag Berlin Heidelberg 2011

1   Introduction

Evolutionary dynamics have been well studied (see [1,6,7,19,21,22]), mainly in the context of homogeneous populations, described by the Moran process [15,17]. In addition, population dynamics have been extensively studied also from the perspective of the strategic interaction in evolutionary game theory, cf. for instance [8,9,10,11,20]. One of the main targets of evolutionary game theory is evolutionary dynamics (see [9,23]). Such dynamics usually examine the propagation of intruders with a given *fitness* to a population, whose initial members (resident individuals) have a different fitness. In fact, "evolutionary stability" is the case where no dissident behaviour can invade and dominate the population. The evolutionary models and the dynamics we consider here belong to this framework. In addition, however, we consider structured populations (i.e. in the form of an undirected graph) and we study how the underlying graph structure affects the evolutionary dynamics. We study in this paper two kinds of evolutionary dynamics. Namely, the "all or nothing" case (where either the intruder overtakes the whole graph or die out) and the "aggregation" case (more similar in spirit to classical evolutionary game theory, where the intruder's fitness aggregates with the population fitness and generates eventually a homogeneous crowd with a new fitness).

In a recent article, Lieberman, Hauert, and Nowak proposed a generalization of the Moran process by arranging individuals on a connected network (i.e. graph) [13] (see also [18]). In this model, vertices correspond to individuals of the population and weighted edges represent the reproductive rates between the adjacent vertices. That is, the population structure is translated into a network (i.e. graph) structure. Furthermore, individuals (i.e. vertices) are partitioned into two types: *aggressive* and *non-aggressive*. The degree of (relative) aggressiveness of an individual is measured by its *relative fitness*; in particular, non-aggressive and aggressive individuals are assumed to have relative fitness 1 and $r \geq 1$, respectively. This modeling approach initiates an ambitious direction of interdisciplinary research, which combines classical aspects of computer science (such as combinatorial structures and complex network topologies), probabilistic calculus (discrete Markov chains), and fundamental aspects of evolutionary game theory (such as evolutionary dynamics).

In the model of [13], one *mutant* (or *invader*) with relative fitness $r \geq 1$ is introduced into a given population of *resident* individuals, each of whom having relative fitness 1. For simplicity, a vertex of the graph that is occupied by a mutant will be referred to as *black*, while the rest of the vertices will be referred to as *white*. At each time step, an individual is chosen for reproduction with a probability proportional to its fitness, while its offspring replaces a randomly chosen neighboring individual in the population. Once u has been selected for reproduction, the probability that vertex u places its offspring into position v is given by the weight w_{uv} of the directed arc $\langle uv \rangle$. This process stops when either all vertices of the graph become black (resulting to a *fixation* of the graph) or they all become white (resulting to *extinction* of the mutants). Several similar models have been previously studied, describing for instance influence propagation in social networks (such as the decreasing cascade model [12,16]), dynamic monopolies [2], particle interactions (such as the voter model, the antivoter model, and the exclusion process), etc. However, the dynamics emerging from these models do not consider different fitnesses for the individuals.

The *fixation probability* f_G of a graph $G = (V, E)$ is the probability that eventually fixation occurs, i.e. the probability that an initially introduced mutant, placed uniformly at random on a vertex of G, eventually spreads over the whole population V, replacing all resident individuals. One of the main characteristics in this model is that at every iteration of the process, a "battle" takes place between aggressive and non-aggressive individuals, while the process stabilizes only when one of the two teams takes over the whole population. This kind of behavior of the individuals can be interpreted as an *all-or-nothing* strategy.

Lieberman et al. [13] proved that the fixation probability for every symmetric directed graph (i.e. when $w_{uv} = w_{vu}$ for every u, v) is equal to that of the complete graph (i.e. the homogeneous population of the Moran process), which tends to $1 - \frac{1}{r}$ as the size n of the population grows. Moreover, exploiting vertices with zero in-degree or zero out-degree ("upstream" and "downstream" populations, respectively), they provided several examples of *directed* graphs with arbitrarily small and arbitrarily large fixation probability [13]. Furthermore, the existence of directions on the arcs leads to examples where neither fixation nor extinction is possible (e.g. a graph with two sources).

In contrast, general *undirected* graphs (i.e. when $\langle uv \rangle \in E$ if and only if $\langle vu \rangle \in E$ for every u, v) appear to have a smoother behavior, as the above process eventually reaches fixation or extinction with probability 1. Furthermore, the coexistence of both directions at every edge in an undirected graph seems to make it more difficult to find *suppressors* or *amplifiers* of selection (i.e. graphs with smaller or greater fixation probability than the complete graph, respectively), or even to derive non-trivial upper and lower bound for the fixation probability on general undirected graphs. This is the main reason why only little progress has been done so far in this direction and why most of the recent work focuses mainly on the exact or numerical computation of the fixation probability for very special cases of undirected graphs, e.g. the star and the path [3,4,5].

Our Contribution. In this paper we overcome this difficulty for undirected graphs and we provide the first class of undirected graphs that act as suppressors of selection in the model of [13], as the number of vertices increases. This is a very simple class of graphs (called *clique-wheels*), where each member G_n has a clique of size $n \geq 3$ and an induced cycle of the same size n with a perfect matching between them. We prove that, when the mutant is introduced to a clique vertex of G_n, then the probability of fixation tends to zero as n grows. Furthermore, we prove that, when the mutant is introduced to a cycle vertex of G_n, then the probability of fixation is at most $1 - \frac{1}{r}$ as n grows (i.e. to the same value with the homogeneous population of the Moran process). Therefore, since the clique and the cycle have the same number n of vertices in G_n, the fixation probability f_{G_n} of G_n is at most $\frac{1}{2}(1 - \frac{1}{r})$ as n increases, i.e. G_n is a suppressor of selection. Furthermore, we provide for the model of [13] the first non-trivial upper and lower bounds for the fixation probability in general undirected graphs. In particular, we first provide a generic upper bound depending on the degrees of some local neighborhood. Second, we present another upper and lower bound, depending on the ratio between the minimum and the maximum degree of the vertices.

As our main contribution, we introduce in this paper the natural alternative of the *all-or-nothing* approach of [13], which can be interpreted as an *aggregation* strategy. In this aggregation model, all individuals interact *simultaneously* and the result is a compromise between the aggressive and non-aggressive individuals. Both these two alternative models for evolutionary dynamics coexist in several domains of interaction between individuals, e.g. in society (dictatorship vs. democracy, war vs. negotiation) and biology (natural selection vs. mutation of species). In particular, another motivation for our models comes from biological networks, in which the interacting individuals (vertices) correspond to cells of an organ and advantageous mutants correspond to viral cells or cancer. Regarding the proposed model of mutual influences, we first prove that it admits a *potential* function. This potential function guarantees that for any graph topology and any initial fitness vector, the system converges to a stable state, where all individuals have the same fitness. Furthermore, we analyze the telescopic behavior of this model for the complete graph. In particular, we prove fast convergence to the stable state, as well as we provide almost tight bounds on the *limit fitness* of the individuals.

Apart from being important on its own, this new evolutionary model enables also the abstract modeling of new control mechanisms over invading populations in networks. We demonstrate this by introducing and analyzing the behavior of two alternative control approaches. In both scenarios we periodically modify the fitness of a small fraction of individuals in the current population, which is arranged on a complete graph with n vertices. In the first scenario, we proceed in phases. Namely, after each modification, we let the system stabilize before we perform the next modification. In the second scenario, we modify the fitness of a small fraction of individuals at each step. In both alternatives, we stop performing these modifications of the population whenever the fitness of every individual becomes sufficiently close to 1 (which is considered to be the "healthy" state of the system). For the first scenario, we prove that the number of *phases* needed for the system to stabilize in the healthy state is logarithmic in $r - 1$ and independent of n. For the second scenario, we prove that the number of *iterations* needed for the system to stabilize in the healthy state is linear in n and proportional to $r \ln(r - 1)$. Due to space limitations we omit the proofs of the results, which can be found in [14].

Notation. In an undirected graph $G = (V, E)$, the edge between vertices $u \in V$ and $v \in V$ is denoted by $uv \in E$, and in this case u and v are said to be *adjacent* in G. If the graph G is directed, we denote by $\langle uv \rangle$ the arc from u to v. For every vertex $u \in V$ in an undirected graph $G = (V, E)$, we denote by $N(u) = \{v \in V \mid uv \in E\}$ the set of neighbors of u in G and by $\deg(u) = |N(u)|$. Furthermore, for any $k \geq 1$, we denote for simplicity $[k] = \{1, 2, \ldots, k\}$.

2   All-or-Nothing vs. Aggregation

In this section we formally define the model of [13] for undirected graphs and we introduce our new model of mutual influences. Similarly to [13], we assume for every edge uv of an undirected graph that $w_{uv} = \frac{1}{\deg u}$ and $w_{vu} = \frac{1}{\deg v}$, i.e. once a vertex u has been chosen for reproduction, it chooses one of its neighbors uniformly at random.

2.1   The Model of Lieberman, Hauert, and Nowak (An All-or-Nothing Approach)

Let $G = (V, E)$ be a connected undirected graph with n vertices. Then, the stochastic process defined in [13] can be described by a Markov chain with state space $\mathcal{S} = 2^V$ (i.e. the set of all subsets of V) and transition probability matrix P, where for any two states $S_1, S_2 \subseteq V$,

$$
P_{S_1,S_2} =
\begin{cases}
\frac{1}{|S_1|r+n-|S_1|} \cdot \sum\limits_{u \in N(v) \cap S_1} \frac{r}{\deg(u)}, & \text{if } S_2 = S_1 \cup \{v\} \text{ and } v \notin S_1 \\[2ex]
\frac{1}{|S_1|r+n-|S_1|} \cdot \sum\limits_{u \in N(v) \setminus S_2} \frac{1}{\deg(u)}, & \text{if } S_1 = S_2 \cup \{v\} \text{ and } v \notin S_2 \\[2ex]
\frac{1}{|S_1|r+n-|S_1|} \left(\sum\limits_{u \in S_1} \frac{r \cdot |N(u) \cap S_1|}{\deg(u)} + \sum\limits_{u \in V \setminus S_1} \frac{|N(u) \cap (V \setminus S_1)|}{\deg(u)} \right), & \text{if } S_2 = S_1 \\[2ex]
0, & \text{otherwise}
\end{cases}
\tag{1}
$$

Notice that in the above Markov chain there are two absorbing states, namely $\emptyset$ and V, which describe the cases where the vertices of G are all white or all black, respectively. Since G is connected, the above Markov chain will eventually reach one of these two absorbing states with probability 1. If we denote by h_v the probability of absorption at state V, given that we start with a single mutant placed on vertex v, then by definition $f_G = \frac{\sum_v h_v}{n}$. Generalizing this notation, let h_S be the probability of absorption at V given that we start at state $S \subseteq V$, and let $h = [h_S]_{S \subseteq V}$. Then, it follows that vector h is the unique solution of the linear system $h = P \cdot h$ with boundary conditions $h_\emptyset = 0$ and $h_V = 1$.

However, observe that the state space $\mathcal{S} = 2^V$ of this Markov chain has size 2^n, i.e. the matrix $P = [P_{S_1,S_2}]$ in (1) has dimension $2^n \times 2^n$. This indicates that the problem of computing the fixation probability f_G of a given graph G is hard, as also mentioned in [13]. This is the main reason why, to the best of our knowledge, all known results so far regarding the computation of the fixation probability of undirected graphs are restricted to regular graphs, stars, and paths [3,4,5,13,18]. In particular, for the case of regular graphs, the above Markov chain is equivalent to a birth-death process with $n - 1$ transient (non-absorbing) states, where the forward bias at every state (i.e. the ratio of the forward probability over the backward probability) is equal to r. In this case, the fixation probability is equal to $\rho = \frac{1}{1 + \sum_{i=1}^{n-1} \frac{1}{r^i}} = \frac{1 - \frac{1}{r}}{1 - \frac{1}{r^n}}$. cf. [18], chapter 8. It is worth mentioning that, even for the case of paths, there is no known exact or approximate formula for the fixation probability [5].

2.2 An Evolutionary Model of Mutual Influences (An Aggregation Approach)

The evolutionary model of [13] constitutes a sequential process, in every step of which only two individuals interact and the process eventually reaches one of two extreme states. However, in many evolutionary processes, all individuals may interact simultaneously at each time step, while some individuals have greater influence to the rest of the population than others. This observation leads naturally to the following model for evolution on graphs, which can be thought as a smooth version of the model presented in [13].

Consider a population of size n and a portion $\alpha \in [0, 1]$ of newly introduced mutants with relative fitness r. The topology of the population is given in general by a directed graph $G = (V, E)$ with $|V| = n$ vertices, where the directed arcs of E describe the allowed interactions between the individuals. At each time step, *every* individual $u \in V$ of the population influences every individual $v \in V$, for which $\langle uv \rangle \in E$, while the degree of this influence is proportional to the fitness of u and to the weight w_{uv} of the arc $\langle uv \rangle$. Note that we can assume without loss of generality that the weights w_{uv} on the arcs are normalized, i.e. for every fixed vertex $u \in V$ it holds $\sum_{\langle uv \rangle \in E} w_{uv} = 1$. Although this model can be defined in general for directed graphs with arbitrary arc weights w_{uv}, we will focus in the following to the case where G is an undirected graph (i.e. $\langle u_i u_j \rangle \in E$ if and only if $\langle u_j u_i \rangle \in E$, for every i, j) and $w_{uv} = \frac{1}{\deg(u)}$ for all edges $uv \in E$.

Formally, let $V = \{u_1, u_2, \ldots, u_n\}$ be the set of vertices and $r_{u_i}(k)$ be the fitness of the vertex $u_i \in V$ at iteration $k \geq 0$. Let $\Sigma(k)$ denote the sum of the fitnesses of all vertices at iteration k, i.e. $\Sigma(k) = \sum_{i=1}^{n} r_{u_i}(k)$. Then the vector $r(k + 1)$ with the fitnesses $r_{u_i}(k + 1)$ of the vertices $u_i \in V$ at the next iteration $k + 1$ is given by $[r_{u_1}(k + 1), r_{u_2}(k + 1), \ldots, r_{u_n}(k + 1)]^T = P \cdot [r_{u_1}(k), r_{u_2}(k), \ldots, r_{u_n}(k)]^T$, i.e.

$$r(k + 1) = P \cdot r(k) \tag{2}$$

In the latter equation, the elements of the square matrix $P = [P_{ij}]_{i,j=1}^{n}$ depend on the iteration k and they are given as follows:

$$P_{ij} = \begin{cases} \frac{r_{u_j}(k)}{\deg(u_j)\Sigma(k)}, & \text{if } i \neq j \text{ and } u_i u_j \in E \\ 0, & \text{if } i \neq j \text{ and } u_i u_j \notin E \\ 1 - \sum_{j \neq i} P_{ij}, & \text{if } i = j \end{cases} \tag{3}$$

Note by (2) and (3) that after the first iteration, the fitness of every individual in our new evolutionary model of mutual influences equals the expected fitness of this individual in the model of [13] (cf. Section 2.1). However, this correlation of the two models is not maintained in the next iterations and the two models behave differently as the processes evolve.

In particular, in the case where G is the complete graph, i.e. $\deg(u_i) = n - 1$ for every vertex u_i, the matrix P becomes

$$P = \begin{bmatrix} 1 - \frac{r_{u_2}(k) + \ldots + r_{u_n}(k)}{(n-1)\Sigma(k)} & \cdots & \frac{r_{u_n}(k)}{(n-1)\Sigma(k)} \\ \frac{r_{u_1}(k)}{(n-1)\Sigma(k)} & \cdots & \frac{r_{u_n}(k)}{(n-1)\Sigma(k)} \\ \cdots & \cdots & \cdots \\ \frac{r_{u_1}(k)}{(n-1)\Sigma(k)} & \cdots & 1 - \frac{r_{u_1}(k) + \ldots + r_{u_{n-1}}(k)}{(n-1)\Sigma(k)} \end{bmatrix} \qquad (4)$$

The system given by (2) and (3) can be defined for every initial fitness vector $r(0)$. However, in the case where there is initially a portion $\alpha \in [0, 1]$ of newly introduced mutants with relative fitness r, the initial condition $r(0)$ of the system in (2) is a vector with αn entries equal to r and with $(1 - \alpha)n$ entries equal to 1. Note that the recursive equation (2) is a *non-linear* equation on the fitness values $r_{u_j}(k)$ of the vertices at iteration k.

Since by (3) the sum of every row of the matrix P equals to one, the fitness $r_{u_i}(k)$ of vertex u_i after the $(k + 1)$-th iteration of the process is a convex combination of the fitnesses of the neighbors of u_i after the k-th iteration. Therefore, in particular, the fitness of every vertex u_i at every iteration $k \geq 0$ lies between the smallest and the greatest initial fitness of the vertices. That is, if r_{min} and r_{max} denote the smallest and the greatest initial fitness in $r(0)$, respectively, then $r_{min} \leq r_{u_i}(k) \leq r_{max}$ for every $u_i \in V$ and every $k \geq 0$.

Degree of influence. Suppose that initially αn mutants (for some $\alpha \in [0, 1]$) with relative fitness $r \geq 1$ are introduced in graph G on a subset $S \subseteq V$ of its vertices. Then, as we prove in Theorem 4, after a certain number of iterations the fitness vector $r(k)$ converges to a vector $[r_0^S, r_0^S, \ldots, r_0^S]^T$, for some value r_0^S. This *limit fitness* r_0^S depends in general on the initial relative fitness r of the mutants, on their initial number αn, as well as on their initial position on the vertices of $S \subseteq V$. The relative fitness r of the initially introduced mutants can be thought as having the "black" color, while the initial fitness of all the other vertices can be thought as having the "white" color. Then, the limit fitness r_0^S can be thought as the "degree of gray color" that all the vertices obtain after sufficiently many iterations, given that the mutants are initially placed at the vertices of S. In the case where the αn mutants are initially placed with *uniform* probability to the vertices of G, we can define the *limit fitness r_0 of G* as $r_0 = \frac{1}{\binom{n}{\alpha n}} \cdot \sum_{S \subseteq V, |S| = \alpha n} r_0^S$. For a given initial value of r, the bigger is r_0 the stronger is the effect of natural selection in G.

Since r_0^S is a convex combination of r and 1, there exists a value $f_{G,S}(r) \in [0, 1]$, such that $r_0^S = f_{G,S}(r) \cdot r + (1 - f_{G,S}(r)) \cdot 1$. Then, the value $f_{G,S}(r)$ is the *degree of influence* of the graph G, given that the mutants are initially placed at the vertices of S. In the case where the mutants are initially placed with uniform probability at the vertices of G, we can define the degree of influence of G as $f_G(r) = \frac{1}{\binom{n}{\alpha n}} \sum_{S \subseteq V, |S| = \alpha n} f_{G,S}(r)$.

Number of iterations to stability. For some graphs G, the fitness vector $r(k)$ reaches *exactly* the *limit fitness vector* $[r_0, r_0, \ldots, r_0]^T$ (for instance, the complete graph with two vertices and one mutant not only reaches this limit in exactly one iteration, but also the degree of influence is exactly the fixation probability of this simple graph). However, for other graphs G the fitness vector $r(k)$ converges to $[r_0, r_0, \ldots, r_0]^T$ (cf. Theorem 4 below), but it never becomes equal to it. In the first case, one can compute (exactly or approximately) the number of iterations needed to reach the limit fitness vector. In the second case, given an arbitrary $\varepsilon > 0$, one can compute the number of iterations needed to come ε-close to the limit fitness vector.

3   Analysis of the All-or-Nothing Model

In this section we present analytic results on the evolutionary model of [13], which is based on the sequential interaction among the individuals. In particular, we first present non-trivial upper and lower bounds for the fixation probability, depending on the degrees of vertices. Then we present the first class of undirected graphs that act as suppressors of selection in the model of [13], as the number of vertices increases.

Recall by the preamble of Section 2.2 that, similarly to [13], we assumed that $w_{uv} = \frac{1}{\deg u}$ and $w_{vu} = \frac{1}{\deg v}$ for every edge uv of an undirected graph $G = (V, E)$. It is easy to see that this formulation is equivalent to assigning to every edge $e = uv \in E$ the weight $w_e = w_{uv} = w_{vu} = 1$, since also in this case, once a vertex u has been chosen for reproduction, it chooses one of its neighbors uniformly at random. A natural generalization of this weight assignment is to consider G as a complete graph, where every edge e in the clique is assigned a non-negative weight $w_e \geq 0$, and w_e is not necessarily an integer. Note that, whenever $w_e = 0$, it is as if the edge e is not present in G. Then, once a vertex u has been chosen for reproduction, u chooses any other vertex v with probability $\frac{w_{uv}}{\sum_{x \neq u} w_{ux}}$.

Note that, if we do not impose any additional constraint on the weights, we can simulate multigraphs by just setting the weight of an edge to be equal to the multiplicity of this edge. Furthermore, we can construct graphs with arbitrary small fixation probability. For instance, consider an undirected star with n leaves, where one of the edges has weight an arbitrary small $\varepsilon > 0$ and all the other edges have weight 1. Then, the leaf that is incident to the edge with weight ε acts as a source in the graph as $\varepsilon \to 0$. Thus, the only chance to reach fixation is when we initially place the mutant at the source, i.e. the fixation probability of this graph tends to $\frac{1}{n+1}$ as $\varepsilon \to 0$. Therefore, it seems that the difficulty to construct strong suppressors lies in the fact that unweighted undirected graphs can not simulate sources. For this reason, we consider in the remainder of this paper only unweighted undirected graphs.

3.1   A Generic Upper Bound Approach

In the next theorem we provide a generic upper bound of the fixation probability of undirected graphs, depending on the degrees of the vertices in some local neighborhood.

Theorem 1. *Let* $G = (V, E)$ *be an undirected graph. For any* $uv \in E$, *let* $Q_u = \sum_{x \in N(u)} \frac{1}{\deg x}$ *and* $Q_{uv} = \sum_{x \in N(u) \setminus \{v\}} \frac{1}{\deg x} + \sum_{x \in N(v) \setminus \{u\}} \frac{1}{\deg x}$. *Then* $f_G \leq \max_{uv \in E} \left\{ \frac{r^2}{r^2 + rQ_u + Q_u Q_{uv}} \right\}$.

3.2 Upper and Lower Bounds Depending on Degrees

In the following theorem we provide upper and lower bounds of the fixation probability of undirected graphs, depending on the minimum and the maximum degree of the vertices.

Theorem 2. *Let* $G = (V, E)$ *be an undirected graph, where* $\delta \leq \deg(u) \leq \Delta$ *for every* $u \in V$. *Then, the fixation probability* f_G *of* G, *when the fitness of the mutant is* r, *is upper (resp. lower) bounded by the fixation probability of the clique for mutant fitness* $r_u = \frac{r\Delta}{\delta}$ *(resp. for mutant fitness* $r_l = \frac{r\delta}{\Delta}$).

3.3 The Undirected Suppressor

In this section we provide the first class of undirected graphs (which we call *clique-wheels*) that act as suppressors of selection as the number of vertices increases. In particular, we prove that the fixation probability of the members of this class is at most $\frac{1}{2}(1 - \frac{1}{r})$, i.e. the half of the fixation probability of the complete graph, as $n \to \infty$. The clique-wheel graph G_n consists of a clique of size $n \geq 3$ and an induced cycle of the same size n with a perfect matching between them. We refer to the vertices of the inner clique as *clique vertices* and to the vertices of the outer cycle as *ring vertices*. The proof of the main results of this section (cf. Lemma 1 and Theorem 3) is technically involved. However, due to space limitations, we omit here the proofs; for a full version see [14].

Denote by h_{clique} (resp. h_{ring}) the probability that all the vertices of G_n become black, given that we start with one black clique vertex (resp. with one black ring vertex). We first provide in the next lemma an upper bound on h_{clique}.

Lemma 1. *For any* $r \in \left(1, \frac{4}{3}\right)$, $h_{clique} \leq \frac{7}{6n\left(\frac{4}{3r} - 1\right)} + o\left(\frac{1}{n}\right)$.

In the next theorem we provide also an upper bound on h_{ring}, thus bounding the fixation probability f_{G_n} of G_n (cf. Theorem 3).

Theorem 3. *For any* $r \in \left(1, \frac{4}{3}\right)$, $h_{ring} \leq (1 + o(1))\left(1 - \frac{1}{r}\right)$. *Therefore, by Lemma 1, the fixation probability of the clique-wheel graph* G_n *is* $f_{G_n} \leq \frac{1}{2}\left(1 - \frac{1}{r}\right) + o(1)$ *as* $n \to \infty$.

4 Analysis of the Aggregation Model

In this section, we provide analytic results on the new evolutionary model of mutual influences. More specifically, in Section 4.1 we prove that this model admits a *potential function* for arbitrary undirected graphs and arbitrary initial fitness vector, which implies that the corresponding dynamic system converges to a stable state. Furthermore, in Section 4.2 we prove fast convergence of the dynamic system for the case of a complete graph, as well as we provide almost tight upper and lower bounds on the limit fitness, to which the system converges.

4.1   Potential and Convergence in General Undirected Graphs

In the following theorem we prove convergence of the new model of mutual influences using a potential function.

Theorem 4. *Let $G = (V, E)$ be a connected undirected graph. Let $r(0)$ be an initial fitness vector of G, and let $r_{\min}$ and $r_{\max}$ be the smallest and the greatest initial fitness in $r(0)$, respectively. Then, in the model of mutual influences, the fitness vector $r(k)$ converges to a vector $[r_0, r_0, \ldots, r_0]^T$ as $k \to \infty$, for some value $r_0 \in [r_{\min}, r_{\max}]$.*

4.2   Analysis of the Complete Graph

The next theorem provides an almost tight analysis for the limit fitness value r_0 and the convergence time to this value, in the case of a complete graph (i.e. a homogeneous population).

Theorem 5. *Let $G = (V, E)$ be the complete graph with n vertices and $\varepsilon > 0$. Let $\alpha \in [0, 1]$ be the portion of initially introduced mutants with relative fitness $r \geq 1$ in G, and let r_0 be the limit fitness of G. Then $|r_u(k) - r_v(k)| < \varepsilon$ for every $u, v \in V$, when $k \geq (n - 2) \cdot \ln(\frac{r-1}{\varepsilon})$. Furthermore, for the limit fitness r_0,*

$$1 + \alpha(r - 1) \ \leq \ r_0 \ \leq \ 1 + \alpha(r - 1) + \frac{\alpha(1 - \alpha)}{1 + \alpha(r - 1)} \cdot \frac{(r - 1)^2}{2} \tag{5}$$

Corollary 1. *Let $G = (V, E)$ be the complete graph with n vertices. Suppose that initially exactly one mutant with relative fitness $r \geq 1$ is placed in G and let r_0 be the limit fitness of G. Then $1 + \frac{r-1}{n} \leq r_0 \leq 1 + \frac{r^2-1}{2n}$.*

5   Invasion Control Mechanisms

As stated in the introduction of this paper, our new evolutionary model of mutual influences can be used to model control mechanisms over invading populations in networks. We demonstrate this by presenting two alternative scenarios in Sections 5.1 and 5.2. In both considered scenarios, we assume that αn individuals of relative fitness r (the rest being of fitness 1) are introduced in the complete graph with n vertices. Then, as the process evolves, we periodically choose (arbitrarily) a small fraction $\beta \in [0, 1]$ of individuals in the current population and we reduce their current fitnesses to a value that is considered to correspond to the healthy state of the system (without loss of generality, this value in our setting is 1). In the remainder of this section, we call these modified individuals as "stabilizers", as they help the population resist to the invasion of the mutants.

5.1 Control of Invasion in Phases

In the first scenario of controlling the invasion of advantageous mutants in networks, we insert stabilizers to the population in phases, as follows. In each phase $k \geq 1$, we let the process evolve until all fitnesses $\{r_v \mid v \in V\}$ become ε-relatively-close to their fixed point $r_0^{(k)}$ (i.e. until they ε-approximate $r_0^{(k)}$). That is, until $\frac{|r_v - r_0^{(k)}|}{r_0^{(k)}} < \varepsilon$ for every $v \in V$. Note by Theorem 4 that, at every phase, the fitness values always ε-approximate such a limit fitness $r_0^{(k)}$. After the end of each phase, we introduce βn stabilizers, where $\beta \in [0, 1]$. That is, we replace βn vertices (arbitrarily chosen) by individuals of fitness 1, i.e. by resident individuals. Clearly, the more the number of phases, the closer the fixed point at the end of each phase will be to 1. In the following theorem we bound the number of phases needed until the system stabilizes, i.e. until the fitness of *every* vertex becomes sufficiently close to 1.

Theorem 6. *Let $G = (V, E)$ be the complete graph with n vertices. Let $\alpha \in [0, 1]$ be the portion of initially introduced mutants with relative fitness $r \geq 1$ in G and let $\beta \in [0, 1]$ be the portion of the stabilizers introduced at every phase. Let $r_0^{(k)}$ be the limit fitness after phase k and let $\varepsilon, \delta > 0$, be such that $\frac{\beta}{2} > \sqrt{\varepsilon}$ and $\delta > \frac{4}{3}\sqrt{\varepsilon}$. Finally, let each phase k run until the fitnesses ε-approximate their fixed point $r_0^{(k)}$. Then, after $k \geq 1 + \ln\left(\frac{\varepsilon + (1+\varepsilon)\frac{1+\alpha}{2}(r-1)}{\delta - \frac{4}{3}\sqrt{\varepsilon}}\right) / \ln\left(\frac{1}{(1+\varepsilon)(1-\frac{\beta}{2})}\right)$ phases, the relative fitness of every vertex $u \in V$ is at most $1 + \delta$.*

5.2 Continuous Control of Invasion

In this section we present another variation of controlling the invasion of advantageous mutants, using our new evolutionary model. In this variation, we do not proceed in phases; we rather introduce *at every single iteration* of the process βn stabilizers, where $\beta \in [0, 1]$ is a small portion of the individuals of the population. For simplicity of the presentation, we assume that at every iteration the βn stabilizers with relative fitness 1 are the same.

Theorem 7. *Let $G = (V, E)$ be the complete graph with n vertices. Let $\alpha \in [0, 1]$ be the portion of initially introduced mutants with relative fitness $r \geq 1$ in G and let $\beta \in [0, 1]$ be the portion of the stabilizers introduced at every iteration. Then, for every $\delta > 0$, after $k \geq \frac{r}{\beta}(n-1) \cdot \ln(\frac{r-1}{\delta})$ iterations, the relative fitness of every vertex $u \in V$ is at most $1 + \delta$.*

Observation 1. *The bound in Theorem 7 of the number of iterations needed to achieve everywhere a sufficiently small relative fitness is independent of the portion $\alpha \in [0, 1]$ of initially placed mutants in the graph. Instead, it depends only on the initial relative fitness r of the mutants and on the portion $\beta \in [0, 1]$ of the vertices, to which we introduce the stabilizers.*

Acknowledgment. Paul G. Spirakis wishes to thank Josep Diaz, Leslie Ann Goldberg, and Maria Serna, for many inspiring discussions on the model of [13].

References

1. Antal, T., Scheuring, I.: Fixation of strategies for an evolutionary game in finite populations. Bulletin of Mathematical Biology 68, 1923–1944 (2006)
2. Berger, E.: Dynamic monopolies of constant size. Journal of Combinatorial Theory, Series B 83, 191–200 (2001)
3. Broom, M., Hadjichrysanthou, C., Rychtar, J.: Evolutionary games on graphs and the speed of the evolutionary process. Proceedings of the Royal Society A 466, 1327–1346 (2010)
4. Broom, M., Hadjichrysanthou, C., Rychtar, J.: Two results on evolutionary processes on general non-directed graphs. Proceedings of the Royal Society A 466, 2795–2798 (2010)
5. Broom, M., Rychtar, J.: An analysis of the fixation probability of a mutant on special classes of non-directed graphs. Proceedings of the Royal Society A 464, 2609–2627 (2008)
6. Broom, M., Rychtar, J., Stadler, B.: Evolutionary dynamics on small order graphs. Journal of Interdisciplinary Mathematics 12, 129–140 (2009)
7. Christine Taylor, A.S., Fudenberg, D., Nowak, M.A.: Evolutionary game dynamics in finite populations. Bulletin of Mathematical Biology 66(6), 1621–1644 (2004)
8. Gintis, H.: Game theory evolving: A problem-centered introduction to modeling strategic interaction. Princeton University Press (2000)
9. Hofbauer, J., Sigmund, K.: Evolutionary games and population dynamics. Cambridge University Press (1998)
10. Imhof, L.A.: The long-run behavior of the stochastic replicator dynamics. Annals of applied probability 15(1B), 1019–1045 (2005)
11. Kandori, M., Mailath, G.J., Rob, R.: Learning, mutation, and long run equilibria in games. Econometrica 61(1), 29–56 (1993)
12. Kempe, D., Kleinberg, J.M., Tardos, É.: Influential Nodes in a Diffusion Model for Social Networks. In: Caires, L., Italiano, G.F., Monteiro, L., Palamidessi, C., Yung, M. (eds.) ICALP 2005. LNCS, vol. 3580, pp. 1127–1138. Springer, Heidelberg (2005)
13. Lieberman, E., Hauert, C., Nowak, M.A.: Evolutionary dynamics on graphs. Nature 433, 312–316 (2005)
14. Mertzios, G., Nikoletseas, S., Raptopoulos, C., Spirakis, P.: Natural models for evolution on graphs. Technical report available at arXiv:1102.3426 (2011)
15. Moran, P.: Random processes in genetics. Proceedings of the Cambridge Philosophical Society 54, 60–71 (1958)
16. Mossel, E., Roch, S.: On the submodularity of influence in social networks. In: Proceedings of the 39th annual ACM Symposium on Theory of Computing (STOC), pp. 128–134 (2007)
17. Norris, J.R.: Markov Chains. Cambridge University Press (1999)
18. Nowak, M.A.: Evolutionary Dynamics: Exploring the Equations of Life. Harvard University Press (2006)
19. Ohtsuki, H., Nowak, M.A.: Evolutionary games on cycles. Proceedings of the Royal Society B: Biological Sciences 273, 2249–2256 (2006)
20. Sandholm, W.H.: Population games and evolutionary dynamics. MIT Press (2011)
21. Taylor, C., Iwasa, Y., Nowak, M.A.: A symmetry of fixation times in evoultionary dynamics. Journal of Theoretical Biology 243(2), 245–251 (2006)
22. Traulsen, A., Hauert, C.: Stochastic evolutionary game dynamics. In: Reviews of Nonlinear Dynamics and Complexity, vol. 2. Wiley, NY (2008)
23. Weibull, J.W.: Evolutionary game theory. MIT Press (1995)

Approximate Judgement Aggregation*

Ilan Nehama

Center for the Study of Rationality
& The Selim and Rachel Benin School of Computer Science and Engineering
The Hebrew University of Jerusalem, Israel
`ilan.nehama@mail.huji.ac.il`

Abstract. In this paper we analyze judgement aggregation problems in which a group of agents independently votes on a set of complex propositions that has some interdependency constraint between them (e.g., transitivity when describing preferences). We consider the issue of judgement aggregation from the perspective of approximation. That is, we generalize the previous results by studying approximate judgement aggregation. We relax the main two constraints assumed in the current literature, Consistency and Independence and consider mechanisms that only approximately satisfy these constraints, that is, satisfy them up to a small portion of the inputs. The main question we raise is whether the relaxation of these notions significantly alters the class of satisfying aggregation mechanisms. The recent works for preference aggregation of Kalai, Mossel, and Keller fit into this framework. The main result of this paper is that, as in the case of preference aggregation, in the case of a subclass of a natural class of aggregation problems termed 'truth-functional agendas', the set of satisfying aggregation mechanisms does not extend non-trivially when relaxing the constraints. Our proof techniques involve boolean Fourier transform and analysis of voter influences for voting protocols.

The question we raise for Approximate Aggregation can be stated in terms of Property Testing. For instance, as a corollary from our result we get a generalization of the classic result for property testing of linearity of boolean functions.

Keywords: approximate aggregation, discursive dilemma, doctrinal paradox, truth-functional agendas, inconsistency index, dependency index, computational social choice.

* The research was supported by a grant from the Israeli Science Foundation (ISF) and by the Google Inter-university center for Electronic Markets and Auctions.

Previous versions of this work were presented at Bertinoro Workshop on Frontiers in Mechanism Design 2010, Third International Workshop on Computational Social Choice, Düsseldorf 2010, and Computation and Economics Seminar at the Hebrew University. The author would like to thank the participants in these workshops for their comments.

Due to space constraint the proofs of all theorems are omitted as well as some discussion on the implications of this work to other fields. The long version of this paper can be found on the author website[33].

1 Introduction

A famous jury paradox shows that aggregating complex decisions might be non-trivial. Assume a jury is faced with a case in which a defendant is accused of murder. The legal doctrine (known by all of them) is that the defendant should be convicted if and only if they are convinced that **a)**The defendant indeed killed the victim and **b)**The defendant is sane. We assume that each of the jurors decides his opinion on the two issues independently and based on this decides whether to convict. Then, the members cast their votes simultaneously and we assume no strategic behavior on their behalf. Kornhauser and Sager[19] noticed that it's possible to have an opinion profile in which, when applying issue-wise aggregation using majority, which seems natural, we get a discrepancy between the majority vote on the conviction question and the conjunction of the majority vote on the two basic questions(whether the defendant killed and whether he is sane)[1]. This discrepancy is termed *The Doctrinal Paradox*. Lately, in [21], List showed that the probability to get such a discrepancy is non-negligible under the uniform distribution and also under other mild relaxations of it (still assuming the voters are i.i.d.).

This insight, that is common to many aggregation problems (e.g., Condorcet paradox for preference aggregation), started the field of 'Judgement Aggregation' and nowadays this field is the subject of a growing body of works in economics, computer science, political science, philosophy, law, and other related disciplines. We find this field highly applicable to agent systems, voting protocols in a network and other frameworks in which one needs to aggregate a lot of opinions in a systematic way without letting the voters deliberate. An aggregation problem in our context concerns a given **Agenda**, which is a set of $\{0,1\}$ vectors of length m (the number of issues), that defines the **consistent** (legal/rational/admissible) opinions that an individual might hold. Given an agenda, Aggregation Theory deals with exploring ways to aggregate opinions of (often many) experts/judges while maintaining two main syntactical properties:

- **Consistency** - always returning an admissible opinion.
  In our example, the aggregated opinion should be to convict iff the aggregated opinion was that indeed the defendant killed and is sane.
- **Independence** - define the aggregated opinion on each issue independently of the votes on other issues.
  This criterion can be seen as respecting the structure of the agenda instead of handling it as a set of several different opinions (in the example above, four) disregarding the structure.

---

[1] For instance, the following profile:

	Killed	Sane	Guilty
25% of the jurors:	✓	✓	✓
33% of the jurors:	✓	✗	✗
42% of the jurors:	✗	✓	✗

Most of these works(e.g., [28,9]) find the set of 'acceptable' aggregation mechanisms (i.e., that satisfy the two criteria) to be very small and undesired (e.g., dictatorships) and hence are considered as impossibility results. A survey of this field can be found in [24,22]. Such impossibility results are quite strong, they show the impossibility of finding any reasonable aggregation mechanism that satisfies the two conditions and hence for (almost) every mechanism there will always be some judgement profile that leads to a breakdown of the mechanism.

In this work we extend the question to 'Approximate Judgement Aggregation'. We relax the above two properties and search for an aggregation mechanism that only *approximately* respects the structure of opinions and *up to a small fraction* of the inputs returns a consistent opinion. More specifically, we are interested in exploring the influence of relaxing the two properties on the set of 'acceptable' aggregation mechanisms.

We quantify being almost consistent by defining δ-**consistency** of an aggregation mechanism F as having a consistent aggregation mechanism G that disagrees with F on at most δ fraction of the inputs[2]. Similarly, we quantify being almost independent by defining δ-**independence** of an aggregation mechanism F as having an independent aggregation mechanism that disagrees with F on at most δ fraction of the inputs. Both terms can be equivalently defined as the failure probability of tests as we show in Section 2. Both definitions use the Hamming distance between mechanisms $d^{\mathbb{X}}(F, G) = \Pr\left[F(X) \neq G(X) \mid X \in \mathbb{X}^n\right]$. It includes two assumptions: uniform distribution over the opinions for each voter and assuming voters draw their opinions independently (**Impartial Culture Assumption**). These assumptions, while certainly unrealistic, are the natural choice in this kind of work and are discussed further in Section 2.

Lately there is a series of works coping with impossibility results in Social Choice Theory using approximations (e.g., [5,14]). In some cases allowing approximation enables significantly better results, while in other cases, hardly anything is gained by allowing it. For example, in [5] the authors deal with preference aggregation and show that when one approximates Dodgson's scoring rule one can achieve several desired properties (monotonicity, homogeneity, and low complexity) that cannot be achieved without this relaxation. On the other hand, in [14] the authors also deal with aggregation of preferences and show that relaxing the strategy-proofness property does not extend the set of satisfying aggregation mechanisms non-trivially and by that they strengthen the classic impossibility result of Gibbard & Satterthwaite. In this work we formalize (as far as we found for the first time) this question of quantifying the influence of relaxing the constraints and query whether one can use this in order to circumvent the impossibility results (as in [5]) or whether we strengthen the impossibility results (as in [14]).

In this paper we study a family of agendas: truth-functional agendas in which each conclusion is defined as conjunction or xor of several premises (up to input & output negation). In a truth-functional agenda the issues are divided into two types: premises and conclusions. Each conclusion j is characterized by a boolean function Φ^j over the premises and an opinion is consistent if the answers to the

---

[2] Formally, $\Pr\left[F(X) \neq G(X) \mid X \in \mathbb{X}^n\right] \leqslant \delta$.

conclusion issues are attained by applying the function Φ^j on the answers to the premise issues.

$$\mathbb{X} = \left\{ x \in \{0,1\}^m \big| x^j = \Phi^j(\text{premises}) \quad \text{for every conclusion issue } j. \right\}$$

For instance the (2-premises) conjunction agenda used in the example above is a truth-functional agenda with two premises and one conclusion and we notate the agenda by $\langle A, B, A \wedge B \rangle$.

For all the agendas we examined, we show that relaxing the two constraints, consistency and independence, does not extend the set of acceptable aggregation mechanisms in a non-trivial way.

We concentrated on two basic agendas: **Conjunction Agenda** $\langle A^1, \ldots, A^m, \wedge_{j=1}^m A^j \rangle$ (i.e., $m+1$ issues where the consistency means that the last one should be a conjunction of the first m) and **Xor Agenda** $\langle A^1, \ldots, A^m, \oplus_{j=1}^m A^j \rangle$ (i.e., $m+1$ issues where the consistency means that the last one should be a parity bit of the first m). For these agendas we prove.

Theorem

1. For any $m \geqslant 2$, $\epsilon > 0$, and $n \geqslant 2$, there exists $\delta(\epsilon, n, m)$ polynomial in n and ϵ (but degrades exponentially in m) s.t. if an aggregation mechanism F over n voters for the m-premises conjunction agenda is δ-independent[3] and δ-consistent[4], then it is ϵ-close to a consistent independent aggregation mechanism $G^{[5]}$.
   Moreover, $\delta = \frac{C}{n} \left(\frac{\epsilon}{8m} \right)^{2m-1}$ (for some constant $C > 0$),
2. For any $m \geqslant 2$, $\epsilon > 0$, and $n \geqslant 2$, there exists $\delta(\epsilon, m)$ linear in ϵ (and degrades quadratically in m) s.t. if an aggregation mechanism F over n voters for the m-premises xor agenda is δ-independent[3] and δ-consistent[4], then it is ϵ-close to a consistent independent aggregation mechanism $G^{[5]}$.
   Moreover, $\delta = \frac{\epsilon}{m(2m+3)}$

Hence, the above theorem can be seen as an impossibility result saying that it is impossible even to find a mechanism that is almost consistent and almost independent besides the trivial answers: independent consistent mechanism and perturbations of them which is (still) a relatively small and undesired collection of mechanisms.

Our results are invariant to negation of issues (which is merely renaming), and hence we can easily generalize the results to other agendas such as $\langle A^1, A^2, A^3, A^1 \wedge A^2 \wedge \overline{A^3} \rangle$, $\langle A^1, A^2, A^1 \vee A^2 \rangle$, and $\langle A^1, A^2, A^3, \overline{A^1 \oplus A^2 \oplus A^3} \rangle$. Using induction we can generalize the result

---

[3] I.e., there exists an independent (not necessarily consistent) aggregation mechanism G that returns the same aggregated opinion as F for at least $(1 - \delta)$ fraction of the profiles.

[4] I.e., F returns a consistent result for at least $(1 - \delta)$ fraction of the profiles.

[5] I.e., F returns the same aggregated opinion as G for at least $(1 - \epsilon)$ fraction of the profiles.

to more complex agendas that include several conclusion issues such as $\langle A^1, A^2, A^3,\ A^1 \vee A^2, A^2 \oplus A^3 \rangle$. The general formulation of the theorem can be in the long version of this paper[33]. We notice that this generalize our result to any agenda of the form $\langle A^1, A^2, \Phi\left(A^1, A^2\right) \rangle$ for **any function** Φ[6] and to **any affine agenda** (I.e., the set of admissible opinions form an affine space).

1.1   Previous Works

There is a long line of works trying to circumvent impossibility results in Aggregation Theory (i.e., results which state that the set of consistent independent aggregation mechanisms is very small and undesired). Most of these works suggest consistent aggregating mechanisms while still trying to stay 'reasonably close' to independence (E.g., [19,18,29,23,7,4,20,8,30]). These classical works are heuristic, sometimes uses the semantics of the agenda, and mainly do not prove bounds on the compliance to the independence property. In [21], List studies the asymptotic probability of getting an inconsistent result in the 2-premises conjunction agenda $\langle A, B, A \wedge B \rangle$ for voter-independent distributions and common (majority-based & supermajority-based) aggregation mechanisms. He mainly studies the conditions for the probability to converge to zero and to one. As far as we found, this is the only work that deals with quantifying, although only asymptotically, the property compliance of an aggregation mechanism for agendas other than the Arovian agenda (preference aggregation).

Another approach is Approximate Aggregation. This line of research started with [15] and was extended in [26,16]. In these works the authors deal with preference aggregation (although without stating the general framework of approximate aggregation) and show that relaxing the transitivity constraint (which is equivalent to consistency for this agenda) does not extend the set of satisfying aggregation mechanisms non-trivially.

Theorem ([16] Theorem 1.3). *There exists an absolute constant C such that the following holds: For any $\epsilon > 0$ and $k \geqslant 3$, if f is an aggregation mechanism for the preference agenda over k candidates that satisfies independence and $C \cdot \left(\epsilon/k^2\right)^3$-consistency, then there exists an aggregation mechanism G that satisfies independence and consistency such that $d(F, G) < \epsilon$.*

This result is neither derived by our results nor derives them because the agendas we deal with and the preference agenda are too different (For instance, the preference agenda cannot be represented as a truth-functional agenda and in some sense it is even far from it).

1.2   Connection to Property Testing

We think it might be useful to phrase the question of approximate aggregation using terminology of property testing. In this field we query a function at a

---

[6] The case of a function that ignores one of the two arguments (or both) is trivial.

small number of (random) points, testing for a global property (in our case, the property is being a consistent independent aggregation mechanism). For example, a corollary of the results we present in this paper (in property testing terms):

For any three binary functions $f, g, h : \{0,1\}^n \to \{0,1\}$, if the probability $\Pr[f(x) \oplus g(y) = h(x \oplus y)]$ is larger than $(1-\epsilon)$ (when the addition is in $\mathbb{Z}_2$ and $\mathbb{Z}_2^n$, respectively), then there exists three binary functions $f', g', h' : \{0,1\}^n \to \{0,1\}$ such that $\Pr[f(x) \neq f'(x)]$, $\Pr[g(x) \neq g'(x)]$, and $\Pr[h(x) \neq h'(x)]$ are smaller than $C\epsilon$ for some constant C independent of n and $\forall x, y : f'(x) \oplus g'(y) = h'(x \oplus y)$.

A special case of this result, $f = g = h$, is the classic result of Blum, Luby, and Rubinfeld ([3,1]) for linear testing of boolean functions. We discuss this connection further and its possible implications in in the long version of this paper[33].

1.3   Techniques

We prove the main theorem by proving the specific case of independent aggregation mechanism for two basic agenda families: the conjunction agendas and the xor agendas. Later we extend these theorems to the general theorem of relaxing both constraints in a agenda-independent way.

We use two different techniques in the proofs. For the conjunction agendas we study influence measures of voters on the issue-aggregating functions[7]. and for the xor agendas we use Fourier analysis of the issue-aggregating functions.

An open question is whether one can find such bounds for any agenda or whether there exists an agenda for which the class of aggregation mechanisms that satisfy consistency and independence expands non trivially when we relax the consistency and independence constraints.

We proceed to describe the structure of the paper. In Section 2 we describe the formal model of aggregation mechanisms. In Section 3 we present the main agendas we deal with, truth-functional agendas, and specifically conjunction agendas and xor agendas. In Section 4 we state the motivation to deal with approximate aggregation. Section 5 concludes.

2   The Model

We define the model similarly to [9,10] (which is Rubinstein and Fishburn's model [32] for the boolean case).

We consider a **committee** of n individuals that needs to decide on m boolean issues[8]. An **opinion** is a vector $x = (x_1, x_2, \ldots, x_m) \in \{0,1\}^m$ denoting an

---

[7] Both the known influence (Banzhaf power index) and a new measure we define: The ignorability of an individual and of a coalition of individuals.

[8] There is some literature on aggregating non-boolean issues, e.g., [32,11], but this is outside the scope of this paper.

answer to each of the issues. An opinion **profile** is a matrix $X \in (\{0,1\}^m)^n$ denoting the opinions of the committee members, so an entry X_i^j denotes the vote of the ith voter for the jth issue. In addition we assume that an **agenda** $\mathbb{X} \subseteq \{0,1\}^m$ of the **consistent** opinions is given.

The basic notion in this field is **Aggregation Mechanism** which is a function that returns an **aggregated opinion** (not necessarily consistent) for every profile[9] : $F : (\{0,1\}^m)^n \to \{0,1\}^m$.

An aggregation mechanism satisfies **Independence** (and we say that the mechanism is **independent**) if for any two consistent profiles X and Y and an issue j, if $X^j = Y^j$ (all individuals voted the same on the jth issue in both profiles) then $(F(X))^j = (F(Y))^j$ (the aggregated opinion for the jth issue is the same for both profiles). This means that F satisfies independence if one can find m boolean functions $f^1, f^2, \ldots, f^m : \{0,1\}^n \to \{0,1\}$ s.t. $F(X) \equiv (f^1(X^1), f^2(X^2), \ldots, f^m(X^m))$[10]. An independent aggregation mechanism satisfies **systematicity** if all issues are aggregated using the same function, i.e., $F(X) = \langle f(X^1), \ldots, f(X^m) \rangle$ for some issue aggregating function f. We will use the notation $\langle f^1, f^2, \ldots, f^m \rangle$ for the independent aggregation mechanism that aggregates the jth issue using f^j.

The main two measures we study in this paper are the **inconsistency index** $IC^{\mathbb{X}}(F)$ and the **dependency index** $DI^{\mathbb{X}}(F)$ of a given aggregation mechanism F and a given agenda $\mathbb{X}$. These measures are relaxations of the **consistency** and **independence** criterion that are usually assumed in current works[11]. We define the measures in the following way:

Definition 1 (Inconsistency Index)
For an agenda $\mathbb{X}$ and an aggregation mechanism F for that agenda, the **inconsistency index** *is defined to be the probability to get an inconsistent result.*[12]

$$IC^{\mathbb{X}}(F) = \Pr\left[F(X) \notin \mathbb{X} \mid X \in \mathbb{X}^n\right].$$

Definition 2 (Dependency Index[13])
For an agenda $\mathbb{X}$ and an aggregation mechanism F for that agenda, the **dependency vector** $DI^{j,\mathbb{X}}(F)$ *is defined as*

$$DI^{j,\mathbb{X}}(F) = \mathop{\mathbb{E}}_{X \in \mathbb{X}^n}\left[\mathop{\Pr}_{Y \in \mathbb{X}^n}\left[(F(X))^j \neq (F(Y))^j \mid X^j = Y^j\right]\right].$$

---

[9] We define the function for all profiles for simplicity but we are not interested in the aggregated opinion in cases one of the voters voted an inconsistent opinion.

[10] Notice this property is a generalization of the IIA property for social welfare functions (aggregation mechanism for the preference agenda) so a social welfare function satisfies IIA iff it satisfies independence as defined here (when the issues are the pair-wise comparisons).

[11] F satisfies consistency iff $IC(F) = 0$ and independence iff $DI(F) = 0$.

[12] In [21] List presented this measure under the name 'Probability of a collective inconsistency' and studies its asymptotical behavior for the conjunction agenda and the issue-wise majority aggregation mechanism.

[13] In [26] Mossel defines similar measure for preference aggregation mechanism called η-IIA. Notice that our definition coincides with his definition for this agenda.

The **dependency index** $DI^{\mathbb{X}}(F)$ *is defined by:* $DI^{\mathbb{X}}(F) = \max\limits_{j=1,\ldots,m} DI^{j,\mathbb{X}}(F)$

In contexts where the agenda is clear we omit the agenda superscript and notate these as $IC(F)$, $DI^j(F)$, and $DI(F)$, respectively.

We define these two indices using local tests and prove that the more natural definition of distance to the class of aggregation mechanisms that satisfy consistency (or independence) is equivalent to the above (up to multiplication by a constant). These definitions include two major assumptions on the opinion profile distribution. First, we assume the voters pick their opinions independently and from the same distribution. Second, we assume a uniform distribution over the (consistent) opinions for each voter (**Impartial Culture Assumption**). The uniform distribution assumption, while certainly unrealistic, is the natural choice for proving 'lower bounds' on $IC(F)$. That is, proving results of the format 'Every aggregation mechanism of a given class has inconsistency index of at least $\gamma(n)$'. In particular, the lower bound, up to a factor δ, applies also to any distribution that gives each preference profile at least a δ fraction of the probability given by the uniform distribution[14]. Note that we cannot hope to get a reasonable bound result for every distribution. For instance, since for every aggregation mechanism we can take a distribution on profiles for which it returns a consistent opinion.

2.1 Binary Functions

Throughout this paper we will identify `True` with 1 and `False` with 0 and use logical operators on bits and bit vectors (using entry-wise semantics).

We define the following measures for the influence of an individual or a coalition of individuals on a function $f : \{0,1\}^n \to \{0,1\}$. Both definitions use the uniform distribution over $\{0,1\}^n$ (which is consistent with the assumption we have on the profile distribution).

- The **Influence**[15] of a voter i on f is defined to be the probability that he can change the result by changing his vote.

$$I_i(f) = \Pr\left[f(x) \neq f(x \oplus e_i)\right]$$

($x \oplus e_i =$ adding to x, e_i(the $i^{\underline{th}}$ elementary vector)=flipping the $i^{\underline{th}}$ bit $0 \leftrightarrow 1$)
- The (zero-)**Ignorability** of a coalition $S \subseteq \{1,\ldots,n\}$ is is defined to be the probability that f returns 1 when one of the members of S voted 0.

$$P_S(f) = \Pr\left[f(x) = 1 \mid \exists i \in S\ x_i = 0\right]$$

---

[14] In successive works we relax this assumption and prove similar results for more general distributions.

[15] In the simple cooperative games regime, this is also called the Banzhaf power index of player i in the game f.

In addition we define a distance function over the binary functions. The distance between two functions $f, g : \{0,1\}^n \to \{0,1\}$ is defined to be the probability of getting a different result (normalized Hamming distance). $d(f,g) = \Pr[f(x) \neq g(x)]$. From this measure we will derive a distance from a function to a set of functions by $d(f, \mathcal{G}) = \min_{g \in \mathcal{G}} d(f,g)$.

3   Agenda Examples

A lot of natural problems can be formulated in the framework of aggregation mechanisms. It is natural to divide the agendas into two major classes **Truth-Functional Agendas** and **Non Truth-Functional Agendas**.

3.1   Truth-Functional Agendas

A (k-premise) truth-functional agenda is defined by a conclusions function ($\Phi : \{0,1\}^k \to \{0,1\}^{m-k}$) from the k premises to the $(m-k)$ conclusions. An opinion is consistent if the answers to the conclusion issues are attained by applying Φ on the answers to the premise issues.

$$\mathbb{X} = \left\{ x \in \{0,1\}^m \middle| x^j = \Phi^j(x_1, \ldots, x_k) \quad j = k+1, \ldots, m \right\}$$

These agendas, due to their structure, seem to be a good point to start our work on approximate aggregation and in this paper we prove results for two families of truth-functional agendas. Later we derive results for a more general family of truth-functional agendas.

Conjunction Agendas: In the m-premises conjunction agenda $\langle A^1, \ldots, A^m, \wedge_{j=1}^m A^j \rangle$ there are $m+1$ issues to decide on and the consistency criterion is defined to be that the last issue is a conjunction of the other issues. For instance the Doctrinal Paradox agenda is the 2-premises conjunction agenda.

Xor Agendas: Similarly, in the m-premises xor agenda $\langle A^1, \ldots, A^m, \oplus_{j=1}^m A^j \rangle$ there are $m+1$ issues to decide on and the consistency criterion is defined to be that the last issue is `True` if the number of true-valued opinions for the first m is even. An equivalent way to define this agenda is constraining the number of `True` answers to be odd.

3.2   Non Truth-Functional Agendas

One can think on a lot of agendas that cannot be represented as a truth-functional agenda. Among such interesting natural agendas that were studied one can find the equivalence agenda[13], the membership agenda [31][25], and the preference agenda described below.

Preference Aggregation: Aggregation of preferences is one of the oldest aggregation frameworks studied. In this framework there are s candidates and each individual holds a full strict order over them. We are interested in Social Welfare Functions which are functions that aggregate n such orders to an aggregated order. As seen in [27,6], this problem can be stated naturally in the aggregation framework we defined by defining $\binom{s}{2}$ issues[16].

4 Motivation

We find the motivation for dealing with the field of approximate judgement aggregation in three different disciplines.

- The consistent characterization are often regarded as 'impossibility results' in the sense that they 'permit' a very restrictive set of aggregation mechanisms. (e.g., Arrow's theorem tells us that there is no 'reasonable' way to aggregate preferences). Extending these theorems to approximate aggregation characterizations sheds light on these impossibility results by relaxing the constraints.
- The questions of Aggregation Theory have often roots in Philosophy, Law, and Political Science. There is a long line of works suggesting consistent aggregating mechanisms while still trying to stay 'reasonably close' to independence. The main general (not agenda-tailored) suggestions are premise-based mechanisms and conclusion-based aggregation for truth-functional agendas (see, among others, [19,18,29,23,7,4]), and a generalization of them to non-truth-functional agendas called sequential priority aggregation([20,8]). Another procedure in the literature is the distance-based aggregation([30]) which is well known for preference aggregation (E.g., Kemeny voting rule[17], Dodgson voting rule[2], and lately a more systematic analysis in [12]). Our work contribute to this discussion by pointing out where one should search for solutions while not leaving the consistency and independence constraints entirely.
- Connections to the Property Testing field. Due to the space constraint it is discussed in the in the long version of this paper[33].

5 Summary and Future Work

In this paper we defined the question of approximate aggregation which is a generalization of the study of aggregation mechanisms that satisfy consistency and independence. We defined measures for the relaxation of the consistency constraint (inconsistency index IC) and for the relaxation of the independence constraint (dependency index DI). To our knowledge, this is the first time this question is stated in its general form.

We proved that relaxing these constraints does not extend the set of satisfying aggregation mechanisms in a non-trivial way for any truth-functional agenda in which every conclusion is either conjunction or xor up to negation of inputs or output. We notice that every conclusion of two premises can be stated as such as

---

[16] The issue $\langle i, j \rangle$ (for $i < j$) represents whether an individual prefers c_i over c_j.

well as any affine agenda. Particulary we calculated the dependency between the extension of this class (ϵ) and the inconsistency index ($\delta(\epsilon)$) (although probably not strictly) for two families of truth-functional agendas with one conclusion. The relation we proved includes dependency on the number of voters (n). In similar works for preference agendas [15,26,16] the relation did not include such a dependency. An interesting question is whether such a dependency is inherent for conjunction agendas or whether it is possible to prove a relation that does not include it.

A major assumption in this paper is the uniform distribution over the inputs which is equivalent to assuming i.i.d uniform distribution over the premises. We think that our results can be extended for other distributions (still assuming voters' opinions are distributed i.i.d) over the space over premises' opinions which seem more realistic.

Immediate extensions for this work can be to extend our result to more complex truth-functional agendas and generalize our results to non-truth-functional agendas to get a result unifying our work and Kalai, Mossel, and Keller's works for the preference agenda.

A major open question is whether one can find an agenda for which relaxing the constraints of independence and consistency extends the class of satisfying aggregation mechanisms in a non-trivial way.

References

1. Bellare, M., Coppersmith, D., Hastad, J., Kiwi, M., Sudan, M.: Linearity testing in characteristic two. In: FOCS 1995: Proceedings of the 36th Annual Symposium on Foundations of Computer Science, p. 432. IEEE Computer Society Press, Washington, DC, USA (1995)
2. Black, D.: The theory of committees and elections. Kluwer Academic Publishers (1957) (reprint at 1986)
3. Blum, M., Luby, M., Rubinfeld, R.: Self-testing/correcting with applications to numerical problems. Journal of Computer and System Sciences 47(3), 549–595 (1993)
4. Bovens, L., Rabinowicz, W.: Democratic answers to complex questions an epistemic perspective. Synthese 150, 131–153 (2006)
5. Caragiannis, I., Kaklamanis, C., Karanikolas, N., Procaccia, A.D.: Socially desirable approximations for dodgson's voting rule. In: Proc. 11th ACM Conference on Electronic Commerce (2010)
6. Dietrich, F., List, C.: Arrows theorem in judgment aggregation. Social Choice and Welfare 29(1), 19–33 (2007)
7. Dietrich, F.: Judgment aggregation (im)possibility theorems. Journal of Economic Theory 126(1), 286–298 (2006)
8. Dietrich, F., List, C.: Judgment aggregation by quota rules. Journal of Theoretical Politics 19(4), 391–424 (2007)
9. Dokow, E., Holzman, R.: Aggregation of binary evaluations for truth-functional agendas. Social Choice and Welfare 32(2), 221–241 (2009)
10. Dokow, E., Holzman, R.: Aggregation of binary evaluations. Journal of Economic Theory 145(2), 495–511 (2010)
11. Dokow, E., Holzman, R.: Aggregation of non-binary evaluations. Advances in Applied Mathematics 45(4), 487–504 (2010)

12. Elkind, E., Faliszewski, P., Slinko, A.: Distance rationalization of voting rules. In: The Third International Workshop on Computational Social Choice, COMSOC 2010 (2010)
13. Fishburn, P., Rubinstein, A.: Aggregation of equivalence relations. Journal of Classification 3(1), 61–65 (1986)
14. Friedgut, E., Kalai, G., Nisan, N.: Elections can be manipulated often. In: FOCS 2008: Proceedings of the 2008 49th Annual IEEE Symposium on Foundations of Computer Science, pp. 243–249. IEEE Computer Society Press, Washington, DC, USA (2008)
15. Kalai, G.: A fourier-theoretic perspective on the condorcet paradox and arrow's theorem. Adv. Appl. Math. 29(3), 412–426 (2002)
16. Keller, N.: A tight quantitative version of Arrow's impossibility theorem. Arxiv preprint arXiv:1003.3956 (2010)
17. Kemeny, J.G.: Mathematics without numbers. Daedalus 88(4), 577–591 (1959)
18. Kornhauser, L.A.: Modeling collegial courts. ii. legal doctrine. Journal of Law, Economics and Organization 8(3), 441–470 (1992)
19. Kornhauser, L.A., Sager, L.G.: Unpacking the court. The Yale Law Journal 96(1), 82–117 (1986)
20. List, C.: A model of path-dependence in decisions over multiple propositions. American Political Science Review 98(03), 495–513 (2004)
21. List, C.: The probability of inconsistencies in complex collective decisions. Social Choice and Welfare 24(1), 3–32 (2005)
22. List, C.: Judgment aggregation: a short introduction (August 2008), http://philsci-archive.pitt.edu/4319/
23. List, C., Pettit, P.: Aggregating sets of judgments: An impossibility result. Economics and Philosophy 18, 89–110 (2002)
24. List, C., Puppe, C.: Judgement aggregation: A survey. In: Anand, P., Puppe, P.P. (eds.) The Handbook of Rational and Social Choice. Oxford University Press, USA (2009)
25. Miller, A.D.: Group identification. Games and Economic Behavior 63(1), 188–202 (2008)
26. Mossel, E.: A quantitative arrow theorem. Probability Theory and Related Fields (forthcoming)
27. Nehring, K.: Arrows theorem as a corollary. Economics Letters 80(3), 379–382 (2003)
28. Nehring, K., Puppe, C.: Consistent judgement aggregation: The truth-functional case. Social Choice and Welfare 31(1), 41–57 (2008)
29. Pettit, P.: Deliberative democracy and the discursive dilemma. Philosophical Issues 11(1), 268–299 (2001)
30. Pigozzi, G.: Belief merging and the discursive dilemma: An argument-based account to paradoxes of judgment aggregation. Synthese 152(2), 285–298 (2006)
31. Rubinstein, A., Kasher, A.: On the question "Who is a J?": A social choice approach. Princeton Economic Theory Papers 00s5, Economics Department. Princeton University (1998)
32. Rubinstein, A., Fishburn, P.C.: Algebraic aggregation theory. Journal of Economic Theory 38(1), 63–77 (1986)
33. Nehama, I.: Approximate judgement aggregation. Discussion Paper Series - DP574R, Center for Rationality and Interactive Decision Theory, Hebrew University, Jerusalem (October 2011), http://ideas.repec.org/p/huj/dispap/dp574.html

Liquidity-Sensitive Automated Market Makers via Homogeneous Risk Measures

Abraham Othman and Tuomas Sandholm

Computer Science Department, Carnegie Mellon University
{aothman,sandholm}@cs.cmu.edu

Abstract. Automated market makers are algorithmic agents that provide liquidity in electronic markets. A recent stream of research in automated market making is the design of *liquidity-sensitive* automated market makers, which are able to adjust their price response to the level of active interest in the market. In this paper, we introduce homogeneous risk measures, the general class of liquidity-sensitive automated market makers, and show that members of this class are (necessarily and sufficiently) the convex conjugates of compact convex sets in the non-negative orthant. We discuss the relation between features of this convex conjugate set and features of the corresponding automated market maker in detail, and prove that it is the curvature of the convex conjugate set that is responsible for implicitly regularizing the price response of the market maker. We use our insights into the dual space to develop a new family of liquidity-sensitive automated market makers with desirable properties.

1   Introduction

Automated market makers are algorithmic agents that provide liquidity in electronic markets. Markets with large event spaces or sparse interest from traders might fail because buyers and sellers have trouble finding one another. Automated market makers can prevent this failure by stepping in and providing a counterparty for prospective traders; instead of making bets with each other, traders place bets with the automated market maker. Automated market makers have been the object of theoretical study into market microstructure [Ostrovsky, 2009; Othman and Sandholm, 2010b] and successfully implemented in practice in large electronic markets [Goel et al., 2008; Othman and Sandholm, 2010a]. A broad introduction to the mechanics of automated market making can be found in Pennock and Sami [2007].

Othman et al. [2010] introduce a *liquidity-sensitive* automated market maker. This market maker is able to adapt its price response to increasing activity within the market; with this market maker bets will not move prices very much when there is lots of money already wagered with the market maker. This is in contrast to traditional market-making agents that provide identical price responses regardless of whether there are tens of dollars or tens of millions of dollars wagered with the market maker.

N. Chen, E. Elkind, and E. Koutsoupias (Eds.): WINE 2011, LNCS 7090, pp. 314–325, 2011.
© Springer-Verlag Berlin Heidelberg 2011

Unfortunately, this liquidity-sensitive market maker does not generalize easily. In Othman et al. [2010] it is referred to as a technique to "continuously channel profits into liquidity", a view echoed by Abernethy et al. [2011]. While this view may be accurate, it is not prescriptive: it offers no insight about how to create other liquidity-sensitive market makers, or of the relation between liquidity-sensitive market makers and the other market makers of the literature.

In this paper, we solve the puzzle of how liquidity-sensitive market makers work, and their relation to other market makers from the literature. We are able to contextualize, generalize, and expand the idea of liquidity-sensitive market makers. In order to do this, we first situate liquidity-sensitive market makers within the same framework as their liquidity-insensitive counterparts. Using a set of desiderata taken from the prediction market and finance literature we introduce a new class of automated market makers, *homogeneous risk measures*, which we argue correctly embody the notion of liquidity sensitivity, and we prove that the market maker of Othman et al. [2010] is a member of this class.

Our principal result is a necessary and sufficient characterization of the complete set of homogeneous risk measures: they are the support functions of compact convex sets in the non-negative orthant. Most intriguingly, this dual view allows us to achieve a synthesis between homogeneous risk measures and the experts algorithm perspective of Chen and Vaughan [2010], another recent view of automated market making. In this perspective, homgeneous risk measures are *unregularized* follow-the-leader algorithms that (generally) put non-unit total weight on the set of experts. We show it is the shape of the convex conjugate set (particularly, that set's curvature) that implicitly acts as a regularizer for the homogeneous risk measure. Furthermore, the bulge of the convex set away from the probability simplex defines notions like the maximum sum of prices. We use these insights to create a new family of liquidity-sensitive automated market makers, the *unit ball market makers*, that have desirable properties: defined costs for any possible bet, defined bounds on sums of prices, and tightly bounded loss.

2   Background

In this section we provide a brief introduction to automated market making, with emphasis on the recent results that guide the remainder of the work.

2.1   Cost Functions and Risk Measures

We consider a general setting in which the future state of the world is exhaustively partitioned into n events, $\{\omega_1, \ldots, \omega_n\}$, so that exactly one of the ω_i will occur. This model applies to a wide variety of settings, including financial models on stock prices and interest rates, sports betting, and traditional prediction markets.

In our notation, $\mathbf{x}$ is a vector and x is a scalar, $\mathbf{1}$ is the n-dimensional vector of all ones, and $\nabla_i f$ represents the i-th element of the gradient of a function f. The non-negative orthant is given by $\mathbb{R}^n_+ \equiv \{\mathbf{x} \mid \min_i x_i \geq 0\}$.

Let U be a convex subset of $\mathbb{R}^n$. Our work concerns functions $C : U \mapsto \mathbb{R}$ which map vector payouts over the events to scalar values. A *state* refers to a vector of payouts. Traders make bets with the market maker by changing the market maker's state. To move the market maker from state $\mathbf{x}$ to state $\mathbf{x}'$, traders pay $C(\mathbf{x}') - C(\mathbf{x})$. For instance, if the state is $x_1 = 5$ and $x_2 = 3$, then the market maker needs to pay out five dollars if ω_1 is realized and pay out 3 dollars if ω_2 is realized. If a new trader wants a bet that pays out one dollar if event ω_1 occurs, then they change the market maker's state to be $\{6, 3\}$, and pay $C(\{6, 3\}) - C(\{5, 3\})$. There are two broad research streams that explore these functions. The prediction market literature, where they are called *cost functions*, and the finance literature, where they are called *risk measures*. We use the terms *cost function* and *risk measure* interchangeably.

The most popular cost function used in Internet prediction markets is Hanson's logarithmic market scoring rule (LMSR), an automated market maker with particularly desirable properties, including bounded loss and a simple analytical form [Hanson, 2003, 2007]. The LMSR is defined as

$$C(\mathbf{x}) = b \log \left(\sum_i \exp(x_i/b) \right)$$

for fixed $b > 0$. b is called the *liquidity parameter*, because it controls the magnitude of the price response of the market maker to bets.[1] For instance, if the LMSR is used with $b = 10$ in our example above, $C(\{6, 3\}) - C(\{5, 3\}) \approx .56$, and so the market maker would quote a price of 56 cents to the agent for their bet. If $b = 1$, the same bet would cost 92 cents.

The *prices* p_i of a differentiable risk measure are given by the gradient of the cost function—the marginal cost on each event: $p_i = \frac{\exp(x_i/b)}{\sum_j \exp(x_j/b)}$. Observe that the prices in the LMSR sum to one. The notion of *sum of prices* is crucial to our work. The market maker's profit cut (or *vigorish* in gambling contexts) can be thought of as the difference between the sum of prices and unity [Othman et al., 2010]. This profit cut serves to compensate the market maker for taking bets with traders, and typical values for the vigorish in real applications are small, ranging from one percent to 20 percent. Since the LMSR and many other cost functions of the literature [Chen and Pennock, 2007; Peters et al., 2007; Agrawal et al., 2009; Abernethy et al., 2011] do not have a profit cut, they can be expected to run at a loss in practice [Pennock and Sami, 2007].

2.2   Link to Online Learning

One of the most intriguing recent developments in automated market making is the link between cost functions and online learning algorithms, particularly between cost functions and online follow-the-regularized-leader algorithms.

---

[1] With $b = 1$, the LMSR is equivalent to the *entropic risk measure* of the finance literature [Föllmer and Schied, 2002].

This link first appeared in a supporting role in Chen et al. [2008], and was significantly expanded in later work by those authors [Chen and Vaughan, 2010; Abernethy et al., 2011]. Any loss-bounded convex risk measure (Section 3 will make this precise) is equivalent to a no-regret follow-the-regularized-leader online learning algorithm. These online learning algorithms are conventionally expressed not as cost functions (or, in the machine learning literature, *potential functions*), but rather in dual space [Shalev-Shwartz and Singer, 2007]. The dual-space formulation is a powerful way of interpreting and constructing automated market makers.

Let Π be the probability simplex. Chen and Vaughan [2010] show that we can write any convex risk measure in terms of a convex optimization over a *follow-the-leader* term and a convex *regularizer* term. This optimization is in fact a conjugacy operation restricted to the probability simplex: $C(\mathbf{x}) = \max_{\mathbf{y} \in \Pi} \mathbf{x} \cdot \mathbf{y} - f(\mathbf{y})$ Here, $\mathbf{x} \cdot \mathbf{y}$ is the follow-the-leader term, and f is a regularizer.

2.3   The OPRS Cost Function

The *Othman-Pennock-Reeves-Sandholm cost function (OPRS)* was originally introduced in Othman et al. [2010] as a liquidity-sensitive extension of the LMSR. The OPRS is defined as $C(\mathbf{x}) = b(\mathbf{x}) \log \left(\sum_i \exp(x_i / b(\mathbf{x})) \right)$, where $b(\mathbf{x}) = \alpha \sum_i x_i$ for $\alpha > 0$. The OPRS can be contrasted with the LMSR, for which $b(\mathbf{x}) \equiv b$. Unlike the LMSR, the OPRS is only defined over the non-negative orthant (for continuity we can set $C(\mathbf{0}) = 0$). Also unlike the LMSR, the sum of prices in the OPRS is always greater than 1.

The OPRS has several desirable properties. These include a concise analytical closed form and *outcome-independent profit*, the ability to (for certain final quantity vectors) book a profit regardless of the realized outcome. Perhaps the most practical property of the OPRS is its scale-invariant liquidity sensitivity: its consistent price reaction over different scales of market activity. (This scale-invariance is a consequence of the OPRS cost function being positive homogeneous.) For large liquid markets, say with millions of dollars, a one-dollar bet will have a much smaller impact on prices than in a less-liquid market. This is not the case for the LMSR, where a one dollar bet moves prices the same amount in both heavily- and lightly-traded markets.

3   Desiderata, Dual Spaces, and an Impossibility Result

This section expands upon the dual-space approach to automated market making [Agrawal et al., 2009; Chen and Vaughan, 2010; Abernethy et al., 2011], particularly as a vehicle for contextualizing and generalizing the OPRS.

3.1   Desiderata and Their Combinations

In this section we introduce five desiderata for cost functions. Each of these properties has been acknowledged as desirable in the market making literature [Agrawal et al., 2009; Othman et al., 2010; Abernethy et al., 2011]. The market makers from the literature satisfy various subsets of these desiderata.

Monotonicity: For all $\mathbf{x}$ and $\mathbf{y}$ such that $x_i \leq y_i$, $C(\mathbf{x}) \leq C(\mathbf{y})$.

Monotonicity prevents simple arbitrages like a trader buying a zero-cost contract that never results in losses but sometimes results in gains.

Convexity: For all $\mathbf{x}$ and $\mathbf{y}$ and $\lambda \in [0,1]$, $C(\lambda\mathbf{x}+(1-\lambda)\mathbf{y}) \leq \lambda C(\mathbf{x})+(1-\lambda)C(\mathbf{y})$.

Convexity can be thought of as a condition that encourages diversification. The cost of the blend of two payout vectors is not greater than the sum of the cost of each individually. Consequently, the market maker is incentivized to diversify away its risk. The acknowledgment of diversification as desirable goes back to the very beginning of the mathematical finance literature [Markowitz, 1952].

Bounded loss: $\sup_{\mathbf{x}} [\max_i (x_i) - C(\mathbf{x})] < \infty$.

A market maker using a cost function with bounded loss can only lose a finite amount to interacting traders, regardless of the traders' actions and the realized outcome.

Translation invariance: For all $\mathbf{x}$ and scalar α, $C(\mathbf{x} + \alpha\mathbf{1}) = C(\mathbf{x}) + \alpha$.

Translation invariance ensures that adding a dollar to the payout of every state of the world will cost a dollar.

Positive homogeneity: For all $\mathbf{x}$ and scalar $\gamma > 0$, $C(\gamma\mathbf{x}) = \gamma C(\mathbf{x})$.

Positive homogeneity ensures a scale-invariant, currency-independent price response, as in the OPRS. From a risk measurement perspective, positive homogeneity ensures that doubling a risk doubles its cost.

A cost function that satisfies all of these desiderata except bounded loss is called a *coherent risk measure*. Coherent risk measures were first introduced in Artzner et al. [1999].

Definition 1. *A* coherent risk measure *is a cost function that satisfies monotonicity, convexity, translation invariance, and positive homogeneity.*

When we relax positive homogeneity from a coherent risk measure, we get a convex risk measure. Convex risk measures were first introduced in Carr et al. [2001] and feature prominently in the prediction market literature [Hanson, 2003; Ben-Tal and Teboulle, 2007; Hanson, 2007; Chen and Pennock, 2007; Peters et al., 2007; Agrawal et al., 2009; Abernethy et al., 2011].

When we instead relax translation invariance from a coherent risk measure, we get what we dub a homogeneous risk measure.

Definition 2. *A* homogeneous risk measure *is a cost function satisfying monotonicity, convexity, and positive homogeneity.*

To our knowledge, the only homogeneous risk measure of the literature that is not also a coherent risk measure is the OPRS.

Proposition 1. *The OPRS is a homogeneous risk measure (for vectors in the non-negative orthant).*

The desiderata are global properties that need to hold over the entire space the cost function is defined over. It is often difficult to verify that a given cost function satisfies these desiderata directly, and inversely, it is difficult to construct new cost functions that satisfy specific desiderata. Remarkably, each of these desiderata have simple representations in Legendre-Fenchel dual space.

3.2   Dual Space Equivalences

The rest of the paper relies on the well-developed theory of convex conjugacy.

Definition 3. *The* Legendre-Fenchel dual *(aka* convex conjugate*) of a convex cost function C is a convex function $f : \mathbb{Y} \mapsto \mathbb{R}$ over a convex set $\mathbb{Y} \subset \mathbb{R}^n$ such that $C(\mathbf{x}) = \max_{\mathbf{y} \in \mathbb{Y}} [\mathbf{x} \cdot \mathbf{y} - f(\mathbf{y})]$. We say that the cost function is "conjugate to" the pair $\mathbb{Y}$ and f. Convex conjugates exist uniquely for convex cost functions defined over $\mathbb{R}^n$ [Rockafellar, 1970; Boyd and Vandenberghe, 2004].*

We will refer to the convex optimization in dual space as the "optimization" or "optimization problem", and the maximizing $\mathbf{y}$ as the "maximizing argument". One way of interpreting the dual is that it represents the "price space" of the market maker, as opposed to a cost function which is defined over a "quantity space" [Abernethy et al., 2011]. The only prices a market maker can assume are those $\mathbf{y} \in \mathbb{Y}$, while the function f serves as a measure of market sensitivity and a way to limit how quickly prices are adjusted in response to bets. As we have discussed, in the prediction market literature "prices" denote the partial derivatives of the cost function [Pennock and Sami, 2007; Othman et al., 2010]. When it is unique, the maximizing argument of the convex conjugate is the gradient of the cost function, and when it is not unique, then the maximizing arguments represent the subgradients of the cost function. A fuller discussion of the relation between convex conjugates and derivatives is available in convex analysis texts [Rockafellar, 1970; Boyd and Vandenberghe, 2004].

Another interpretation of the dual space is from online learning, specifically online regularized follow-the-leader algorithms [Chen and Vaughan, 2010]. We discussed the literature relating to this link in Section 2.2. Here, the set $\mathbb{Y}$ represents the allowable weights we can assign to experts, and the function f is a regularizer that determines how quickly we adjust the weight between experts in response to returns which are the same as payouts in this interpretation. Generally speaking when the set $\mathbb{Y}$ exceeds the probability simplex $\varPi$, then the weights placed on the experts will not be guaranteed to sum to unity.

With these interpretations in mind, we proceed to show the power of the dual space: we can represent homogeneous risk measures with a compact convex set in the non-negative orthant. The relations between convex and monotonic cost functions, convex and positive homogeneous cost functions, and their respective duals are a consequence of well-known results in the convex analysis literature [Rockafellar, 1966, 1970].

Proposition 2. *A risk measure is convex and monotonic if and only if the set $\mathbb{Y}$ is exclusively within the non-negative orthant.*

Proposition 3. *A risk measure is convex and positive homogeneous if and only if its convex conjugate has compact* $\mathbb{Y}$ *and has* $f(\mathbf{y}) = 0$ *for every* $\mathbf{y} \in \mathbb{Y}$.

In the literature this latter result relates *indicator sets* (here, the set $\mathbb{Y}$) to *support functions* (here, the cost function). Since $f(\mathbf{y}) = 0$ for all $\mathbf{y} \in \mathbb{Y}$, the cost function conjugacy is defined only by the set $\mathbb{Y}$. Consequently, we will abuse terminology slightly and refer to the cost functions as conjugate to the convex compact set alone. A necessary and sufficient condition on the set of homogeneous risk measures follows.

Corollary 1. *A cost function is a homogeneous risk measure if and only if it is conjugate to a compact convex set in the non-negative orthant.*

The following results can be derived from convex analysis and the work of Abernethy et al. [2011].

Proposition 4. *A risk measure is convex, monotonic, and translation invariant if and only if the set* $\mathbb{Y}$ *lies exclusively on the probability simplex.*

Proposition 5. *A risk measure is convex and has bounded loss if and only if the set* $\mathbb{Y}$ *includes the probability simplex.*

The only market maker that satisfies all five of our desiderata is max.

Proposition 6. *The only coherent risk measure with bounded loss is* $C(\mathbf{x}) = \max_i x_i$.

The max market maker corresponds to an order-matching, risk-averse cost function that either charges agents nothing for their transactions, or exactly as much as they could be expected to gain in the best case. For instance, a trader wishing to move the max market maker from state $\{5, 3\}$ to state $\{7, 3\}$ would be charged 2 dollars, exactly as much as they would win if the first event happened—which means taking the bet is a dominated action. On the other hand, a trader wishing to move the market maker from state $\{5, 3\}$ to state $\{5, 5\}$ pays nothing! These two small examples suggest that max is a poor risk measure in practice, and therefore Proposition 6 should be viewed as an impossibility result.

Combining all of our dual-space equivalences, we have that the conjugate of max is defined exclusively on the whole probability simplex, where it is identically 0.[2]

There are now two ways to smooth out the price response of max: One way is to use a regularizer, so that price estimates do not immediately jump to the axes (i.e., zero or one). This corresponds to a regularized online follow-the-leader algorithm, which is a convex risk measure [Chen and Vaughan, 2010]. We introduce a different approach, to expand the shape of the valid price, so that the shape of the space itself serves as an implicit regularizer over the price estimates. This will generally result in prices that are not probability distributions, and as we explore in the next section, this approach leads to homogeneous risk measures.

---

[2] In dual price space, the maximizing argument to the max cost function can always be represented as one of the axes. In the online learning view, max represents an unregularized follow-the-leader algorithm, putting all of its probability weight on the current best expert (i.e., the event with largest current payout).

4   Shaping the Dual Space

Recall that only the convex conjugate set $\mathbb{Y}$ of a homogenenous risk measure is responsible for determining the market maker's behavior, because the conjugate function f takes value zero everywhere in that set. In this section, we explore two features of the conjugate set that produce desirable properties: its *curvature* and its *divergence* from the probability simplex.

4.1   Curvature

We would like for our cost function to always be differentiable (outside of $\mathbf{0}$, where a derivative of a positive homogeneous function will not generally exist). The OPRS is differentiable in the non-negative orthant (again, excepting $\mathbf{0}$) while max is differentiable only when the maximum is unique. In this section, we show that only curved conjugate sets produce homogeneous risk measures that are differentiable.[3]

Definition 4. *A closed, convex set $\mathbb{Y}$ is strictly convex if its boundary does not contain a non-degenerate line segment. Formally, let $\partial\mathbb{Y}$ denote the boundary of the set. Let $0 \leq \lambda \leq 1$ and $\mathbf{x}, \mathbf{x}' \in \partial\mathbb{Y}$. Then $\lambda\mathbf{x} + (1-\lambda)\mathbf{x}' \in \partial\mathbb{Y}$ holds only for $\mathbf{x} = \mathbf{x}'$.*

Since strictly convex sets are never linear on their boundary they can be thought of as sets with curved boundaries.

Proposition 7. *A homogeneous risk measure is differentiable on $\mathbb{R}^n \backslash \mathbf{0}$ if and only if its conjugate set is strictly convex.*

4.2   Divergence from Probability Simplex

The amount of divergence from the probability simplex governs the market maker's divergence from translation-invariant prices (i.e., prices that sum to unity). Recall that max is the homogeneous risk measure that is defined only over the probability simplex.

Proposition 8. *Let $\mathbb{Y}$ be the dual set of a differentiable homogeneous risk measure. Then the maximum sum of prices (the most a trader would ever need to spend for a unit guaranteed payout) is given by $\max_{\mathbf{y} \in \mathbb{Y}} \sum_i y_i$, and the minimum sum of prices (the most the market maker would ever pay for a unit guaranteed payout) is given by $\min_{\mathbf{y} \in \mathbb{Y}} \sum_i y_i$.*

---

[3] It might be argued that what we are really interested in, particularly if we claim that curved sets act as a regularizer in the price response, is whether or not curved sets also imply *continuous* differentiability of the cost function. Continuous differentiability would mean that prices both exist and are continuous in the quantity vector. These conditions are in fact the same for convex functions defined over an open interval (such as $\mathbb{R}^n \backslash \mathbf{0}$), because for such functions differentiability implies continuous differentiability [Rockafellar, 1970].

Given any (efficiently representable) convex set corresponding to a differentiable homogeneous risk measure, the extreme price sums can be solved for in polynomial time, since it is a convex optimization over a convex set.

It was shown in Othman et al. [2010] that the OPRS achieved its maximum sum of prices for quantity vectors that are scalar multiples of $\mathbf{1}$. A corollary of the above result is that this property holds for every homogeneous risk measure. (Other vectors may also achieve the same sum of prices.)

Corollary 2. *In a homogeneous risk measure every vector that is a positive multiple of $\mathbf{1}$ achieves the maximum sum of prices.*

In addition to maximum prices, the shape of the convex set also determines the worst-case loss of the resulting market maker. The notion of worst-case loss is closely related to our desideratum of bounded loss—a market maker with unbounded worst-case loss does not have bounded loss, and a market maker with finite worst-case loss has bounded loss.

Definition 5. *The* worst-case loss *of a market maker is given by $\max_i x_i - C(\mathbf{x}) + C(\mathbf{x^0})$ where $\mathbf{x^0} \in \mathbb{R}^n_+$ is some initial quantity vector the market maker selects.*

In homogeneous risk measures, the amount of liquidity sensitivity is proportional to the market's state. Since in practice there is some latent level of interest in trading on the event before the market's initiation, it is desirable to seed the market initially to reflect a certain level of liquidity. It is desirable to have a tight bound on that worst-case loss, reflecting that in practice, market administrators are likely to have bounds on how much the market maker could lose in the worst case. Tight bounds on worst-case loss assure the administrator that that bound will be satisfied with maximum liquidity injected at the market's initiation.

Proposition 9. *Let $\mathbb{Y}$ be a convex set conjugate to a homogeneous risk measure that includes the unit axes but does not exceed the unit hypercube. Then the worst-case loss of the risk measure is* tightly *bounded by the initial cost of the market's starting point.*

By bringing $\mathbf{x^0}$ as close as desired to $\mathbf{0}$, we have the following corollary, which is a generalization of a similar result for the OPRS.

Corollary 3. *Let $\mathbb{Y}$ be a convex set conjugate to a homogeneous risk measure that includes the unit axes. Then the worst-case loss of the risk measure can be set arbitrarily small.*

A bound on prices also emerges from this result.

Corollary 4. *Let $\mathbb{Y}$ be a convex set conjugate to a homogeneous risk measure that includes the unit axes but does not exceed the unit hypercube. Then the maximum price on any event is 1.*

5   A New Family of Liquidity-Sensitive Market Makers

We proceed to use our theoretical results constructively, to create a family of homogeneous risk measures with desirable properties that the OPRS, the only prior homogeneous risk measure, lacks. These include tight bounds on minimum sum of prices and worst-case losses, and definition over all of $\mathbb{R}^n$. Our new family of market makers is parameterized (in much the same way as the OPRS) by the maximum sum of prices. The OPRS is not a member of this new family.

Our scheme is to take as our dual set the intersection of two unit balls in different $\mathcal{L}^p$ norms, one ball at $\mathbf{0}$ and the other ball at $\mathbf{1}$. For $1 < p < \infty$, the intersection of the two balls is a strictly convex set that includes the unit axes but does not exceed the unit hypercube. (At $p = 1$, we get the probability simplex, which is not strictly convex. At $p = \infty$ we get the unit hypercube, which is also not strictly convex.) Let $|| \cdot ||_p$ denote the $\mathcal{L}^p$ norm. Then we can define the vectors in the intersections of the unit balls, $\mathcal{U}(p)$, as

$$\mathcal{U}(p) \equiv \{\mathbf{y} \mid \mathbf{y} \in \mathbb{R}^n, ||\mathbf{y} - \mathbf{1}||_p \leq 1, ||\mathbf{y}||_p \leq 1\}$$

This set gives us a cost function $C(\mathbf{x}) = \max_{\mathbf{y} \in \mathcal{U}(p)} \mathbf{x} \cdot \mathbf{y}$. We dub this the *unit ball market maker*. Since we can easily test whether a vector is within both unit balls (i.e., within $\mathcal{U}(p)$), the optimization problem for the cost function can be solved in polynomial time.

This family of market makers is parameterized by the $\mathcal{L}^p$ norm that defines which vectors in dual space are in the convex set. By choosing the value of p correctly, we can engineer a market maker with the desired maximum sum of prices. The outer boundary of the set is defined by the unit ball from $\mathbf{0}$ in $\mathcal{L}^p$ space. Its boundary along $\mathbf{1}$ is given by the k that solves $\sqrt[p]{nk^p} = 1$. Solving for k we get $k = n^{-1/p}$, and so the maximum sum of prices is $nk = n\left(n^{-1/p}\right) = n^{1-1/p}$. For prices that are at most $1 + v$, we can set $1 + v = n^{1-1/p}$. Solving this equation for p yields $p = \frac{\log n}{\log n - \log(1+v)}$. Given any target maximum level of vigorish, this formula provides the exponent of the unit ball market maker to use. Considering that only small divergences away from unity are natural to the setting, the p we select for our $\mathcal{L}^p$ norm should be quite small. The norm increases in the maximum sum of prices, and for larger n the same norm produces larger sums of prices.

One of the advantages of the unit ball market maker is that it is defined over all of $\mathbb{R}^n$, as opposed to just the non-negative orthant. Its behavior in the positive orthant is to charge agents more than a dollar for a dollar guaranteed payout, because the outer boundary is diverges outwards from the probability simplex. Its behavior in the negative orthant, where its points on the inner boundary are selected in the maximization, is to pay less than a dollar for a dollar guaranteed payout. Its behavior in all other orthants is equivalent to max, as the unit axes are selected as maximizing arguments. Finally, if we restrict the unit ball market maker to only the non-negative orthant (like the OPRS), the sum of prices is tightly bounded between 1 and $n^{1-1/p}$.

6 Conclusions and Future Work

Using five desiderata that have appeared in the finance and prediction market literature, we contextualized a new class of cost functions, which we dubbed *homogeneous risk measures*. We showed that the OPRS [Othman et al., 2010] is a member of this class, because it is convex, monotonic, and positive homogeneous. We proved only the max cost function satisfies all five of our desiderata, but it does not have a differentiable price response. To produce a differentiable price response, one can add a regularizer, leading to the regularized online learning algorithms explored by Chen and Vaughan [2010]. Another approach is to curve the conjugate dual space, relaxing it from the probability simplex. We discussed how the properties of the convex set induce desirable properties in its conjugate homogeneous risk measure. Finally, using our insights, we developed a new family of homogeneous risk measures, the *unit ball market makers*, with desirable properties.

Our work centered on cost functions that are positive homogeneous, because these are the only cost functions that display identical relative price responses at different levels of liquidity. However, another direction is to explore cost functions that display some characteristics of liquidity sensitivity (more muted price responses at high levels of liquidity) without necessarily being homogeneous.

Finally, we are attracted to the work of Agrawal et al. [2009] because it provides a framework to simply add functionality to handle limit orders (orders of the form "I will pay no more than p for the payout vector $\mathbf{x}$") into a cost function market maker. That framework relies on convex optimization and so would also be able to run in polynomial time, a significant gain over naïve implementations of limit orders within cost function market makers. However, that work relied heavily on simplifications to the optimization that could be made because of translation invariance, so it is unclear how to embed a market maker whose convex conjugate is defined over more than the probability simplex into a limit order framework.

Acknowledgements. We thank Peter Carr, Yiling Chen, Geoff Gordon, Dilip Madan, Dave Pennock, Steve Shreve, and Kevin Waugh for helpful discussions and suggestions. This work was supported by NSF grants CCF-1101668, IIS-0905390, IIS-0964579, and by the Google Fellowship in Market Algorithms.

References

Abernethy, J., Chen, Y., Vaughan, J.W.: An optimization-based framework for automated market-making. In: ACM Conference on Electronic Commerce, EC (2011)

Agrawal, S., Delage, E., Peters, M., Wang, Z., Ye, Y.: A unified framework for dynamic pari-mutuel information market design. In: ACM Conference on Electronic Commerce (EC), pp. 255–264 (2009)

Artzner, P., Delbaen, F., Eber, J., Heath, D.: Coherent measures of risk. Mathematical finance 9(3), 203–228 (1999)

Ben-Tal, A., Teboulle, M.: An old-new concept of convex risk measures: The optimized certainty equivalent. Mathematical Finance 17(3), 449–476 (2007)

Boyd, S., Vandenberghe, L.: Convex Optimization. Cambridge University Press (2004)

Carr, P., Geman, H., Madan, D.: Pricing and hedging in incomplete markets. Journal of Financial Economics 62(1), 131–167 (2001)

Chen, Y., Pennock, D.M.: A utility framework for bounded-loss market makers. In: Proceedings of the 23rd Annual Conference on Uncertainty in Artificial Intelligence (UAI), pp. 49–56 (2007)

Chen, Y., Vaughan, J.W.: A new understanding of prediction markets via no-regret learning. In: ACM Conference on Electronic Commerce (EC), pp. 189–198 (2010)

Chen, Y., Fortnow, L., Lambert, N., Pennock, D.M., Wortman, J.: Complexity of combinatorial market makers. In: ACM Conference on Electronic Commerce (EC), pp. 190–199 (2008)

Föllmer, H., Schied, A.: Stochastic Finance. Studies in Mathematics, vol. 27. De Gruyter (2002)

Goel, S., Pennock, D., Reeves, D., Yu, C.: Yoopick: a combinatorial sports prediction market. In: Proceedings of the National Conference on Artificial Intelligence (AAAI), pp. 1880–1881 (2008)

Hanson, R.: Combinatorial information market design. Information Systems Frontiers 5(1), 107–119 (2003)

Hanson, R.: Logarithmic market scoring rules for modular combinatorial information aggregation. Journal of Prediction Markets 1(1), 1–15 (2007)

Markowitz, H.: Portfolio Selection. The Journal of Finance 7(1), 77–91 (1952)

Ostrovsky, M.: Information aggregation in dynamic markets with strategic traders. In: ACM Conference on Electronic Commerce (EC), pp. 253–254 (2009)

Othman, A., Sandholm, T.: Automated market-making in the large: the Gates Hillman prediction market. In: ACM Conference on Electronic Commerce (EC), pp. 367–376 (2010a)

Othman, A., Sandholm, T.: When Do Markets with Simple Agents Fail? In: International Conference on Autonomous Agents and Multi-Agent Systems (AAMAS), pp. 865–872 (2010)

Othman, A., Pennock, D.M., Reeves, D.M., Sandholm, T.: A practical liquidity-sensitive automated market maker. In: ACM Conference on Electronic Commerce (EC), pp. 377–386 (2010)

Pennock, D., Sami, R.: Computational Aspects of Prediction Markets. In: Algorithmic Game Theory, ch. 26, pp. 651–674. Cambridge University Press (2007)

Peters, M., So, A.M.-C., Ye, Y.: Pari-Mutuel Markets: Mechanisms and Performance. In: Deng, X., Graham, F.C. (eds.) WINE 2007. LNCS, vol. 4858, pp. 82–95. Springer, Heidelberg (2007)

Rockafellar, R.T.: Level sets and continuity of conjugate convex functions. Transactions of the American Mathematical Society 123(1), 46–63 (1966)

Rockafellar, R.T.: Convex Analysis. Princeton University Press (1970)

Shalev-Shwartz, S., Singer, Y.: A primal-dual perspective of online learning algorithms. Machine Learning 69, 115–142 (2007)

Manipulating Stochastically Generated Single-Elimination Tournaments for Nearly All Players

Isabelle Stanton and Virginia Vassilevska Williams

Computer Science Department
UC Berkeley
{isabelle,virgi}@eecs.berkeley.edu

Abstract. We study the power of a tournament organizer in manipulating the outcome of a balanced single-elimination tournament by fixing the initial seeding. This problem is known as *agenda control for balanced voting trees*. It is not known whether there is a polynomial time algorithm that computes a seeding for which a given player can win the tournament, even if the match outcomes for all pairwise player match-ups are known in advance. We approach the problem by giving a sufficient condition under which the organizer can *always* efficiently find a tournament seeding for which the given player will win the tournament. We then use this result to show that for most match outcomes generated by a natural random model attributed to Condorcet, the tournament organizer can very efficiently make a large constant fraction of the players win, by manipulating the initial seeding.

1 Introduction

The study of election manipulation is an integral part of social choice theory. Results such as the Gibbard-Satterthwaite theorem [8,13] show that all voting protocols that meet certain rationality criteria are manipulable. The seminal work of Bartholdi, Tovey and Trick [1,2] proposes to judge the quality of voting systems using computational complexity: a protocol may be manipulable, but it may still be good if manipulation is computationally expensive. This idea is at the heart of computational social choice.

The particular type of election manipulation that we study in this paper is called *agenda control* and was introduced in [2]: there is an election organizer who has power over some part of the protocol, say the order in which candidates are considered. The organizer would like to exploit this power to fix the outcome of the election by making their favorite candidate win. [2] focused on plurality and Condorcet voting, agenda control by adding, deleting, or partitioning candidates or voters. We study the balanced binary cup voting rule, also called *balanced voting tree*, or a *balanced single-elimination* (SE) tournament: the number of candidates is a power of 2 and at each stage the remaining candidates are paired up and their votes are compared. The losers are eliminated and the winners move

N. Chen, E. Elkind, and E. Koutsoupias (Eds.): WINE 2011, LNCS 7090, pp. 326–337, 2011.

on to the next round until only one candidate remains. The power of the election organizer is to pick the pairing of the players in each round. We assume that the organizer knows all the votes in advance, i.e. for any two candidates, they know which candidate is preferred. In this case, picking the pairings for each round is equivalent to picking the initial tournament seeding.

Single-elimination is prevalent in sports tournaments such as Wimbledon or March Madness. In this setting, a tournament organizer may have some information, say from prior matches or betting experts, about the winner in any possible match. The organizer creates a *seeding* of the players through which they are distributed in the tournament bracket. The question is, can the tournament organizer abuse this power to determine the winner of the tournament?

There is significant prior work on this problem. Lang et al. [10] showed that if the tournament organizer only has probabilistic information about each match, then the agenda control problem is NP-hard. Vu et al [17,18] showed that the problem is NP-hard even when the probabilities are in $\{0, 1, 1/2\}$ and that it is NP-hard to obtain a tournament bracket that approximates the maximum probability that a given player wins within any constant factor. Vassilevska Williams [16] showed that the agenda control problem is NP-hard even when the information is deterministic but some match-ups are disallowed. [16] also gave conditions under which the organizer can always make their favorite player win the tournament with advance knowledge of each match outcome. It is still an open problem whether the agenda control problem in this deterministic setting can be solved in polynomial time.

The binary cup is a complete binary voting tree. Other related work has studied more general voting trees [9,7], and manipulation by the players themselves by throwing games to manipulate SE tournaments [12].

The match outcome information available to the tournament organizer can be represented as a weighted or unweighted tournament graph, a graph such that for every two nodes u, v exactly one of (u, v) or (v, u) is an edge. An edge (u, v) signifies that u beats v, and a weight p on an edge (u, v) means that u will beat v with probability p. With this representation, the agenda control problem becomes a computational problem on tournament graphs.

The tournament graph structure which comes from real world sports tournaments or from elections is not arbitrary. Although the graphs are not necessarily transitive, stronger players typically beat weaker ones. Some generative models have been proposed in order to study real-world tournaments. In this work, we study a standard model in social choice theory attributed to Condorcet (see, e.g., Young [19]). The model was more recently studied by Braverman and Mossel [3]. We refer to this model as the *Condorcet Random (CR)* model[1].

The CR model has an underlying total order of the players and the outcome of every match is probabilistic. There is some *global* probability $p < 1/2$ with which a weaker player beats a stronger player. This probability represents outside factors which do not depend on the players' abilities.

---

[1] A previous version of this paper referred to the model as the Braverman-Mossel model.

Vassilevska Williams [16] has shown that when $p \geq 16\sqrt{\ln(n)/n}$, with high probability, the model generates a tournament graph T such that there is always a poly-time computable seeding for which *any* given player is a single-elimination tournament winner, provided all match outcomes occur as T predicts.

This result was initially surprising as the CR model is often considered to be a good model of the real world. Recent work by Russell [11] confirmed the theoretical results of [16] by giving experimental evidence that in real world instances (from tennis, basketball and hockey tournaments) one can either quickly find a winning seeding for any player, or decide that it is not possible. Russell's work uses a variant of the generalized CR model that we will define later.

The result from [16], however, was meaningful only for large n. For instance, when $n = 512$, the noise parameter $p \geq 16\sqrt{\ln(n)/n}$ is close to $1/2$, and the result is not at all surprising since then all players are essentially indistinguishable. A natural question emerges: can we still make almost all players win with a much smaller noise value? A second question is, can we relax the CR model to allow a different error probability for each pair of players, and what manipulation is then possible? We address both questions.

Finally, the CR model has been previously considered in fault-tolerant and parallel computing. For instance, Feige et al. [6] consider comparison circuits that are incorrect with probability p and develop algorithms to sort this noisy data. In particular, one of their results uses tournaments for finding the maximum in parallel. In a sense, their algorithm provides a better mechanism for finding 'the' winner (the top player in the underlying total order) in the CR model, although this mechanism may not satisfy the other nice properties of SE tournaments.

Contributions. We continue the study begun in [16] on whether one can compute a winning SE tournament seeding for a *king* player when the match outcomes are known in advance. A king is a player K such that for any other player a, either K beats a, or K beats some other player who beats a. Kings are very strong players, yet the agenda control problem for SE tournaments is not known to be polynomial-time solvable even for kings. We show that in order for a winning seeding to exist for a king, it is sufficient for the king to be among the top third of the players when sorted by the number of potential matches they can win. Before our work only much stricter conditions were known, e.g. that it is sufficient if the king beats half of the players. Our more general result allows us to obtain better results for the Condorcet random model as well.

There are $\log n$ rounds in an SE tournament over n players, so a necessary condition for a player to be a winner is that it can beat at least $\log n$ players. We consider a generalization of the Condorcet random model in which the error probabilities $p(i, j)$ can vary but are all lower-bounded by a global parameter p. The expected outdegree of the weakest player i in such a tournament is $\sum_j p(i, j) \geq p(n - 1)$, and it needs to be at least $\log n$ in order for i to win an SE tournament. Thus, we focus on the case where p is $\Omega(\log n/n)$, as this is a necessary condition for all players to be winners.

We consider tournaments generated with noise $p = \Omega(\log n/n)$. The ranking obtained by sorting the players in nondecreasing order of the number of matches they can win is known to be a constant factor approximation to the Slater ranking [14,4], and is hence a good notion of ranking in itself. We show that for almost all tournaments generated by the CR model, one can efficiently compute a seeding so that essentially the top half of the players can be made SE winners. We also show that there is a trade-off between the amount of noise and the number of players that can be made winners: as the level of noise increases, the tournament can be fixed for a larger constant fraction and eventually for all of the players. While this result does not answer the question of whether it is computationally difficult to manipulate an SE tournament in general, it does show that for tournaments we might expect to see in practice, manipulation can be quite easy.

1.1   Condorcet Random Model – Formal Definition

The premise of the Condorcet random (CR) model is that there is an implicit ranking π of the players by intrinsic abilities so that $\pi(i) < \pi(j)$ means i has strictly better abilities than j. For ease of notation, we will assume that π is the identity permutation (if not, rename the players), so that $\pi(i)$ is i. When i and j play a match there may be outside influences so that even if $i < j$, j might beat i. The CR model allows that weaker players can beat stronger players, but only with probability $p < 1/2$. Here, p is a global parameter and if $i < j$, i beats j with probability $1 - p$. A random tournament graph generated in the CR model, a *CR tournament*, is defined as follows: for every i, j with $i < j$, add edge (i, j) independently with probability $1 - p$ and otherwise add (j, i). In other words, a CR tournament is initially a completely transitive tournament where each edge is independently reversed with probability p.

We generalize the CR model to the GCR model, in which j beats i with probability $p(j, i)$, where $p \le p(j, i) \le 1/2$ for all i, j with $i < j$, i.e. the error probabilities can differ but are all lower-bounded by a global p. A random tournament graph generated in the GCR model (*GCR tournament*) is defined as follows: for every i, j with $i < j$, add edge (i, j) independently with probability $1 - p(j, i)$ and otherwise add (j, i).

Unless noted otherwise, all graphs in the paper are tournament graphs over n vertices, where n is a power of 2, and all SE tournaments are balanced. In Table 1, we define the notation used in the rest of this paper. For the definitions, let $a \in V$ be any node, $X \subset V$ and $Y \subset V$ such that X and Y are disjoint. Given a player $\mathcal{A}$, A denotes $N^{out}(\mathcal{A})$ and B denotes $N^{in}(\mathcal{A})$.

The outcome of a round-robin tournament has a natural graph representation as a tournament graph. The nodes of a tournament graph represent the players, and a directed edge (a, b) represents a win of a over b.

We will use the concept of a *king* in a graph. Although the definition makes sense for any graph, it is particularly useful for tournaments, as the highest outdegree node is always a king. We also define a *superking*, as in [16].

Definition 1. *A king in $G = (V, E)$ is a node $\mathcal{A}$ such that for every other $x \in V$ either $(\mathcal{A}, x) \in E$ or there exists $y \in V$ such that $(\mathcal{A}, y), (y, x) \in E$.*

Table 1. A summary of the notation used in this paper

Notation	
$N^{out}(a) = \{v \vert (a,v) \in E\}$	$N_X^{out}(a) = N^{out}(a) \cap X$
$N^{in}(a) = \{v \vert (v,a) \in E\}$	$N_X^{in}(a) = N^{in}(a) \cap X$
$out(a) = \vert N^{out}(a) \vert$	$out_X(a) = \vert N_X^{out}(a) \vert$
$in(a) = \vert N^{in}(a) \vert$	$in_X(a) = \vert N_X^{in}(a) \vert$
$\mathcal{H}^{in}(a) = \{v \vert v \in N^{in}(a), out(v) > out(a)\}$	
$\mathcal{H}^{out}(a) = \{v \vert v \in N^{out}(a), out(v) > out(a)\}$	
$\mathcal{H}(a) = \mathcal{H}^{in}(a) \cup \mathcal{H}^{out}(a)$	
$E(X,Y) = \{(u,v) \vert (u,v) \in E, u \in X, v \in Y\}$	

Definition 2. *A superking in $G = (V, E)$ is a node $\mathcal{A}$ such that for every other $x \in V$ either $(\mathcal{A}, x) \in E$ or there exist $\log n$ nodes $y_1, \ldots, y_{\log n} \in V$ such that $\forall i, (\mathcal{A}, y_i), (y_i, x) \in E$.*

2   Kings That Are also SE Winners

Being a king in the tournament graph is not a sufficient condition for a player to also be able to win an SE tournament. For instance, a player may be a king by beating only 1 player who, in turn, beats all the other players. This king beats less than $\log n$ players, so it cannot win an SE tournament. [16] considered the question of how strong a king player needs to be in order for there to always exist a winning SE tournament seeding for which they win the SE tournament.

Theorem 1. *[16] Let $G = (V, E)$ be a tournament graph and let $\mathcal{A} \in V$ be a king. One can efficiently construct a winning single-elimination tournament seeding for $\mathcal{A}$ if either $\mathcal{H}^{in}(\mathcal{A}) = \emptyset$, or $out(\mathcal{A}) \geq n/2$.*

We generalize the above result by giving a condition which completely subsumes the one in Theorem 1.

Theorem 2 (Kings with High Outdegree). *Let G be a tournament graph on n nodes and $\mathcal{A}$ be a king. If $out(\mathcal{A}) \geq \vert \mathcal{H}^{in}(\mathcal{A}) \vert + 1$, then one can efficiently compute a winning single-elimination seeding for $\mathcal{A}$.*

To see that the above theorem implies Theorem 1, note that if $out(\mathcal{A}) \geq n/2$, then $\vert \mathcal{H}^{in}(a) \vert \leq n/2 - 1 \leq out(\mathcal{A}) - 1$. Also, if $\mathcal{H}^{in}(\mathcal{A}) = \emptyset$ and $n \geq 2$, then $out(\mathcal{A}) \geq 1 \geq 1 + \vert \mathcal{H}^{in}(\mathcal{A}) \vert$.

Theorem 2 is more general than Theorem 1. In Figure 1 we have an example of a tournament where node $\mathcal{A}$ satisfies the requirements of Theorem 2, but not those of Theorem 1. Here, $\vert \mathcal{H}^{in}(\mathcal{A}) \vert = \frac{n}{4}$ and $\vert N^{out}(\mathcal{A}) \vert = \frac{n}{4} + 1$. The purpose of node a is just to guarantee that $\mathcal{A}$ is a king. The example requires that each node in $N^{in}(\mathcal{A}) \setminus \mathcal{H}^{in}(\mathcal{A})$ has lower outdegree than $\mathcal{A}$; it suffices to use an outdegree-balanced[2] tournament for this set.

---

[2] An outdegree-balanced tournament is a tournament in which every vertex has outdegree equal to half the graph; such a tournament can easily be constructed inductively.

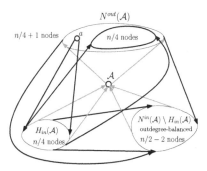

Fig. 1. An example for which Theorem 1 does not apply, but Theorem 2 does apply

The intuition behind the proof of Theorem 2 is partially inspired by our recent results in [15]. There we show that a large fraction of highly ranked nodes can be tournament winners, provided a matching exists from the lower ranked to the higher ranked players. In this paper, we are working with a king node, and are able to weaken the matching requirement. Instead, we carefully construct matchings that maintain that $\mathcal{A}$ is a king over the graph, while eliminating the elements of $\mathcal{H}^{in}(\mathcal{A})$ until we reduce the problem to the case of Theorem 1.

We will need a technical lemma from prior work relating the indegree and outdegree of two nodes in order to prove Theorem 2. By definition, if a node $\mathcal{A}$ is a king then for every other node b, $N^{out}(\mathcal{A}) \cap N^{in}(b) \neq \emptyset$. The following lemma is useful for showing a node is a king.

Lemma 1 ([16]). *Let a be a given node, $A = N^{out}(a), B = N^{in}(a), b \in B$. Then $out(a) - out(b) = in_A(b) - out_B(b)$. In particular, $out(a) \geq out(b)$ if and only if $out_B(b) \leq in_A(b)$.*

Proof of Theorem 2: We will design the matching for each consecutive round r of the tournament. In the induced graph before the rth round, let $\mathcal{H}_r$ be the subset of $\mathcal{H}^{in}(\mathcal{A})$ that is still live, A_r be the current outneighborhood of $\mathcal{A}$ and B_r be the current inneighborhood of $\mathcal{A}$. We will keep the invariant that if $B_r \setminus \mathcal{H}_r \neq \emptyset$, we have $|A_r| \geq |\mathcal{H}_r| + 1$, $\mathcal{A}$ is a king and the subset of nodes from the inneighborhood of $\mathcal{A}$ that have larger outdegree than $\mathcal{A}$ is contained in $\mathcal{H}_r$.

We now assume that the invariant is true for round $r - 1$. We will show how to construct round r. If $\mathcal{H}_r = \emptyset$ we are done by reducing the problem to Theorem 1, so assume that $|\mathcal{H}_r| \geq 1$. We begin by taking a maximal matching M_r from A_r to $\mathcal{H}_r$. Since $|A_r| \geq |\mathcal{H}_r| + 1$, $A_r \setminus M_r \neq \emptyset$ i.e. M_r cannot match all of A_r. Now, let M'_r be a maximal matching from $A_r \setminus M_r$ to $B_r \setminus \mathcal{H}_r$.

If $A_r \setminus (M'_r \cup M_r) \neq \emptyset$, there is some node a' leftover to match $\mathcal{A}$ to. Otherwise, pick any $a' \in M'_r \cap A_r$. Remove the edge matched to a' from M'_r and match a' with $\mathcal{A}$. To complete the matching, create maximal matchings within $\bar{A}_r = A_r \setminus (M'_r \cup M_r) \setminus \{a'\}$, $\bar{B}_r = B_r \setminus \mathcal{H}_r \setminus M'_r$ and $\mathcal{H}_r \setminus M_r$. Either 0 or 2 of $|\bar{A}_r|, |\bar{B}_r|, |\mathcal{H}_r \setminus M_r|$ can be odd and so there are at most 2 unmatched nodes that can be matched against each other. Let M be the union of these matchings.

We will now show that the invariants still hold. Notice that $\mathcal{A}$ is still a king on the sources of the created matching M. Now, consider any node b from $B_r \setminus \mathcal{H}_r$ which is a source in M. We have two choices. The first is that b survived by beating another node of B_r so it lost at least one outneighbor from B_r. Since M'_r was maximal, b may have lost at most one of its inneighbors (a'). Hence,

$$out_{B_{r+1}}(b) + 1 \le (out_{B_r}(b) - 1 + 1) \le in_{A_r}(b) - 1 \le in_{A_{r+1}}(b).$$

By Lemma 1 this means that $out(b) \le out(\mathcal{A})$. The second choice is if b survived by beating a leftover node $\bar{a}$ from A_r. This can only happen if $A_r \setminus (M'_r \cup M_r) \ne \emptyset$. Thus, $\bar{a}$ was in $A_r \setminus (M'_r \cup M_r)$. However, since M'_r was maximal, $\bar{a}$ must lose to b, and so all inneighbors of b from A_r move on to the next round, and $out(b) \le out(\mathcal{A})$. Thus $\mathcal{A}$ has outdegree at least as high as all nodes in $B_{r+1} \setminus \mathcal{H}_{r+1}$.

Now we consider A_{r+1} vs $\mathcal{H}_{r+1}$. We have

$$|A_{r+1}| \ge \lfloor (|A_r| + |M'_r| + |M_r| - 1)/2 \rfloor, \text{ and}$$

$$|\mathcal{H}_{r+1}| \le \lceil (|\mathcal{H}_r| - |M_r|)/2 \rceil = \lfloor (|\mathcal{H}_r| + 1 - |M_r|)/2 \rfloor.$$

Since $|\mathcal{H}_r| \ge 1$ we must have $|M_r| \ge 1$. If either $|M_r| \ge 2$, $|A_r| \ge |\mathcal{H}_r| + 2$, or $|M'_r| \ge 1$ then it must be that $|A_{r+1}| \ge \lfloor (|\mathcal{H}_r| + 2)/2 \rfloor \ge |\mathcal{H}_{r+1}| + 1$. Also, if $|\mathcal{H}_r|$ is even then $|A_{r+1}| \ge |\mathcal{H}_r|/2 = 1 + \lfloor (|\mathcal{H}_r| - 1)/2 \rfloor \ge |\mathcal{H}_{r+1}| + 1$, and the invariant is satisfied for round $r + 1$.

On the other hand, assume that $|M_r| = 1, |M'_r| = 0, |A_r| = |\mathcal{H}_r| + 1$ and $|\mathcal{H}_r|$ is odd. This necessarily implies that $|B_r \setminus \mathcal{H}_r| \le 1$. Since $|A_r| = |\mathcal{H}_r| + 1$ is even, $|B_r|$ must be odd and so $|B_r \setminus \mathcal{H}_r|$ must be even. $|B_r \setminus \mathcal{H}_r|$ can only be 0. This means $|\mathcal{H}_r| = n_r/2 - 1$ (where n_r is the current number of nodes). We can conclude that $\mathcal{A}$ is a king with outdegree at least half the graph and the tournament can be efficiently fixed so that $\mathcal{A}$ wins by Theorem 1. $\square$

Theorem 2 implies the following corollaries.

Corollary 1. *Let $\mathcal{A}$ be a king in a tournament graph. If $|\mathcal{H}^{in}(\mathcal{A})| \le (n-3)/4$, then one can efficiently compute a winning SE tournament seeding for $\mathcal{A}$.*

Corollary 2. *Let $\mathcal{A}$ be a king in a tournament graph. If $|\mathcal{H}(\mathcal{A})| \le n/3 - 1$, then one can efficiently compute a winning SE tournament seeding for $\mathcal{A}$.*

The proof of Corollary 1 follows by the fact that if $|\mathcal{H}^{in}(\mathcal{A})| = k$, then $out(\mathcal{A}) \ge (n - k)/3$. Corollary 2 simply states that any player in the top third of the bracket who is a king is also a tournament winner.

Proof of Corollary 2: Let $K = |\mathcal{H}(\mathcal{A})|$. Then the outdegree of $\mathcal{A}$ is at least $(n - K - 1)/2$. Let $h = |\mathcal{H}^{in}(\mathcal{A})|$. By Theorem 2, a sufficient condition for $\mathcal{A}$ to be able to win an SE tournament is that $out(\mathcal{A}) \ge h + 1$. Hence it is sufficient that $n - K - 1 \ge 2h + 2$, or that $2h + K \le n - 3$. Since $2h + K \le 3K$, it is sufficient that $3K \le n - 3$, and since $K \le (n - 3)/3$ we have our result. $\square$

3   Condorcet Random Model

We can now apply our results to graphs generated by the CR Model. From prior work we know that if $p \geq C\sqrt{\ln n/n}$ for $C > 4$, then with probability at least $1 - 1/\text{poly}(n)$, any node in a tournament graph generated by the CR model can win an SE tournament. However, since p must be less than $1/2$, this result only applies for $n \geq 512$. Moreover, even for $n = 8192$ the relevant value of p is $> 13\%$ which is a very high noise rate. We consider how many players can be efficiently made winners when p is a slower growing function of n. We show that even when $p \geq C \ln n/n$ for a large enough constant C, a constant fraction of the top players in a CR tournament can be efficiently made winners.

Theorem 3 (CR Model Winners for Lower p). *For any given constant $C > 16$, there exists a constant n_C so that for all $n > n_C$ the following holds. Let $p \geq C \ln n/n$, and G be a tournament graph generated by the CR model with error p. With probability at least $1 - 3/n^{C/8-2}$, any node v with $v \leq n/2 - 5C\sqrt{n}\ln n$ can win an SE tournament.*

This result applies for $n \geq 256$ and also reduces the amount of noise needed. For example, if $C = 17$ then when $n = 8192$, it is only necessary that $p < 2\%$, as opposed to $> 13\%$. This is a significant improvement. The proof of Theorem 3 uses Theorem 2 and Chernoff-Hoeffding bounds.

Theorem 4 (Chernoff-Hoeffding). *Let $X_1, \ldots, X_n$ be random variables with $X = \sum_i X_i$, $E[X] = \mu$. Then for $0 \leq D < \mu$, $Pr[X \geq \mu + D] \leq \exp(-D^2/(4\mu))$ and $Pr[X < \mu - D] \leq \exp(-D^2/(2\mu))$.*

Proof of Theorem 3: Let C be given. Consider player j. The expectation of the number n_j of outneighbors of j in G is

$$E[n_j] = (1 - p)(n - j) + (j - 1)p = n(1 - p) - p - j(1 - 2p).$$

This is exactly where we use the CR model. Our result is not directly applicable to the GCR model because this is only a lower bound on the expectation of n_j in that model. We will show that with high probability, all n_j are concentrated around their expectations and that all players $j \leq n/2$ are kings.

Showing that each n_j is concentrated around its expectation is a standard application of the Chernoff bounds and a union bound. Therefore, for $C > 16$ and $n > 2$, we have $2/n^{C^2/4} < 1/n^C$. Hence, with probability at least $1 - 1/n^{C-1}$ for every j, $|E[n_j] - n_j| \leq C\sqrt{n}\ln n$.

We assume n is large enough so that $n >> \sqrt{n}\ln n$ and that $p \leq 1/4$ so that $1 \geq (1 - 2p) \geq 1/2$. Now fix $j \leq n/2$. By the concentration result, this implies

$$n_j \geq 3n/4 - 1 - j - C\sqrt{n}\ln n \geq n/4 - 1 - C\sqrt{n}\ln n \geq \varepsilon n,$$

where $\varepsilon = 1/8$ works. The probability that j is a king is quite high: the probability that some node z has no inneighbor from $N^{out}(j)$ is at most

$$n(1 - p)^{n_j} \leq n(1 - C \ln n/n)^{(n/(C \ln n)) \cdot C\varepsilon \ln n} \leq 1/n^{\varepsilon C-1}.$$

By a union bound, the probability that some node j is not a king is at most $1/n^{\varepsilon C-2}$. Therefore, we can conclude that the probability that all the n_j are concentrated around their expectations and all nodes $j \leq n/2$ are kings is at least $1 - (1/n^{C-1} + 1/n^{\varepsilon C-2})$.

We now need to upper bound $|\mathcal{H}^{in}(j)|$. We are interested in how many nodes with $i < j + 2C\sqrt{n \ln n}/(1 - 2p)$ appear in $N^{in}(j)$: if we have an upper bound on them, we can apply Theorem 2 to get a bound on j. First, consider how small $n_j - n_i$ can be for any i:

$$n_j - n_i \geq (i - j)(1 - 2p) - 2C\sqrt{n \ln n}.$$

So for $i \geq j + 2C\sqrt{n \ln n}/(1 - 2p)$, $n_j \geq n_i$ with high probability. The expected number of nodes $i < j$ that appear in $N^{in}(j)$ is $(1-p)(j-1)$. By the Chernoff bound, the probability that at least $(1-p)(j-1) + C\sqrt{j \ln n}$ of the $j-1$ nodes less than j are in $N^{in}(j)$ is $\leq exp(-C^2 j \ln n/4j) = n^{-C^2/4}$. Therefore, with probability at least $1 - 1/n^{C^2/4}$, the number of such i is at most $(1-p)(j-1) + C\sqrt{j \ln n}$. By a union bound, this holds for all j with probability at least $1 - 1/n^{C^2/4-1}$. Now, we can say with high probability that $|\mathcal{H}^{in}(j)|$ is at most

$$(1-p)(j-1) + C\sqrt{j \ln n} + \frac{2C\sqrt{n \ln n}}{1 - 2p} \leq (1-p)(j-1) + 5C\sqrt{n \ln n}.$$

By Theorem 2, for there to be a winning seeding for j, it is sufficient that $\mathcal{H}^{in}(j) < n_j$ or that

$$(1-p)(j-1) + 5C\sqrt{n \ln n} < n(1-p) - p - j(1-2p) - C\sqrt{n \ln n}.$$

Rearranging the above equation, it is sufficient if

$$j < n/2 + \frac{pn}{(2(2 - 3p))} + \frac{(1 - 2p)}{(2 - 3p)} - 24C\sqrt{n \ln n}/5,$$

and so for all $j \leq n/2 - 5C\sqrt{n \ln n}$, there is a winning seeding for j with probability at least

$$1 - (2/n^{C-1} + 1/n^{\varepsilon C-2}) \geq 1 - 3/n^{C/8-2}.$$

$\square$

3.1 Improving the Result for the GCR Model through Perfect Matchings

Next, we show that there is a trade-off between the constant in front of $\log n/n$ and the fraction of nodes that can win an SE tournament. The proofs are based on the following result of Erdős and Rényi [5]. Let $B(n,p)$ denote a random bipartite graph on n nodes in each partition such that every edge between the two partitions appears with probability p.

Theorem 5 (Erdős and Rényi [5]). *Let c_n be any function of n, then consider $G = B(n,p)$ for $p = (\ln n + c_n)/n$. The probability that G contains a perfect matching is at least $1 - 2/e^{c_n}$.*

For the particular case $c_n = \Theta(\ln n)$, G contains a perfect matching with probability at least $1 - 1/\text{poly}(n)$.

Lemma 2. *Let $C \geq 64$ be a constant. Let $n \geq 16$ and G be a GCR tournament for $p = C \ln n/n$. With probability at least $1 - 2/n^{C/32-1}$, G is such that one can efficiently construct a winning SE tournament seeding for the node ranked 1.*

Proof. We will call the top ranked node s. We will show that with high probability s has outdegree at least $n/4$ and that every node in $N^{in}(s)$ has at least $\log n$ inneighbors in $N^{out}(s)$. This makes s a superking, and by [16], s can win an SE tournament.

The probability that s beats any node j is $> 1/2$, the expected outdegree of s is $> (n-1)/2$. By a Chernoff bound, the probability that s has outdegree $< n/4$ is at most $exp(-(n-1)/16) << 1/n^{C/32-1}$. Given that the outdegree of s is at least $n/4$, the expected number of inneighbors in $N^{out}(s)$ of any particular node y in $N^{in}(s)$ is at least $(n/4) \cdot (C \ln n/n) = (C/4) \ln n$.

We can show that each node in $N^{in}(s)$ has at least $\log n$ inneighbors from $N^{out}(s)$ by using a Chernoff bound and union bound. By a Chernoff bound, the probability that y has less than $(C/8) \ln n$ inneighbors from $N^{out}(s)$ is at most $exp(-(C/32) \ln n) = 1/n^{C/32}$. By a union bound, the probability that some $y \in N^{in}(s)$ has less than $(C/8) \ln n$ inneighbors from $N^{out}(s)$ is at most $1/n^{C/32-1}$. Therefore, s is a superking is with probability at least $1 - 2/n^{C/32-1}$ where $n \geq 16$, $n/4 \geq \log n, C > 64$, and $(C/8) \ln n \geq \log n$.   □

Lemma 2 concerned itself only with the player who is ranked highest in intrinsic ability. The next theorem shows that as we increase the noise factor, we can fix the tournament for an increasingly large set of players. As the noise level increases, we can argue recursively that there exists a matching from $\frac{n}{2} + 1 \ldots n$ to $1 \ldots \frac{n}{2}$, and from $\frac{3n}{4} + 1 \ldots n$ to $\frac{n}{2} + 1 \ldots \frac{3n}{4}$ and so forth. These matchings form each successive round of the tournament, eliminating all the stronger players.

Theorem 6. *Let $n \geq 16$, $i \geq 0$ be a constant and $p \geq 64 \cdot 2^i \ln n/n \in [0, 1]$. With probability at least $1 - 1/\text{poly}(n)$, one can efficiently construct a winning SE seeding for any of the top $1 + n(1 - 1/2^i)$ players in a GCR tournament.*

Proof. Let G be a GCR tournament for $p = C2^i \ln n/n$, $C \geq 64$. Let S be the set of all $n/2^{i-1}$ players j with $j > n(1 - 1/2^{i-1})$. Let s be a node with $1 + n(1 - 1/2^{i-1}) \leq s \leq 1 + n(1 - 1/2^i)$. The probability that s wins an SE tournament on the subtournament of G induced by S is high: there is a set X of at least $n/2^i - 1$ nodes that are after s. By Lemma 2, s wins an SE tournament on $X \cup \{s\}$ with high probability $1 - \frac{2}{(n/2^i)^{C/32-1}}$.

In addition, by Theorem 5, with probability at least $1 - \frac{2}{(n/2^i)^{C-1}}$, there is a perfect matching from $X \cup \{s\}$ to $S \setminus (X \cup \{s\})$. For every $1 \leq k \leq i-1$, consider

$$A_k = \{x \mid 1 + n(1 - 1/2^k) \leq x\}, \text{ and}$$

$$B_k = \{x \mid 1 + n(1 - 1/2^{k-1}) \leq x \leq n(1 - 1/2^k)\}.$$

Then $A_{k-1} = A_k \cup B_k$, $A_k \cap B_k = \emptyset$, and $|A_k| = |B_k| = n/2^k$. Hence $p \geq C \ln |A_k|/|A_k|$ for all $k \leq i - 1$. By Theorem 5, the probability that there is no perfect matching from A_k to B_k for a particular k is at most $2/(n/2^k)^{C2^{i-k}-1}$. This value is maximized for $k = i$, and it is $2/(n/2^i)^{C-1}$. Thus by a union bound, with probability at least $1 - 2i/(n/2^i)^{C-1} = 1 - 1/\mathrm{poly}(n)$, there is a perfect matching from A_k to B_k, for every k.

Thus, with probability at least $1 - 1/\mathrm{poly}(n)$, s wins an SE tournament in G with high probability, and the full bracket seeding can be constructed by taking the unions of the perfect matchings from A_k to B_k and the bracket from S.   □

For the CR model we can strengthen the bound from Theorem 3 by combining the arguments from Theorems 3 and 6.

Theorem 7. *There exists a constant n_0 such that for all $n > n_0$ the following holds. Let $i \geq 0$ be a constant, and $p = 64 \cdot 2^i \ln n/n \in [0, 1]$. With probability at least $1 - 1/\mathrm{poly}(n)$, one can efficiently construct a winning seeding for any of the top $n(1 - 1/2^{i+1}) - (80/2^{i/2})\sqrt{n \ln n}$ players in a CR tournament.*

As an example, for $p = 256 \ln n/n$, Theorem 7 says that any of the top $7n/8 - 40\sqrt{n \ln n}$ players are winners while Theorem 6 only gives $3n/4+1$ for this setting of p in the GCR model.

Proof. As in Theorem 6, for every $1 \leq k \leq i$, consider

$$A_k = \{x \mid 1 + n(1 - 1/2^k) \leq x\}, \text{ and}$$

$$B_k = \{x \mid 1 + n(1 - 1/2^{k-1}) \leq x \leq n(1 - 1/2^k)\}.$$

Then $A_{k-1} = A_k \cup B_k$, $A_k \cap B_k = \emptyset$, and $|A_k| = |B_k| = n/2^k$. By the argument from Theorem 6, w.h.p. there is a perfect matching from A_k to B_k, for all k.

Consider A_i. By Theorem 3, with probability $1 - 1/\mathrm{poly}(n)$, we can efficiently fix the tournament for any of the first $n/2^{i+1} - 80\sqrt{(n/2^i)\ln(n/2^i)}$ nodes in A_i. Combining the construction with the perfect matchings between A_k and B_k, we can efficiently construct a winning tournament seeding for any of the top

$$n - \frac{n}{2^i} + \frac{n}{2^{i+1}} - 80\sqrt{\frac{n}{2^i} \ln(\frac{n}{2^i})} \geq n(1 - \frac{1}{2^{(i+1)}}) - \frac{80}{2^{i/2}}\sqrt{n \ln n} \text{ nodes.}$$

□

4   Conclusions

In this paper, we have shown a tight bound (up to a constant factor) on the noise needed to fix an SE tournament for a large fraction of players when the match outcomes are generated by the CR model. As this model is believed to be a good model for real-world tournaments, this result shows that many tournaments in practice can be easily manipulated. In some sense, this sidesteps the question of whether it is NP-hard to fix a tournament in general by showing that it is easy on examples that we care about.

Acknowledgments. The authors are grateful for the detailed comments from the anonymous reviewers of WSCAI. The first author was supported by the NDSEG and NSF Graduate Fellowships and NSF Grant CCF-0830797. The second author was supported by the NSF under Grant #0963904 and under Grant #0937060 to the CRA for the CIFellows Project. Any opinions, findings, and conclusions or recommendations expressed in this material are those of the authors and do not necessarily reflect the views of the NSF or the CRA.

References

1. Bartholdi, J., Tovey, C., Trick, M.: The computational difficulty of manipulating an election. Social Choice Welfare 6(3), 227–241 (1989)
2. Bartholdi, J., Tovey, C., Trick, M.: How hard is it to control an election. Mathematical and Computer Modeling, 27–40 (1992)
3. Braverman, M., Mossel, E.: Noisy sorting without resampling. In: SODA, pp. 268–276 (2008)
4. Coppersmith, D., Fleischer, L., Rudra, A.: Ordering by weighted number of wins gives a good ranking for weighted tournaments. In: SODA, pp. 776–782 (2006)
5. Erdős, P., Rényi, A.: On random matrices. Publications of the Mathematical Institute Hungarian Academy of Science 8, 455–561 (1964)
6. Feige, U., Peleg, D., Laghavan, P., Upfal, E.: Computing with unreliable information. In: STOC, pp. 128–137 (1990)
7. Fischer, F., Procaccia, A.D., Samorodnitsky, A.: On voting caterpillars:approximating maximum degree in a tournament by binary trees. In: COMSOC (2008)
8. Gibbard, A.: Manipulation of voting schemes: a general result. Econometrica 41 (1973)
9. Hazon, N., Dunne, P.E., Kraus, S., Wooldridge, M.: How to rig elections and competitions. In: COMSOC (2008)
10. Lang, J., Pini, M.S., Rossi, F., Venable, K.B., Walsh, T.: Winner determination in sequential majority voting. In: IJCAI (2007)
11. Russell, T.: A computational study of problems in sports. University of Waterloo PhD Disseration (2010)
12. Russell, T., Walsh, T.: Manipulating tournaments in cup and round robin competitions. In: Algorithmic Decision Theory (2009)
13. Satterthwaite, M.A.: Strategy-proofness and arrow's conditions: Existence and correspondence theorems for voting procedures and social welfare functions. Journal of Economic Theory 10 (1975)
14. Slater, P.: Inconsistencies in a schedule of paired comparisons. Biometrika 48(3/4), 303–312 (1961)
15. Stanton, I., Vassilevska Williams, V.: Rigging tournament brackets for weaker players. In: IJCAI (2011)
16. Vassilevska Williams, V.: Fixing a tournament. In: AAAI, pp. 895–900 (2010)
17. Vu, T., Altman, A., Shoham, Y.: On the complexity of schedule control problems for knockout tournaments. In: AAMAS (2009)
18. Vu, T., Hazon, N., Altman, A., Kraus, S., Shoham, Y., Wooldridge, M.: On the complexity of schedule control problems for knock-out tournaments. In: JAIR (2010)
19. Young, H.P.: Condorcets theory of voting. The American Political Science Review 82(4), 1231–1244 (1988)

Computing Nash Equilibria of Action-Graph Games via Support Enumeration

David R.M. Thompson[1], Samantha Leung[2], and Kevin Leyton-Brown[1]

[1] University of British Columbia
{daveth,kevinlb}@cs.ubc.ca
[2] Cornell University
samlyy@cs.cornell.edu

Abstract. The support-enumeration method (SEM) for computation of Nash equilibrium has been shown to achieve state-of-the-art empirical performance on normal-form games. Action-graph games (AGGs) are exponentially smaller than the normal form on many important classes of games. We show how SEM can be extended to the AGG representation, yielding an exponential improvement in worst-case runtime. Empirically, we demonstrate that our AGG-optimized SEM algorithm substantially outperforms the original SEM, and also outperforms state-of-the-art AGG-optimized algorithms on most problem distributions.

1 Introduction

The canonical representation of simultaneous-move, perfect information games is the normal form. Because this representation grows exponentially in the number of players, it is impractical for modeling interactions that involve more than a handful of players. There is an extensive literature on compactly representing interesting games [19,9,12]. Action-graph games [8] (AGGs) unify these past representations and can furthermore represent many additional families of games in polynomial space. Computing solution concepts (e.g., Nash equilibrium) of compactly represented games is an active area of research. Although in some cases novel algorithms have been proposed [4,20], a particularly fruitful approach has been to take existing, normal-form-based algorithms, and modify them to operate efficiently with a compact representation [2,1]. Indeed, the state-of-the-art Nash-equilibrium-finding algorithms for AGGs are normal-form algorithms with AGG-optimized expected-utility calculations [8].

One algorithm that has not been modified to work with any compact representation also has very strong empirical performance: the support enumeration method (SEM) [18]. This algorithm also has other useful properties, such as returning the equilibrium in which agents randomize as little as possible, and being able to find all equilibria. A key reason that SEM has not been extended to work with compact game representations is that it operates very differently from other equilibrium-finding algorithms, and hence existing techniques cannot be applied to it directly. In this paper we show how SEM can be extended to work with AGGs (and hence with other game families compactly encodable as AGGs, such as graphical games and congestion games). Specifically, we show how three of SEM's subroutines can be made exponentially faster. Experimentally, we

N. Chen, E. Elkind, and E. Koutsoupias (Eds.): WINE 2011, LNCS 7090, pp. 338–350, 2011.

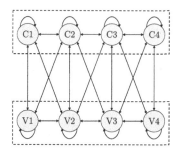

Fig. 1. An action graph for the "ice cream game" [8]. In the ice cream game, each player has either chocolate (C) or vanilla (V) ice cream to sell and must choose one of four possible locations to set up his stand. The action-graph encodes that a player's payoff depends only on his location, and the type and number of competitors within one step of his location.

show that these optimizations dramatically improve SEM's performance, rendering it competitive with and often stronger than other state-of-the-art algorithms for computing equilibria of AGGs.

2   Technical Background

In this section we summarize the two strands of related work that this paper brings together: AGGs and SEM.

2.1   Action-Graph Games

Action-graph games [8] achieve compactness by exploiting several kinds of structure in the payoffs. The first kind of structure is anonymity, which means that an agent's payoff depends only on his own action and the "configuration" induced by the other agents–a tuple of counts of how many agents played each action. Anonymity means that an agent's payoff does not depend on who played which action. This yields representational savings because, instead of storing a payoff for every pure-strategy profile, AGGs only need to store one for every configuration. The second kind of structure is context-specific independence, which means that a given agent only cares about the actions of others who take *a specific subset* of their actions. (This is a strengthening of (strict) independence, as captured by Graphical Games [9], which says that the agent never cares about certain others' actions, regardless of which actions they choose.) The context-specific aspect is that the actions (and hence agents) that can affect a given agent's payoff depend on which action that agent chooses. These independencies are encoded in an action graph, a directed graph where each node corresponds to an action, and payoff for playing an action only depends on the counts on its neighboring nodes. Instead of storing an action's payoff for every configuration, AGGs only need to store one for every "projected configuration," a tuple of counts on the neighbors of a vertex (denoted $C^{(v)}$ for vertex v). See Figure 1 for an example.

$$\sum_{a_{-i} \in S_{-i}} p(a_{-i})u_i(a_i, a_{-i}) = v_i \qquad \forall i \in N, \forall a_i \in S_i \tag{1}$$

$$\sum_{a_{-i} \in S_{-i}} p(a_{-i})u_i(a_i, a_{-i}) \le v_i \qquad \forall i \in N, \forall a_i \in A_i \setminus S_i \tag{2}$$

$$\sum_{a_i \in S_i} p_i(a_i) = 1 \qquad \forall i \in N \tag{3}$$

$$p_i(a_i) \ge 0 \qquad \forall i \in N, \forall a_i \in S_i \tag{4}$$

Fig. 2. The Test-Given-Support (TGS) feasibility program for n-player games. For any given support profile $S \in \prod_{i \in N} 2^{A_i}$, we can construct a TGS feasibility program where any feasible solution p, v is a Nash equilibrium with support S, where the players randomize according to the probabilities in p and get the payoffs specified by v. The constraints on line (1) specify that each player is indifferent between all the actions in his support. Those on line (2) specify that each player weakly prefers the actions in his support. The remaining lines specify that each mixed strategy is a probability distribution. (Note that this formulation allows for actions in the support to be played with zero probability. This doesn't adversely affect SEM's behavior; if such a Nash equilibrium existed, SEM would have found it already.)

Formally, an action-graph game is a 4-tuple (N, A, G, u) where N is the set of agents $(1, 2, ..., n)$; $A = \Pi_{i \in N} A_i$ is the set of action profiles (where $m = \max_{i \in N} |A_i|$); $G = (V, E)$ is a directed graph with vertices V (where $V = \bigcup_{i \in N} A_i$) and edges E; and $u = (u_1, u_2, ..., u_{|V|})$ is a tuple of utility functions, where $u_v : C^{(v)} \mapsto \mathbb{R}$.

Note that there is a more complicated family of AGGs, AGGs with function nodes (described in detail in [8]). The algorithms described in this paper also work (and have the same asymptotic performance) with function nodes, but we omit the description of these games for notational clarity and simplicity.

The other advantage of AGGs, besides their ability to represent games compactly, is that they can be reasoned about efficiently. In particular, given a mixed-strategy profile it is possible in polynomial time to compute a given agent i's expected utility for playing action a_i, using dynamic programming [8]. The dynamic program proceeds through n iterations, where at the kth iteration, it computes the marginal distribution over the projected configurations $C^{(a_i)}$ given the strategies of the first k agents.

2.2   Support-Enumeration Method

The support-enumeration method (SEM) is a brute-force-search method of finding Nash equilibria. However, rather than searching through all mixed strategy profiles, it searches through support profiles (specifying which actions each agent plays with positive probability) and tests whether there is a Nash equilibrium with that particular support. This test can be performed using the polynomial feasibility program given in Figure 2. Though several algorithms have been proposed for searching in the space of supports to find Nash equilibria [14,5,13], we will focus on the most recent SEM variant, due to Porter, Nudelman and Shoham [18]. This variant introduces two important features designed to improve empirical performance. First, it uses heuristics to order its exploration of the space of supports, searching from smallest to largest, breaking ties

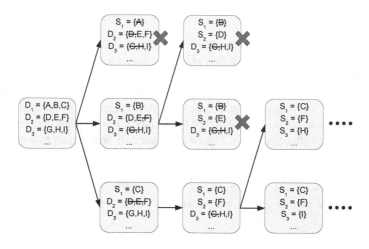

Fig. 3. Porter *et al*'s [18] tree-search works by instantiating strategies from agents' supports, one agent at a time, and removing strategies that are strictly dominated conditional on the agents playing within their given supports. The search backtracks whenever an agent has an empty support or the TGS feasibility program is infeasible.

in favor of more balanced support profiles. This order can speed up equilibrium finding for several reasons. First, there are fewer small support-size profiles to search through. Second, the corresponding feasibility programs have fewer variables and smaller constraints. And third, in games of interest (see, e.g., [11]), Nash equilibria with small, balanced supports are common [18]. Second, instead of simply iterating through the complete set of support profiles of a given size, SEM explores this space by tree search (see Figure 3 for an example). This search works by, at each level, selecting a support for a single additional player. At the leaves of the tree, the support is specified for every agent, and that support profile can be tested using TGS for the existence of a Nash equilibrium. The advantage of using search comes from pruning: after an agent's support is selected, SEM performs iterative removal of strictly dominated strategies (IRSDS), conditional on agents only playing actions in their supports. This has the effect of eliminating many support profiles from consideration. The search backtracks whenever an agent has an empty support or the TGS feasibility program is infeasible.

There are several reasons to be interested in the SEM algorithm. One is that it is the only known method for finding small-support equilibria. Another is that it has been shown to achieve better empirical performance than the previous state-of-the-art algorithms for the sample equilibrium problem, simplicial subdivision (SimpDiv) [10] and the global Newton method (GNM) [7], on many game families of interest. Notably, most of these games had pure-strategy Nash equilibria (PSNEs), which SEM finds in polynomial time. However, even on games without PSNEs—where SEM has exponential worst-case running time—SEM is still often faster than SimpDiv and GNM.

3   SEM for AGGs

Observe that we can trivially make a version of SEM that takes AGGs as input, simply replacing the normal form game (NFG) utility lookups $(u_i(a))$ with the AGG equivalents $(u_i(a) = u_{a_i}(c^{(a_i)}))$, where $c^{(a_i)}$ is the projected configuration given a). We denote this algorithm NFG-SEM, because its behavior is exactly the same as that of SEM for normal-form games. However, because AGGs can be exponentially smaller than NFGs, we show below that NFG-SEM's asymptotic worst-case performance, as a function of the length of its input, can be exponentially worse than that of SEM for the induced normal form of the same game. Specifically, we show that NFG-SEM's inner-loop operations—iterative removal of strictly dominated strategies and the TGS feasibility program—are at least worst-case exponential in the AGG input length (denoted ℓ). The outer-loop search over supports also requires exponential time, even for games with PSNEs.

However, we can do better if we construct a version of SEM that explicitly takes AGG structure into account. We present such an extension of SEM, denoted AGG-SEM[1], and its asymptotic analysis. Overall, we show that AGG-SEM's worst-case performance is exponentially faster than that of NFG-SEM.

3.1   Conditional Dominance

Because SEM makes extensive use of iterative removal of strictly dominated strategies, efficiently identifying dominated strategies is critical. For normal-form games, testing whether or not some pure strategy a_i is dominated by some other a_i' is straightforward: one can exhaustively search through through the pure strategy profiles of the other agents, looking for the existence of some a_{-i} to which a_i is a weakly better response. This trivial algorithm only requires time linear in the size of a normal-form game. However, it can require time exponential in the size of an action-graph game.

Lemma 1. *NFG-SEM's dominance check has a worst-case running time of* $\Theta(2^\ell)$.

Proof. Consider the family of action-graph games with two actions per player and no edges. For this family, there are at most $2n$ nodes, the payoff table for each of which contains only a single value. Thus, ℓ is $\Theta(n)$, while $|A_{-i}| = 2^{n-1}$ and therefore is $\Theta(2^\ell)$. In the worst case, exhaustive search iterates over every $a_{-i} \in A_{-i}$ to confirm that a_i is not a best response to any action profile. $\square$

However, we can do better: a straightforward, polynomial-time algorithm for AGG dominance checking can be derived from Jiang et al's [8] dynamic-programming algorithm. To determine whether or not a_i is dominated by a_i', we do not need to search through the entirety of A_{-i}; we only need to search over the set of possible projected configurations on the joint neighborhoods of a_i and a_i'. This adaptation guarantees polynomial runtime. However, empirically we observed that it often gave rise to poor performance, compared to exhaustive search over A_{-i}. We attribute this to stopping conditions: the

---

[1] Pseudo-code for AGG-SEM can be found in the extended version of this paper, which is available on the authors' websites.

exhaustive search can stop as soon as it finds any case where a_i is a better response, while the dynamic-programming algorithm must build up all configurations first before it ever encounters a better response, effectively performing a breadth-first search. Based on this insight, we created a depth-first-tree-search-based algorithm that combines the best of both approaches: like exhaustive search, it can find a better response without needing to compute the entire set of projected configurations; like our adaptation above, it exploits AGG structure and so needs only to evaluate a polynomial number of projected configurations. It works as follows. At each level, the search fixes the action of some agent, giving a search tree that potentially includes every A_{-i}. However, we also perform multiple-path pruning: a search refinement in which previously visited nodes are recorded, and the search backtracks whenever it re-encounters a node along a different search path [17]. In our case, the algorithm backtracks whenever it encounters a previously visited projected configuration, based on a lookup from a trie map.

Lemma 2. *AGG-SEM's dominance check has a worst-case running time of $O(nm\ell^3)$.*

Proof. The search traverses a tree with a depth of n and a branching factor of m. However, at every level, at most ζ^2 nodes are expanded (where ζ denotes the largest set of possible projected configurations for any node), because there are at most ζ^2 distinct projected configurations on the neighborhood of a_i, a_i'. In the worst case, when it traverses the whole tree, the search must follow each of m arcs from $O(\zeta^2)$ nodes at each of n levels, or $O(nm\zeta^2)$ arcs. For each arc, the search may perform a trie-map lookup and insert; these operations each require runtime that grows like the maximum in-degree of the graph, ι, and so the total cost is $O(nm\zeta^2\iota)$. Because ℓ is $\Omega(\zeta + \iota)$, $nm\zeta^2\iota$ is $O(nm\ell^3)$. □

3.2 TGS Feasibility Program

SEM's asymptotic performance is dominated by the Test-Given-Support feasibility program. (Polynomial feasibility is NP-hard; e.g., polynomial constraints generalize 0–1 integrality constraints [22].) For NFG-SEM, this complexity obstacle is particularly severe: directly representing the TGS feasibility program requires space exponential in the size of the AGG. Thus, unless $P = NP$, TGS requires doubly exponential time.

Lemma 3. *The NFG-SEM TGS feasibility program has worst-case size of $\Theta(nm2^\ell)$.*

Proof sketch, similar to Lemma 1's proof. TGS can have $|A_{-i}|$ terms in each constraint, and this quantity is $\Theta(2^\ell)$ in the worst case. There are $O(nm)$ such constraints. □

The essential challenge is in the expected utility constraints (lines 1 and 2 of Figure 2) which can be exponentially long. We already have Jiang et al's [8] dynamic-programming algorithm for computing expected utility given a specific mixed-strategy profile. Now we want to compute expected utility symbolically, without specifying the probabilities beforehand. This can be accomplished by "unrolling" the dynamic program: every update in the dynamic program is expressed as a polynomial equality constraint in the TGS program. This set of new constraints is polynomial in the size of the AGG.

Lemma 4. *The AGG-SEM TGS feasibility program has worst-case size of $O(n^2 m^2 \ell^2)$.*

Proof. For each $j \in \{1, \ldots, n\}$ we introduce $O(\zeta)$ new constraints, each corresponding to the probability of a projected configuration given the strategies of the first j agents. This gives $O(n\zeta)$ constraints. Each contains at most $O(\zeta m)$ terms, corresponding to the possible projected configurations and actions that could lead to some new projected configuration when another agent is added. Since ζ is $O(\ell)$, the output requires $O(nm\ell^2)$ space. It must be run once for each agent i and for each action in A_i: $O(nm)$ times in total. Thus, the TGS feasibility program requires $O(n^2 m^2 \ell^2)$ space.   □

Although this optimization speeds up the worst case exponentially, it is not guaranteed to be helpful on average. This is because the symbolic representation of the TGS system is made exponentially smaller by replacing each exponentially long expected-utility constraint with multiple small constraints. How this change affects runtime in the average case depends on the (black-box) feasibility solver.

3.3 Asymptotic Analysis of SEM for AGGs

We are now ready to demonstrate that asymptotically, AGG-SEM is exponentially faster than NFG-SEM. We assume that both algorithms make use of a polynomial feasibility solver with worst-case runtime $O(2^x)$ where x is the length of the feasibility program.

Theorem 1. *Assume that we have access to a polynomial feasibility solver with worst-case runtime $O(2^x)$, where x is the length of the feasibility program. Then NFG-SEM requires doubly exponential time time to find a sample Nash equilibrium in an AGG.*

Proof sketch. In the worst case, AGGs can have $\Omega(2^\ell)$ support profiles (as in Lemma 1), even for symmetric games. Thus, the search must traverse a tree with $O(2^\ell)$ leaf nodes where TGS is solved, and $O(2^\ell)$ interior nodes where iterative removal of strictly dominated strategies (IRSDS) is performed. Solving TGS takes $O(2^{nm2^\ell})$ runtime in the worst case. This expression is $O(2^x)$ where x is $O(nm2^\ell)$ by Lemma 3, and so dominates IRSDS. Thus the total runtime is $O(2^{nm2^\ell + \ell})$.   □

Theorem 2. *AGG-SEM requires (only) exponential time to find a sample Nash equilibrium in an AGG.*

Proof sketch. AGG-SEM still searches $O(2^\ell)$ nodes, but TGS now requires $O(2^{n^2 m^2 \ell^2})$ time (Lemma 4), which again dominates IRSDS. The total runtime is $O(2^{n^2 m^2 \ell^2 + \ell})$.   □

Given complexity results known in the literature, it is unsurprising that AGG-SEM requires exponential time in the worst case. In particular, finding even PSNEs of AGGs in polynomial time would imply P=NP: AGGs generalize graphical games, and finding a PSNE of an arbitrary graphical game is NP-hard [6]. Further, finding a PSNE of a symmetric AGG with unbounded m is also known to be NP-hard [4].

3.4 Further Speedups for k-Symmetric Games

We now show that the search over supports can be sped up in the case of AGGs with k-symmetry, i.e., where the players can be partitioned into k classes such that all players in a class are identical. (We describe the algorithm for the case of $k = 1$, or full symmetry. The generalization is straightforward.) We saw in the proof of Theorem 1 that symmetry does not help for NFG-SEM. Here we strengthen that result, showing that NFG-SEM can take exponential time even when PSNEs exist in k-symmetric AGGs with bounded m and k.

Theorem 3. *NFG-SEM requires exponential time time to find a sample PSNE in a k-symmetric AGG with bounded m and k.*

Proof sketch. For games with PSNEs, we never need to solve TGS: any support profile that survives IRSDS is a Nash equilibrium. The tree search must still expand $O(2^\ell)$ nodes to explore all $O(2^\ell)$ pure support profiles. At each interior node, IRSDS is called, requiring $O(n^2 m^3)$ calls to the conditional dominance test, which requires $O(2^\ell)$ time (by Lemma 1). For bounded m, the total runtime to find a PSNE is $O(n^2 2^{2\ell})$.   □

Next, we show that we *can* achieve an improvement on such games for AGG-SEM. This optimization works by skipping any support profile that is a permutation of a previously explored support profile. At every stage of the tree search, we explore a support S_i iff $S_i \succeq S_j$ where j is any player with support selected higher in the tree, and where $\succ$ is the order in which supports are explored at each level of the tree.

Lemma 5. *AGG-SEM's search evaluates* poly(n) *support profiles in the worst case, even for games without PSNEs, given a k-symmetric AGG with bounded k and m.*

Proof. Every distinct support profile can be identified by a vector of $O(k2^m)$ integers in the range $[0, n]$, where each element indicates how many agents of a given class have a given support. There are at most $O(n^{k2^m})$ such vectors. For bounded k and m, this quantity is poly(n).   □

Theorem 4. *AGG-SEM requires* poly(ℓ) *time to find a sample PSNE in a k-symmetric AGG with bounded m and k.*

Proof sketch. For bounded m and k, AGG-SEM's search expands polynomially many nodes (by Lemma 5), each of which requires running IRSDS. IRSDS performs $O(n^2 m^3)$ conditional dominance tests, requiring $O(nm\ell^3)$ time (by Lemma 2). Thus, AGG-SEM has poly(ℓ) runtime on such games.   □

4 Experimental Evaluation

So far, our analysis has concentrated on the worst case. However, improvements to the worst case do not necessarily imply improvements on instances of interest. As we are motivated by developing *practical* methods for computing Nash equilibria, we conducted an experimental evaluation to compare the performance of NFG-SEM and AGG-SEM.

4.1 Experimental Setup

We sampled 50 instances from each of 11 different game distributions (see Table 1). Nine distributions were from GAMUT [11]; each had $n = 10$ players, $m = 10$ actions

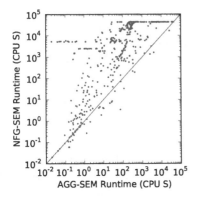 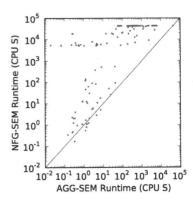

Fig. 4. Scatterplot contrasting the runtimes of AGG-SEM and NFG-SEM. The left plot shows runtimes for computing a sample Nash equilibria; the right plot shows runtimes for computing all pure strategy Nash equilibria.

per player, and action graphs with in-degree at most five. The remaining two distributions were over position auction games with $n = 10$ players and up to $m = 11$ actions per player (though weakly dominated actions, which occurred frequently, were omitted by the generator) [21]. On each game, we compared AGG-SEM to three other algorithms: NFG-SEM, and the two existing state-of-the-art Nash-equilibrium-finding algorithms: GNM, the global Newton method [7], and SimpDiv, simplicial subdivision [10], both using Gambit implementations [15] extended to work efficiently with AGGs by [8]. All algorithms were given error tolerance of 10^{-10}. For AGG-SEM and NFG-SEM, we used minos [16] to solve the TGS feasibility problems.

We performed a blocking mean-of-means test [3] (with $p \leq 0.01$) to compare mean runtimes across game distributions. In three distributions (see Table 1), we were not able to conclude that differences were significant because of high runtime variation. Such problems can be overcome by obtaining additional data; thus, for these distributions we generated an additional 150 instances (i.e., 200 total). In the end, we were able to identify a significantly faster algorithm for every distribution.

Our experiments were performed on machines with dual Intel Xeon 3.2GHz CPUs, 2MB cache and 2GB RAM, running Suse Linux 11.1 (Linux kernel 2.6.27.48-0.3-pae). Each run was limited to 12 CPU hours; we report runs that did not complete as having taken 12 hours. In total, our experiments required about 420 CPU days.

4.2   Results

Overall, we found that AGG-SEM provided a substantial performance improvement over NFG-SEM, outperforming it on the vast majority of instances; see Figure 4. As we expected, AGG-SEM was not faster on absolutely every instance. Nevertheless, AGG-SEM achieved significantly faster mean performance in every game distribution. Its largest speedup over NFG-SEM was 280× (on D1), its smallest speedup was 1.45× (on D4), and its median speedup was 10× (on D10). While the biggest speedups were on games where AGG-SEM could leverage k-symmetry, ranging from 280× (on D1) to

$7\times$ (on D2), we still achieved substantial speedups on asymmetric games (those with n player classes), ranging from $10\times$ (on D10) to $1.45\times$ (on D4). AGG-SEM stochastically dominated NFG-SEM overall (see Figure 5), but on a per-distribution basis, it only stochastically dominated on four distributions (D3 and D9–11). AGG-SEM's failure to stochastically dominate on the remaining seven distributions was due to the fact that all contained instances that both methods solved extremely quickly (in less than one second), but that NFG-SEM finished more quickly. Considering runtimes over a second, AGG-SEM stochastically dominated NFG-SEM on every distribution.

AGG-SEM outperformed SimpDiv and GNM on 9 of the 11 game distributions (see Table 1), and furthermore stochastically dominated GNM overall (see Figure 5). Comparing the runtimes of AGG-SEM and SimpDiv, we found that the two were correlated: they both solved many (338) of the same instances in under 600s, with SimpDiv having better mean runtime on these instances ($\mu = 12.71$s vs $\mu = 108.32$s). On the remaining instances, SEM finished far more often (87.74% vs 23.11%). Thus, SimpDiv was only fastest on D2 (Ice-cream games), which contained almost exclusively instances that were easy for both algorithms. AGG-SEM only stochastically dominated SimpDiv on D10 and D11, which contained no instances that were easy for simpDiv. AGG-SEM's runtime was less correlated with that of GNM than it was with SimpDiv. For example, GNM solved every instance in D4 (GFP position auctions), which contained instances that were not solved by either SimpDiv or AGG-SEM. Overall, however, GNM had the worst performance (and was stochastically dominated by AGG-SEM on every distribution but D3 and D4). Although AGG-SEM had the fastest overall average runtime, there were at least a few instances for which each of AGG-SEM, SimpDiv and GNM was hundreds of times faster than the others. The best practical approach may thus be a portfolio of all three algorithms, following e.g. [23].

Like Porter, *et. al.*, [18], we found that in many (7 of 11) distributions, most games (over 90%) had PSNEs. AGG-SEM finished on every such game. Thus the four distributions (D4 and D9-11) in which PSNEs were least common were also those on which AGG-SEM was mostly likely to time out. (We verified that AGG-SEM terminated on every game with PSNEs by checking the support size AGG-SEM was considering when it timed out. In every case, it ruled out all supports of size 1 for every agent before running out of time.)

One advantage of SEM over other Nash-equilibrium-finding algorithms is its ability to find all Nash equilibria (or all equilibria with support sizes not more than some constant). This is particularly useful when we want to understand the range of possible outcomes. For example, in [21] one of the goals was to identify the minimum and maximum revenue possible in equilibrium of position auction games. At the time, only empirical bounds on revenue were possible, because there was no algorithm available for finding all the PSNEs of an AGG. We have since tested equilibrium enumeration by searching for all PSNEs on a representative subset of these games (20 from each distribution). We found that AGG-SEM was significantly faster than NFG-SEM (see Figure 4) at enumerating the set of all equilibria. Notably, for every position auction game, AGG-SEM was able to find all PSNEs in under one CPU minute. (These games each have ten bidders, eight positions and eleven bid increments per bidder.)

Table 1. Mean runtimes. * denotes significantly slower than fastest solver, $\alpha = 0.01$. Capped runs count as 43200s. † denotes distributions where, due to high variance, more instances (200) were necessary for statistical significance.

No.	Game type	Player Classes	Mean runtime (CPU s)			
			AGG-SEM	NFG-SEM	GNM	SimpDiv
D1	Coffee shop	1	**18.00**	5032.38*	3309.73*	362.63*†
D2	Ice cream	3	131.64*	957.42*	151.59*	**0.39**
D3	Job market	1	**249.02**	6070.61*	372.96*†	1536.45*†
	Position Auctions:					
D4	GFP	n	7519.90*	10878.19*	**75.73**	10750.93*
D5	Weighted GSP	n	**45.10**	96.78*	723.19*	734.56*†
	Random AGGs:					
D6	Random graph	1	**68.02**	7005.34*	10580.58*	5188.02*
D7	Road graph	1	**441.11**	32103.15*	41814.79*	9507.58*
D8	Small-world graph	1	**596.75**	31750.79*	28195.09*	4665.58*
	Random Graphical Games:					
D9	Random graph	n	**11953.48**	20469.50*	24337.47*	27002.81*
D10	Road graph	n	**3244.50**	32052.36*	43200.00*	43200.00*
D11	Small-world graph	n	**11356.47**	29861.96*	43200.00*	40677.67*
	Overall:		**3244.28**	16520.09*	18265.93*	13176.94*

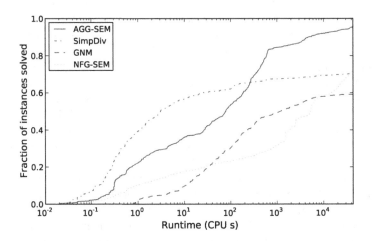

Fig. 5. Runtime CDFs for our four algorithms

5   Conclusion

We have showed that the support enumeration method can be extended to games compactly represented as AGGs. Our approach outperforms the original SEM algorithm for such games both asymptotically and in practice. Theoretically, we showed that SEM's worst-case runtime can be reduced exponentially. Our work in this vein may also be of independent interest, as it shows novel ways of exploiting AGG structure. In particular, the polynomial-time algorithm for removing dominated strategies could be useful, e.g. as a preprocessing step for other equilibrium-finding algorithms. Empirically, we observed that our new algorithm was substantially (often orders of magnitude) faster, and that it almost always outperformed current state-of-the-art algorithms. Beyond this, our algorithm offers substantial advantages over existing algorithms, such as the ability to enumerate equilibria and to identify pure-strategy Nash equilibria or prove their non-existence.

We envision several extensions to AGG-SEM. One promising direction is to search for specific types of (e.g., symmetric or social-welfare-maximizing) equilibria, for example by replacing the depth-first search with branch-and-bound search. Performance could also be improved by using good heuristics to choose the order in which supports are instantiated, or even by exploring the space of supports using stochastic local search rather than tree search.

References

1. Bhat, N., Leyton-Brown, K.: Computing Nash equilibria of Action-Graph Games. In: UAI (2004)
2. Blum, B., Shelton, C., Kohler, D.: A continuation method for Nash equilibria in structured games. JAIR 25, 457–502 (2006)
3. Chernick, M.R.: Bootstrap Methods, A practitioner's guide. Wiley (1999)
4. Daskalakis, C., Schoenebeck, G., Valiant, G., Valiant, P.: On the complexity of Nash equilibria of action-graph games. In: SODA (2009)
5. Dickhaut, J., Kaplan, T.: A program for finding Nash equilibria. Mathematica J. 1, 87–93 (1991)
6. Gottlob, G., Greco, G., Scarcello, F.: Pure nash equilibria: hard and easy games. In: TARK (2003)
7. Govindan, S., Wilson, R.: A global Newton method to compute Nash equilibria. J. Economic Theory 110, 65–86 (2003)
8. Jiang, A.X., Leyton-Brown, K., Bhat, N.A.R.: Action-graph games. GEB 71, 141–173 (2011)
9. Kearns, M.J., Littman, M.L., Singh, S.P.: Graphical models for game theory. In: UAI (2001)
10. van der Laan, G., Talman, A.J.J., van Der Heyden, L.: Simplicial variable dimension algorithms for solving the nonlinear complementarity problem on a product of unit simplices using a general labelling. Mathematics of Operations Research 12, 377–397 (1987)
11. Leyton-Brown, K., Nudelman, E., Wortman, J., Shoham, Y.: Run the GAMUT: A comprehensive approach to evaluating game-theoretic algorithms. In: AAMAS (2004)
12. Leyton-Brown, K., Tennenholtz, M.: Local-effect games. In: IJCAI (2003)
13. Lipton, R.J., Markakis, E.: Nash Equilibria Via Polynomial Equations. In: Farach-Colton, M. (ed.) LATIN 2004. LNCS, vol. 2976, pp. 413–422. Springer, Heidelberg (2004)
14. Mangasarian, O.: Equilibrium points of bimatrix games. J. Society for Industrial and Applied Mathematics 12, 778–780 (1964)
15. McKelvey, R.D., McLennan, A.M., Turocy, T.L.: Gambit: Software tools for game theory (2006), http://econweb.tamu.edu/gambit
16. Murtagh, B., Saunders, M.: MINOS (2010), http://www.sbsi-sol-optimize.com
17. Poole, D.L., Mackworth, A.K.: Artificial Intelligence. Cambridge University Press (2011)
18. Porter, R.W., Nudelman, E., Shoham, Y.: Simple Search Methods for Finding a Nash Equilibrium. GEB 63, 642–662 (2009)

19. Rosenthal, R.: A class of games possessing pure-strategy Nash equilibria. Int. J. Game Theory 2, 65–67 (1973)
20. Roughgarden, T., Papadimitriou, C.: Computing correlated equilibria in multi-player games. JACM 37, 49–56 (2008)
21. Thompson, D.R.M., Leyton-Brown, K.: Computational analysis of perfect-information position auctions. In: ACM-EC (2009)
22. Vohra, R.V.: Advanced Mathematical Economics. Routledge (2005)
23. Xu, L., Hutter, F., Hoos, H.H., Leyton-Brown, K.: SATzilla: portfolio-based algorithm selection for SAT. JAIR 32, 565–606 (2008)

Heavy Traffic Approximation of Equilibria in Resource Sharing Games

Yu Wu, Loc Bui, and Ramesh Johari

MS&E, Stanford University
{yuwu,locbui,ramesh.johari}@stanford.edu

Abstract. We consider a model of priced resource sharing that combines both *queueing behavior* and *strategic behavior*. We study a priority service model where a single server allocates its capacity to agents in proportion to their payment to the system, and users from different classes act to minimize the sum of their cost for processing delay and payment. As the exact processing time of this system is hard to compute, we introduce the notion of *heavy traffic equilibrium* as an approximation of the Nash equilibrium, derived by considering the asymptotic regime where the system load approaches capacity. We discuss efficiency and revenue, and in particular provide a bound for the price of anarchy of the heavy traffic equilibrium.

Keywords: resource sharing, discriminatory processor sharing, equilibrium, heavy traffic approximation.

1 Introduction

A range of resource sharing systems, such as computing or communication services, exhibit two distinct characteristics: *queueing behavior* and *strategic behavior*. Queueing behavior arises because jobs or flows are served with the limited capacity of system resources. Strategic behavior arises because these jobs or flows are typically generated by self-interested, payoff-maximizing users. Analysis of strategic behavior in queueing systems has a long history, dating to the seminal work of Naor [16]; see the book by Hassin and Haviv [7] for a comprehensive survey. The interaction of queueing and strategic behaviors has become especially important recently, with the rise of paid resource sharing systems such as cloud computing platform. For example, [1] and [4] discussed systems with multiple service providers, modelled as first-come-first-serve queues, that compete in both price and response time for potential buyers.

In this paper we consider a particular queueing model where a single server is shared among multiple jobs, and the service capacity allocated to each job depends on its priority level. The particular scheduling policy we consider is known in the literature as the *discriminatory processor sharing* (DPS) policy [12]. In the DPS model, the server shares its capacity *in proportion* to the priority level of all jobs currently in the system. This service allocation rule is a special case of a more general scheduling policy for queueing networks known as *proportionally fair* resource sharing [10, 14];

N. Chen, E. Elkind, and E. Koutsoupias (Eds.): WINE 2011, LNCS 7090, pp. 351–362, 2011.

such scheduling policies have been studied extensively in the context of networked resource sharing (see [9, 19] and references therein). A survey of the DPS literature can also be found in [2].

We consider a DPS system in steady state, and study a *job level* game where every individual job is a single strategic user. This user chooses a payment β; which corresponds to the priority level of that user. The user also incurs a cost proportional to total processing time. The users' goal is to choose priority levels to minimize the sum of expected processing cost and payment. (We also briefly discuss a *class level* game, where every *class* is a single user.) A central difficulty in analysis of equilibria arises because exact computation of the steady state processing time of a single job, given the priority choices of other jobs, is intractable. Since the queueing behavior computation itself involves numerical complexity, equilibrium characterization in closed form for the strategic behavior is essentially impossible. Thus obtaining structural insight into the games is a significant challenge.

In this paper, we propose an alternate approximate approach to equilibrium characterization that is amenable to analysis, computable in closed form, and provably exact in an appropriate asymptotic regime where the load on the system increases, known as the *heavy traffic* regime [11, 18]. The heavy traffic asymptotic regime is widely used in analysis of queueing systems and is especially valuable to study systems with many users. Asymptotics yield two benefits. *First*, they significantly simplify stochastic analysis. The *second* key benefit of asymptotics is that we are also able to simplify our game theoretic analysis. Informally, an important reason is that when the number of users grows large, no single user has a large impact on the whole system; this effect allows us to simplify calculation of equilibria.

Our main contributions are as follows.

(1) An approximate notion of equilibrium. Using an approximation to the processing time derived via the heavy traffic asymptotic regime, we suggest a natural corresponding notion of equilibrium that we call *heavy traffic equilibrium* (HTE). In an HTE, users minimize the sum of their payment and heavy traffic processing time cost, rather than their true expected processing time cost. We show that under mild conditions, HTE exists and is unique, and that it can be computed in closed form in terms of system parameters. It is thus both simple to compute, and asymptotically accurate when the system approaches heavy traffic.

(2) Economic analysis: parameter sensitivity, efficiency, and revenue. A significant benefit of our approach is that since we can compute the equilibrium in closed form, it is straightforward to carry out analysis on efficiency and revenue. We study how the system behavior changes when cost or arrival rate parameters are scaled, and more importantly, we investigate social efficiency and system revenue of HTE under different system parameters, and give a bound for the price of anarchy of HTE. We obtain some intriguing insights: in particular, we show that within a particular class of pricing schemes, and for a wide range of parameter choices, the incentives of the revenue maximizing service provider become aligned with minimization of total system processing cost.

We believe our work makes significant progress on two fronts. First, the DPS queueing model is important in its own right as a benchmark model for analysis of priority pricing for shared resource services. Our analysis provides extensive insight into this queueing system with strategic behavior. Second, and perhaps of greater longer term interest, our approximation methodology suggests a broader research program for understanding strategic behavior in queueing systems: by exploiting large system asymptotics, we can simplify *both* the complexity of the stochastic system, as well as the complexity of the economic system.

The reader is referred to the companion technical report for the proofs of all theorems in this paper [20].

2 Resource Sharing Game

We consider a queueing game in which K classes of jobs share a single server of unit capacity. Class i $(i = 1, \cdots, K)$ jobs arrive according to a Poisson process with arrival rate λ_i and have i.i.d. exponentially distributed service requirements (measured in units of service, e.g., processing cycles) with mean $1/\mu_i$. Throughout this paper, we assume for simplicity that $\mu_i = \mu$ for all classes. Let $\lambda = \sum_k \lambda_k$ denote the total arrival rate to the system. Also, let $\rho_i = \lambda_i/\mu$ be the *load* of class i, and define the *system load* as $\rho = \sum_k \rho_k = \sum_k \lambda_k/\mu$. To ensure stability, we assume $\rho < 1$. It is well known that under this condition, the resulting queueing system is ergodic and possesses a unique steady state distribution [13]. Waiting and being served in the system induce a cost c_i per unit time for users of class i. Without loss of generality we assume $c_1 > c_2 > \cdots > c_K$: if two classes i and j have the same cost $c_i = c_j$, then they can be merged to one class with arrival rate $\lambda_i + \lambda_j$.

We assume that the server allocates its capacity according to the *discriminatory processor sharing* (DPS) policy. Under this policy, each job is associated with a priority level. If there are currently N jobs in the system and job ℓ has chosen priority level β_ℓ, then the fraction of service capacity allocated to job k is $\beta_\ell / \sum_{m=1}^N \beta_m$.

Upon arrival, without observing the state of the system, each job chooses a priority level β. Throughout the paper, we assume $\beta \geq \underline{\beta}$, where $\underline{\beta} > 0$ is a sufficiently small minimum priority level required of any participant in the system. We consider a family of pricing rules for priority that we refer to as α-*fair pricing rules*, where $\alpha > 0$. Formally, we assume that if a job chooses priority level β, then the system manager charges that job a price β^α, where $\alpha > 0$. Varying α allows us to study a range of pricing schemes. In particular, as $\alpha \to 0$, jobs face a strongly diminishing marginal cost to higher choices of β; while as $\alpha \to \infty$, jobs face a strongly increasing marginal cost with higher choices of β.

The pricing rules we consider are closely related to α-*fair allocation rules* studied in the networking literature [15]. In an α-fair allocation system, one unit of resource is allocated to N users, whose utility functions are characterized by α: $U^{(\alpha)}(x) = x^{1-\alpha}/(1 - \alpha)$ if $\alpha \neq 1$, and $U^{(\alpha)}(x) = \log(x)$ if $\alpha = 1$. Users make payments for use of the system; let w_ℓ be the payment of user ℓ; the payments determine users' weights in the system. Formally, suppose the payment vector of users is w and the allocation vector is x; then the resource manager solves the following optimization problem:

$$\max_{x} \sum_{\ell=1}^{N} w_\ell U^{(\alpha)}(x_\ell) \quad s.t. \sum_{\ell=1}^{N} x_\ell \leq 1,$$

The solution of this problem is $x_\ell = w_\ell^{1/\alpha}/(\sum w_m^{1/\alpha})$. A well-known example of an α-fair allocation rule is the *proportionally fair* allocation rule, obtained when $\alpha = 1$ [10]: resource is allocated proportional to payment. Now suppose that the α-fair pricing rule is used in our model, so that $w_\ell = \beta_\ell^\alpha$. Then the α-fair allocation rule reduces to the discriminatory processor sharing policy described above—i.e., allocation of server capacity in proportion to the priority levels β_ℓ.

In this paper we will generally be interested in scenarios where all jobs of the same class i choose the same priority level. In an abuse of notation we denote by β_i the priority level chosen by all class i jobs, and in this case we succinctly denote $(\beta_1, \cdots, \beta_K)$ by $\boldsymbol{\beta}$. We refer to $\boldsymbol{\beta}$ as the *class priority vector*.

Let $V(\beta; \boldsymbol{\beta})$ be the expected processing time for a job with priority β that arrives to the system in steady state, with the class priority vector given by $\boldsymbol{\beta}$. Observe that with this notation, a class i job with priority level β_i has expected processing time $V(\beta_i; \boldsymbol{\beta})$. For convenience we define $W_i(\boldsymbol{\beta}) = V(\beta_i; \boldsymbol{\beta})$. The total cost of a user is $cV(\beta; \boldsymbol{\beta}) + \beta^\alpha$, where c is the user's unit time cost and β is its priority level.

We frequently make use of *Little's law*, which provides a relationship between steady state expected processing times and steady state queue lengths [13]. In particular, let N_i denote the steady state number of class i jobs in the system. In a system consisting of K classes (λ_i, β_i), $i = 1, \ldots, K$, Little's law establishes that in steady state, for every class i we have $E[N_i] = \lambda_i W_i$.

2.1   Nash Equilibrium

We consider two types of games for this system: the *job level game* and the *class level game*. In the job level game, each *job* is an individual user, aiming to minimize its expected total cost by choosing its own priority level β. Although jobs from the same class are allowed to choose different priority levels, because jobs of the same class share the same parameters *ex ante*, we restrict our attention only to *symmetric* equilibria of the job level game; these are equilibria where jobs from the same class choose the same priority levels. Such an equilibrium can be characterized by a class priority vector $(\beta_1, \cdots, \beta_K)$.

Definition 1. *A* **job level Nash equilibrium** *consists of a class priority vector* $\boldsymbol{\beta} = (\beta_1, \cdots, \beta_K)$ *such that for all* $i = 1, \cdots, K$,

$$\beta_i = \arg\min_{\beta \geq \underline{\beta}} [c_i V(\beta; \boldsymbol{\beta}) + \beta^\alpha], \; \forall \, i = 1, \cdots, K. \tag{1}$$

In the class level game, each class is regarded as a single user and chooses a priority level for all of its jobs, therefore the equilibrium is again characterized by a class priority vector.

Definition 2. *A* **class level Nash equilibrium** *consists of a class priority vector* $\boldsymbol{\beta} = (\beta_1, \cdots, \beta_K)$ *such that for all* $i = 1, \cdots, K$,

$$\beta_i = \arg\min_{\beta \geq \underline{\beta}} [c_i W_i(\beta_1, \cdots, \beta_{i-1}, \beta, \beta_{i+1}, \cdots, \beta_K) + \beta^\alpha], \; \forall \, i = 1, \cdots, K. \tag{2}$$

We emphasize that, although jobs from the same class choose the same priority in both the symmetric job level equilibrium and the class level equilibrium, these two equilibria are not identical. The difference is that in the class level game, changing the priority level of a whole class i causes an externality within the class itself, while by contrast, in the job level game, a single job alters its priority level in isolation. In this paper, we mainly study the job level game, but also briefly discuss how our study can be adapted to the class level game.

2.2 Characterizing Processing Times

Nash equilibria of both the job level and class level games require characterization of the processing times V and W_i, which is in general quite complex. For the K class DPS model, Fayolle et al. [6] show that the expected steady state processing time W_i for each class i can be determined by a linear system.

Theorem 1. *[6] In a K-class DPS model with class priority vector β, $(W_1(\beta),$ $\cdots, W_K(\beta))$ is the unique solution of the following system of equations:*

$$\mu W_k(\beta) - \sum_{i=1}^{K} \frac{\lambda_i \beta_i}{\beta_i + \beta_k}[W_k(\beta) + W_i(\beta)] = 1, \ k = 1, \cdots, K. \tag{3}$$

On the other hand, computing the job level processing time $V(\beta; \beta)$ can be reduced to computing the class level processing time $W_i(\beta)$ as stated by the following theorem.

Theorem 2. *Let N_i be the steady state number of class i jobs in a K-class DPS system with class priority vector β. Then the steady state processing time of a job with priority β is*

$$V(\beta; \beta) = U_0(\beta; \beta) + \sum_{i=1}^{K} U_i(\beta; \beta) E[N_i], \tag{4}$$

and
$$\begin{cases} U_i(\beta; \beta) = \dfrac{\beta_i}{\beta_i + \beta} U_0(\beta; \beta), \ i = 1, \cdots, K; \\[2mm] U_0(\beta; \beta) = \left[\mu - \displaystyle\sum_{i=1}^{K} \frac{\lambda_i \beta_i}{\beta_i + \beta}\right]^{-1}. \end{cases} \tag{5}$$

The values of $E[N_i]$ can be obtained by applying Little's law to the solution of the system of linear equations (3). We conclude, therefore, that solving for $V(\beta; \beta)$ in (4) can be reduced to computing $W_i(\beta)$. In general, explicitly solving (3) requires the inversion of a nontrivial $K \times K$ matrix with complexity $O(K^3)$, thus in general there is no closed form expression for either W_i or V.

Nevertheless, when $K = 1$ or $K = 2$, we are able to solve for W_i and V in closed form. The solution of (4) with $K = 1$ is first established in [8, 7] and will be used frequently later:

$$V(\beta; \hat{\beta}) = \frac{1}{\mu(1 - \rho)} \cdot \frac{\beta(1 - \rho) + \hat{\beta}}{\hat{\beta}(1 - \rho) + \beta}. \tag{6}$$

When $K = 2$, the solution for W_i is given by [6], and the solution for V directly follows. Both solutions are lengthy and omitted for brevity.

2.3   Existence of NE

Existence of Nash equilibrium can be guaranteed when $\alpha \geq 1$, by exploiting convexity of the job cost function in (1).

Theorem 3. *There exists a Nash equilibrium for the job level game when $\alpha \geq 1$.*

When $\alpha < 1$, the payment term β^α is strictly concave, therefore the convexity of the objective function is not guaranteed; establishing existence of Nash equilibrium in this regime remains an open question.

As usual, this existence result is nonconstructive, since it uses a fixed point theorem. In general, given the implicit equations that define the processing times in (3), there is no closed form characterization of the Nash equilibrium, and no tractable approach for computation is available. Although we could resort to some heuristics (e.g., best response dynamics) to approach NE, each step of such an algorithm requires computing a range of processing times with fixed parameters, and as established above each such computation has complexity $O(K^3)$. Further, there is no guarantee that such dynamics will converge.

3   Heavy Traffic Approximation: Job Level

In the remainder of the paper we consider an alternate approach to the equilibrium analysis, by approximating the processing time. We aim to overcome the complexity of computing the processing times by exploiting a *heavy traffic* approximation, i.e., an approximation where the load approaches service capacity. Such an approximation is relevant for large systems such as cloud computing services, where providers will typically not want to provision significant excesses of capacity relative to demand.

Since $W_i(\boldsymbol{\beta}) = V(\beta_i; \boldsymbol{\beta})$, our focus is the job level processing time $V(\beta; \boldsymbol{\beta})$.

3.1   Approximating the Processing Time

In heavy traffic, a phenomenon known as *state space collapse* gives us a simplified solution for the steady state distribution of the system [18]; informally, state space collapse refers to the fact that the numbers of jobs of each class in the system become perfectly correlated when the system is heavily loaded.

In a slight abuse, whenever we write $\rho \to 1$, we mean that we consider a sequence of systems such that $(\rho_1, \cdots, \rho_K)$ converges to some $(\overline{\rho}_1, \cdots, \overline{\rho}_K)$ with $\sum_{i=1}^{K} \overline{\rho}_i = 1$. Moreover, we emphasize that both $V(\beta; \boldsymbol{\beta})$ and $W_i(\boldsymbol{\beta})$ depend on ρ, though we suppress this dependence for notational brevity. Let N_i denote the steady state number of type i jobs in the system. Then we have the following result on the joint steady state distribution of $(N_1, \cdots, N_K)$ for a DPS system in heavy traffic.

Theorem 4. *[17] Let N_i be the steady state number of class i jobs in a K-class DPS system with class priority vector $\boldsymbol{\beta}$. Then as $\rho \to 1$, we have*

$$(1 - \rho)(N_1, \cdots, N_K) \xrightarrow{d.} Z \cdot (\frac{\overline{\rho}_1}{\beta_1}, \cdots, \frac{\overline{\rho}_K}{\beta_K}), \tag{7}$$

where "$\xrightarrow{d.}$" denotes convergence in distribution, and Z is an exponentially distributed random variable with parameter $\overline{\gamma}(\beta) = \sum_{i=1}^{K} \overline{\rho}_i/\beta_i$.

Convergence of the joint distribution directly implies convergence of marginal distributions, so $(1 - \rho)N_i \xrightarrow{d.} Z\overline{\rho}_i/\beta_i$ for each i. Moreover, the second moment of N_i is shown to be uniformly bounded [17], so the N_i's are uniformly integrable. It follows from [3] that in this case convergence in distribution implies convergence in mean, and hence,

$$(1 - \rho)E[N_i] \to E[Z]\frac{\overline{\rho}_i}{\beta_i} = \frac{\overline{\rho}_i}{\beta_i\overline{\gamma}(\beta)} \quad \text{as} \quad \rho \to 1. \tag{8}$$

Taking advantage of this approximation of $E[N_i]$, we are now able to approximate $V(\beta; \boldsymbol{\beta})$. Substituting (5) and (8) into (4) yields

$$\lim_{\rho \to 1}(1 - \rho)V(\beta; \boldsymbol{\beta}) = \lim_{\rho \to 1} U_0(\beta; \boldsymbol{\beta})\left[\sum_{i=1}^{K} \frac{\overline{\rho}_i}{\overline{\gamma}(\beta)(\beta_i + \beta)} + (1 - \rho)\right]$$

$$= \lim_{\rho \to 1} \frac{(1 - \rho)\overline{\gamma}(\beta)\prod_{i=1}^{K}(\beta_i + \beta) + \sum_{i=1}^{K} \overline{\rho}_i \prod_{j \neq i}(\beta_j + \beta)}{\mu\overline{\gamma}(\beta)[(1 - \rho)\prod_{i=1}^{K}(\beta_i + \beta) + \beta\sum_{i=1}^{K} \rho_i \prod_{j \neq i}(\beta_j + \beta)]} = \frac{1}{\mu\beta\overline{\gamma}(\beta)}. \tag{9}$$

In the light of the above approximation, we have the following definition.

Definition 3. *The* **heavy traffic processing time** *for a job with priority level* β *in a system with* K *classes with class priority vector* $\boldsymbol{\beta}$ *is defined as*

$$V^{HT}(\beta; \boldsymbol{\beta}) = \frac{1}{(1 - \rho)} \cdot \frac{1}{\mu\beta\gamma(\beta)}, \quad \text{where} \quad \gamma(\beta) = \frac{1}{\rho}\sum_{i=1}^{K} \frac{\rho_i}{\beta_i}. \tag{10}$$

We note that $V^{HT}(\beta; \boldsymbol{\beta})$ has a closed form, and is easy to compute. Moreover, it is asymptotically exact in the heavy traffic regime: it is straightforward to show that as $\rho \to 1$, $\gamma(\beta) \to \overline{\gamma}(\beta)$, and hence, $(1 - \rho)[V^{HT}(\beta; \boldsymbol{\beta}) - V(\beta; \boldsymbol{\beta})] \to 0$. Next, we will approximate the Nash equilibrium based on this approximation of processing time.

3.2 Heavy Traffic Equilibrium

Recall from Definition 1 that $\boldsymbol{\beta} = (\beta_1, \cdots, \beta_K)$ is a job-level Nash equilibrium if (1) holds, i.e.,

$$\beta_i = \arg\min_{\beta > \underline{\beta}} c_i V(\beta; \boldsymbol{\beta}) + \beta, \quad i = 1, \cdots, K.$$

For general K it is quite hard to solve for pure Nash equilibrium because: (i) computing $V(\beta; \boldsymbol{\beta})$ requires matrix inversion to solve the linear system (3), which can only be done numerically; and (ii) even if we are able to solve $V(\beta; \boldsymbol{\beta})$ numerically and obtain optimality conditions for each player (which cannot be done in closed form), we would still need to solve a possibly *nonlinear* system with K equations and K unknowns to compute the Nash equilibrium.

In this section, we propose a novel concept of equilibrium which can be used to approximate the Nash equilibrium, yet can be computed in closed form. We approximate

$V(\beta; \boldsymbol{\beta})$ by $V^{HT}(\beta; \boldsymbol{\beta})$ in the objective function, and based on this approximation we define a concept of equilibrium that we call *heavy traffic equilibrium* (HTE) for job level games, as follows.

Definition 4. *A* **heavy traffic equilibrium** *of the job level game consists of a set of priorities* $\boldsymbol{\beta} = (\beta_1, \cdots, \beta_K)$ *such that*

$$\beta_i = \arg\min_{\beta \geq \underline{\beta}} \left(c_i V^{HT}(\beta; \boldsymbol{\beta}) + \beta^{\alpha} \right), \ i = 1, \cdots, K.$$

We can explicitly compute the heavy traffic equilibrium.

Theorem 5. *A heavy-traffic equilibrium always exists, and it is unique. Moreover, it can be calculated in closed form:*

$$\beta_i = c_i^{\frac{1}{\alpha+1}} [\alpha(1-\rho)\rho^{-1}S_1]^{-\frac{1}{\alpha}}, \tag{11}$$

where $S_1 = \sum_{i=1}^K \lambda_i c_i^{-\frac{1}{\alpha+1}}$.

We have two remarks on this result. First, this closed form expression allows us to carry out the analysis on sensitivity, efficiency, and revenue of the HTE (see Section 5). Second, the HTE is easily computable with complexity $O(K)$. In comparison, the complexity for computing the exact processing time with fixed parameters is $O(K^3)$, and as discussed computing exact NE is intractable.

We have observed above that the difference between the heavy traffic processing time and the exact processing time approaches zero as $\rho \to 1$, when scaled by a factor $1 - \rho$. Using this approximation, we can also prove an approximation theorem for the heavy traffic equilibrium: we show that deviating by any constant factor from the HTE is not profitable as $\rho \to 1$.

Theorem 6. *Consider a sequence of systems indexed by* n *such that classes have the same service capacity* μ, *and the loads of the systems* $\rho^{(n)} \to 1$ *as* $n \to \infty$. *Let* $\boldsymbol{\beta}^{(n)}$ *be the unique HTE of the n-th system, then for any* $\delta \geq 0$,

$$\lim_{n \to \infty} (1 - \rho^{(n)}) \left[c_i V_{(n)}(\beta_i^{(n)}; \boldsymbol{\beta}^{(n)}) + (\beta_i^{(n)})^{\alpha} - c_i V_{(n)} \left(\delta\beta_i^{(n)}; \boldsymbol{\beta}^{(n)} \right) - \left(\delta\beta_i^{(n)} \right)^{\alpha} \right] \leq 0.$$

Here V *is subscripted by* (n) *to indicate that the processing time is computed in system* n *with load* $\rho^{(n)}$.

In the theorem, we consider deviations by a multiplicative constant factor rather than by an additive constant because (11) implies that, as $\rho \to 1$, the heavy traffic equilibrium increases without bound; as a result, it is straightforward to check that any additive constant deviation has no beneficial effect as ρ approaches 1. Note that the waiting time is only asymptotically exact up to a $1 - \rho$ scaling, thus the same is true for this approximation theorem as well. Indeed, this is what we give up by studying heavy traffic: while we gain analytical tractability, the "resolution" to which we can study deviations is scaled by $1 - \rho$. This tradeoff is systematic throughout the study of large scale queueing models even without strategic behavior.

4 Heavy Traffic Approximation: Class Level

Based on the heavy traffic processing time approximation results in Theorem 4, one can also propose a similar heavy traffic equilibrium (HTE) concept for class level games. However, although the processing time approximation allows us to greatly simplify the computation of best response strategy, we are not able to obtain the closed form expression for this class level HTE. The main reason is that, unlike job level games, a class level game has an *intra-class externality* behavior: in a class level game, a class chooses one priority level for *all* its jobs simultaneously; while in a job level game, a job can choose any priority level regardless of other jobs belonging to its class.

Nevertheless, the intra-class externality effect will become negligible in a regime where the number of classes is large; this is a particularly useful regime for computing services where the number of users grows large. This observation motivates us to consider a limiting model in which the number of classes approaches infinity and any single class becomes infinitesimal. Thus, it can be connected to the job level game model. We refer the reader to the companion technical report for an investigation of this limiting class level game model in which we show that the heavy traffic equilibrium exists, is unique, can be computed in closed form, and is the limit of the finite class equilibrium as the number of class goes to infinity [20].

5 Sensitivity, Efficiency and Revenue

The tractability of heavy traffic equilibrium allows us to analytically study parameter sensitivity, as well as efficiency and revenue at the HTE equilibrium. We only study the job level HTE in this section; a similar investigation can be easily extended to the class level HTE. Throughout this section, we let β^* denote the job level HTE.

5.1 Sensitivity

In this subsection, we analyze the sensitivity of the HTE, i.e., how the equilibrium behaves with respect to changes in system parameters. These observations follow directly from (11).

Sensitivity with respect to c. If all c_i are scaled by a constant $\zeta > 0$, then every β_i^* is scaled by $\zeta^{\frac{1}{\alpha}}$. This is rather intuitive since the objective function is the sum of expected processing cost and β_i^α, and the expected processing cost does not change any V_i. Therefore the equilibrium is the same up to a scaling factor.

Sensitivity with respect to ρ. The ratio $\beta_i^*/\beta_j^* = (c_i/c_j)^{\frac{1}{\alpha+1}}$ is independent of ρ, i.e., changing ρ will change each β_i^* but will not affect β_i^*/β_j^* for any i, j. Therefore the ratio between service capacity allocated to any pair of jobs, as well as the ratio between the heavy traffic processing times of a pair of jobs, are invariant to the load of the system.

Sensitivity with respect to α. When $\alpha \to 0$, every $\beta_i^* \to \infty$; when $\alpha \to \infty$, every $\beta_i^* \to 1$. This is due to the fact that as $\alpha \to 0$, jobs face a strongly diminishing marginal cost to higher choices of β, and hence, prefer to choose higher β at the equilibrium; while the effect is reversed as $\alpha \to \infty$.

5.2   Efficiency

In HTE, efficiency is characterized by the expected total cost incurred to the system in one unit of time:

$$C = \sum_{i=1}^{K} \lambda_i c_i V^{HT}(\beta_i^*; \beta^*) = \left(\frac{\rho}{1-\rho}\right)\left(\sum_{i=1}^{K} \lambda_i c_i^{\frac{\alpha}{\alpha+1}}\right)\left(\sum_{i=1}^{K} \lambda_i c_i^{-\frac{1}{\alpha+1}}\right)^{-1}. \quad (12)$$

We call C the *system processing cost* (a more efficient system has a lower value of C). Given fixed λ_i and c_i $(i = 1, \cdots, K)$, the efficiency depends on the system parameter α and the load ρ as follows.

Dependence of C on ρ. We note that C is proportional to $\rho/(1-\rho)$, and hence is increasing in ρ. This is because a larger load ρ implies a busier system, and therefore the processing time is longer. (Note that we fixed λ, so varying ρ is equivalent to varying μ.)

Dependence of C on α. It is well known that the system optimal scheduling policy is the c-μ rule [5]: classes are given strict priority in descending order of $c_i \mu_i$ (or equivalently in this paper, in descending order of c_i, since we assume that all μ_i are the same). That is, for any $1 \le i, j \le K$, class j jobs are preempted by class i jobs if $c_i \mu_i > c_j \mu_j$. Jobs with the same value of $c\mu$ are served in first-in-first-out (FIFO) scheme. Since $\beta_i^*/\beta_j^* = (c_i/c_j)^{\frac{1}{\alpha+1}}$, for $c_i > c_j$, the ratio β_i^*/β_j^* is higher with smaller α, so we expect higher α lead to less efficient equilibria. This intuition is analytically stated in the following theorem.

Theorem 7. *The HTE system processing cost C is increasing in $\alpha > 0$.*

We note that even when α approaches 0, the HTE does not approach social optimum. In fact, for any i, j such that $c_i > c_j$, we have that $\beta_i^*/\beta_j^* = (c_i/c_j)^{\frac{1}{\alpha+1}}$, and hence $1 < \beta_i^*/\beta_j^* < c_i/c_j$. On the other hand, with the c-μ rule, if $c_i > c_j$, then class i jobs completely preempt class j jobs, which can be interpreted as the case where $\beta_i^*/\beta_j^* = \infty$. Therefore, it is clear that the HTE can never be as efficient as the c-μ rule, for any choice of α. However, we can upper bound the *price of anarchy* (PoA) of the HTE, as stated in the following theorem. The PoA is the ratio C/C^{opt}, where C^{opt} is the minimum expected system processing cost (achieved by the c-μ rule).

Theorem 8. *The price of anarchy (PoA) of the HTE is upper-bounded by:*

$$\frac{C}{C^{opt}} < \frac{\sum_{i=1}^{K-1}(\lambda_i/\lambda_K)(c_i/c_K)^{\frac{\alpha}{\alpha+1}} + 1}{\sum_{i=1}^{K-1}(\lambda_i/\lambda_K)(c_i/c_K)^{-\frac{1}{\alpha+1}} + 1} < \left(\frac{\lambda - \lambda_K}{\lambda_K}\right)\left(\frac{c_1}{c_K}\right)^{\frac{\alpha}{\alpha+1}} + 1.$$

Note that the upper bound can be made arbitrarily large through appropriate parameter choices; further, this is tight, in the sense that there exist systems where the PoA of HTE is in fact arbitrarily large. For example, let $\lambda_i = \lambda$ for all i, and set $c_K = 1$, $c_i = m$ for $i = 1, \cdots, K - 1$, and choose μ so that $\rho = 1 - m^{-2}$. Then it can be shown that $\frac{C}{C^{opt}} = \Omega\left((K-1)m^{\frac{\alpha}{\alpha+1}}\right)$ as $m \to \infty$ (see the proof of the theorem for details).

We also note that the PoA bound in increasing in α, which matches the intuition that a scheme closer to strict priority in descending cost order yields higher social welfare.

If we let $\alpha \to 0$, then the PoA is asymptotically bounded by λ/λ_K. In that case, if the arrival rates of all classes are the same, then the PoA is bounded by K. We can also let $\lambda_1, \cdots, \lambda_{K-1} \to 0$ to make the PoA approach 1, but this is not surprising since in this case the system essentially consists of only one class.

5.3   Revenue

The revenue of the server is the sum of expected payments in one unit of time:

$$\mathcal{R} = \sum_{i=1}^{K} \lambda_i (\beta_i^*)^\alpha = \left(\frac{\rho}{\alpha(1-\rho)}\right) \left(\sum_{i=1}^{K} \lambda_i c_i^{\frac{\alpha}{\alpha+1}}\right) \left(\sum_{i=1}^{K} \lambda_i c_i^{-\frac{1}{\alpha+1}}\right)^{-1}. \quad (13)$$

Given fixed λ_i and c_i $(i = 1, \cdots, K)$, the revenue depends on the system parameter α and the load ρ as follows.

Dependence of $\mathcal{R}$ on ρ. The revenue is proportional to $\rho/(1-\rho)$, therefore the revenue is increasing in ρ. Heavier traffic will induce greater congestion, and hence, jobs have to invest more in their purchase of priority in order to keep the same performance.

Dependence of $\mathcal{R}$ on α. The revenue depends on α in three terms, and it seems that in general the effect of changing α in the last two terms is significantly smaller than that of changing α in the first term $\rho/(\alpha(1-\rho))$. Hence we would expect that the revenue is in general decreasing in α. The next result shows this intuition holds if c_1/c_K is not too high.

Theorem 9. *The revenue $\mathcal{R}$ is decreasing in $\alpha > 0$ if $c_1/c_K < e^4$.*

On the other hand, $\mathcal{R}$ could be increasing in α in some cases. For instance, if $K = 2$ and c_1/c_2 is large enough, then $\partial\mathcal{R}/\partial\alpha$ is positive around $\alpha = 1$ (see the proof of theorem for details). To explain this special scenario where the monotonicity does not hold, we first note that a smaller α in general induces a higher revenue because jobs have incentive to purchase higher priority (as a response to the stronger diminishing marginal cost effect). However, in the HTE, significant asymmetry in costs will result in significant asymmetry in equilibrium priorities. Therefore when c_1/c_K is large, the optimal priorities already exhibit significant differences even when α is not small, and thus in equilibrium at small α, jobs have lower incentive to increase their priorities compared to what they do with mutually comparable costs.

With both (12) and (13), it is quite surprising to see that in the HTE,

$$\text{total cost of all jobs} = \mathcal{C} = \alpha\mathcal{R} = \alpha \cdot \text{total revenue of the system.}$$

Thus, we obtain an interesting insight: *another interpretation of α is the users' equilibrium cost per unit revenue.* We have shown that the user's total cost is increasing in α, (i.e., the system efficiency is decreasing in α), and the system revenue is decreasing in α under some mild conditions. Therefore, in a wide range of regimes, from the standpoint of system manager, smaller α is more favorable in terms of both efficiency and revenue. Note that smaller α is somewhat more "unfair," however, as it approaches a strict priority system.

Acknowledgments. This work was supported by the National Science Foundation under grants CMMI-0948434, CNS-0904609, CCF-0832820, and CNS-0644114, and by the Defense Advanced Research Projects Agency under the ITMANET program.

References

1. Allon, G., Gurvich, I.: Pricing and dimensioning competing large-scale service providers. Manufacturing & Service Operations Management 12, 449–469 (2010)
2. Altman, E., Avrachenkov, K., Ayesta, U.: A survey on discriminatory processor sharing. Queueing Systems 53(1-2), 53–63 (2006)
3. Billingsley, P.: Weak Convergence of Measures: Applications in Probability. Society for Industrial Mathematics, Philadelphia (1987)
4. Chen, Y., Maglaras, C., Vulcano, G.: Design of an aggregated marketplace under congestion effects: Asymptotic analysis and equilibrium characterization (2010) Working Paper
5. Cox, D.R., Smith, W.L.: Queues. Methuen and Wiley, London and New York (1961)
6. Fayolle, G., Mitrani, I., Iasnogorodski, R.: Sharing a processor among many job classes. Journal of the ACM 27(3), 519–532 (1980)
7. Hassin, R., Haviv, M.: To queue or not to queue: Equilibrium behavior in queueing systems. Kluwer Academic Publishers (2003)
8. Haviv, M., van der Wal, J.: Equilibrium strategies for processor sharing and queues with relative priorities. Probability in the Engineering and Informational Sciences 11(4), 403–412 (1997)
9. Kang, W., Kelly, F., Lee, N., Williams, R.: State space collapse and diffusion approximation for a network operation under a fair bandwidth sharing policy. The Annals of Applied Probability 19(5), 1719–1780 (2009)
10. Kelly, F.P., Maulloo, A., Tan, D.: Rate control in communication networks: shadow prices, proportional fairness and stability. Journal of the Operational Research Society 49(3), 237–252 (1998)
11. Kingman, J.F.C.: On queues in heavy traffic. Journal of the Royal Statistical Society. Series B (Methodological) 24(2), 383–392 (1962)
12. Kleinrock, L.: Time-shared systems: A theoretical treatment. Journal of ACM 14(2), 242–261 (1967)
13. Kleinrock, L.: Queueing Systems: Theory, vol. 1. Wiley–Interscience, New York (1975)
14. Massoulié, L., Roberts, J.: Bandwidth sharing and admission control for elastic traffic. Telecommunication Systems 15(1-2), 185–201 (2000)
15. Mo, J., Walrand, J.: Fair end-to-end window-based congestion control. IEEE/ACM Trans. Netw. 8, 556–567 (2000)
16. Naor, P.: The regulation of queue size by levying tolls. Econometrica 37(1), 15–24 (1969)
17. Rege, K., Sengupta, B.: Queue-length distribution for the discriminatory processor-sharing queue. Operations Research 44(4), 653–657 (1996)
18. Reiman, M.I.: Open queueing networks in heavy traffic. Mathematics of Operations Research 9(3), 441–458 (1984)
19. Verloop, I.M., Ayesta, U., Nunez-Queija, R.: Heavy-traffic analysis of a multiple-phase network with discriminatory processor sharing. Operations Research 59(3), 648–660 (2011)
20. Wu, Y., Bui, L., Johari, R.: Heavy traffic approximations of equilibria in resource sharing games. Technical Report, arXiv:1109.6166 (2011)

An NTU Cooperative Game Theoretic View
of Manipulating Elections

Michael Zuckerman[1], Piotr Faliszewski[2], Vincent Conitzer[3],
and Jeffrey S. Rosenschein[1]

[1] School of Computer Science and Engineering, The Hebrew Univ. of Jerusalem, Israel
{michez,jeff}@cs.huji.ac.il
[2] AGH University of Science and Technology, Kraków, Poland
faliszew@agh.edu.pl
[3] Department of Computer Science, Duke University Durham, NC 27708, USA
conitzer@cs.duke.edu

Abstract. Social choice theory and cooperative (coalitional) game theory have
become important foundations for the design and analysis of multiagent systems.
In this paper, we use cooperative game theory tools in order to explore the coalition formation process in the coalitional manipulation problem. Unlike earlier
work on a cooperative-game-theoretic approach to the manipulation problem [2],
we consider a model where utilities are not transferable. We investigate the issue of stability in coalitional manipulation voting games; we define two notions
of the core in these domains, the α-core and the β-core. For each type of core,
we investigate how hard it is to determine whether a given candidate is in the
core. We prove that for both types of core, this determination is at least as hard as
the coalitional manipulation problem. On the other hand, we show that for some
voting rules, the α- and the β-core problems are no harder than the coalitional
manipulation problem. We also show that some prominent voting rules, when applied to the truthful preferences of voters, may produce an outcome not in the
core, even when the core is not empty.

1 Introduction

Voting constitutes a natural methodology for a group of agents to make a joint decision
in spite of (possibly) conflicting preferences. Unfortunately, voting and elections are
not a universal, perfect solution to preference aggregation problems. For example, the
classic result of Gibbard and Satterthwaite [9,13] says that in sufficiently general settings, any reasonable voting rule may lead to a situation where some voter(s) are better
off by casting votes different from their true preferences (this is called *manipulation* or
strategic voting). One of the most influential ideas regarding the computational study
of elections, due to Bartholdi, Orlin, Tovey, and Trick [4,3], was to study the computational complexity of computing manipulative votes. The rationale behind it was that
if planning a manipulation were computationally hard, then in practice voters would
not be able to vote strategically (see the surveys of Faliszewski, Hemaspaandra, and
Hemaspaandra [7] and of Faliszewski and Procaccia [8] for a detailed overview of this
approach and for two viewpoints regarding its applicability).

N. Chen, E. Elkind, and E. Koutsoupias (Eds.): WINE 2011, LNCS 7090, pp. 363–374, 2011.
© Springer-Verlag Berlin Heidelberg 2011

Formally, in the *coalitional* manipulation problem, introduced by Conitzer, Sand-holm, and Lang [5], the voters are divided into two groups, the manipulators and the nonmanipulators. The votes of the nonmanipulators are assumed to be known, and the problem is to determine whether the manipulators can use their votes to achieve a given goal. The goal is either to ensure that some preferred candidate wins (the *construc-tive* variant) or to ensure that some despised candidate does not win (the *destructive* variant). Further, we obtain different flavors of the problem depending on whether the votes are weighted or not, which voting rule is used, etc. Some results on coalitional manipulation can be found in [17,15,14,16]; see also surveys [8,7].

Let us focus on unweighted constructive coalitional manipulation. If all the manip-ulators have identical preferences, this is exactly the problem that they want to be able to solve—they would first try to ensure their most preferred candidate's victory; then, if that were impossible, they would try the second best one, third best one, and so on, until they would either find a successful manipulation or determine that they cannot do better than electing the truthful winner.

Nonetheless, generally manipulators do *not* have identical preferences. In this case, the manipulators may still work together, but it is much less clear which candidate they should try to promote (even ignoring computational considerations). While they may all agree that they would prefer a different winner than the truthful one, deciding *which* candidate to support is a whole new game that they need to play among themselves. (To push our scenario to the limit, consider a situation where all the voters are manipulators.)

In this paper, we take the viewpoint of *cooperative game theory* to solve such games among the manipulators, and study computational aspects of the relevant solution con-cepts. As in most of the literature on voting, we assume that the agents do not have the ability to make or receive payments, so that we are in the *nontransferable utility (NTU)* case of cooperative game theory. (Recently, Bachrach, Elkind, and Faliszewski [2] stud-ied a similar problem in the *transferable utility* setting, and obtained results linking coalitional manipulation [5] and bribery [6] problems with their cooperative game-theoretic model.) Moreover, in this setting, what one (sub)coalition of manipulators can achieve depends on the actions (votes) of the manipulators outside the coalition. We consider two different ways of addressing this—via the standard notions of the α- and the β-core [12].

2   Preliminaries

Let us now define the basic notions of (computational) social choice theory and coali-tional game theory, as used in this paper.

An election E is a triple $(C, V, \mathcal{P})$, where $C = \{c_1, \ldots, c_m\}$ is the set of candidates, $V = \{1, \ldots, n\}$ is the set of voters, and $\mathcal{P} = (P_1, \ldots, P_n)$ is a preference profile of voters in V. That is, each voter i, $1 \leq i \leq n$, is associated with preference order P_i from $\mathcal{P}$. A preference order is a linear order over the candidates in C. We will sometimes write $\succ_i$ instead of P_i. We write $L(C)$ to denote the set of all linear orders over C. Let U be some subset of V. By $\mathcal{P}_U$ we mean $(P_i)_{i \in U}$ and by $\mathcal{P}_{-U}$ we mean $(P_i)_{i \notin U}$. Using standard notational conventions, we have $\mathcal{P} = (\mathcal{P}_U, \mathcal{P}_{-U})$.

A *voting rule* $\mathcal{R}$ is a function that, given election E as input, returns one of the can-didates, denoted $\mathcal{R}(E)$, as the election's winner. (We assume ties are resolved by some simple tie-breaking rule and that the manipulators, unwilling to rely on tie-breaking, always seek unique winners; we point the reader to the work of Obraztsova, Elkind, and Hazon [11], and Obraztsova and Elkind [10] for a detailed discussion of tie-breaking in voting manipulation.) We focus on the following (families of) voting rules:

Positional Scoring Rules. Let $s = (s_1, \ldots, s_m)$ be a vector of non-negative integers, such that $s_1 \geq s_2 \geq \ldots \geq s_m$. For each voter, a candidate receives s_1 points if it is ranked first by the voter, s_2 points if it is ranked second, etc. The score of a candidate is the total number of points the candidate receives. The winner is the candidate with the maximum score. The scoring rules which we will consider here are k-approval, where $s = (1, \ldots, 1, 0, \ldots, 0)$ ($s_1 = \ldots = s_k = 1; s_{k+1} = \ldots = s_m = 0$), Plurality, where $s = (1, 0, \ldots, 0)$, and Borda, where $s = (m - 1, m - 2, \ldots, 0)$.

Plurality with Runoff. In this rule, a first round eliminates all candidates except the two with the highest plurality scores. The second round determines the winner between these two by their pairwise election.

Simplified Bucklin. A candidate c's Bucklin score is the smallest number k such that more than half of the votes rank c among the top k candidates. The winner is the candidate that has the smallest Bucklin score.[1]

We use the term *(coalitional) manipulation* to refer to a situation where a voter (a group of voters) casts votes not according to his (their) true preferences, but rather to obtain some goal. It is one of the best-studied forms of strategic behavior in elections (see the surveys [8,7]). The definition below is taken from the paper of Bachrach, Elkind, and Faliszewski [2], which itself is inspired by the definition of Conitzer, Sandholm, and Lang [5].

Definition 1. *For any voting rule $\mathcal{R}$, an instance $I = (E, S, c)$ of the $\mathcal{R}$-COALITIONAL MANIPULATION problem is given by an election $E = (C, V, \mathcal{P})$, a set of manipulators $S, S \cap V = \emptyset$, and a distinguished candidate $c \in C$. It is a "yes"-instance if there is a vector $\mathcal{P}_S = (P_i)_{i \in S} \in (L(C))^{|S|}$ such that $\mathcal{R}(\mathcal{P}, \mathcal{P}_S) = c$, and a "no"-instance otherwise.*

Note that in the traditional definition of coalitional manipulation the manipulators, unlike honest voters, do not have preferences over the candidates; they simply want to get a particular candidate elected. Thus, this definition eliminates the problem of deciding which candidates the manipulators should support by making it external to the setting.

3 Manipulation Model

The goal of this paper is to study the unweighted constructive coalitional manipulation problem in the setting in which not all manipulators have identical preferences. Specifically, our focus is on the computational complexity of deciding which candidates the manipulators may support in a stable way, without breaking the coalition.

---

[1] The Nonsimplified Bucklin rule additionally breaks ties by the number of votes that rank a candidate among the top k candidates. In computational social choice it is common to focus on Simplified Bucklin instead of its full variant.

Formally, we consider the following setting. We are given an election $E = (C, V, \mathcal{P})$ where some of the voters are truthful and some are willing to manipulate. We denote the set of possible manipulators by M and we will refer to them as *colluders*. In our model the remaining voters, those in $H = V \setminus M$, are *honest voters* who vote truthfully. We assume that the manipulators have a way of communicating with one another (this is a standard assumption, though typically it is not mentioned explicitly). We ask which candidate the colluders should support. Naturally, the answer to this question depends strongly on the attitudes that the colluders have towards one another, and on the way they expect one another to behave. We consider two settings, depending on how the colluders react to the breaking out of the coalition by some subset M' of M: (a) the pessimistic model, where players in M' want to succeed irrespective of the reaction of players in $M \setminus M'$, and (b) the adaptive model, where players in M' can pick their votes depending on the votes of players in $M \setminus M'$.

Definition 2. *Let $\mathcal{R}$ be a voting rule and let $E = (C, V, \mathcal{P})$ be an election with colluder set M. Fix a candidate $c \in C$ and a coalition $S \subseteq M$.*

1. *We say that c is feasible for S if there is a profile $\mathcal{P}'_S$ such that $\mathcal{R}(\mathcal{P}'_S, \mathcal{P}_{-S}) = c$.*
2. *We say that c is α-feasible for a coalition S if there is a preference profile $\mathcal{P}'_S$ such that for all preference profiles $\mathcal{P}''_{M \setminus S}$ it holds that $\mathcal{R}(\mathcal{P}'_S, \mathcal{P}''_{M \setminus S}, \mathcal{P}_{-M}) = c$.*
3. *We say that $c \in C$ is β-feasible for a coalition $S \subseteq M$ if for all preference profiles $\mathcal{P}'_{M \setminus S}$ there exists a preference profile $\mathcal{P}''_S$ such that $\mathcal{R}(\mathcal{P}''_S, \mathcal{P}'_{M \setminus S}, \mathcal{P}_{-M}) = c$.*

We denote the set of all candidates that are feasible (respectively, α-feasible, β-feasible) for S by $F(S)$ (respectively, $F_\alpha(S)$, $F_\beta(S)$).

As a side remark, it is easy to see that $F(S)$ is never empty as it always contains $\mathcal{R}(E)$; but $F_\alpha(S)$ and $F_\beta(S)$ may be empty.

Using our feasibility notions, we can adapt the notions of α- and β-core to our setting [12].

Definition 3. *Let $E = (C, V, \mathcal{P})$ be an election, M be a subset of V, $\mathcal{R}$ be a voting rule, and c be a candidate. We say that c belongs to the α-core (respectively, β-core) if $c \in F(M)$ and there is no candidate $c' \in C$ and non-empty coalition $W' \subseteq M$ such that (a) $c' \in F_\alpha(W')$ (respectively, $c' \in F_\beta(W')$), and (b) each voter in W' prefers c' to c.*

One can see that these notions are related to the notion of strong Nash equilibrium (SNE) in the game played by the colluders. Recall that an SNE is a Nash equilibrium in which no coalition, taking the actions of its complements as given, can cooperatively deviate in a way that benefits all of its members [1]. In the context of SNE, we have the following scenario: the colluders have agreed upon some voting profile. The deviating coalition can privately communicate; when a coalition coordinates a deviation, the remaining players are unaware of it, so they stick to their agreed-upon strategies. In this model, it is relatively easy for a sub-coalition to break off since the deviating colluders know in advance how the rest of the colluders will vote. On the other hand, the α-core and the β-core model an election in which it is non-trivial for a subgroup of manipulators to break off from the coalition. (Consider, e.g., an election on choosing an acceptable debt level for a country and the manipulators being MPs from the ruling party. Even if they personally disagreed with some particular proposed debt level, they would need to obtain very strong support before breaking off from the coalition.) Here, if a subcoalition breaks off then it has to be ready to face every possible reaction of the colluders they abandon.

The difference between the α-core and the β-core is that in the case of the β-core the splitting subcoalition knows that it will know the remaining colluders' votes before having to cast their own. (In the parliamentary example from the previous paragraph, this means that MPs leaving the ruling party know they will be asked about their vote on the debt issue only after the ruling party will be.)

The above intuitions are supported by the observation that, by definition, if $\mathcal{P}'_M$ is SNE then $c = \mathcal{R}(\mathcal{P}_H, \mathcal{P}'_M) \in \beta$-core, and β-core $\subseteq \alpha$-core. (To see this, note that if for $c' \in C \setminus \{c\}$, and $W' \subseteq M$ such that everybody in W' prefers c' to c, $c' \in F_\beta(W')$ then W' can deviate from $\mathcal{P}'_M$ to make c' a winner, and so $\mathcal{P}'_M$ is not a SNE; and if for some candidate c' and set $W' \subseteq M$ it holds that $c' \in F_\alpha(W')$ then $c' \in F_\beta(W')$.) That is, settings where the α-core is an appropriate model are more stable than those where colluders do not counteract deviations by other colluders.

It is interesting that there are prominent voting rules and preference profiles for which the non-empty α-core/β-core does not contain the truthful winner.

Example 4. Consider a Borda election with candidate set $C = \{a, b, c, d, e, f\}$, and the following preference profiles (the numbers in parentheses are the serial numbers of the voters): (1) $f \succ d \succ b \succ c \succ a \succ e$, (2) $a \succ f \succ b \succ c \succ d \succ e$, (3) $b \succ f \succ d \succ c \succ a \succ e$, and (4) $d \succ b \succ f \succ a \succ e \succ c$.

Here we have one honest voter (1), and the rest are the colluders. In the above election, f wins. Now consider the set $W_{bf} = \{3, 4\}$. If (3) votes $b \succ e \succ c \succ a \succ d \succ f$, and (4) votes $b \succ a \succ e \succ c \succ d \succ f$, then no matter how voter (2) votes, b wins the election. Hence $b \in F_\alpha(W_{bf}) \Rightarrow f \notin \alpha$-core. We claim that $b \in \beta$-core (and hence $b \in \alpha$-core). Indeed, b is the Condorcet winner among the colluders. Hence, if the two colluders who prefer b to some candidate $x \in \{a, c, d, e, f\}$, vote $b \succ e \succ c \succ a \succ d \succ f$ and $b \succ a \succ e \succ c \succ d \succ f$, then no matter how the third colluder votes, b wins the election. Hence, no other candidate is β-feasible for the coalition that prefers him, which implies b is in the β-core.

We also have a similar example for Maximin, but we omit it due to space restrictions.[2]

To study our core notions computationally, we need to define an appropriate decision problem.

Definition 5. *For a voting rule $\mathcal{R}$, an instance $I = (E, M, c)$ of the $\mathcal{R}$-α-CORE (respectively, $\mathcal{R}$-β-Core) problem, is given by an election $E = (C, V, \mathcal{P})$, a set of colluders $M \subseteq V$ and a candidate $c \in C$. It is a "yes"-instance if c is in the α-core (respectively, β-core) of the corresponding game among the manipulators, and a "no"-instance otherwise.*

---

[2] See `ftp://ftp.cs.huji.ac.il/users/jeff/wine11zuckermanfull.pdf` for a version of this paper containing the example.

4   Main Results

In this section we present our main computational results. Namely, we provide a computational analysis of the notions of the α- and β-cores. We first provide a connection between our notions of the core and the standard problem of coalitional manipulation.

Theorem 6. *Given an election $E = (C, V, \mathcal{P})$, a set $M \subseteq V$, a voting rule $\mathcal{R}$ and a candidate $c \in C$, the problem of determining whether c is in the α- or β-core is at least as hard as the corresponding coalitional manipulation problem.*

Proof. We show a reduction from the coalitional manipulation problem. Given a voting rule $\mathcal{R}$ and an instance $I = (E, S, c)$, where $E = (C, V, \mathcal{P})$, of the $\mathcal{R}$-Coalitional Manipulation problem, we construct the core problem in the following way. We set $E' = (C, V', \mathcal{P}')$ where $V' = V \cup S$, $\mathcal{P}' = \mathcal{P} \cup \mathcal{P}'_S$, and $\mathcal{P}'_S$ is a profile of voters in S where c is ranked first in all the ballots, and the rest of the candidates are ranked in some arbitrary order. We set $M = S$. There exists a manipulation by voters in S making c win in the coalitional manipulation problem if and only if c is in the core (either α or β) of the core problem (which is defined by the instance (E', M, c)), since here c is in the core if and only if c is feasible for the coalition M (as there is no non-empty coalition W' where the voters in W' prefer c' to c). □

However, could it be that the α- or the β-core problems are strictly harder than the coalitional manipulation problem? We will see that for many voting rules, the answer is "no". Due to Theorem 6, we focus on those voting rules for which the coalitional manipulation problem is polynomial-time solvable.

Theorem 7. *Let R be a positional scoring rule with scoring vector $s = (s_1, \ldots, s_m)$ such that the R-Coalitional Manipulation problem is polynomial-time solvable. Then there exists a polynomial-time algorithm solving the R-α-Core problem.*

Proof. Suppose we are given an instance $I = (E, M, c)$ of the R-α-Core problem, where $E = (C, V, \mathcal{P})$. Suppose w.l.o.g. that $s_m = 0$. The algorithm first checks whether $c \in F(M)$ by solving the R-Coalitional Manipulation problem. Then it goes over all the other candidates. For each candidate $x \neq c$ it checks what is the maximal set $W_{xc} \subseteq M$ of the colluders who prefer x to c. If $W_{xc} \neq \emptyset$, it builds a "profile"[3] $\mathcal{Q}_{M \setminus W_{xc}}$ for $M \setminus W_{xc}$ s.t. the score of x is $s(\mathcal{Q}_{M \setminus W_{xc}}, x) = 0$, and the score of any other candidate $y \neq x$ is $s(\mathcal{Q}_{M \setminus W_{xc}}, y) = s_1 \cdot |M \setminus W_{xc}|$. Finally, it solves the R-Coalitional Manipulation problem $(C, V \setminus W_{xc}, \mathcal{P}_H \cup \mathcal{Q}_{M \setminus W_{xc}}), x, |W_{xc}|$ to see whether the votes in W_{xc} can be cast in order to make x win the election. Next we prove that we manage to make x win if and only if x is α-feasible for W_{xc}.

Claim. Let $\mathcal{Q}_{W_{xc}}$ be a profile. x is the winner under $\mathcal{P}_H \cup \mathcal{Q}_{M \setminus W_{xc}} \cup \mathcal{Q}_{W_{xc}}$ (where $\mathcal{Q}_{M \setminus W_{xc}}$ is as defined above) if and only if for any (real) profile $\mathcal{P}'_{M \setminus W_{xc}}$ x wins under $\mathcal{P}_H \cup \mathcal{P}'_{M \setminus W_{xc}} \cup \mathcal{Q}_{W_{xc}}$.

---

[3] We used the double quotes for the word "profile" since it is not a real profile, but rather a score specification. Nevertheless, we use the above notation for simplicity, as if it were a real profile.

Proof. We first prove the "only if" part. Suppose that x wins under $\mathcal{P}_H \cup \mathcal{Q}_{M \setminus W_{xc}} \cup \mathcal{Q}_{W_{xc}}$. Let $\mathcal{P}'_{M \setminus W_{xc}}$ be any profile. For all y, $y \neq x$, it holds that $s(\mathcal{P}'_{M \setminus W_{xc}}, y) \leq s_1 \cdot |M \setminus W_{xc}|$. Also, $s(\mathcal{P}'_{M \setminus W_{xc}}, x) \geq 0$. Therefore, for each y, $y \neq x$, $s(\mathcal{P}_H \cup \mathcal{P}'_{M \setminus W_{xc}} \cup \mathcal{Q}_{W_{xc}}, x) \geq s(\mathcal{P}_H \cup \mathcal{Q}_{M \setminus W_{xc}} \cup \mathcal{Q}_{W_{xc}}, x) > s(\mathcal{P}_H \cup \mathcal{Q}_{M \setminus W_{xc}} \cup \mathcal{Q}_{W_{xc}}, y) = s(\mathcal{P}_H \cup \mathcal{Q}_{W_{xc}}, y) + s_1 \cdot |M \setminus W_{xc}| \geq s(\mathcal{P}_H \cup \mathcal{P}'_{M \setminus W_{xc}} \cup \mathcal{Q}_{W_{xc}}, y).[4] And so, x will win under $\mathcal{P}_H \cup \mathcal{P}'_{M \setminus W_{xc}} \cup \mathcal{Q}_{W_{xc}}$.

Now we prove the "if" part. Suppose that for any profile $\mathcal{P}'_{M \setminus W_{xc}}$, x is the winner under $\mathcal{P}_H \cup \mathcal{P}'_{M \setminus W_{xc}} \cup \mathcal{Q}_{W_{xc}}$. Let y, $y \neq x$ be a candidate. By our assumption, x is the winner under $\mathcal{P}_H \cup \mathcal{P}'_{M \setminus W_{xc}} \cup \mathcal{Q}_{W_{xc}}$, where $\mathcal{P}'_{M \setminus W_{xc}}$ is built as follows: y is ranked on top of each preference in $\mathcal{P}'_{M \setminus W_{xc}}$, x is ranked at the bottom of each preference in $\mathcal{P}'_{M \setminus W_{xc}}$, and the other candidates are ranked arbitrarily. From this construction we get: $s(\mathcal{P}_H \cup \mathcal{Q}_{M \setminus W_{xc}} \cup \mathcal{Q}_{W_{xc}}, x) = s(\mathcal{P}_H, x) + s(\mathcal{Q}_{W_{xc}}, x) = s(\mathcal{P}_H \cup \mathcal{P}'_{M \setminus W_{xc}} \cup \mathcal{Q}_{W_{xc}}, x) > s(\mathcal{P}_H \cup \mathcal{P}'_{M \setminus W_{xc}} \cup \mathcal{Q}_{W_{xc}}, y) = s(\mathcal{P}_H, y) + s_1 \cdot |M \setminus W_{xc}| + s(\mathcal{Q}_{W_{xc}}, y) = s(\mathcal{P}_H \cup \mathcal{Q}_{M \setminus W_{xc}} \cup \mathcal{Q}_{W_{xc}}, y)$. This is true for every $y \neq x$. Therefore, x is the winner under $\mathcal{P}_H \cup \mathcal{Q}_{M \setminus W_{xc}} \cup \mathcal{Q}_{W_{xc}}$. **(End of proof of claim.)** □

If we find a candidate $x \neq c$ and a non-empty coalition W_{xc} such that x is α-feasible for W_{xc}, then c is not in the α-core. Otherwise, c is in the α-core. □

Theorem 8. *Let $k \in \mathbb{N}$, $1 \leq k \leq |C| - 1 = m - 1$. There is a polynomial-time algorithm solving the k-approval-β-Core problem.*

Due to space limitations, the proof of Theorem 8 is omitted.[5]

Theorem 9. *There exists a polynomial-time algorithm solving the Simplified-Bucklin-α-Core problem.*

Proof. Suppose we are given an instance $I = (E, M, c)$ of the Bucklin-α-Core problem, where $E = (C, V, \mathcal{P})$. The algorithm first checks whether $c \in F(M)$ by solving the Bucklin-Coalitional Manipulation problem (see [15] for the algorithm for this). It then goes over all the other candidates; for each candidate $x \neq c$ it computes the maximal set $W_{xc} \subseteq M$ of colluders who prefer x to c. If $W_{xc} \neq \emptyset$, we need to check whether x is α-feasible for W_{xc}. We first introduce some new notation. Let $B(a, k, \mathcal{Q})$ denote the number of votes in the profile $\mathcal{Q}$ that rank the candidate a in the k first places. Let d be the minimal integer such that $B(x, d, \mathcal{P}_H) + |W_{xc}| > \frac{1}{2} \cdot |V|$. Let $\mathcal{P}''_{M \setminus W_{xc}}$ be a "profile" of $M \setminus W_{xc}$ such that $B(x, d, \mathcal{P}''_{M \setminus W_{xc}}) = 0$; and for all $y \neq x$, $B(y, 1, \mathcal{P}''_{M \setminus W_{xc}}) = |M \setminus W_{xc}|$.[6] In the following claim we prove that it is enough to check whether x can win vs. $\mathcal{P}''_{M \setminus W_{xc}}$ to determine whether x is α-feasible for W_{xc}.

---

[4] We use here a strong inequality as we prove our results for the unique winner model. However, our results can be modified to work for the co-winner model as well.

[5] Again, see ftp://ftp.cs.huji.ac.il/users/jeff/winellzuckermanfull.pdf for a version of this paper containing the proof.

[6] Of course, if $m > 2$ then $\mathcal{P}''_{M \setminus W_{xc}}$ cannot be a real profile; rather it is just a score specification. As before, we use the notation as if it were a real profile, for simplicity.

Claim. x is the winner under $\mathcal{P}_H \cup \mathcal{P}''_{M \setminus W_{xc}} \cup \mathcal{P}^x_{W_{xc}}$ where $\mathcal{P}''_{M \setminus W_{xc}}$ is as defined above, and $\mathcal{P}^x_{W_{xc}}$ is some fixed (real) profile for W_{xc} which ranks x in the first position in all the votes, if and only if for all profiles $\mathcal{P}'_{M \setminus W_{xc}}$ x is the winner under $\mathcal{P}_H \cup \mathcal{P}'_{M \setminus W_{xc}} \cup \mathcal{P}^x_{W_{xc}}$.

Proof. Suppose x is the winner under $\mathcal{P}_H \cup \mathcal{P}''_{M \setminus W_{xc}} \cup \mathcal{P}^x_{W_{xc}}$. Let $\mathcal{P}'_{M \setminus W_{xc}}$ be any profile. Let d be as defined above. Since for all $y \neq x$, $B(y, d, \mathcal{P}'_{M \setminus W_{xc}}) \leq |M \setminus W_{xc}|$, we have that $g'_y := \lfloor \frac{1}{2} |V| \rfloor - B(y, d, \mathcal{P}_H \cup \mathcal{P}'_{M \setminus W_{xc}}) \geq \lfloor \frac{1}{2} |V| \rfloor - B(y, d, \mathcal{P}_H) - |M \setminus W_{xc}| = \lfloor \frac{1}{2} |V| \rfloor - B(y, d, \mathcal{P}_H \cup \mathcal{P}''_{M \setminus W_{xc}}) =: g''_y$. Therefore, $k'_y := \min\{g'_y, |W_{xc}|\} \geq \min\{g''_y, |W_{xc}|\} =: k''_y$. Since x is the winner under $\mathcal{P}_H \cup \mathcal{P}''_{M \setminus W_{xc}} \cup \mathcal{P}^x_{W_{xc}}$, we have for all $y \neq x$, $B(y, d, \mathcal{P}^x_{W_{xc}}) \leq k''_y \leq k'_y$. So, by definition of k'_y, $B(y, d, \mathcal{P}_H \cup \mathcal{P}'_{M \setminus W_{xc}} \cup \mathcal{P}^x_{W_{xc}}) \leq \lfloor \frac{1}{2} |V| \rfloor$. On the other hand, $B(x, d, \mathcal{P}_H \cup \mathcal{P}'_{M \setminus W_{xc}} \cup \mathcal{P}^x_{W_{xc}}) \geq B(x, d, \mathcal{P}_H \cup \mathcal{P}^x_{W_{xc}}) = B(x, d, \mathcal{P}_H) + |W_{xc}| > \frac{1}{2} |V|$ (where the last inequality follows from the definition of d). And so, x wins under $\mathcal{P}_H \cup \mathcal{P}'_{M \setminus W_{xc}} \cup \mathcal{P}^x_{W_{xc}}$.

For the opposite direction, suppose that for every profile $\mathcal{P}'_{M \setminus W_{xc}}$, x is the winner under $\mathcal{P}_H \cup \mathcal{P}'_{M \setminus W_{xc}} \cup \mathcal{P}^x_{W_{xc}}$, where $\mathcal{P}^x_{W_{xc}}$ is some fixed profile of W_{xc} with x ranked at the top of each vote. Let y, $y \neq x$, be any candidate. $\mathcal{P}^x_{W_{xc}}$ makes x win also vs. the profile $\mathcal{P}^{y,x}_{M \setminus W_{xc}}$ where y is ranked first and x is ranked last in all the preferences. Since the score of x under $\mathcal{P}_H \cup \mathcal{P}^{y,x}_{M \setminus W_{xc}} \cup \mathcal{P}^x_{W_{xc}}$ is d, it follows that the score of y under $\mathcal{P}_H \cup \mathcal{P}^{y,x}_{M \setminus W_{xc}} \cup \mathcal{P}^x_{W_{xc}}$ is greater than d. Therefore, $B(y, d, \mathcal{P}^x_{W_{xc}}) \leq \min\{|W_{xc}|, \lfloor \frac{1}{2} |V| \rfloor - B(y, d, \mathcal{P}_H) - |M \setminus W_{xc}|\}$. Hence, the score of y under $\mathcal{P}_H \cup \mathcal{P}''_{M \setminus W_{xc}} \cup \mathcal{P}^x_{W_{xc}}$ is $> d$. On the other hand, by definition of d and $\mathcal{P}^x_{W_{xc}}$, $B(x, d, \mathcal{P}_H \cup \mathcal{P}''_{M \setminus W_{xc}} \cup \mathcal{P}^x_{W_{xc}}) = B(x, d, \mathcal{P}_H \cup \mathcal{P}^x_{W_{xc}}) = B(x, d, \mathcal{P}_H) + |W_{xc}| > \frac{1}{2} \cdot |V|$. So, the score of x under $\mathcal{P}_H \cup \mathcal{P}''_{M \setminus W_{xc}} \cup \mathcal{P}^x_{W_{xc}}$ is d. Hence, x is the winner under $\mathcal{P}_H \cup \mathcal{P}''_{M \setminus W_{xc}} \cup \mathcal{P}^x_{W_{xc}}$. **(End of proof of claim.)** □

We now resume the proof of Theorem 9. With the above claim in hand, we can compute in polynomial time whether x is α-feasible for the coalition W_{xc}, in the following way. First compute d as the minimal integer such that $B(x, d, \mathcal{P}_H) + |W_{xc}| > \frac{1}{2} |V|$. Then define for each $y \neq x$, $g''_y = \lfloor \frac{1}{2} |V| \rfloor - B(y, d, \mathcal{P}_H) - |M \setminus W_{xc}|$, and $k''_y = \min\{g''_y, |W_{xc}|\}$. If $\sum_{y \neq x} k''_y < (d-1)|W_{xc}|$, then there does not exist a profile $\mathcal{P}'''_{W_{xc}}$ making x win under $\mathcal{P}_H \cup \mathcal{P}''_{M \setminus W_{xc}} \cup \mathcal{P}'''_{W_{xc}}$ (see [15] for details of the algorithm). Otherwise, we build the profile $\mathcal{P}^x_{W_{xc}}$ as follows. We first put x on top of all the preferences of $\mathcal{P}^x_{W_{xc}}$. Then for all $i = 1, \ldots, m - 1$ we put the candidate c_i in the next k''_{c_i} available places in the votes of $\mathcal{P}^x_{W_{xc}}$, such that $B(c_i, d, \mathcal{P}^x_{W_{xc}}) \leq k''_{c_i}$, until we fill all the critical places. By Claim 4, if we have found a manipulation, then it works for all profiles $\mathcal{P}'_{M \setminus W_{xc}}$, and if we have not found a manipulation, then there does not exist a manipulation that works for all the profiles $\mathcal{P}'_{M \setminus W_{xc}}$. That is, we have found a manipulation if and only if $x \in F_\alpha(W_{xc})$. If we have found some $x \neq c$ such that $W_{xc} \neq \emptyset$ and $x \in F_\alpha(W_{xc})$, then c is not in the α-core. Otherwise, c is in the α-core. □

Theorem 10. *There exists a polynomial-time algorithm solving the Simplified-Bucklin-β-Core problem.*

The proof of Theorem 10 is omitted due to space limitations.[7]

Theorem 11. *Let R be the Plurality with Runoff voting rule. There exists a polynomial-time algorithm solving the R-β-Core problem.*

Proof. Suppose we are given an instance $I = (E, M, c)$ of the R-β-Core problem, where R is the Plurality with Runoff voting rule, and $E = (C, V, \mathcal{P})$. The algorithm first checks whether $c \in F(M)$ by solving the R-Coalitional Manipulation problem (see [17] for the algorithm for this). Then it goes over all the other candidates. For each candidate $x \neq c$ it computes the maximal set $W_{xc} \subseteq M$ of the colluders who prefer x to c. If $W_{xc} \neq \emptyset$, we need to check whether x is β-feasible for W_{xc}. To do so, we iterate over all candidates $a \neq x$, and for each a, we check whether W_{xc} can make x win vs. the profile $\mathcal{P}^{a,x}_{M \backslash W_{xc}}$ where everybody ranks a first, x last, and the other candidates in some arbitrary order. Then we iterate over all pairs of candidates (a, b), and for each pair (a, b) and for all $i = 1, \dots, |M \backslash W_{xc}| - 1$, we check whether W_{xc} can make x win vs. the profile $\mathcal{P}^{i,a,b,x}_{M \backslash W_{xc}}$ where i voters in $M \backslash W_{xc}$ rank a first, the $|M \backslash W_{xc}| - i$ remaining voters rank b first, all these voters rank x last, and the other candidates are ranked in some arbitrary order. In the next claim we prove that it is enough to check these profiles to determine whether x is β-feasible for W_{xc}.

Claim. If:

1. for each $a \neq x$, there exists a profile $\mathcal{P}''_{W_{xc}}$ such that x is the winner under $\mathcal{P}_H \cup \mathcal{P}^{a,x}_{M \backslash W_{xc}} \cup \mathcal{P}''_{W_{xc}}$, and
2. for each pair (a, b) (where $x \notin \{a, b\}$) and for each $i, 1 \leq i \leq |M \backslash W_{xc}| - 1$, there exists a profile $\mathcal{P}'''_{W_{xc}}$ such that x is the winner under $\mathcal{P}_H \cup \mathcal{P}^{i,a,b,x}_{M \backslash W_{xc}} \cup \mathcal{P}'''_{W_{xc}}$,

then x is β-feasible. Otherwise x is not β-feasible.

Proof. It is clear that if (1) there is a candidate $a \neq x$, such that for each profile $\mathcal{P}'_{W_{xc}}$ x is not a winner under $\mathcal{P}_H \cup \mathcal{P}^{a,x}_{M \backslash W_{xc}} \cup \mathcal{P}'_{W_{xc}}$, or (2) there is a pair of candidates (a, b) and $i, 1 \leq i \leq |M \backslash W_{xc}| - 1$, such that for each profile $\mathcal{P}'_{W_{xc}}$ x is not a winner under $\mathcal{P}_H \cup \mathcal{P}^{i,a,b,x}_{M \backslash W_{xc}} \cup \mathcal{P}'_{W_{xc}}$, then by definition, x is not β-feasible for W_{xc}. For the other direction, suppose that for each $a \neq x$, there exists a profile $\mathcal{P}''_{W_{xc}}$ such that x is the winner under $\mathcal{P}_H \cup \mathcal{P}^{a,x}_{M \backslash W_{xc}} \cup \mathcal{P}''_{W_{xc}}$, and for each (a, b) and i there exists a profile $\mathcal{P}'''_{W_{xc}}$ such that x is the winner under $\mathcal{P}_H \cup \mathcal{P}^{i,a,b,x}_{M \backslash W_{xc}} \cup \mathcal{P}'''_{W_{xc}}$. Let $\mathcal{P}'_{M \backslash W_{xc}}$ be any profile for $M \backslash W_{xc}$. We will show that there exists a profile $\mathcal{P}'_{W_{xc}}$ such that x is the winner of the election under $\mathcal{P}_H \cup \mathcal{P}'_{M \backslash W_{xc}} \cup \mathcal{P}'_{W_{xc}}$. Let $y, z \in C$ be the winners of the first round under the partial profile $\mathcal{P}_H \cup \mathcal{P}'_{M \backslash W_{xc}}$. Denote by $\gamma(\mathcal{P}, b)$ the plurality score of candidate b under the profile $\mathcal{P}$. Suppose w.l.o.g. that

$$\gamma(\mathcal{P}_H \cup \mathcal{P}'_{M \backslash W_{xc}}, y) \geq \gamma(\mathcal{P}_H \cup \mathcal{P}'_{M \backslash W_{xc}}, z). \tag{1}$$

Denote by $N(\mathcal{P}, a, b)$ the number of votes in the profile $\mathcal{P}$ who prefer a over b. We divide the proof into 3 cases:

---

[7] See `ftp://ftp.cs.huji.ac.il/users/jeff/wine11zuckermanfull.pdf` for a version of this paper containing the proof.

Case 1. $x \notin \{y, z\}$. Let us consider the profile $\mathcal{P}_{M \setminus W_{xc}}^{i,y,z,x}$, where $i = |M \setminus W_{xc}| - \gamma(\mathcal{P}'_{M \setminus W_{xc}}, z)$. That is, we choose i such that $\gamma(\mathcal{P}_{M \setminus W_{xc}}^{i,y,z,x}, z) = \gamma(\mathcal{P}'_{M \setminus W_{xc}}, z)$. Let $\mathcal{P}_{W_{xc}}'''$ be the profile such that x is the winner under $\mathcal{P}_H \cup \mathcal{P}_{M \setminus W_{xc}}^{i,y,z,x} \cup \mathcal{P}_{W_{xc}}'''$. Let x and b be the two candidates who proceed to the second round under $\mathcal{P}_H \cup \mathcal{P}_{M \setminus W_{xc}}^{i,y,z,x} \cup \mathcal{P}_{W_{xc}}'''$. Here we have 3 cases:

Case 1.a. $b = y$. We may assume that $\mathcal{P}_{W_{xc}}'''$ has in the first places only x's and y's (if it does not, then we can change it appropriately, and the winners in the first and the second round will not change). Then under $\mathcal{P}_H \cup \mathcal{P}'_{M \setminus W_{xc}} \cup \mathcal{P}_{W_{xc}}'''$ the only possible winners of the first round are y, z and x. Denote for brevity, $\mathcal{Q}^1 = \mathcal{P}_H \cup \mathcal{P}'_{M \setminus W_{xc}} \cup \mathcal{P}_{W_{xc}}'''$, and $\mathcal{Q}^2 = \mathcal{P}_H \cup \mathcal{P}_{M \setminus W_{xc}}^{i,y,z,x} \cup \mathcal{P}_{W_{xc}}'''$. Since x is ranked last by all the voters in $\mathcal{P}_{M \setminus W_{xc}}^{i,y,z,x}$, we have $\gamma(\mathcal{Q}^1, x) \geq \gamma(\mathcal{Q}^2, x)$. Also, by the definition of $\mathcal{P}_{M \setminus W_{xc}}^{i,y,z,x}$, we have $\gamma(\mathcal{Q}^1, y) \leq \gamma(\mathcal{Q}^2, y)$ and $\gamma(\mathcal{Q}^1, z) = \gamma(\mathcal{Q}^2, z)$. As mentioned before, x is one of the winners of the first round under $\mathcal{Q}^2$. It follows that also under the profile $\mathcal{Q}^1$ x will be one of the winners of the first round. By assumption (1), and by the inequality $\gamma(\mathcal{P}_{W_{xc}}''', y) \geq \gamma(\mathcal{P}_{W_{xc}}''', z)$, we have $\gamma(\mathcal{Q}^1, y) \geq \gamma(\mathcal{Q}^1, z)$, and so the second winner of the first round is y. Now, $N(\mathcal{P}'_{M \setminus W_{xc}}, x, y) \geq 0 = N(\mathcal{P}_{M \setminus W_{xc}}^{i,y,z,x}, x, y)$. So, since x beats y in the second round under the profile $\mathcal{Q}^2$, x also beats y in the second round under $\mathcal{Q}^1$. So, we found a profile $(\mathcal{P}_{W_{xc}}''')$, such that x is the winner under $\mathcal{Q}^1 = \mathcal{P}_H \cup \mathcal{P}'_{M \setminus W_{xc}} \cup \mathcal{P}_{W_{xc}}'''$.

Case 1.b. $b = z$. Here we may assume that $\mathcal{P}_{W_{xc}}'''$ contains in the first places only x's and z's. Here we also have that the only possible winners of the first round under $\mathcal{Q}^1$ are x, y and z. Again, we have $\gamma(\mathcal{Q}^1, x) \geq \gamma(\mathcal{Q}^2, x)$, $\gamma(\mathcal{Q}^1, y) \leq \gamma(\mathcal{Q}^2, y)$ and $\gamma(\mathcal{Q}^1, z) = \gamma(\mathcal{Q}^2, z)$. So, since under $\mathcal{Q}^2$ the winners of the first round are x and z, we have that under $\mathcal{Q}^1$ the winners are also x and z. x beats z in the second round under $\mathcal{Q}^2$, and $N(\mathcal{P}'_{M \setminus W_{xc}}, x, z) \geq 0 = N(\mathcal{P}_{M \setminus W_{xc}}^{i,y,z,x}, x, z)$. Hence, x beats z in the second round under $\mathcal{Q}^1$ as well.

Case 1.c. $b \notin \{y, z\}$. We may assume that $\mathcal{P}_{W_{xc}}'''$ contains in the first places only x's and b's. The only possible winners of the first round under $\mathcal{Q}^1$ are y, z, x and b. We have the following inequalities: $\gamma(\mathcal{P}'_{M \setminus W_{xc}}, x) \geq 0 = \gamma(\mathcal{P}_{M \setminus W_{xc}}^{i,y,z,x}, x)$, $\gamma(\mathcal{P}'_{M \setminus W_{xc}}, b) \geq 0 = \gamma(\mathcal{P}_{M \setminus W_{xc}}^{i,y,z,x}, b)$, $\gamma(\mathcal{P}'_{M \setminus W_{xc}}, y) \leq \gamma(\mathcal{P}_{M \setminus W_{xc}}^{i,y,z,x}, y)$, and $\gamma(\mathcal{P}'_{M \setminus W_{xc}}, z) = \gamma(\mathcal{P}_{M \setminus W_{xc}}^{i,y,z,x}, z)$. And so, since x and b are the winners of the first round under $\mathcal{Q}^2$, they are also the winners of the first round under $\mathcal{Q}^1$. In the second round, x beats b under the profile $\mathcal{Q}^2$, and $N(\mathcal{P}'_{M \setminus W_{xc}}, x, b) \geq 0 = N(\mathcal{P}_{M \setminus W_{xc}}^{i,y,z,x}, x, b)$. It follows that x beats b in the second round under the profile $\mathcal{Q}^1$ as well.

Case 2. $x = y$. So x and z are the two winners of the first round under the partial profile $\mathcal{P}_H \cup \mathcal{P}'_{M \setminus W_{xc}}$. Recall that $\mathcal{P}_{M \setminus W_{xc}}^{z,x}$ is a profile where everybody in $M \setminus W_{xc}$ ranks z first, x last, and the other candidates in some arbitrary order. Let $\mathcal{P}_{W_{xc}}''$ be the profile such that x is the winner under $\mathcal{P}_H \cup \mathcal{P}_{M \setminus W_{xc}}^{z,x} \cup \mathcal{P}_{W_{xc}}''$. Denote, for shortness, $\mathcal{Q}^3 = \mathcal{P}_H \cup \mathcal{P}'_{M \setminus W_{xc}} \cup \mathcal{P}_{W_{xc}}''$ and $\mathcal{Q}^4 = \mathcal{P}_H \cup \mathcal{P}_{M \setminus W_{xc}}^{z,x} \cup \mathcal{P}_{W_{xc}}''$. Let x and d be the two candidates who are the winners of the first round under $\mathcal{Q}^4$.

Case 2.a. $d = z$. Here we can assume that $\mathcal{P}_{W_{xc}}''$ contains only x's and z's in the first places. Hence, under $\mathcal{Q}^3$, x and $z = d$ are the winners of the first round.

Case 2.b. $d \neq z$. We can assume that $\mathcal{P}''_{W_{xc}}$ contains only x's and d's in the first places. Therefore, under $\mathcal{Q}^3$ the only possible winners of the first round are x, z and d. Since under $\mathcal{Q}^4$ x and d proceed to the second round, and due to the fact that in $\mathcal{P}^{z,x}_{M \setminus W_{xc}}$ z is ranked on top of all the preferences, we have $\gamma(\mathcal{Q}^3, d) \geq \gamma(\mathcal{Q}^4, d) > \gamma(\mathcal{Q}^4, z) \geq \gamma(\mathcal{Q}^3, z)$, and $\gamma(\mathcal{Q}^3, x) \geq \gamma(\mathcal{Q}^4, x) > \gamma(\mathcal{Q}^4, z) \geq \gamma(\mathcal{Q}^3, z)$. Therefore, under $\mathcal{Q}^3$, x and d are the winners of the first round.

Case 2 (continued). Now, as x is the winner of the second round under $\mathcal{Q}^4$, and since $N(\mathcal{P}'_{M \setminus W_{xc}}, x, d) \geq 0 = N(\mathcal{P}^{z,x}_{M \setminus W_{xc}}, x, d)$, we have $N(\mathcal{Q}^3, x, d) \geq N(\mathcal{Q}^4, x, d) > N(\mathcal{Q}^4, d, x) \geq N(\mathcal{Q}^3, d, x)$. Hence, x beats d in the second round under $\mathcal{Q}^3$, and so x wins the election.

Case 3. $x = z$. This case is handled similarly to Case 2. (**End of proof of claim.**)

$\square$

If we found $x \neq c$ such that $W_{xc} \neq \emptyset$ and $x \in F_\beta(W_{xc})$, then by definition $c \notin \beta$-core. Otherwise, $c \in \beta$-core.

$\square$

5 Conclusions and Future Work

In this paper we have provided a computational analysis of the following question: given a coalition of manipulative voters, which candidate should they manipulate in favor of, given that they might not have identical preferences. To perform our analysis we have used the notions of α- and β-core, which—under various assumptions—describe the sets of candidates that the manipulators can support in a stable manner (i.e., without running the risk of breaking the coalition).

Our main results are the following. The complexity of determining membership in the α- and β-cores is at least as high as the complexity of the constructive coalitional manipulation problem for the same rule. On the other hand, for the several prominent voting rules for which coalitional manipulation is easy, we have also provided polynomial time algorithms for determining membership in the α- and β-cores. One direction for future work is to extend the above research to other voting rules. Another interesting direction is to try to find voting rules that produce an outcome in the core, if the core is non-empty, when applied to the truthful preferences of the voters. Yet another direction is to investigate the computational complexity of finding a voting profile of the colluders which is a strong Nash equilibrium, if one exists.

Acknowledgments. Michael Zuckerman and Jeffrey S. Rosenschein were partially funded by Israel Science Foundation grant #898/05, Israel Ministry of Science and Technology grant #3-6797, and, along with Vincent Conitzer, United States-Israel Binational Science Foundation grant #2006-216. Piotr Faliszewski was Supported by AGH University of Technology Grant no. 11.11.120.865, by Polish Ministry of Science and Higher Education grant N-N206-378637, and by Foundation for Polish Science's program Homing/Powroty. Vincent Conitzer acknowledges NSF IIS-0812113, IIS-0953756, and CCF-1101659, as well as an Alfred P. Sloan fellowship, for support.

References

1. Aumann, R.: Acceptable points in general cooperative n-person games. Contributions to the Theory of Games 4, 287–324 (1959)
2. Bachrach, Y., Elkind, E., Faliszewski, P.: Coalitional voting manipulation: A game-theoretic perspective. In: The Twenty-Second International Joint Conference on Artificial Intelligence (IJCAI 2011), pp. 49–54 (2011)
3. Bartholdi, J., Orlin, J.: Single Transferable Vote resists strategic voting. Social Choice and Welfare 8, 341–354 (1991)
4. Bartholdi, J., Tovey, C.A., Trick, M.A.: The computational difficulty of manipulating an election. Social Choice and Welfare 6, 227–241 (1989)
5. Conitzer, V., Sandholm, T., Lang, J.: When are elections with few candidates hard to manipulate? JACM 54(3) (June 2007)
6. Faliszewski, P., Hemaspaandra, E., Hemaspaandra, L.: How hard is bribery in elections? Journal of AI Research 35, 485–532 (2009)
7. Faliszewski, P., Hemaspaandra, E., Hemaspaandra, L.: Using complexity to protect elections. Communications of the ACM 53(11), 74–82 (2010)
8. Faliszewski, P., Procaccia, A.: AI's war on manipulation: Are we winning? AI Magazine 31(4), 53–64 (2010)
9. Gibbard, A.: Manipulation of voting schemes. Econometrica 41, 587–602 (1973)
10. Obraztsova, S., Elkind, E.: On the complexity of voting manipulation under randomized tie-breaking. In: The Twenty-Second International Joint Conference on Artificial Intelligence (IJCAI 2011), pp. 319–324 (2011)
11. Obraztsova, S., Elkind, E., Hazon, N.: Ties matter: Complexity of voting manipulation revisited. In: The Tenth International Joint Conference on Autonomous Agents and Multiagent Systems (AAMAS 2011), pp. 71–78 (2011)
12. Peleg, B., Sudholter, P.: Introduction to the Theory of Cooperative Games. Kluwer Academic Publishers, Boston (2003)
13. Satterthwaite, M.: Strategy-proofness and Arrow's conditions: Existence and correspondence theorems for voting procedures and social welfare functions. Journal of Economic Theory 10, 187–217 (1975)
14. Xia, L., Conitzer, V., Procaccia, A.D.: A scheduling approach to coalitional manipulation. In: Proceedings of the 11th ACM Conference on Electronic Commerce (EC 2010), pp. 275–284 (2010)
15. Xia, L., Zuckerman, M., Procaccia, A.D., Conitzer, V., Rosenschein, J.S.: Complexity of unweighted coalitional manipulation under some common voting rules. In: The Twenty-First International Joint Conference on Artificial Intelligence (IJCAI 2009), Pasadena, California, pp. 348–353 (July 2009)
16. Zuckerman, M., Lev, O., Rosenschein, J.S.: An algorithm for the coalitional manipulation problem under maximin. In: The Tenth International Joint Conference on Autonomous Agents and Multiagent Systems (AAMAS 2011), Taipei, Taiwan, pp. 845–852 (2011)
17. Zuckerman, M., Procaccia, A.D., Rosenschein, J.S.: Algorithms for the coalitional manipulation problem. AI Journal 173(2), 392–412 (2009)

The Price of Civil Society*

Russell Buehler[1], Zachary Goldman[2], David Liben-Nowell[1], Yuechao Pei[2],
Jamie Quadri[3], Alexa Sharp[3], Sam Taggart[3], Tom Wexler[3], and Kevin Woods[3]

[1] Carleton College, Northfield MN, USA
[2] Denison University, Granville OH, USA
[3] Oberlin College, Oberlin OH, USA

Abstract. Most work in algorithmic game theory assumes that players
ignore costs incurred by their fellow players. In this paper, we consider
superimposing a social network over a game, where players are concerned
with minimizing not only their own costs, but also the costs of their
neighbors in the network. We aim to understand how properties of the
underlying game are affected by this alteration to the standard model.
The new social game has its own equilibria, and the *price of civil society*
denotes the ratio of the social cost of the worst such equilibrium relative
to the worst Nash equilibrium under standard selfish play. We initiate
the study of the price of civil society in the context of a simple class
of games. Counterintuitively, we show that when players become less
selfish (optimizing over both themselves and their friends), the resulting
outcomes may be worse than they would have been in the base game. We
give tight bounds on this phenomenon in a simple class of load-balancing
games, over arbitrary social networks, and present some extensions.

1 Introduction

The world of traditional game-theoretic analysis is a cold one. Each individual
cares only about himself; he pays no heed to his neighbor's happiness. He makes
strategic decisions with exclusive regard for his own direct preferences about
the state of the world. He is not a good friend. Over the past decade, research
on the *price of anarchy*—beginning with the seminal work of Koutsoupias and
Papadimitriou [11], and progressing through landmark results like Roughgarden
and Tardos's work on selfish routing [17], among many others—has flourished in
the algorithmic game theory community. The price of anarchy measures the cost
of this cold world: relative to a centrally planned optimum, how much worse are
the outcomes that arise from purely selfish decision-makers?

But this picture of human decision-makers is unrealistically bleak. Humans
maintain long-term dyadic relationships with nonrelatives; in other words, we
have *friends*. The presence of this "friend" relationship is unusual among species,
and it is a long-standing matter of research in the social sciences to explain its

---

* This work was supported by NSF grant CCF-0728779 and was underwritten in part
by grants in support of undergraduate research from Carleton, Denison, and Oberlin.
Thanks to Dev Gupta, Jason Hartline, and Thanh Nguyen for helpful discussions.

N. Chen, E. Elkind, and E. Koutsoupias (Eds.): WINE 2011, LNCS 7090, pp. 375–382, 2011.

origins. (See Silk [18], for example.) Our daily lives are influenced deeply by what is called *civil society* by political scientists: that is, everything about society that is not "the state" or "the market." Civil society is the church, the book club, the Association for Computing Machinery, the knitting circle. It is clear that our bonds with others affect our decisions; our relationships formed through civil society change our preferences relative to the solipsistic baseline. And, generally speaking, this effect is positive. (See Putnam's *Bowling Alone* [15], for example.)

Our goal in this paper is to understand the way in which superimposing a social network on a strategic situation affects the quality of the resulting equilibria. To this end, we consider augmenting a "base game" with a social network. Each individual cares about both her own happiness and that of her neighbors in the social network. When the social network has no edges, we have "the market" and the classic Nash equilibrium. When the social network is the complete graph, we have "the state," or at least individual agents myopically striving to optimize the welfare of society as a whole. Our interest lies in exploring the middle ground between selfishness and altruism. Specifically, we wish to analyze the following question: *How much better are the equilibria when, rather than acting in a purely selfish manner, players act on behalf of their friends as well as themselves?* That is, what benefits accrue from the presence of civil society?

Perhaps surprisingly, we observe that, by becoming *more* socially concerned, players can often end up at worse outcomes than standard Nash equilibria. This phenomenon has been observed in multiple related settings (e.g. [4, 9]), but our observation adds to the collection of such examples, suggesting that this effect is ubiquitous rather than pathological. Thus we are forced to address a different question: *How much worse are the equilibria when, rather than acting in a purely selfish manner, players act on behalf of their friends as well as themselves?*

We take a particular class of load-balancing game as our base games, and we examine the effect of varying the social network structure. We present tight constant bounds on the degradation of equilibrium quality in these games. Thus, in these games, we show that caring about friends can make the world worse, but only to a limited degree.

The price of civil society. Let Γ denote an n-player game, and let G denote an undirected graph on the players, where an edge reflects friendship between its endpoints. The *social game* Γ_G has the same n players and the same actual cost functions, but each player seeks to optimize the sum of her own cost *and* the costs of her neighbors in G. We are interested in the relative cost of the worst Nash equilibrium in the social game Γ_G compared to the cost of the worst Nash equilibrium in the base game Γ. We call this ratio the *price of civil society*.

The complexity of this model lies in the structure of the social network. To isolate the effects of this structure, we limit our attention to *load-balancing games with identical linear machines*. Even in this simplistic setting, our model exhibits interesting and nontrivial behavior. Our main result is that, while the price of civil society can exceed 1, it does not exceed 5/4, and furthermore this bound is tight. The extremal example is a small game, and so we may wonder if this deterioration is merely a symptom of the discrete nature of atomic load-balancing.

We prove that it is not. As the number of players tends to infinity, and the game converges to its nonatomic version, the price of civil society becomes 9/8: smaller than in the atomic case, but still greater than 1. We also consider generalizations of this game and restrictions to particular classes of social networks.

Related work. Chen and Kempe [4] consider a routing game in which players optimize a linear combination of their own latency and the average latency of *all* players. The authors study the change in equilibria as a function of the weight given to the "altruistic" term, allowing for interpolation between totally selfish and totally altruistic behavior. Our model similarly interpolates between these extremes, but our "middle ground" reflects a player who cares about a select group of other players, rather than caring to a limited degree about all players.

Meier, Oswald, Schmid, and Wattenhofer [13] study a virus inoculation game in which players care about their neighbors' costs in addition to their own. Unlike our model, however, their game is intrinsically linked to the social network: changing the graph also changes the game. The primary goal of the present work is to study varied network topologies for a *fixed* game, thereby allowing us to isolate the impact of social structure on competitive play.

Ashlagi, Krysta, and Tennenholtz [1] introduce "social context games" which consider players who optimize general functions of their friends' utilities (rather than just the sum). Their work focuses on the existence of equilibria for a variety of functions, whereas we are concerned with the quality of these outcomes.

Hayrapetyan, Tardos, and Wexler [9] consider games in which players form coalitions, each modeled as a single player. Their model is closely related to the special case of our game in which the social network is a collection of cliques.

The price of anarchy was introduced by Koutsoupias and Papadimitriou [11] in the context of load-balancing games. Subsequent work has explored numerous variations of these games, including restricted classes of latency functions; symmetric or asymmetric access; unit or weighted jobs; mixed or pure equilibria; worst or best Nash equilibria; and sum or makespan objective functions. See, for example, [2, 3, 5–8, 10, 12, 19] and the references therein.

2   Model and Notation

For a game with player set N, a social network is given by an undirected graph with vertices N and edges representing (symmetric) friendships between players. The *perceived cost* to player i under a strategy profile $\mathbf{s}$ is the cost i incurs plus the total costs incurred by all of i's neighbors. The social cost $\mathrm{SC}(\mathbf{s})$ of profile $\mathbf{s}$ is the sum of the actual (not perceived) costs experienced by each player.

We define OPT as a socially optimal strategy profile, that is, one with minimum social cost. An outcome is a pure Nash equilibrium (NE) when no individual player can decrease her actual cost by unilaterally switching to a different strategy; it is a *civil society Nash equilibrium* (CSNE) when no player can decrease her perceived cost by switching. Let WNE and WCSNE, respectively, denote the worst among the NE and the worst among the CSNE, measured by social cost.

The *price of anarchy* (POA) measures the extent to which the cost of any Nash equilibrium can exceed that of the optimal solution: it is SC(WNE)/SC(OPT). The *price of civil society* (POCS) measures the extent to which the cost of a civil society Nash equilibrium can exceed that of the worst pure Nash equilibrium. That is, the POCS is SC(WCSNE)/SC(WNE). Note that we are comparing worst social-game equilibria to worst base-game equilibria; it could also be interesting to consider *best* equilibria in one or both games.

In this paper, we consider a simple base game in which each player has access to a common set M of resources. Each player selects a single resource, and incurs a cost equal to the total number of players who pick that resource. Given an outcome $\mathbf{s}$, the *load* $\ell_j = \ell_j(\mathbf{s})$ on a resource j is the number of players who choose j. The social cost, given by $\sum_j \ell_j^2$, is completely determined by the load vector $L = (\ell_1, \dots, \ell_m)$. This game is equivalent to load balancing with identical linear latencies. Such games are known to have pure Nash equilibria [14, 16]. Furthermore, it can be shown that for any social network G, these games also have civil society equilibria. In particular, a potential function for this game is $\Phi(\mathbf{s}) = \sum_{j \in M} \left[\binom{\ell_j + 1}{2} + e_j \right]$, where e_j is the number of edges with both endpoints choosing j under $\mathbf{s}$: one can check that if a player switches to a new resource, the improvement in her perceived cost is exactly the decrease in Φ, and so repeated best-response moves must eventually terminate at a CSNE.

We also consider the nonatomic version of this game. Informally, we picture a continuum of infinitesimally small players. More formally, let $\text{POCS}(n, m)$ denote the maximum POCS over all n-player, m-resource atomic games, and then define the nonatomic POCS to be $\sup_m \left(\lim \sup_{n \to \infty} \text{POCS}(n, m) \right)$.

3   Main Results

Consider an n-player, m-resource instance of our game. By convexity, the socially optimal solution assigns players to resources as evenly as possible. Furthermore, the only pure Nash equilibria are of the same form. Therefore WNE = OPT and thus POA = 1. Likewise, in the nonatomic version of this game, all resources have exactly the same load in both the optimal solution and the unique equilibrium. However, superimposing certain social networks on these games may cause the POCS to exceed 1; that is, some networks lead to social games with worse stable outcomes than without the network. We consider both the atomic and nonatomic settings, and prove that the POCS in these settings is 5/4 and 9/8, respectively.

We begin with some useful lemmas. Let δ be the size of a player in a game; that is, let $\delta = 1$ for atomic games and let $\delta \to 0$ for nonatomic games.

Lemma 1. *For any CSNE and any two resources 1 and 2, $\ell_1 \le 2\ell_2 + \delta$.*

Proof. Consider a player i using resource 1 in the CSNE. Suppose that she is friends with all players using resource 2 and with no one using resource 1: this arrangement is the case in which she will be most averse to switching to resource 2. As it stands, her cost is ℓ_1, each of her ℓ_2/δ friends' costs are ℓ_2, and so her total perceived cost is $\ell_1 + \ell_2^2/\delta$. If she were to switch to resource 2, her cost would be $\ell_2 + \delta$, each of her ℓ_2/δ friends' costs would be $\ell_2 + \delta$, and so her total perceived cost would be $\ell_2 + \delta + \ell_2^2/\delta + \ell_2$. Because she does not want to switch in a CSNE, we have that $\ell_1 + \ell_2^2/\delta \leq \ell_2 + \delta + \ell_2^2/\delta + \ell_2$. Simplifying, we get the desired $\ell_1 \leq 2\ell_2 + \delta$. This inequality holds for any social network. □

Given load vector $L = (\ell_1, \cdots, \ell_m)$, define $\Phi_m(L) = (\sum_i \ell_i^2)/[(\sum_i \ell_i)^2/m]$. This quantity is the ratio of the social cost of L to that of a "fractional" Nash equilibrium (all loads are identical, even if that amount is not an integer), and thus is an upper bound on the POCS of this instance.

Lemma 2. *For a given machine i and given bounds a and b with $0 \leq a \leq b$, suppose that we maximize $\Phi_m(L)$ over all ℓ_i such that $a \leq \ell_i \leq b$. Then this maximum will be achieved at either $\ell_i = a$ or $\ell_i = b$.*

Proof. We note that $\Phi_m(L)$ is not generally a convex function of ℓ_i. However, it is continuously differentiable for $\ell_i \geq 0$, the derivative is zero at the unique point $x_i = (\sum_{j \neq i} \ell_j^2)/(\sum_{j \neq i} \ell_j)$, and the derivative is negative for $0 \leq \ell_i < x_i$ and positive for $\ell_i > x_i$. Thus the maximum occurs at either $\ell_i = a$ or $\ell_i = b$. □

3.1   Atomic Games

We begin with an example illustrating a nontrivial price of civil society. Consider four players, two resources, and social network $\mathcal{K}_{1,3}$, the complete bipartite graph on 1 and 3 vertices. Assigning the three leaf players to resource 1 and the root player to resource 2 yields a CSNE of cost 10. All NE have a cost of 8; thus, the POCS of this instance is $5/4$. We will show that this example is in fact the worst possible, and therefore that the POCS for atomic games is $5/4$. We first prove this result for any CSNE in which the load on each resource is at least 2:

Lemma 3. *Let L be the load vector for a CSNE with minimum load at least 2. Then $\Phi_m(L) \leq 49/40 < 5/4$.*

Proof. Combining Lemmas 1 and 2, $\Phi_m(L)$ is maximized when all resources have load either ℓ or $2\ell + 1$ for some $\ell \geq 2$. Let p and $\gamma \cdot p$ denote the number of resources with loads ℓ and $2\ell + 1$, respectively. Then $\Phi_m(L)$ is maximized when $\gamma = \ell/(2\ell + 1)$ and $\ell = 2$, giving an upper bound of $49/40 < 5/4$, as desired. □

Unfortunately, if some resources have a load of 0 or 1 in a CSNE with load vector L, then $\Phi_m(L)$ may exceed $5/4$: the POCS is still bounded by $5/4$, but $\Phi_m(L)$ uses a fractional Nash equilibrium, and more careful analysis is needed.

Theorem 1. *The price of civil society of the atomic game is $5/4$.*

Proof. We have shown that that the POCS can be as large as $5/4$. We must finish showing that $5/4$ is an upper bound. Let L be the load vector of a WCSNE. By Lemma 3, it only remains to consider the case in which the minimum load of L is either 0 or 1. If the minimum load in L is 0, then, by Lemma 1, every resource has load either 0 or 1. Such a configuration is a NE, and thus the POCS is 1.

If the minimum load in L is 1, Lemma 1 implies that the max load is 3. Let p, q, and r denote the number of resources with loads 1, 2, and 3, respectively. The social cost of this outcome is $p + 4q + 9r$; to determine the POCS of this instance, it remains to determine SC(WNE) and optimize over p, q, and r.

Note that a Nash equilibrium cannot have both load-1 and load-3 resources: players will move from a load-3 to a load-1 resource until either the load-1 or load-3 resources are exhausted. Thus a Nash equilibrium either has no load-1 resources (if $p \leq r$) or no load-3 resources (if $r \geq p$). Suppose that $p \leq r$, so at NE, there are $(q + 2p)$ load-2 and $(r - p)$ load-3 resources. The social cost of this configuration is thus $4(q + 2p) + 9(r - p) = -p + 4q + 9r$, and so POCS $= (p + 4q + 9r)/(-p + 4q + 9r)$. This expression is maximized when $p = r$ and $q = 0$, and evaluates to $5/4$. Similar analysis shows that if $r \leq p$, the same bound holds.   □

3.2 Nonatomic Games

In the atomic case, the worst instance for POCS had only a few players. We now consider nonatomic games, with an infinite number of infinitesimally small players. We show that while the price of civil society generally decreases as the number of players grows, it does not improve below $9/8$. Thus, while a portion of our atomic POCS $= 5/4$ example can be attributed to an integrality issue (which tends to zero as the number of players grows) the remainder is due purely to the presence of socially conscious agents (the effects of which persist even in the nonatomic case).

We start with a nonatomic example whose POCS is $9/8$. Let there be three resources and $n \to \infty$ players, with social network a complete tripartite graph with parts of size $n/4$, $n/4$, and $n/2$. The unique Nash equilibrium places $n/3$ players on each resource, for a social cost of $3(n/3)^2 = n^2/3$. There is a CSNE that places the $n/2$-player part on resource 1 and the other parts on resources 2 and 3, for a social cost of $3n^2/8$, yielding a POCS of $9/8$.

Theorem 2. *The price of civil society of the nonatomic game is $9/8$.*

Proof. In light of the previous example, we need only prove an upper bound. It suffices to show $\Phi_m(L) \leq 9/8$ for a WCSNE with load vector L. Lemmas 1 and 2 together imply that the maximum $\Phi_m(L)$ occurs when each ℓ_i is either ℓ or 2ℓ, for some $\ell > 0$. Let p and q denote the number of resources with loads ℓ and 2ℓ, respectively. Then $\Phi_m(L)$ is maximized at $q = p/2$, where the ratio is $9/8$.   □

4 Extensions and Future Work

The POCS of $5/4$ for the discrete game was only achievable on a complete bipartite graph. These graphs have *no* triadic closure: friends of friends are never friends. But real social networks exhibit a high degree of triadic closure [20]. What happens if we force the graph to be more like a friendship-based social network?

We first look at a graph with *complete* triadic closure (i.e., all connected components are cliques). Here, loads in a CSNE must be as balanced as possible (or else at least one clique would have more people using some overloaded resource than some underloaded one). Therefore every CSNE is a NE, and the POCS is 1: there is no degradation because of the social network.

What happens with graphs between these extremes? What "intermediate" social structures should we examine? Here is one option. Precisely define the triadic closure of a graph to be the probability Δ that a path of length two, chosen uniformly at random, is part of a triangle in the graph [20]. What happens as we vary Δ in our social network? Unfortunately, graphs with Δ approaching 1 can have the worst possible POCS of $5/4$ (using many copies of the original worst-case example). However, for a fixed number of resources, restricting the allowable triadic closure can provide improved bounds on the worst possible POCS. We have seen that with no constraints on Δ, the POCS can be as large as $5/4$, while if Δ is 1, the POCS is 1. Many intermediate results are possible: e.g., with two resources, we can show that if $\Delta > 7/11$, the POCS is at most $10/9$.

Another way to generalize our results is to change our base game beyond a load-balancing game with unit-weight jobs and machines with identical linear latencies. We have proven several such results that we simply state here without proof. (The proof methods are substantially similar to those in Section 3.)

First we consider *related machines*, where machine j has latency function $a_j \ell$ under load ℓ. For two machines with atomic players, we can show that the price of civil society is $14/11 \approx 1.27$, a mild worsening over our earlier bound of $5/4 = 1.25$ for identical machines. The extremal example has two jobs on a machine of latency 3ℓ and one job on a machine of latency 2ℓ. For nonatomic players, the POCS remains $9/8 = 1.125$, the same as for identical machines. Unsurprisingly, for broader classes of latency functions, the POCS increases substantially: for general convex and increasing latencies, the POCS reaches n, the number of players.

Another generalization is *weighted* load balancing, where jobs may have different sizes. For two identical linear machines, the POCS increases to $34/25 = 1.36$. The extremal example has one machine with two jobs of weight 4 and one machine with two jobs of weight 1. For more than two identical linear machines, we can prove that the POCS lies in a narrow range slightly greater than 1.5.

More generally, similar analysis can be applied to arbitrary base games. We suspect that for some classes of games, social networks may cause substantially greater degradation in the resulting equilibria, while in others, social structure may only improve outcomes. In what games are social networks most harmful?

References

1. Ashlagi, I., Krysta, P., Tennenholtz, M.: Social Context Games. In: Proc. Workshop on Internet and Network Economics, pp. 675–683 (2008)
2. Awerbuch, B., Azar, Y., Epstein, A.: The Price of Routing Unsplittable Flow. In: Proc. Symposium on Theory of Computing, pp. 57–66 (2005)

3. Caragiannis, I., Flammini, M., Kaklamanis, C., Kanellopoulos, P., Moscardelli, L.: Tight Bounds for Selfish and Greedy Load Balancing. In: Bugliesi, M., Preneel, B., Sassone, V., Wegener, I. (eds.) ICALP 2006. LNCS, vol. 4051, pp. 311–322. Springer, Heidelberg (2006)
4. Chen, P.-A., Kempe, D.: Altruism, Selfishness, and Spite in Traffic Routing. In: Proc. Conference on Electronic Commerce, pp. 140–149 (2008)
5. Christodoulou, G., Koutsoupias, E.: On the Price of Anarchy and Stability of Correlated Equilibria of Linear Congestion Games. In: Brodal, G.S., Leonardi, S. (eds.) ESA 2005. LNCS, vol. 3669, pp. 59–70. Springer, Heidelberg (2005)
6. Christodoulou, G., Koutsoupias, E.: The Price of Anarchy of Finite Congestion Games. In: Proc. Symposium on Theory of Computing, pp. 67–73 (2005)
7. Fotakis, D., Kontogiannis, S., Spirakis, P.: Selfish Unsplittable Flows. Theoretical Computer Science 348(2), 226–239 (2005)
8. Gairing, M., Lücking, T., Mavronicolas, M., Monien, B., Rode, M.: Nash Equilibria in Discrete Routing Games with Convex Latency Functions. Journal of Computer Systems Science 74(7), 1199–1225 (2008)
9. Hayrapetyan, A., Tardos, É., Wexler, T.: The Effect of Collusion in Congestion Games. In: Proc. Symposium on Theory of Computing, pp. 89–98 (2006)
10. Kothari, A., Suri, S., Tóth, C.D., Zhou, Y.: Congestion Games, Load Balancing, and Price of Anarchy. In: Proc. Workshop on Combinatorial and Algorithmic Aspects of Networking (2004)
11. Koutsoupias, E., Papadimitriou, C.H.: Worst-Case Equilibria. In: Proc. Symposium on Theoretical Aspects of Computer Science, pp. 404–413 (1999)
12. Lücking, T., Mavronicolas, M., Monien, B., Rode, M.: A New Model for Selfish Routing. Theoretical Computer Science 406(3), 187–206 (2008)
13. Meier, D., Oswald, Y.A., Schmid, S., Wattenhofer, R.: On the Windfall of Friendship: Inoculation Strategies on Social Networks. In: Proc. Conference on Electronic Commerce, pp. 294–301 (2008)
14. Monderer, D., Shapley, L.S.: Potential Games. Games and Economic Behavior 14, 124–143 (1996)
15. Putnam, R.: Bowling Alone: The Collapse and Revival of American Community. Simon & Schuster (2000)
16. Rosenthal, R.W.: A Class of Games Possessing Pure-Strategy Nash Equilibria. International Journal of Game Theory 2, 65–67 (1973)
17. Roughgarden, T., Tardos, É.: How Bad is Selfish Routing? Journal of the ACM (2002)
18. Silk, J.: Cooperation Without Counting: The Puzzle of Friendship. In: Hammerstein, P. (ed.) Genetic and Cultural Evolution of Cooperation, pp. 37–54. MIT Press (2003)
19. Suri, S., Tóth, C.D., Zhou, Y.: Selfish Load Balancing and Atomic Congestion Games. Algorithmica 47(1), 79–96 (2007)
20. Watts, D.J., Strogatz, S.H.: Collective Dynamics of 'Small-World' Networks. Nature 393, 440–442 (1998)

The Robust Price
of Anarchy of Altruistic Games

Po-An Chen[1], Bart de Keijzer[2], David Kempe[1], and Guido Schäfer[2,3]

[1] Department of Computer Science, University of Southern California, USA
{poanchen,dkempe}@usc.edu
[2] Algorithms, Combinatorics and Optimization, CWI Amsterdam, The Netherlands
{b.de.keijzer,g.schaefer}@cwi.nl
[3] Dept. of Econometrics and Operations Research, VU University Amsterdam,
The Netherlands

Abstract. We study the inefficiency of equilibria for several classes of
games when players are (partially) altruistic. We model altruistic behavior by assuming that player i's perceived cost is a convex combination of
$1-\alpha_i$ times his direct cost and α_i times the social cost. Tuning the parameters α_i allows smooth interpolation between purely selfish and purely
altruistic behavior. Within this framework, we study altruistic extensions
of cost-sharing games, utility games, and linear congestion games. Our
main contribution is an adaptation of Roughgarden's *smoothness* notion
to altruistic extensions of games. We show that this extension captures
the essential properties to determine the *robust price of anarchy* of these
games, and use it to derive mostly tight bounds.

1 Introduction

Many large-scale decentralized systems involve the interactions of large numbers
of individuals acting to benefit themselves. Thus, such systems are naturally
studied from the viewpoint of game theory, with an eye on the social efficiency
of stable outcomes. Traditionally, "stable outcomes" have been associated with
pure Nash equilibria of the corresponding game. The notions of *price of anarchy*
[9] and *price of stability* [2] provide natural measures of the system degradation,
by capturing the degradation of the worst and best Nash equilibria, respectively,
compared to the socially optimal outcome. However, the predictive power of
such bounds has been questioned on (at least) two grounds: First, the adoption of Nash equilibria as a prescriptive solution concept implicitly assumes that
players are able to reach such equilibria, a very suspect assumption for computationally bounded players. In response, recent work has begun analyzing the
outcomes of natural response dynamics [3,15], as well as more permissive solution concepts such as mixed, correlated or coarse correlated equilibria. (This
general direction of inquiry has become known as "robust price of anarchy".)
Second, the assumption that players seek only to maximize their own utility
is at odds with altruistic behavior routinely observed in the real world. While
modeling human incentives and behavior accurately is a formidable task, several

N. Chen, E. Elkind, and E. Koutsoupias (Eds.): WINE 2011, LNCS 7090, pp. 383–390, 2011.
© Springer-Verlag Berlin Heidelberg 2011

papers have proposed natural models of altruism and analyzed its impact on the outcomes of games [4,5,6,10].

The goal of this paper is to begin a thorough investigation of the effects of relaxing both of the standard assumptions simultaneously, i.e., considering the combination of weaker solution concepts and notions of partially altruistic behavior by players. We formally define the *altruistic extension* of an n-player game in the spirit of past work on altruism (see [10, p. 154] and [4,5,8]): player i has an associated altruism parameter α_i, and his cost (or payoff) is a convex combination of $(1 - \alpha_i)$ times his direct cost (or payoff) and α_i times the social cost (or social welfare). By tuning the parameters α_i, this model allows smooth interpolation between pure selfishness ($\alpha_i = 0$) and pure altruism ($\alpha_i = 1$). To analyze the degradation of system performance in light of partially altruistic behavior, we extend the notion of *robust price of anarchy* [15] to games with altruistic players, and show that a suitably adapted notion of *smoothness* [15] captures the properties of a system that determine its robust price of anarchy. We use our framework to analyze the robust price of anarchy of three fundamental classes of games.

1. In a *cost-sharing game* [2], players choose subsets of resources, and all players choosing the same resource share its cost evenly. Using our framework, we derive a bound of $n/(1 - \hat{\alpha})$ on the robust price of anarchy of these games, where $\hat{\alpha}$ is the maximum altruism level of a player. This bound is tight for uniformly altruistic players.

2. We apply our framework to *utility games* [16], in which players choose subsets of resources and derive utility of the chosen set. The total welfare is determined by a submodular function of the union of all chosen sets. We derive a bound of 2 on the robust price of anarchy of these games. In particular, the bound remains at 2 regardless of the (possibly different) altruism levels of the players. This bound is tight.

3. We revisit and extend the analysis of *atomic congestion games* [14], in which players choose subsets of resources whose costs increase (linearly) with the number of players using them. Caragiannis et al. [4] recently derived a tight bound of $(5 + 4\alpha)/(2 + \alpha)$ on the pure price of anarchy when all players have the *same* altruism level α.[1] Our framework makes it an easy observation that their proof in fact bounds the robust price of anarchy. We generalize their bound to the case when different players have different altruism levels, obtaining a bound in terms of the maximum and minimum altruism levels. This partially answers an open question from [4]. For the special case of symmetric singleton congestion games (which corresponds to selfish scheduling on machines), we extend our study of non-uniform altruism and obtain an improved bound of $(4 - 2\alpha)/(3 - \alpha)$ on the price of anarchy when an α-fraction of the players are entirely altruistic and the remaining players are entirely selfish.

Notice that many of these bounds on the robust price of anarchy reveal a counter-intuitive trend: at best, for utility games, the bound is independent of the level of

---

[1] The altruism model of [4] differs from ours in a slight technicality discussed in Section 2 (Remark 1). Therefore, various bounds we cite here are stated differently in [4].

altruism, and for congestion games and cost-sharing games, it actually *increases* in the altruism level, unboundedly so for cost-sharing games. Intuitively, this phenomenon is explained by the fact that a change of strategy by player i may affect many players. An altruistic player will care more about these other players than a selfish player; hence, an altruistic player accepts more states as "stable". This suggests that the best stable solution can also be chosen from a larger set, and the price of stability should thus decrease. Our results on the price of stability lend support to this intuition: for congestion games, we derive an upper bound on the price of stability which decreases as $2/(1+\alpha)$; similarly, for cost-sharing games, we establish an upper bound which decreases as $(1-\alpha)H_n + \alpha$.

The increase in the price of anarchy is not a universal phenomenon, demonstrated by *symmetric singleton* congestion games. Caragiannis et al. [4] showed a bound of $4/(3+\alpha)$ for pure Nash equilibria with uniformly altruistic players, which decreases with the altruism level α. Our bound of $(4-2\alpha)/(3-\alpha)$ for mixtures of entirely altruistic and selfish players is also decreasing in the fraction of entirely altruistic players. We also extend an example of Lücking et al. [11] to show that symmetric singleton congestion games may have a mixed price of anarchy arbitrarily close to 2 for arbitrary altruism levels. In light of the above bounds, this establishes that pure Nash equilibria can result in strictly lower price of anarchy than weaker solution concepts.

Proofs are omitted from this short paper; they are available in the full version.

2 Altruistic Games and the Robust Price of Anarchy

Let $G = (N, \{\Sigma_i\}_{i \in N}, \{C_i\}_{i \in N})$ be a finite strategic game, where $N = [n]$ is the set of players, Σ_i the strategy space of player i, and $C_i : \Sigma \to \mathbb{R}$ the cost function of player i, mapping every joint strategy $s \in \Sigma = \Sigma_1 \times \cdots \times \Sigma_n$ to the player's direct cost. Unless stated otherwise, we assume that every player i wants to minimize his individual cost function C_i. We also call such games *cost-minimization games*. A *social cost* function $C : \Sigma \to \mathbb{R}$ maps strategies to social costs. We require that C is *sum-bounded*, that is, $C(s) \leq \sum_{i=1}^{n} C_i(s)$ for all $s \in \Sigma$. We study *altruistic extensions* of strategic games equipped with sum-bounded social cost functions. Our definition is based on one used (among others) in [5], and similar to ones given in [4,6,10].

Definition 1. *Let $\alpha \in [0,1]^n$. The α-altruistic extension of G (or simply α-altruistic game) is defined as the strategic game $G^\alpha = (N, \{\Sigma_i\}_{i \in N}, \{C_i^\alpha\}_{i \in N})$, where for every $i \in N$ and $s \in \Sigma$, $C_i^\alpha(s) = (1-\alpha_i)C_i(s) + \alpha_i C(s)$.*

Thus, the perceived cost that player i experiences is a convex combination of his direct (selfish) cost and the social cost; we call such a player α_i-*altruistic*. When $\alpha_i = 0$, player i is entirely selfish; thus, $\alpha = \mathbf{0}$ recovers the original game. A player with $\alpha_i = 1$ is entirely altruistic. Given an altruism vector $\alpha \in [0,1]^n$, we let $\hat{\alpha} = \max_{i \in N} \alpha_i$ and $\check{\alpha} = \min_{i \in N} \alpha_i$ denote the maximum and minimum altruism levels, respectively. When $\alpha_i = \alpha$ (a scalar) for all i, we call such games *uniformly α-altruistic games*.

Remark 1. In a recent paper, Caragiannis et al. [4] model uniformly altruistic players by defining the perceived cost of player i as $(1-\xi)C_i(s)+\xi(C(s)-C_i(s))$, where $\xi \in [0,1]$. It is not hard to see that in the range $\xi \in [0, \frac{1}{2}]$ this definition is equivalent to ours by setting $\alpha = \xi/(1-\xi)$ or $\xi = \alpha/(1+\alpha)$.

The most general equilibrium concept we consider is coarse (correlated) equilibria.

Definition 2 (Coarse equilibrium). *A* coarse (correlated) equilibrium *of a game G is a probability distribution σ over $\Sigma = \Sigma_1 \times \cdots \times \Sigma_n$ with the following property: if s is a random variable with distribution σ, then for each player i, and all $s_i^* \in \Sigma_i$:*

$$\mathbf{E}_{s\sim\sigma}\left[C_i(s)\right] \leq \mathbf{E}_{s_{-i}\sim\sigma_{-i}}\left[C_i(s_i^*, s_{-i})\right], \tag{1}$$

where σ_{-i} is the projection of σ on $\Sigma_{-i} = \Sigma_1 \times \cdots \times \Sigma_{i-1} \times \Sigma_{i+1} \times \cdots \times \Sigma_n$.

It includes several other solution concepts, such as correlated equilibria, mixed Nash equilibria and pure Nash equilibria.

The *price of anarchy (PoA)* [9] and *price of stability (PoS)* [2] quantify the inefficiency of equilibria for classes of games: Let $S \subseteq \Sigma$ be a set of strategy profiles for a cost-minimization game G with social cost function C, and let s^* be a strategy profile that minimizes C. We define $\text{PoA}(S,G) = \sup \{C(s)/C(s^*) : s \in S\}$ and $\text{PoS}(S,G) = \inf \{C(s)/C(s^*) : s \in S\}$. The *coarse* (or *correlated, mixed, pure*) *PoA* (or *PoS*) of a class of games $\mathcal{G}$ is the supremum over all games in $\mathcal{G}$ and all strategy profiles in the respective set of equilibrium outcomes. Notice that the PoA and PoS are defined with respect to the *original* social cost function C, not accounting for the altruistic components. This reflects our desire to understand the overall performance of the system (or strategic game), which is not affected by different *perceptions* of costs by individuals.[2]

Roughgarden [15] introduced the notion of (λ, μ)-*smoothness* of strategic games with sum-bounded social cost functions and showed that it provides a generic template for proving bounds on the PoA as well as the outcomes of no-regret sequences [3].

The smoothness approach cannot be applied directly to our altruistic games because the social cost function C that we consider here is in general not sum-bounded in terms of C_i^α (which is a crucial prerequisite in [15]). However, we are able to generalize the (λ, μ)-smoothness notion to altruistic games, thereby preserving many of its applications. For notational convenience, we define $C_{-i}(s) = C(s) - C_i(s)$.

---

[2] If all players have a uniform altruism level $\alpha_i = \alpha \in [0, 1]$ and the social cost function C is equal to the sum of all players' direct costs, then for every strategy profile $s \in \Sigma$, the sum of the perceived costs of all players is equal to $(1-\alpha+\alpha n)C(s)$. In particular, bounding the PoA (or PoS) with respect to C is equivalent to bounding the PoA (or PoS) with respect to total perceived cost in this case.

Definition 3. G^α *is* (λ, μ, α)-*smooth iff for any two strategy profiles* $s, s^* \in \Sigma$,

$$\sum_{i=1}^{n} C_i(s_i^*, s_{-i}) + \alpha_i(C_{-i}(s_i^*, s_{-i}) - C_{-i}(s)) \leq \lambda C(s^*) + \mu C(s).$$

Most of the results in [15] following from (λ, μ)-smoothness carry over to our altruistic setting using the generalized (λ, μ, α)-smoothness notion. The following result allows a calculation of the PoA.[3]

Proposition 1. *Let* G^α *be an* α-*altruistic game that is* (λ, μ, α)-*smooth with* $\mu < 1$. *Then, the coarse (and thus also correlated, mixed, and pure) price of anarchy of* G^α *is at most* $\frac{\lambda}{1-\mu}$.

For many important classes of games, the bounds obtained by (λ, μ, α)-smoothness arguments are actually tight, even for pure Nash equilibria. This motivates defining the *robust PoA* as the best bound that can be proved using the smoothness technique.

Definition 4. *Let* G^α *be an* α-*altruistic game. Its* robust PoA *is defined as* $RPoA_G(\alpha) = \inf\{\frac{\lambda}{1-\mu} : G^\alpha$ *is* (λ, μ, α)-*smooth with* $\mu < 1\}$. *For a class* $\mathcal{G}$ *of games, we define* $RPoA_{\mathcal{G}}(\alpha) = \sup\{RPoA_G(\alpha) : G \in \mathcal{G}\}$.

We study the robust PoA of three classes of games: they are all described by a set E of *resources* (or *facilities*), and strategy sets $\Sigma_i \subseteq 2^E$ for each player, from which the player can choose a subset $s_i \in \Sigma_i$ of resources. Given a joint strategy s, we define $x_e(s) = |\{i \in N : e \in s_i\}|$ as the number of players that use resource $e \in E$ under s. We also use $U(s)$ to refer to the union of all resources used under s, i.e., $U(s) = \bigcup_{i \in N} s_i$.

3   Cost-Sharing Games

A *cost-sharing game* is given by $G = (N, E, \{\Sigma_i\}_{i \in N}, \{c_e\}_{e \in E})$, where $c_e \geq 0$ is the cost of facility $e \in E$. The cost of each facility is shared evenly among all players using it, i.e., the direct cost of player i is $C_i(s) = \sum_{e \in s_i} c_e / x_e(s)$. The social cost function is $C(s) = \sum_{i=1}^{n} C_i(s) = \sum_{e \in U(s)} c_e$.

It is well-known that the pure PoA of cost-sharing games is n [13]. We show that it can get significantly worse when there is altruism. Also we provide an upper bound on the pure PoS when altruism is uniform.

Theorem 1. *For* α-*altruistic cost-sharing games, the robust PoA is* $\frac{n}{1-\tilde{\alpha}}$ *(where* $n/0 = \infty$).

While Theorem 1 shows that the PoA gets worse with increasing altruism, this does not happen for the price of stability.

Proposition 2. *The pure PoS of uniformly* α-*altruistic cost-sharing games is at most* $(1 - \alpha)H_n + \alpha$.

---

[3] All results in this section continue to hold for altruistic extensions of *payoff-maximization games* G: One needs only replace C by Π and μ by $-\mu$ in Definition 3, and replace $\frac{\lambda}{1-\mu}$ by $\frac{1+\mu}{\lambda}$ and $\mu < 1$ by $\mu > -1$ in Definition 4.

4   Utility Games

A *utility game* [16] $G = (N, E, \{\Sigma_i\}_{i \in N}, \{\Pi_i\}_{i \in N}, V)$ is a payoff maximization game, in which Π_i is the payoff function of player i, and V is a submodular[4] and non-negative function on E. Every player i strives to maximize his payoff function Π_i. The social welfare function $\Pi : \Sigma \to \mathbb{R}$ to be maximized is $\Pi(s) = V(U(s))$, and thus depends on the union of the players' chosen resources, evaluated by V. The payoff function of every player i is assumed to satisfy $\Pi_i(s) \geq \Pi(s) - V(U(s) \setminus s_i)$ for every strategy profile $s \in \Sigma$. Intuitively, this means that the payoff of a player is at least his contribution to the social welfare. Moreover, it is assumed that $\Pi(s) \geq \sum_{i=1}^{n} \Pi_i(s)$ for every $s \in \Sigma$; see [16] for a justification of these assumptions. Vetta [16] proved a bound of 2 on the pure PoA for utility games with non-decreasing V; Roughgarden [15] showed that this bound is achieved via a (λ, μ)-smoothness argument. We extend it to altruistic extensions of these games.

Theorem 2. *The robust PoA of α-altruistic utility games is 2.*

5   Congestion Games

In an *atomic congestion game* $G = (N, E, \{\Sigma_i\}_{i \in N}, \{d_e\}_{e \in E})$, we have a *delay function* $d_e : \mathbb{N} \to \mathbb{R}$ associated with each facility $e \in E$. Player i's cost is $C_i(s) = \sum_{e \in s_i} d_e(x_e(s))$, and the social cost is $C(s) = \sum_{i=1}^{n} C_i(s)$. We focus on *linear* congestion games, i.e., the delay functions are of the form $d_e(x) = a_e x + b_e$, where a_e, b_e are non-negative rational numbers. Pure Nash equilibria of altruistic extensions of linear congestion games always exist [8]; this may not be the case for arbitrary (non-linear) congestion games.[5]

The PoA of linear congestion games is known to be $\frac{5}{2}$ [7]. Recently, Caragiannis et al. [4] extended this result to linear congestion games with uniformly altruistic players. Applying the transformation outlined in Remark 1, their result can be stated as follows:

Theorem 3 ([4]). *The pure PoA of uniformly α-altruistic linear congestion games is at most $\frac{5+4\alpha}{2+\alpha}$.*

The proof in [4] implicitly uses a smoothness argument in the framework we define here for altruistic games. Thus, without any additional work, our framework allows the extension of Theorem 3 to the robust PoA. Caragiannis et al. [4] also showed that the bound of Theorem 3 is asymptotically tight. A simpler example (deferred to the full version of this paper) proves tightness of this bound (not

---

[4] A function $f : 2^E \to \mathbb{R}$ is called *submodular* iff $f(A \cup \{x\}) - f(A) \geq f(B \cup \{x\}) - f(B)$ for any $A \subseteq B \subseteq E$, $x \in E$.

[5] Hoefer and Skopalik [8] established the existence of Nash Equilibria for several subclasses of atomic congestion games. For the generalization of arbitrary player-specific cost functions, Milchtaich [12] showed existence for (symmetric) singleton congestion games and Ackermann et al. [1] for matroid congestion games.

only asymptotically). Thus, the robust PoA is exactly $\frac{5+4\alpha}{2+\alpha}$. We give a refinement of Theorem 3 to non-uniform altruism distributions, obtaining a bound in terms of the maximum and minimum altruism levels.

Theorem 4. *The robust PoA of α-altruistic linear congestion games is at most* $\frac{5+2\hat{\alpha}+2\check{\alpha}}{2-\hat{\alpha}+2\check{\alpha}}$.

We turn to the pure price of stability of α-altruistic congestion games. Clearly, an upper bound on the pure price of stability extends to the mixed, correlated and coarse price of stability.

Proposition 3. *The pure PoS of uniformly α-altruistic linear congestion games is at most* $\frac{2}{1+\alpha}$.

Symmetric Singleton Congestion Games. *Symmetric singleton congestion games* are an important special case of congestion games. They are defined as $G = (N, E, \{\Sigma_i\}_{i \in N}, \{d_e\}_{e \in E})$: every player chooses one facility (also called *edge*) from $E = [m]$, and all strategy sets are identical, i.e., $\Sigma_i = E$ for every i. In *singleton linear congestion games*, the focus here, delay functions are also assumed to be linear, of the form $d_e(x) = a_e x + b_e$.

Caragiannis et al. [4] prove the following theorem (stated using the transformation from Remark 1). It shows that the pure PoA does not always increase with the altruism level; the relationship between α and the PoA is thus rather subtle.

Theorem 5 (Caragiannis et al. [4]). *The pure PoA of uniformly α-altruistic singleton linear congestion games is* $\frac{4}{3+\alpha}$.

We show that even the mixed PoA (and thus also the robust PoA) will be at least 2 regardless of the altruism levels of the players, by generalizing a result of Lücking et al. [11, Theorem 5.4]. This implies that the benefits of higher altruism in singleton congestion games are only reaped in pure Nash equilibria, and the gap between the pure and mixed PoA increases in α.

Proposition 4. *For every $\alpha \in [0, 1]^n$, the mixed PoA for α-altruistic singleton linear congestion games is at least 2.*

As a first step to extend the analysis to non-uniform altruism, we analyze the case when all altruism levels are in $\{0, 1\}$, i.e., each player is either completely altruistic or completely selfish.[6] Then, the system is entirely characterized by the fraction α of altruistic players (which coincides with the average altruism level). The next theorem shows that in this case, too, the pure PoA *improves* with the overall altruism level.

Theorem 6. *Assume that an α fraction of the players are completely altruistic, and the remaining $(1 - \alpha)$ fraction are completely selfish. Then, the pure PoA of the altruistic singleton linear congestion game is at most* $\frac{4-2\alpha}{3-\alpha}$.

---

[6] This model relates naturally to *Stackelberg scheduling games* (see, e.g., [6]).

Acknowledgements. We thank anonymous reviewers for useful feedback, and for pointing out the similarities between our work on congestion games and that of [4].

References

1. Ackermann, H., Röglin, H., Vöcking, B.: Pure Nash Equilibria in Player-Specific and Weighted Congestion Games. In: Spirakis, P.G., Mavronicolas, M., Kontogiannis, S.C. (eds.) WINE 2006. LNCS, vol. 4286, pp. 50–61. Springer, Heidelberg (2006)
2. Anshelevich, E., Dasgupta, A., Kleinberg, J., Tardos, E., Wexler, T., Roughgarden, T.: The price of stability for network design with fair cost allocation. In: Proc. 45th Symposium on Foundations of Computer Science (2004)
3. Blum, A., Hajiaghayi, M.T., Ligett, K., Roth, A.: Regret minimization and the price of total anarchy. In: Proc. 40th Annual ACM Symposium on Theory of Computing, pp. 373–382 (2008)
4. Caragiannis, I., Kaklamanis, C., Kanellopoulos, P., Kyropoulou, M., Papaioannou, E.: The impact of altruism on the efficiency of atomic congestion games. In: Proc. 5th Symposium on Trustworthy Global Computing (2010)
5. Chen, P.-A., David, M., Kempe, D.: Better vaccination strategies for better people. In: Proc. 12th ACM Conference on Electronic Commerce (2010)
6. Chen, P.-A., Kempe, D.: Altruism, selfishness, and spite in traffic routing. In: Proc. 10th ACM Conference on Electronic Commerce, pp. 140–149 (2008)
7. Christodoulou, G., Koutsoupias, E.: The price of anarchy of finite congestion games. In: Proc. 37th Annual ACM Symposium on Theory of Computing (2005)
8. Hoefer, M., Skopalik, A.: Altruism in atomic congestion games. In: Proc. 17th European Symposium on Algorithms (2009)
9. Koutsoupias, E., Papadimitriou, C.: Worst-case equilibria. In: Proc. 16th Annual Symposium on Theoretical Aspects of Computer Science, March 4-6, pp. 404–413 (1999)
10. Ledyard, J.: Public goods: A survey of experimental resesarch. In: Kagel, J., Roth, A. (eds.) Handbook of Experimental Economics, pp. 111–194. Princeton University Press (1997)
11. Lücking, T., Mavronicolas, M., Monien, B., Rode, M.: A new model for selfish routing. Theoretical Computer Science 406, 187–206 (2008)
12. Milchtaich, I.: Congestion games with player-specific payoff functions. Games and Economic Behavior 13(1), 111–124 (1996)
13. Nisan, N., Roughgarden, T., Tardos, É., Vazirani, V.V.: Algorithmic Game Theory. Cambridge University Press (2007)
14. Roughgarden, T.: Selfish Routing and the Price of Anarchy. MIT Press (2005)
15. Roughgarden, T.: Intrinsic robustness of the price of anarchy. In: Proc. 41st Annual ACM Symposium on Theory of Computing, pp. 513–522 (2009)
16. Vetta, A.: Nash equilibria in competitive societies, with applications to facility location, traffic routing and auctions. In: Proc. 43rd Symposium on Foundations of Computer Science (2002)

Revenue Enhancement in Ad Auctions

Michal Feldman[1,4,*], Reshef Meir[1,2], and Moshe Tennenholtz[2,3]

[1] Hebrew University of Jerusalem
mfeldman@huji.ac.il, reshef.meir@mail.huji.ac.il
[2] Microsoft Research Herzlia
moshet@microsoft.com
[3] Technion-Israel Institute of Technology
[4] Microsoft Research New England

Abstract. We consider the revenue of the Generalized Second Price (GSP) auction, which is one of the most widely used mechanisms for ad auctions. While the standard model of ad auctions implies that the revenue of GSP in equilibrium is at least as high as the revenue of VCG, the literature suggests that it is not strictly higher due to the selection of a natural equilibrium that coincides with the VCG outcome. We propose a randomized modification of the GSP mechanism, which eliminates the low-revenue equilibria of the GSP mechanism under some natural restrictions. The proposed mechanism leads to a higher revenue to the seller.

1 Introduction

Ad auctions are perhaps the most widely studied economic setup in the literature on on-line markets. As ad auctions generate revenues of billions of dollars per year to publishers, every subtle feature of their design may have tremendous effect. One of the two most widely discussed mechanisms in the study of ad auctions is the Generalized Second Price (GSP) auction, versions of which are those typically used in practice. The other is the classical Vickrey-Clarke-Groves (VCG) auction, which is known to be *incentive compatible*. That is, in VCG bidders are incentivized to report their true evaluations. We consider the original ad auctions model introduced in the seminal work of Varian [12] and Edelman et al. [3], where bidders' valuations per click are fixed and independent of the ad slot. Previous work characterized a special family of equilibria of GSP auctions in that setting, termed *envy free* or *Symmetric* Nash Equilibria (SNE), and showed that the SNE leading to the lowest revenue for the seller, termed Lower Equilibrium (LE), coincides with the truth-revealing equilibrium of VCG.

While the above results suggest that GSP may lead to higher revenue than VCG, other arguments for revenue comparison between these mechanisms have been discussed. Kuminov and Tennenholtz model user behavior explicitly as glancing through the ads in a sequence [7]. Interestingly, in this setting the VCG outcome coincides with the GSP equilibrium that leads to the *highest* revenue. Closer to the study of the standard ad auctions setting, Edelman and Schwarz consider equilibrium selection in GSP by comparing the revenue in the standard static game to a dynamic variant of the

---

* At the time of research the author was affiliated with Microsoft Research Herzlia, Israel.

N. Chen, E. Elkind, and E. Koutsoupias (Eds.): WINE 2011, LNCS 7090, pp. 391–398, 2011.
© Springer-Verlag Berlin Heidelberg 2011

game [4]. They suggest that the GSP equilibrium that leads to the highest revenue is less natural than the one that coincides with the VCG outcome since it generates too high revenue compared to the revenue obtained in the dynamic model. Recently, Lucier et al. [8] performed a detailed analysis of the revenue in the GSP auction, under both complete and incomplete information, using the VCG revenue as a baseline. In particular, they outlined the conditions under which the revenue in non-envy-free equilibria can be lower or higher than the revenue in SNE.

Thompson and Leyton-Brown [11] computed the revenue of GSP in equilibrium using several models from the literature. They found that while the expected revenue of GSP in Varian's model was slightly higher than the VCG baseline, in most models the revenue was profoundly affected by equilibrium selection.

The above suggests that GSP has attracted much attention from both researchers and practitioners, but it is unclear whether it has revenue advantage over VCG. Hence, one may wish to consider natural modifications for GSP that increase the auctioneer/publisher revenue. Notice that GSP is by now a standard practice and modifications to it should conform to having a relatively similar structure in the way bidders are assigned to ad slots and the way they are assigned payments; i.e., an advertiser's payment should be bounded by his bids and have some intuitive relations with other (less successful) bids. This is not a "mathematical" requirement, but a practical one, given the way advertisers perceive ad auctions.

Several modifications to the GSP mechanism have already been suggested in the literature. The most common one is the addition of *reserve prices*. Indeed, field experiments (and to some extent also theory) suggest that reserve prices can substantially increase the revenue in GSP auctions [10]. A different modification deals with allocation efficiency. When some assumptions in Varian's model are violated, the GSP mechanism may not have efficient equilibria. Blumrosen et al. modify the GSP mechanism to guarantee the existence of an efficient equilibrium in a more general model [1].

Revenue of other incentive compatible and envy-free mechanisms has also been studied. Hartline and Yan [6] analyzed the *maximal* revenue attainable under envy-freeness constraints, which are perceived as a relaxation of incentive compatibility (IC), in a wide variety of single-parameter domains (including ad auctions). They show that optimal envy-free revenue is a good proxy of the optimal IC revenue, and use it as a baseline to evaluate the revenue of other IC mechanisms. These results highlight another angle of the roles of truthfulness and envy-freeness in revenue analysis, but lie outside the scope of our model. This is mainly since Hartline and Yan assume truthful bidding (even with non-IC pricing) whereas the envy-free outcomes of the GSP auction are the result of *strategic bidding*. Thus GSP and its variations have multiple equilibria with different revenues, as discussed above.

Our contribution. In this paper we take the study of the GSP revenue to the next stage, by suggesting natural modifications to the GSP mechanism which result in revenue boosting for many natural click-through rates. Recall that in the GSP mechanism every winning agent pays the *second* price, i.e., the bid of the bidder directly below her bid. Our revised mechanism selects randomly between GSP and a variant of it, in which an agent pays the *third* price, i.e., the bid that follows the bid below her bid.[1]. We show conditions under which the combined mechanism admits an ex-post envy-free equilibrium that achieves revenue that is arbitrarily close to the revenue obtained in the highest revenue equilibrium of GSP, while eliminating the low revenue equilibria outcomes. More generally, we introduce the family of m-price auctions by generalizing the GSP auctions as well as the random selection among them, study their ex-post envy-free equilibria, and prove that by random selection between a pair of such mechanisms we can boost the revenue of GSP. Proofs are omitted due to space constraints, they appear in the full version of this paper [5].

2 Model and Preliminaries

2.1 Ad Auctions

In an ad auction there are s slots to allocate, and n bidders, each with valuation v_i per click, for $i \in \{1, \ldots, n\}$. Every slot $1 \leq j \leq s$ is associated with a click-through rate (CTR) $x_j > 0$, where $x_j \geq x_{j+1}$. For mathematical convenience, we define $x_j = 0$ for every $j > s$. Throughout the paper we make the simplifying assumption that CTRs are strictly decreasing, i.e., $x_j > x_{j+1}$.

The mechanism receives as input a bid b_i from every bidder and determines an allocation of the slots to the bidders and a payment per click, p_i, for every bidder. We denote the slot allocated to bidder i by $\pi(i)$. A bidder i that has been allocated slot $\pi(i)$ gains v_i per click (regardless of the slot), and is charged p_i per click. Thus, his total utility is given by $u_i = (v_i - p_i)x_{\pi(i)}$. A mechanism that assigns better slots to higher bids is called *efficient*. A mechanism that never assigns lower payments to higher bids is called *monotone*. We restrict our attention to efficient and monotone mechanisms.

Nash equilibria. Let $v_1 > \ldots > v_n$ and $\mathbf{b} = (b_1, \ldots, b_n)$, where v_i and b_i are the private valuation and the submitted bid of bidder i, respectively. Let f be an efficient and monotone auction mechanism, and let $p_1, \ldots, p_n$ be the payments assigned by f according to the bids. We say that the bids are in Nash equilibrium (NE), if no bidder wants to deviate by changing her bid. Let $p_i^f(b')$ be the payment assigned by f to bidder i if she changes her bid to b' (and all other bidders keep their current bids). From the efficiency of f, in order to get slot $j < \pi(i)$, bidder i must bid like the bidder currently occupying the slot, i.e., $b' = b_j$. Similarly, in order to get slot $j > \pi(i)$, bidder i must bid below bidder j, i.e., $b' = b_{j+1}$. The stability requirements of NE can be divided into three parts. For every bidder i the following should hold:

bid neutral. Bidder i has no incentive to change her bid if this deviation does not change the slot assigned to i.

up-Nash. Bidder i does not want to get a better slot:

$$\forall j < \pi(i), \quad (v_i - p_i)x_{\pi(i)} \geq (v_i - p_i^f(b_{\pi^{-1}(j)}))x_j. \tag{1}$$

---

[1] k-price auctions have been shown to lead to some intriguing results in the more classical single-items setting; see [9]

down-Nash. Bidder i does not want to get a worse slot:

$$\forall j > \pi(i), \quad (v_i - p_i)x_{\pi(i)} \geq (v_i - p_i^f(b_{\pi^{-1}(j+1)}))x_j. \tag{2}$$

The first requirement is usually handled by mechanisms that ignore b_i when setting the payment p_i (i.e., b_i is only used to decide on the allocation and on p_j for $j \neq i$).

Efficiency. We say that an equilibrium outcome is *efficient*, if $b_i \geq b_{i+1}$ for all i. Note that if both the mechanism f and the outcome **b** are efficient, then every bidder i gets slot i, i.e., $\pi(i) = i$. Efficient equilibria guarantee that the *social welfare* (i.e. the sum of utilities of all bidders and the auctioneer) is maximized.

Envy freeness. A different notion of stability than Nash is captured by the *envy-freeness* requirement. An outcome is envy free if no bidder is interested in swapping slots (and payments) with any other bidder. Formally, it takes the following form:

$$\forall j \neq \pi(i), \quad (v_i - p_i)x_{\pi(i)} \geq (v_i - p_i^f(b_{\pi^{-1}(j+1)}))x_j. \tag{3}$$

It is easy to verify that (3) entails requirements (1) and (2). Thus, any envy-free outcome in a bid-neutral mechanism is also a NE. In fact, envy freeness is more restricting than (1), and thus we are left with a subset of the original set of NE.

Envy-free equilibria have been thoroughly studied in the literature of games in general, and ad auctions in particular. They are also known as *symmetric Nash equilibria* (SNE), due to the symmetry of up-Nash and down-Nash constraints. Varian [12] and Lucier et al. [8] further studied properties of SNEs in the GSP mechanism. For example, it is shown that every SNE is efficient, which is not true for arbitrary NE.

When a randomized mechanism is in use, we must distinguish between outcomes that are *envy free in expectation* from outcomes that are *envy free ex-post*. The latter definition means no bidder wants to change slots, even after the randomization has taken place and the outcome is known. We will be interested in this stronger interpretation of envy freeness.

The revenue interval. Suppose we are using some auction mechanism f. If f has multiple NE outcomes, then the auctioneer might end up with different revenues for the different equilibria. We define R_f^U (resp. R_f^L) as the highest (resp. lowest) revenue generated by mechanism f in some NE. We use a similar notation to denote the highest and lowest revenues in restricted subsets of NE, replacing R with ER (for *efficient NE*) or with SR (for *symmetric NE*). Clearly $[SR_f^L, SR_f^U], [ER_f^L, ER_f^U] \subseteq [R_f^L, R_f^U]$.

The revenue interval raises the natural question of *equilibrium selection*. Clearly, the auctioneer would like the bidders to end up playing an equilibrium with high revenue. However, the auctioneer is not a player in this game. The players are the bidders, and given an efficient allocation their joint incentive is basically the opposite - to end up paying the lowest possible amount.

2.2 VCG and GSP

The VCG mechanism sorts the bidders by their bids, and allocates the j'th slot, $j = 1, \ldots, s$, to the j'th highest bidder. Each bidder j is charged $p_j = \sum_{k \geq j+1} b_k(x_{k-1} - x_k)$ per click (we define $b_i = 0$ for all $i > n$). Note that VCG is efficient and monotone.

It is well known that under the VCG mechanism, reporting the true values (i.e., $b_i = v_i$) is a dominant strategy, and in particular it is a Nash equilibrium. We denote the revenue in the truthful equilibrium of VCG by R^T_{VCG}.

The allocation of the GSP mechanism is efficient, i.e., identical to that of VCG. The charge of bidder $j = 1, \ldots, s$ equals the bid of the next bidder; i.e., $p_j = b_{j+1}$. For mathematical convenience, we define $b_{j+1} = 0$ for $j \geq n$. GSP is clearly efficient and monotone. Varian [12] focuses on the analysis of envy-free equilibria (i.e., SNEs) in the GSP auction, due to their many attractive properties. For example, Varian shows that every SNE is efficient. The SNE requirement (3) in GSP takes the following form:

$$\forall j \neq i, (v_i - b_{i+1})x_i \geq (v_i - b_{j+1})x_j. \tag{4}$$

2.3   Known Properties of GSP

Before presenting some known properties of the VCG and GSP mechanisms, we put forward the following basic definitions. Let $g_1, \ldots, g_m \in \mathbb{R}_+$ be the elements of a monotonically nonincreasing series. We sometimes refer to such series as *functions* (of the form $g : \{1, \ldots, m\} \to \mathbb{R}$). We say that g is *convex* if it has a decreasing marginal loss, i.e., $g_i - g_{i+1} \geq g_j - g_{j+1}$ for every $i < j$. Similarly, if g has an *increasing* marginal loss then it is *concave*. Notice that linear functions are both convex and concave.

A special case of convexity is when the marginal loss decreases exponentially fast. We say that g is *β-separated* (for some $0 < \beta < 1$) if $g_{i+1} \leq \beta g_i$ for every i. If the above holds with equality (rather than inequality), g is said to be *exponential*. Lastly, a series/function g is said to be *log-concave*, if for every $i < j \leq m$, $\frac{g_i}{g_{i-1}} \geq \frac{g_j}{g_{j-1}}$.

Distinguished equilibria. Of particular interest are the two equilibria of GSP that reside on the boundaries of the SNE set, referred to as *Lower Equilibrium* (LE) and *Upper Equilibrium* (UE). We denote the LE and UE profiles by $\mathbf{b}^L = \{b^L_i\}_{i \in N}$ and $\mathbf{b}^U = \{b^U_i\}_{i \in N}$, respectively. The bids in the LE, for every $2 \leq i \leq s$, are given by

$$b^L_i x_{i-1} = v_i(x_{i-1} - x_i) + b^L_{i+1}x_i = \sum_{s+1 \geq t \geq i} v_t(x_{t-1} - x_t).$$

A similar recursive formula is derived for the UE. Note that since $x_{i-1} > x_i$, no two bidders submit the same bid in LE (or in UE). Interestingly, the LE outcome coincides with that of VCG in terms of the revenue (although bids may be different).

Proposition 1 (Varian [12]). *The payments of all bidders in the LE of GSP are the same as under the truthful bidding in VCG. In particular, it follows that* $SR^L_{GSP} = R^T_{VCG}$.

The revenue interval of GSP. Since SNE is always efficient, we have that

$$[SR^L_{GSP}, SR^U_{GSP}] \subseteq [ER^L_{GSP}, ER^U_{GSP}] \subseteq [R^L_{GSP}, R^U_{GSP}],$$

where in the most general case, these inclusions may be strict.

Lucier et al. [8] study the conditions on the CTR function under which the boundaries of these sets become close or equal. They show that if $x_i > x_{i+1}$ for all i (as we assume here), then the first inclusion becomes an equality.

Algorithm 1. THE RANDOM NEXT-PRICE MECHANISM ($M(q)$)

Collect bids from all agents. Allocate slots according to bids in decreasing order.
  w.p. q, each bidder i pays b_{i+1}. // *GSP is applied*
  w.p. $1 - q$, each bidder i pays b_{i+2}. // *GTP is applied*

Justifying the Lower equilibrium. We argue that when faced with the equilibrium selection problem of GSP, bidders are likely to play the lower envy-free equilibrium $\mathbf{b}^L$. The concept of envy freeness itself is well justified in various settings. In addition, in our ad auction setting, bidders have a particular interest in an efficient allocation, which makes the set of SNEs even more prominent. Within this set, there are two natural focal points, introduced in the previous paragraph. However, observe that if all bidders change their bids from b_i to b_i^L, then the allocation remains the same, and every bidder i pays the same, or strictly less (if $b_{i+1}^L < b_{i+1}$). Thus every SNE is Pareto dominated by $\mathbf{b}^L$.

in conclusion: if bidders try to influence the outcome equilibrium, they are most likely to play the lowest envy-free equilibrium (LE in the GSP case), and the revenue of the auctioneer will be $SR_{GSP}^L = ER_{GSP}^L$.

3   Generalized Next Price Auctions

Consider a modified GSP mechanism, termed m-price auction (or just m-price), in which $p_i = b_{i+m-1}$. For example, the 2-Price auction is GSP, as $p_i = b_{i+1}$. We argue that the SNE outcomes of the auction are essentially the same for all $m \geq 2$.

Lemma 1. *Let* $\mathbf{b}$ *be a (sorted) bid vector, and let* k, m *s.t.* $2 \leq k < m \leq n + 1 - s$. $\mathbf{b}$ *is an SNE of m-price if and only if* $\mathbf{b}'$ *is an SNE of k-price, where* $b_i' = b_{i+k-m}$.[2]

This is simply since the bid vectors in both mechanisms induce exactly the same allocation (in decreasing order of valuations), and exactly the same payments. In particular, from the perspective of the bidders the outcome is the same whether GSP or any other m-price auction is used (although they will submit different bids). Clearly the revenue of the auctioneer is the same as well.

We can now easily derive the lower and upper bounds on SNE bids in m-price auctions, which we denote by $b_i^L(m), b_i^U(m)$. For every $m \geq 2$, $b_i^L(m) = b_{i+2-m}^L$, and $b_i^U(m) = b_{i+2-m}^U$. We will focus on 2-price (i.e., GSP) and 3-price auctions; we refer to the latter as the *generalized third-price* auction (GTP).

3.1   Boosting Revenue via Randomization

Due to Lemma 1, it seems that there is no benefit in using the m-price auction rather than the original GSP. Quite surprisingly, it turns out that combining these mechanisms enables us to improve the revenue interval. In particular, we introduce a randomized mechanism that boosts the auctioneer's revenue. This mechanism, denoted $M(q)$, runs GSP w.p. q (where $q \in [0, 1]$) and otherwise runs GTP (see Algorithm 1).

---

[2] If $i + k - m < 1$ then b_i' can be completed arbitrarily, as long as $b_i' > b_{i+1}'$.

We identify natural conditions under which the expected revenue of the $M(q)$ mechanism exceeds that of GSP and VCG, or even approaches the highest possible revenue of the GSP auction (in any Nash equilibrium).

Theorem 2. *Suppose that the CTR function is convex and log-concave, and that for every $2 \geq i \geq s$ it holds that $\frac{v_i}{v_{i-1}} \leq \frac{x_i}{x_{i-1}}$. For any $0 < q < 1$ it holds that $SR^L_{M(q)} > SR^L_{GSP}$ (and thus $SR^L_{M(q)} > R^T_{VCG}$).*

Theorem 3. *When the CTR function is exponential, $SR^L_{M(q)} > R^T_{VCG}$. Further,*
$$\lim_{q \to 1} SR^L_{M(q)} = SR^U_{GSP} = ER^U_{GSP}.$$

In the remainder of this section, we sketch the proof of both theorems. The crucial part lies in showing that under the conditions we required (in either theorem), the upper-revenue bidding profile of the GSP auction is also an SNE in GTP, and in particular $b^U_i \geq b^L_i(3)$ for all i. On the other hand, the lower-revenue profile ($\mathbf{b}^L$) is not envy-free if GTP is applied. The expected payment of bidder i is therefore

$$\mathbb{E}[p_i] \geq q b^L_{i+1}(3) + (1-q)b^L_{i+2}(3) \stackrel{\text{(Lemma 1)}}{=} q b^L_i + (1-q)b^L_{i+1} > b^L_{i+1} = p^L_i.$$

That is, every bidder is paying (in expectation) strictly more in the $M(q)$ mechanism than in GSP. When the CTRs are exponential, we show that $\mathbf{b}^U$ and $\mathbf{b}^L(3)$ coincide (i.e. this is the unique ex-post envy free bidding profile). Thus

$$\mathbb{E}[p_i] = q b^L_{i+1}(3) + (1-q)b^L_{i+2}(3) = q b^U_{i+1} + (1-q)b^U_{i+2} = q p^U_i + (1-q)p^U_{i+1},$$

and as q gets closer to 1, the revenue gets arbitrarily close to SR^U_{GSP}, i.e., to the highest revenue possible in the GSP mechanism *in envy-free equilibrium*.

Finally, since exponential CTR is in particular strictly decreasing, it follows (according to Lucier et al. [8]) that $SR^U_{GSP} = ER^U_{GSP}$, as required.

4   Discussion

We propose a randomized modification to the GSP auction, and analyze the set of ex-post envy-free equilibria in this new mechanism. We show natural conditions under which the revenue to the auctioneer strictly increases compared to the revenue in the "natural" envy-free equilibrium of the original GSP auction (which equals to the revenue in VCG under truthful bidding). When the CTRs are exponentially decreasing, our mechanism eliminates all ex-post envy-free equilibria, except the one leading to the maximal possible equilibrium revenue of the GSP auction. We note that convex and even exponential CTRs are common in the real world, as shown in Figure 1.[3]

Future research directions include the extension of our analysis to generalizations of the basic model (e.g., by considering the "ad's quality," as in [12]), and a study of the randomized mechanism in a model with incomplete information. Field experiments with the randomized mechanism (in the spirit of Ostrovsky and Schwarz [10]) will help to determine the practical value of our approach.

---

[3] Stats taken from Atlas Institutes rank report [2]. We present the data on the "click potential" attribute, which corresponds to the actual CTR in our model.

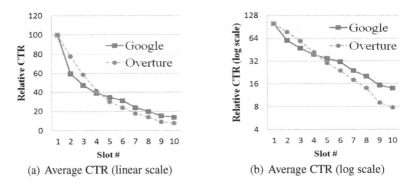

(a) Average CTR (linear scale)   (b) Average CTR (log scale)

Fig. 1. The average click-through rate for ads positioned in any of the first ten slots are shown in Fig. 1(a) (numbers are normalized so that the CTR of slot 1 is 100). We can see that the shape of the CTR function is convex in both Google and Overture. In Fig. 1(b) we see the same data in log scale. Interestingly, the Overture CTR function is very close to exponential (with $\beta \cong 1.3$).

References

1. Blumrosen, L., Hartline, J.D., Nong, S.: Position auctions and non-uniform conversion rates. In: Proceedings of the 4th ACM EC Workshop on Ad Auctions (2008)
2. Brooks, N.: The Atlas rank report: How search engine rank impacts traffic, http://tiny.cc/wj8f3
3. Edelman, B., Ostrovsky, M., Schwarz, M.: Internet advertising and the generalized second price auction: Selling billions of dollars worth of keywords. American Economic Review 97(1), 242–259 (2007)
4. Edelman, B., Schwarz, M.: Optimal auction design and equilibrium selection in sponsored search auctions. American Economic Review 100, 597–602 (2010)
5. Feldman, M., Meir, R., Tennenholtz, M.: Revenue enhancement in ad auctions (2011) (manuscript), http://www.cs.huji.ac.il/ reshef24/auctions. wine11.pdf
6. Hartline, J.D., Yan, Q.: Envy, truth, and profit. In: EC 2011: Proceeding of the 12th ACM Conference on Electronic Commerce, pp. 243–252 (2011)
7. Kuminov, D., Tennenholtz, M.: User modeling in position auctions: re-considering the GSP and VCG mechanisms. In: AAMAS 2009: Proceedings of the 8th International Joint Conference on Autonomous Agents and Multiagent Systems, pp. 273–280 (2009)
8. Lucier, B., Leme, R.P., Tardos, É.: On revenue in the generalized second price auction. In: Proceedings of the 7th ACM EC Workshop on Ad Auctions (2011)
9. Monderer, D., Tennenholtz, M.: K-price auctions: revenue inequalities, utility equivalence, and competition in auction design. Economic Theory 24(2), 255–270 (2002)
10. Ostrovsky, M., Schwarz, M.: Reserve prices in internet advertising auctions: A field experiment. In: EC 2011: Proceedings of the 12th ACM Conference on Electronic Commerce, pp. 59–60 (2011)
11. Thompson, D.R.M., Leyton-Brown, K.: Computational analysis of perfect-information position auctions. In: EC 2009: Proceedings of the 10th ACM Conference on Electronic Commerce, pp. 51–60 (2009)
12. Varian, H.R.: Position auctions. International Journal of Industrial Organization 25(6), 1163–1178 (2007)

Bilinear Games: Polynomial Time Algorithms for Rank Based Subclasses*

Jugal Garg[1,**], Albert Xin Jiang[2], and Ruta Mehta[1]

[1] Indian Institute of Technology, Bombay
{jugal,ruta}@cse.iitb.ac.in
[2] University of British Columbia
jiang@cs.ubc.ca

Abstract. Motivated by the sequence form formulation of Koller et al. [16], this paper considers *bilinear games*, represented by two payoff matrices (A, B) and compact polytopal strategy sets. Bilinear games are very general and capture many interesting classes of games including bimatrix games, two player Bayesian games, polymatrix games, and two-player extensive form games with perfect recall as special cases, and hence are hard to solve in general. For a bilinear game, we define its *best response polytopes* (BRPs) and characterize its Nash equilibria as the *fully-labeled* pairs of the BRPs. Rank of a game (A, B) is defined as $rank(A + B)$. In this paper, we give polynomial-time algorithms for computing Nash equilibria of (i) rank-1 games, (ii) FPTAS for constant-rank games, and (iii) when $rank(A)$ or $rank(B)$ is constant.

1   Introduction

One fundamental class of computational problems in game theory is the computation of Nash equilibria (NE) of finite games. The recent results [4,6] established that the problem is PPAD-complete, even for games with only two players. In light of these negative results, one direction is to identify tractable subclasses of games.

A two-player normal form game can be represented by two payoff matrices, say A and B, one for each player, and hence is also known as bimatrix game. For bimatrix games, polynomial time NE computation algorithms are known for many subclasses, including zero-sum games [5], (quasi-) concave games [17], and games with low rank payoff matrices [18]. A line of work focuses on games of small *rank*, defined as $rank(A + B)$ by Kannan and Theobald [14]. They gave a fully polynomial time approximation scheme (FPTAS) for fixed rank games. Such a rank-based approach has wide applicability: even when the game does not have small rank, one can solve a small-rank approximation of the game (using e.g. singular-value decomposition) to compute an approximate Nash equilibrium of the original game. Recently, Adsul et al. [1] gave a polynomial time algorithm for computing an *exact* Nash equilibrium for rank-1 games.

Specifying the two payoff matrices of a bimatrix game requires a polynomial number of entries in the numbers of pure strategies available to the players. This is adequate

---

* Refer [9] for the full version. We thank Milind Sohoni for helpful comments and corrections.
** Work done while the author was an intern at Microsoft Research India.

N. Chen, E. Elkind, and E. Koutsoupias (Eds.): WINE 2011, LNCS 7090, pp. 399–407, 2011.

when the set of pure strategies are explicitly given. However, there are cases where the natural description gives the set of pure strategies implicitly. For example, the normal form representation of a two player extensive-form game may have exponentially many strategies in the size of the extensive-form description [8]. In such a case, even if the resulting bimatrix game has a fixed rank, the above results may not be applied for efficient computation. Koller, Megiddo and von Stengel [16] converted an arbitrary two-player, perfect-recall, extensive form game into a payoff-equivalent two-player game with continuous strategy sets. In this derived formulation, which they call the *sequence form*, there is a pair of payoff matrices A and B, one for each player. Further, their strategy sets turn out to be compact polytopes in Euclidean space of polynomial dimension. Given a pair of strategies (x, y), utilities of the players are $x^T A y$ and $x^T B y$ respectively. The sequence form requires only a polynomial number of bits to specify.

Motivated by the sequence form of Koller *et al.*, we define *bilinear games*, which are two-player, non-cooperative, single shot games represented by two payoff matrices, say A and B, of dimension $M \times N$ and two polytopal compact strategy sets $X = \{x \in \mathbb{R}^M \mid Ex = e,\ x \geq 0\}$ and $Y = \{y \in \mathbb{R}^N \mid Fy = f,\ y \geq 0\}$. If $(x, y) \in X \times Y$ is the played strategy, then $x^T A y$ and $x^T B y$ are the utilities derived by player one and player two respectively. In other words, the payoffs are bilinear functions of strategies, hence the name *bilinear games*. The scope of bilinear games is large enough to capture many interesting classes of games (see Section 2.1 of [9]) besides bimatrix and two-player extensive form games with perfect recall, like for two-player Bayesian games [12,19], polymatrix games [11], and various classes of optimization duels [13], the proposed polynomial-sized payoff-equivalent formulations turn out to be bilinear games.

Since, the hardness results of bimatrix games apply to bilinear games, the only hope is to design efficient algorithms or FPTAS for special subclasses. A natural approach is to try to adapt algorithms for bimatrix games to bilinear games. However, a technical challenge is that the polytopal strategy sets of bilinear games are much more complex than the bimatrix case; in particular the number of vertices may be exponential, while the set of mixed strategies is just a simplex. Recently, we were pointed to *Constrained games* [3], similar to bilinear games. The linear programming approach of [3] also works to solve zero-sum bilinear games (see also [7,13,15,19]).

Our Contribution. We extend Kannan and Theobald's [14] rank-based hierarchy for bimatrix games to bilinear games, by defining the rank of a bilinear game with payoff matrices (A, B) as the rank of $(A + B)$. In Section 3, we show that in spite of a very general structure of the strategy sets in bilinear games, the basic approach given by Adsul et al. [1] to compute a NE of a rank-1 bimatrix game can be generalized to compute a NE of a rank-1 bilinear game. In Section 4, we discuss two FPTAS algorithms for the fixed rank bilinear games, which are generalization of the algorithms by Kannan and Theobald [14] for the bimatrix games. Finally, in Section 5, we obtain a polynomial time algorithm for the case when the rank of either A or B is a constant, and the rank of E and F are also constant. Since a bimatrix game can be thought of as a bilinear game with E and F being a single row of 1s, this algorithm improves upon a result by Lipton et al. [18] and Kannan et al. [14] for bimatrix games, where they require *both* A and B of constant rank. This algorithm can also enumerate extreme equilibria of a bilinear game, in time polynomial under the above assumptions and exponential in general.

2   Bilinear Games and Nash Equilibria

Notations. We consider a vector x as a column vector by default and for the row vector, we use transpose (*i.e.*, x^T). Let $x \in \mathbb{R}^n$ be a vector and $c \in \mathbb{R}$ be a scalar, $x \leq c$ denotes $\forall i \leq n$, $x_i \leq c$. For a given matrix A, it's i^{th}-row and j^{th}-column are denoted by A_i and A^j respectively, and let $|A| = \max_{ij} |A_{ij}|$.

Bilinear games are two-player non-cooperative, single shot games, represented by two $M \times N$ dimensional payoff matrices A and B, one for each player, and two compact polytopal strategy sets. Let $S_1 = \{1, \ldots, M\}$ be the set of rows and $S_2 = \{1, \ldots, N\}$ be the set of columns of the matrices. Let $E \in \mathbb{R}^{k_1 \times M}$ and $F \in \mathbb{R}^{k_2 \times N}$ be the matrices, and $e \in \mathbb{R}^{k_1}$ and $f \in \mathbb{R}^{k_2}$ be the vectors. The strategy set of the first-player is $X = \{x \in \mathbb{R}^M \mid Ex = e, x \geq 0\}$ and the second-player is $Y = \{y \in \mathbb{R}^N \mid Fy = f, y \geq 0\}$. Sets X and Y are assumed to be compact. From a strategy profile $(x, y) \in X \times Y$, the payoffs obtained by the first and the second player are $x^T A y$ and $x^T B y$ respectively.

Definition 1. *An* $(x, y) \in X \times Y$ *is a NE of the game* (A, B) *iff* $x^T A y \geq x'^T A y$, $\forall x' \in X$ *and* $x^T B y \geq x^T B y'$, $\forall y' \in Y$, *i.e., no player gains by unilateral deviation.*

As a corollary of Glicksberg's [10] result that there always exists a Nash equilibrium in a game if strategy spaces are convex and compact, and utility function for each player i is continuous in all players' strategies and quasi-concave in i's strategy, we have

Proposition 1. *Every bilinear game has at least one Nash equilibrium.*

A bilinear game is completely represented by a six-tuple (A, B, E, F, e, f) in general. However, for ease of notation we represent it by (A, B) fixing (E, F, e, f). Given a strategy $y \in Y$ of the second-player, the objective of the first player is to play $x \in X$ such that $x^T (Ay)$ is maximized, *i.e.,* solve the following LP [16].

$$
\begin{array}{lll}
max : x^T(Ay) & & min : e^T p \\
s.t. \quad Ex = e & \xrightarrow{\text{Dual}} & s.t. \quad E^T p \geq Ay \\
\quad\quad x \geq 0 & &
\end{array}
\tag{1}
$$

At the optimal point (x, p) of (1), we get $x_i > 0 \Rightarrow A_i y = p^T E^i$, $\forall i \in S_1$. This with a similar condition for the second-player, characterize the NE as follows: A strategy pair $(x, y) \in X \times Y$ is a Nash equilibrium of the game (A, B), iff,

$$
\begin{array}{l}
\exists p \in \mathbb{R}^{k_1} \text{ s.t. } Ay \leq E^T p \text{ and } \forall i \in S_1, \ x_i > 0 \ \Rightarrow \ A_i y = p^T E^i \\
\exists q \in \mathbb{R}^{k_2} \text{ s.t. } x^T B \leq q^T F \text{ and } \forall j \in S_2, \ y_j > 0 \ \Rightarrow \ x^T B^j = q^T F^j
\end{array}
\tag{2}
$$

At a NE, a player plays a strategy with positive weight only if it gives the maximum payoff with respect to (w.r.t.) the opponent's strategy in *some sense*. Such strategies are called the *best response* strategies. Using this fact, we define *best response polytopes* (BRPs) (similar to bimatrix game [20]), where x, y, p and q are vector variables.

$$
\begin{array}{l}
P = \{(y, p) \in \mathbb{R}^{N+k_1} | A_i y - p^T E^i \leq 0, \ \forall i \in S_1; \ y_j \geq 0, \ \forall j \in S_2; \ Fy = f\} \\
Q = \{(x, q) \in \mathbb{R}^{M+k_2} | x_i \geq 0, \ \forall i \in S_1; \ x^T B^j - q^T F^j \leq 0, \ \forall j \in S_2; \ Ex = e\}
\end{array}
\tag{3}
$$

The polytopes P and Q are called BRPs of the first and the second player respectively. Since $|S_1| = M$ and $|S_2| = N$, number the inequalities from 1 to M, and $M + 1$ to

$M + N$ in both the polytopes. Let the *label* $L(v)$ of a point v be the set of indices of the tight inequalities at v. If for a pair $(v, w) \in P \times Q$, $L(v) \cup L(w) = \{1, \ldots, M + N\}$, then it is called a *fully-labeled pair*. The next lemma follows directly using (2).

Lemma 1. *A strategy profile* (x, y) *is a NE of the game* (A, B) *iff* $((y, p), (x, q)) \in P \times Q$ *is a fully-labeled pair, for some p and q.*

A game is called non-degenerate if both the polytopes are non-degenerate. Note that a fully-labeled pair of a non-degenerate game has to be a vertex-pair. Lemma 1 implies that a $((y, p), (x, q)) \in P \times Q$ corresponds to a NE if and only if it satisfies the following linear complementarity conditions (LCP).

$$x^T (Ay - E^T p) = 0 \text{ and } (x^T B - q^T F)y = 0 \tag{4}$$

Clearly, $x^T (Ay - E^T p) \le 0$ and $(x^T B - q^T F)y \le 0$ over $P \times Q$ and hence $x^T (Ay - E^T p) + (x^T B - q^T F)y \le 0 \Rightarrow x^T (A + B)y - e^T p - f^T q \le 0$ and equality holds iff (x, y) is a NE (using (4)). This gives the following QP formulation which captures all the NE of game (A, B) at its optimal points.

$$\begin{aligned} \max: \; & x^T (A + B)y - e^T p - f^T q \\ \text{s.t. } & ((y, p), (x, q)) \in P \times Q \end{aligned} \tag{5}$$

Some proofs, a section on examples, and *symmetric bilinear games* are omitted due to space constraints. Interested readers are referred to [9] for details.

3   Rank-1 Games and Polynomial Time Algorithm

The approach used in this section is motivated by the paper [1]. Given a rank-1 game (A, B), $\exists \alpha \in \mathbb{R}^M, \exists \beta \in \mathbb{R}^N$ such that $A + B = \alpha \cdot \beta^T$. In that case $B = -A + \alpha \cdot \beta^T$. Let $G(\alpha) = (A, -A + \alpha \cdot \beta^T)$ be a parametrized game for a fixed $A \in \mathbb{R}^{M \times N}$ and $\beta \in \mathbb{R}^N$. For any game $G(\alpha)$ the BRP $P(\alpha)$ of the first-player is fixed to P (of (3)), since A is fixed. However, the BRP of second-player $Q(\alpha)$ changes with the parameter. Consider the following polytope with x, q as vector variables and λ as a scalar variable:

$$Q' = \{(x, \lambda, q) \in \mathbb{R}^{M+1+k_2} \mid x_i \ge 0, \; \forall i \in S_1; \atop x^T (-A^j) + \lambda \beta_j - q^T F^j \le 0, \; \forall j \in S_2; \; Ex = e\} \tag{6}$$

It is easy to see that $Q(\alpha)$ is the projection of $\{(x, \lambda, q) \in Q' \mid \lambda = x^T \alpha\}$ on (x, q)-space, and hence Q' covers $Q(\alpha), \forall \alpha \in \mathbb{R}^M$. Number the equations of Q' in a similar way as the equations of Q. Let $\mathcal{N}$ be the set of fully-labeled pairs of $P \times Q'$, i.e., $\mathcal{N} = \{(v, w) \in P \times Q' \mid L(v) \cup L(w) = \{1, \ldots, M + N\}\}$.

$$((y, p), (x, \lambda, q)) \in \mathcal{N} \Leftrightarrow x^T (Ay - E^T p) = 0, \; (x^T (-A) + \lambda \beta^T - q^T F)y = 0 \tag{7}$$

Lemma 2. *Let* $(v, w) \in \mathcal{N}$, $v = (y, p)$ *and* $w = (x, \lambda, q)$.

- *For all α such that $\lambda = x^T \alpha$, (x, y) is a NE of $G(\alpha)$.*
- *For every NE (x, y) of a game $G(\alpha)$, $\exists (v, w) \in \mathcal{N}$, where $\lambda = x^T \alpha$.*

Next we discuss the structure of $\mathcal{N}$. The polytopes P and Q' are assumed to be non-degenerate, and let $k_1 = k_2 = k$ for simplicity. Therefore, $\forall (v, w) \in \mathcal{N}, |L(v)| \leq N$ and $|L(w)| \leq M+1$, and $\mathcal{N} \subset$ 1-skeleton of $P \times Q'$. Further, if $(v, w) \in \mathcal{N}$ is a vertex pair then $|L(v) \cap L(w)| = 1$. Let the label in the intersection be called the *duplicate label* of (v, w). Relaxing the inequality corresponding to the duplicate label at (v, w) in P and Q' respectively gives its two adjacent edges in $\mathcal{N}$. Therefore, every vertex of $\mathcal{N}$ has degree two. This implies that $\mathcal{N}$ is a set of cycles and infinite paths. We show that $\mathcal{N}$ forms a single infinite path. The next lemma follows directly from Equations (3,6,7).

Lemma 3. *For all $(v, w) = ((y, p), (x, \lambda, q)) \in P \times Q'$, we have $\lambda(\beta^T y) - e^T p - f^T q \leq 0$, and the equality holds iff $(v, w) \in \mathcal{N}$.*

$$LP(\delta) \quad - \quad \max: \delta(\beta^T y) - e^T p - f^T q$$
$$\text{s.t. } ((y, p), (x, \lambda, q)) \in P \times Q'; \quad \lambda = \delta$$

For an $a \in \mathbb{R}$, let $OPT(a)$ be the set of optimal solutions of $LP(a)$ and $\mathcal{N}(a) = \{(v, w) \in \mathcal{N} \mid w = (x, \lambda, q) \text{ and } \lambda = a\}$.

Lemma 4. *For an $a \in \mathbb{R}, \mathcal{N}(a) \neq \emptyset$ and $OPT(a) = \mathcal{N}(a)$*

Proof. Clearly, $\mathcal{N}(a) \neq \emptyset$ as points on $\mathcal{N}$ corresponding to NE of $G([a, \ldots, a])$ are in $\mathcal{N}(a)$ (Lemma 2). And, $OPT(a) = \mathcal{N}(a)$ follows directly from Lemma 3. □

Lemma 5. *The set $\mathcal{N}$ forms an infinite path, with λ being monotonic on it.*

Proof. Let $\mathcal{C}$ be a cycle in $\mathcal{N}$. Clearly, $\exists a \in \mathbb{R}$, s.t. either $\lambda = a$ contains $\mathcal{C}$ or cuts it exactly at two points. Contradiction to $\mathcal{N}(a)$ being convex. Since, $\forall a \in \mathbb{R}, \mathcal{N}(a)$ is convex, if there are two paths in $\mathcal{N}$, then λ should be monotonic on both and its range on the paths should be mutually disjoint. Therefore if the range of λ covered by one is $(-\infty, a]$, then the other covers $(a, \inf)$, which contradicts closeness of the paths. □

3.1 Algorithm

Let (A, B) be a given rank-1 game s.t. $A + B = \gamma \cdot \beta^T, \gamma_{min} = \min_{x \in X} \sum_{i \in S_1} \gamma_i x_i$ and $\gamma_{max} = \max_{x \in X} \sum_{i \in S_1} \gamma_i x_i$. Let $H_\gamma : \lambda - \sum_{i \in S_1} \gamma_i x_i = 0$ and H_γ^-, H_γ^+ be its half-spaces. The γ_{min} and γ_{max} exists since X is a bounded polytope. Clearly, $\mathcal{N} \cap H_\gamma =$ NE of (A, B) (Lemma 2), and corresponding $\lambda \in [\gamma_{min}, \gamma_{max}]$. Therefore, all the points in $\mathcal{N} \cap H_\gamma$ lies between $\mathcal{N}(\gamma_{min}) \in H_\gamma^-$ and $\mathcal{N}(\gamma_{max}) \in H_\gamma^+$. As λ is monotonic on $\mathcal{N}$ (Lemma 5), the following algorithm does binary search on $\mathcal{N}$ between $\mathcal{N}(\gamma_{min})$ and $\mathcal{N}(\gamma_{max})$ to find a point in the intersection (similar to *BinSearch* of [1]).

S_1 Initialize $a_1 = \gamma_{min}$ and $a_2 = \gamma_{max}$. If the edge containing $\mathcal{N}(a_1)$ or $\mathcal{N}(a_2)$ intersects H_γ, then output the intersection and exit.

S_2 Let $a = \frac{a_1 + a_2}{2}$. Let $\overline{u, v}$ be the edge containing $\mathcal{N}(a)$.

S_3 If $\overline{u, v}$ intersects H_γ, then output the intersection and exit.

S_4 Else if $\overline{u, v} \in H_\gamma^-$, then set $a_1 = a$ else set $a_2 = a$ and continue from step S_2.

Let $Z = \max\{|A|, |E|, |F|, |e|, |f|, |\gamma|, |\beta|\}, l = M + N + k_1 + k_2$, and $\Delta = l! Z^l$.

Theorem 1. *The above algorithm finds a NE of* (A, B) *in polynomial time.*

Proof. Clearly, if λ is not constant on an edge of $\mathcal{N}$, then it changes by at least $\frac{1}{\Delta^2}$ on the edge. Therefore, if $a_2 - a_1 < \frac{1}{\Delta^2}$ the algorithm terminates. The proof follows. □

4   FPTAS for Rank-k Games

The approximation notion in bilinear games may be defined in a similar way to that of bimatrix games given by Kannan et al. [14]. Let $x_{max} = \max_{x \in X} \sum_i x_i$, $y_{max} = \max_{y \in Y} \sum_j y_j$ and $D = |A + B|$. Clearly the total payoff derived from a strategy profile $(x, y) \in X \times Y$ is at most $x_{max} D y_{max}$.

Definition 2. *For a strategy profile* $(x, y) \in X \times Y$, *let* $u = \max_{x' \in X} x'^T A y$ *and* $v = \max_{y' \in Y} x^T B y'$. *Then* (x, y) *is an* ϵ-*approximate NE of the game* (A, B) *if* $u + v - x^T(A + B)y \le \epsilon(x_{max} D y_{max})$.

Next we define a stronger notion of approximate NE, called *relative* ϵ-approximate NE.

Definition 3. *For a strategy profile* $(x, y) \in X \times Y$, *let* $u = \max_{x' \in X} x'^T A y$ *and* $v = \max_{y' \in Y} x^T B y'$. *Then* (x, y) *is a* relative ϵ-*approximate NE of* (A, B) *if* $u + v - x^T(A + B)y \le \epsilon(u + v)$, *i.e., the total error is relatively small.*

Since the value of $u + v$ is at most $x_{max} D y_{max}$, if (x, y) is relative ϵ-approximate NE, then it is also ϵ-approximate NE. As scaling A, B, E, F, e and f by a positive value does not change the set of (relative) ϵ-approximate NE, we assume them to be integer matrices. Next we discuss two FPTAS to solve QP of (5), one for each definition of approximation. The approaches used in these algorithms are generalization of [14].

4.1   FPTAS for Approximate NE

We show that the result by Vavasis [21] can be applied to get an ϵ-approximate NE.

Proposition 2. *[21] Let* $f(x) = \frac{1}{2}x^T Q x + q^T x$, *and* $\min\{f(x) : Ax \le b\}$ *be a quadratic optimization problem with compact polytope* $\{x \in \mathbb{R}^n : Ax \le b\}$, *and let the rank of* Q *be a constant. If* x^* *and* $x^\#$ *minimizes and maximizes* $f(x)$ *in the feasible region, respectively, then one can find in time* $poly(\mathcal{L}, \frac{1}{\epsilon})$ *a point* x° *satisfying* $f(x^\circ) - f(x^*) \le \epsilon(f(x^\#) - f(x^*))$.

Consider the QP formulation of (5) to capture all NE of a rank-k game (A, B).

Theorem 2. *For every* $\epsilon > 0$, *an* ϵ-*approximate NE of the game* (A, B), *can be computed in time* $poly(\mathcal{L}, \frac{1}{\epsilon})$, *where* $\mathcal{L}$ *is the bit length of the game.*

Proof. The objective function of (5) can be transformed to the form $\min : \frac{1}{2}x^T Q x + q^T x$, where $rank(Q) = 2k$. To apply Proposition 2 on this QP, we need to bound its feasible set. The only variables to bound are ps and qs. Since, maximum absolute value of a co-ordinate of any vertex in the polytope is at most $l!Z^l$, impose $-l!Z^l \le p \le l!Z^l$ and $-l!Z^l \le q \le l!Z^l$ (with polynomial increase in $\mathcal{L}$). The minimum and the maximum objective values of this QP are zero (Lemma 3) and at most $2x_{max}Dy_{max}$. Let $((y^\circ, p^\circ), (x^\circ, q^\circ))$ be the solution given by Vavasis algorithm for $\frac{\epsilon}{2}$, then from Proposition 2 we get,

$$e^T p^\circ + f^T q^\circ - x^{\circ T}(A + B)y^\circ \leq \epsilon(x_{max}Dy_{max})$$

From the primal-dual formulation of (1) it is clear that $\max_{x'\in X} x'^T Ay^\circ \leq e^T p^\circ$ and $\max_{y'\in Y} x^{\circ T} By' \leq f^T q^\circ$. Therefore, we get $\max_{x'\in X} x'^T Ay^\circ + \max_{y'\in Y} x^{\circ T} By' - x^{\circ T}(A + B)y^\circ \leq \epsilon(x_{max}Dy_{max})$. □

4.2 FPTAS for Relative Approximate NE

Let the rank of a game (A, B) be k, then $A + B = \sum_{i=1}^k \alpha(i)\beta(i)^T$, where $\forall i$, $\alpha(i) \in \mathbb{R}^M$ and $\beta(i) \in \mathbb{R}^N$. We assume that the game is such that $\alpha(i)$s and $\beta(i)$s are positive vectors. For all $i \leq k$, let $w_i = \min_{x\in X} x^T\alpha(i)$ and $w_i' = \max_{x\in X} x^T\alpha(i)$, similarly let $z_i = \min_{y\in Y} \beta(i)^T y$ and $z_i' = \max_{y\in Y} \beta(i)^T y$. Note that w_i, w_i', z_i and z_i' can be represented by $poly(\mathcal{L}, M, N)$ bits, since X and Y are compact. Given an $\epsilon > 0$, consider the sub-intervals $[w_i, (1 + \epsilon)w_i]$, $[(1 + \epsilon)w_i, (1 + \epsilon)^2 w_i]$, and so on of $[w_i, w_i']$ and similarly of $[z_i, z_i']$. All combinations of these intervals form a grid in $2k$-dimensional box $\mathcal{B} = \times_i[w_i, w_i'] \times_i [z_i, z_i']$. For a fixed hyper-cube $\times_i[u_i, (1+\epsilon)u_i] \times_i [v_i, (1 + \epsilon)v_i]$ of the grid, consider the following LP based on the QP of (5),

min: $e^T p + f^T q$
s.t. $Ay \leq E^T p$, $Fy = f$, $y \geq 0$; $\forall i$, $v_i \leq \beta(i)^T y \leq (1+\epsilon)v_i$;
    $x^T B \leq q^T F$; $Ex = e$; $x \geq 0$; $\forall i$, $u_i \leq x^T\alpha(i) \leq (1+\epsilon)u_i$;

Algorithm. Run the above LP for each hyper-cube of the grid, and output an optimal point of the one giving the best approximation. As the number of hyper-cubes in the grid is $poly(\mathcal{L}, 1/\log(1 + \epsilon))$, the running time is $poly(M, N, \mathcal{L}, 1/\log(1 + \epsilon))$.

Theorem 3. Let (A, B) be a rank-k game, and $A + B = \sum_{i=1}^k \alpha(i)\beta(i)^T$, s.t. $\alpha(i)$s and $\beta(j)$s are positive. Then given an $\epsilon > 0$, a relative $(1 - 1/(1 + \epsilon)^2)$-approximate NE can be computed in time $poly(\mathcal{L}, 1/\log(1 + \epsilon))$, where $\mathcal{L}$ is the input bit length.

Proof. Clearly, for a $(x, y) \in X \times Y$ such that $\forall i, x^T\alpha(i) \in [u_i, (1+\epsilon)u_i]$ and $\beta(i)^T y \in [v_i, (1+\epsilon)v_i]$, we have $\sum_{i=1}^k u_i v_i \leq x^T(A+B)y \leq (1+\epsilon)^2 \sum_{i=1}^k u_i v_i$. Using this we can show that the optimal of LP, for the hyper-cube containing the point corresponding to a NE, is a relative $(1 - 1/(1 + \epsilon)^2)$-approximate NE. The proof follows. □

5 Games with a Low Rank Matrix

In this section we show that if rank of even one payoff matrix (A or B) is constant, then Nash equilibrium computation can be done in polynomial time. Recall the best response polytopes P and Q (3) for the bilinear game (A, B).

Lemma 6. Given a bilinear game (A, B), there exists a vertex pair $((y, p), (x, q)) \in P \times Q$ such that (x, y) is a NE of (A, B).

Lemma 7. *Let* $k_1 = k_2 = k$ *and* $rank(A) = l$. P *has at most* $O(N^{l+k})$ *vertices.*

Proof. As $Fy = f$ gives k linearly independent (l.i.) equalities, P is of dimension N. Note that, $\exists! S \subset S_1$, $|S| = rank([A \ -E]) \le l+k$ such that $\forall i \in S_1 \backslash S$, $A_i y - p^T E^i \le 0$ are not needed in defining the polytope P. At a vertex, if d inequalities of S are tight then rest $N - d$ must be of type $y_j = 0$. Therefore, the total number of vertices are at most $\sum_{i=0}^{l+k} \binom{l+k}{i} \binom{N}{i} \le 2^{l+k} N^{l+k}$. □

Theorem 4. *If rank of either A or B is constant then a NE of a bilinear game (A, B) can be computed in polynomial time, assuming k to be a constant.*

Proof. The proof follows from the fact that, whether a vertex of P or Q corresponds to a NE of (A, B), can be checked in polynomial time. □

As the set of bimatrix games is the bilinear games with $k_1 = k_2 = 1$, Theorem 4 strengthens the results by Lipton et al. [18] (Corollary 4), and Kannan et al. [14] (Theorem 3.2), where they require both A and B of constant rank.

In fact Theorem 4 gives a polynomial time algorithm to enumerate all the extreme equilibria of a bilinear game with a fixed rank matrix, and an exponential time enumeration algorithm for any bilinear game. A similar (exponential time) algorithm was given by Avis et al. [2] to enumerate all NE of a bimatrix game.

References

1. Adsul, B., Garg, J., Mehta, R., Sohoni, M.: Rank-1 bimatrix games: A homeomorphism and a polynomial time algorithm. In: STOC, pp. 195–204 (2011)
2. Avis, D., Rosenberg, G.D., Savani, R., von Stengel, B.: Enumeration of Nash equilibria for two-player games. Economic Theory 42, 9–37 (2010)
3. Charnes, A.: Constrained games and linear programming. Proceedings of the National Academy of Sciences of the USA 39, 639–641 (1953)
4. Chen, X., Deng, X.: Settling the complexity of 2-player Nash-equilibrium. In: FOCS, pp. 261–272 (2006)
5. Dantzig, G.B.: Linear Programming and Extensions. Princeton Univ. Press (1963)
6. Daskalakis, C., Goldberg, P.W., Papadimitriou, C.H.: The complexity of computing a Nash equilibrium. In: STOC, pp. 71–78 (2006)
7. Daskalakis, C., Papadimitriou, C.H.: On a Network Generalization of the Minmax Theorem. In: Albers, S., Marchetti-Spaccamela, A., Matias, Y., Nikoletseas, S., Thomas, W. (eds.) ICALP 2009. LNCS, vol. 5556, pp. 423–434. Springer, Heidelberg (2009)
8. Fudenberg, D., Tirole, J.: Game Theory. MIT Press (1991)
9. Garg, J., Jiang, A.X., Mehta, R.: Bilinear games: Polynomial time algorithms for rank based subclasses. arXiv:1109.6182 (2011)
10. Glicksberg, I.L.: A further generalization of the Kakutani fixed point theorem, with application to Nash equilibrium points, vol. 3(1), pp. 170–174. AMS (1952)
11. Howson Jr., J.: Equilibria of polymatrix games. Management Science (1972)
12. Howson Jr, J., Rosenthal, R.: Bayesian equilibria of finite two-person games with incomplete information. Management Science, 313–315 (1974)
13. Immorlica, N., Kalai, A.T., Lucier, B., Moitra, A., Postlewaite, A., Tennenholtz, M.: Dueling algorithms. In: STOC (2011)
14. Kannan, R., Theobald, T.: Games of fixed rank: A hierarchy of bimatrix games. Economic Theory, 1–17 (2009)

15. Koller, D., Megiddo, N., von Stengel, B.: Fast algorithms for finding randomized strategies in game trees. In: STOC, pp. 750–759 (1994)
16. Koller, D., Megiddo, N., von Stengel, B.: Efficient computation of equilibria for extensive two-person games. Games and Economic Behavior 14, 247–259 (1996)
17. Kontogiannis, S., Spirakis, P.: Exploiting Concavity in Bimatrix Games: New Polynomially Tractable Subclasses. In: Serna, M., Shaltiel, R., Jansen, K., Rolim, J. (eds.) APPROX 2010, LNCS, vol. 6302, pp. 312–325. Springer, Heidelberg (2010)
18. Lipton, R., Markakis, E., Mehta, A.: Playing large games using simple strategies. In: EC, pp. 36–41. ACM Press, New York (2003)
19. Ponssard, J.P., Sorin, S.: The LP formulation of finite zero-sum games with incomplete information. International Journal of Game Theory 9, 99–105 (1980)
20. von Stengel, B.: Equilibrium computation for two-player games in strategic and extensive form. In: Nisan, N., Roughgarden, T., Tardos, E., Vazirani, V. (eds.) Algorithmic Game Theory, ch. 3, pp. 53–78 (2007)
21. Vavasis, S.: Approximation algorithms for indefinite quadratic programming. Mathematical Programming 57, 279–311 (1992)

Extending Characterizations of Truthful Mechanisms from Subdomains to Domains*

Angelina Vidali

Theory and Applications of Algorithms Research Group
University of Vienna, Austria
angelina.vidali@univie.ac.at

Abstract. The already extended literature in Combinatorial Auctions, Public Projects and Scheduling demands a more systematic classification of the domains and a clear comparison of the results known. Connecting characterization results for different settings and providing a characterization proof using another characterization result as a black box without having to repeat a tediously similar proof is not only elegant and desirable, but also greatly enhances our intuition and provides a classification of different results and a unified and deeper understanding. We consider whether one can extend a characterization of a subdomain to a domain in a black-box manner. We show that this is possible for n-player stable mechanisms if the only truthful mechanisms for the subdomain are the affine maximizers. We further show that if the characterization of the subdomain involves a combination of affine maximizers and threshold mechanisms, the threshold mechanisms for the subdomain cannot be extended to truthful mechanisms for the union of the subdomain with a (very slight) affine transformation of it. We also show that for every truthful mechanism in a domain there exists a corresponding truthful mechanism for any affine transformation of the domain. We finally plug in as a black box to our theorems the characterization of additive 2-player combinatorial auctions that are decisive and allocate all items (which essentially is the domain for scheduling unrelated machines) and show that the 2-player truthful mechanisms of any domain, which is strictly a superdomain of it are only the affine maximizers. This gives a unique characterization proof of the decisive 2-player mechanisms for: Combinatorial Public Projects, Unrestricted domains, as well as for Submodular and Subadditive Combinatorial Auctions that allocate all items.

1 Introduction

Our Results and Motivation. Suppose that you are in a conference committee and need to choose only one of the following two papers for acceptance: Both give a characterization for n-player stable (/2-player that allocate all items) Combinatorial Auctions, the first for the case when the players have additive

---

* This work has been funded by the Vienna Science and Technology Fund WWTF grant ICT10-002.

N. Chen, E. Elkind, and E. Koutsoupias (Eds.): WINE 2011, LNCS 7090, pp. 408–414, 2011.

valuations and the second for the case when the players have sub-modular valuations. Which one will you accept? Is one of the results stronger, in the sense that the two papers can be merged and the weaker result can be derived as a corollary?

In this paper we address and only partially answer the following questions: Which domains have the same characterization? Can we classify different domains in terms of how difficult it is to characterize them or how rich are the mechanisms allowed? Does a characterization for a "more difficult" domain automatically imply a proof for domains that are lower in this hierarchy? Can we establish a bijection between the mechanisms involved in the characterization of different domains? This work gives some explanations we would have liked to find, back when we started working on characterization results and wondered what do the results about other slightly different domains tell us about the domain we were primarily interested in.

Roberts [11](1979) gives an elegant proof, which shows that the only truthful mechanisms for the Unrestricted domain are the affine maximizers. He also gets the Gibbard-Sattherwhaite Theorem (1973) for voting systems as a corollary. For more "restricted" multi-parameter domains, there exist truthful mechanisms other than affine maximizers (see e.g. [15,10,4]). An important question, posed in [18,14], is to determine how much we need to restrict the domain in order to admit truthful mechanisms different than the affine maximizers. Here we show that for the case of 2 players, this transition domain is the additive combinatorial auctions domain: We show that if we slightly enrich the possible valuations, the Threshold mechanisms involved in the characterization [4] seize to be truthful and the only truthful mechanisms that remain are the affine maximizers.

A crucial observation is: the more "unrestricted" the domain of valuations, the fewer the possible truthful mechanisms. An intuitive explanation for this is that in larger domains there are more inputs that need to satisfy the conditions for truthfulness. On the other hand, *this intuition may be misleading*: Given that a sub-domain admits as truthful mechanisms only the affine maximizers does not immediately imply that the domain also admits the same mechanisms; there may be other mechanisms which when restricted to sub-domain are exactly the affine maximizers. In particular, we do not know whether this is possible for non-stable mechanisms. We also do not know if this is possible for domains where the possible truthful mechanisms are richer than combinations of affine maximizers and threshold mechanisms.

Related Work. The starting point of characterization attempts goes back to Robert's [11] result. Many papers tried to extend this very elegant proof [16,10,19], while others tried different proof techniques [15,4,8,16]. (As the literature in combinatorial auctions is vast we refer the reader to [18] Chapter 11 and the references within and mention here only some recent results.) Computational complexity impossibility results for maximal in range mechanisms (mechanisms obtained by restricting the possible allocations among which an affine maximizer chooses its allocation) where shown in [2,7]. Dobzinski [6] shows that every universally truthful randomized mechanism for combinatorial auctions with submodular valuations that provides an approximation ratio of $m^{\frac{1}{2}-\epsilon}$ must use exponentially many value queries. Krysta and Ventre show that if verification is introduced sub-modular combinatorial auctions become tractable [13]. Nisan and Ronen introduced the truthful scheduling unrelated machines problem [17,4,12,10,20]. A characterization of decisive truthful 2-player mechanisms in terms of affine minimizers and threshold mechanisms was given in [4] and it also implies a characterization for additive combinatorial auctions, which we will use here as a black box. We will alternatively use as a black box the characterization of n-player stable (/2-player that allocate all items) subadditive combinatorial auctions by Dobzinski and Sundararajan [10].

1.1 Definitions and Preliminaries

A mechanism is *decisive* when (for fixed values of the other players) a player can enforce any outcome (allocation), by declaring very high or very low values. A mechanism is called *stable* if the following holds: *For fixed valuations* v_{-i}, *the allocation* a_i *of player* i *determines uniquely the allocation* a_{-i} *of the other players.* (In other words: Fix v_{-i}, then for all v_i for which player i has allocation a_i the allocation a_{-i} is the same.) *Stability can be assumed without loss of generality for n-player unrestricted domains, combinatorial public projects and 2-player auctions where all items are allocated.* It is too restrictive for combinatorial auctions with $n \geq 3$ players (see [15] Example 4), however all known characterization results [11,15,19,10,4,16,10] heavily rely on stability, or characterize domains where stability can be assumed essentially without loss of generality. Stability is implied by S-MON or IIA (see [15,1,10] for a discussion on these conditions and proofs).

We will denote any finite domain of valuations D as a set of matrices [3]. We have one matrix for each valuation function $v = (v_1, \ldots, v_n) : A \to \mathbb{R}^n$ that belongs to the domain. This matrix has one column for each alternative $a \in A$ and one row for each player. The valuation v_i of player i is a vector of numbers with one coordinate for each possible alternative. Let V_i set of all possible such vectors for player i. (The domain is the set of all possible inputs of the mechanism.) Under this notation the domain of unrestricted valuations [11] contains all possible matrices with real values. We will say that S_i *is a subdomain of* V_i if $S_i \subseteq V_i$. We will say that D *is a subdomain of* D' if $D \subseteq D'$.

2 Our Results

Derivation of the Characterization of a Domain from the Characterization of one of Its Sub-Domains. Suppose we know which mechanisms are truthful for a given domain, does this tell us which mechanisms are truthful for any super-domain of it? The first reaction may be: we can read the proofs and produce (tediously) similar ones. But then the mechanism for the bigger domain has to satisfy truthfulness for a superset of the input space. Are then the mechanisms for the bigger domain a subset of the mechanisms for the sub-domain?

We have to be careful: it is true that if a mechanism is truthful for the bigger domain, then its restriction to the smaller domain is a truthful mechanism for the smaller domain (for which we assumed that we know a characterization). However it then remains to describe the mechanism for the additional inputs we allowed by enlarging the domain.

We need Lemma 1 in order to avoid assuming decisiveness in Theorem 1. It shows that shows that by truthfulness the range of the mechanism for the bigger domain is the same as the range of it's restriction to the subdomain. In other words *if the characterization that you plug in Theorem 1 or 4 as a black box does not require decisiveness then the characterization you obtain for the superdomain does not require decisiveness either.* Lemma 2 is the core argument for proving Theorem 1.

Lemma 1. *Let S_i be the domain of additive valuations, or any super-domain of it, and $S_i \subseteq V_i$. For for fixed v_{-i}, consider a truthful social choice function $f(\cdot, v_{-i}) : S_1 \times \ldots \times V_i \times \ldots \times S_n \to A$, and restrict it to the sub-domain $S_1 \times \ldots \times S_n$. If the range of the restricted function is a set of alternatives A, then A is also the range of $f(\cdot, v_{-i})$.*

Lemma 2. *Start with an affine maximizer M defined for the domain of valuations $S_1 \times \ldots \times S_n$ and then consider the bigger domain $S_1 \times \ldots \times V_i \times \ldots \times S_n$, where $S_i \subseteq V_i$. There is a unique way to extend M (preserving truthfulness) to an n-player stable (/2-player that allocates all items) mechanism for the bigger domain, namely an affine maximizer defined by the same λ, γ as M.*

Note that if we did not require the mechanism to be truthful, then there would have been many possibilities to extend the mechanism to a mechanism that would not be an affine maximizer for the whole domain.

Theorem 1. *Let V be a sub-domain of the domain of unrestricted valuations and superdomain of the domain of additive valuations. If the only possible truthful n-player stable mechanisms for V are affine maximizers, then the same holds for every super-domain of V.*[1]

Plugging in as a black box the characterization of n-player stable scalable (if you multiply all entries of the input matrix by the same number the allocation remains the same) mechanisms for subadditive combinatorial auctions [10] we get:

Corollary 1. *The only truthful n-player stable (/2-player that allocate all items) mechanisms with at least 3 outcomes for any superdomain of Subadditive Combinatorial Auctions that satisfy scalability are affine maximizers. This proves that the only truthful scalable mechanisms for (a) Unrestricted domains as well as for (b) stable Combinatorial Auctions (with general valuations) are affine maximizers.*

---

[1] The proof of Theorem 1 for the 2-player case, goes along exactly the same lines as the proof of Lemma 3.1 [5] by Dobzinski. (The statement of that Lemma involves a different setting, with which we don't deal with in this paper, that of two-player multi-unit auctions.)

Affine Transformations of Domains. Note that the next theorem holds for any choice of the domain D, and not only for the domain of additive valuations. It implies that if we characterize all possible mechanisms for a domain of valuations D then the same characterization holds for all domains we get by translating D.

If D is the matrix representation of a domain we denote by $\lambda D + c$ the following affine transformation of D: Multiply the valuations of each player i by a positive constant λ_i and add a matrix of constants c, with one row c^i for each player and one column for each possible allocation. For example the following is an affine transformation of 2-player combinatorial auctions:

$$\begin{pmatrix} c_\emptyset^1 & \lambda_1 v_1(\{1\}) + c_{\{1\}}^1 & \lambda_1 v_1(\{2\}) + c_{\{2\}}^1 & \lambda_1 v_1(\{1,2\}) + c_{\{1,2\}}^1 \\ \lambda_2 v_2(\{1,2\}) + c_{\{1,2\}}^2 & \lambda_2 v_2(\{2\}) + c_{\{2\}}^2 & \lambda_2 v_2(\{1\}) + c_{\{1\}}^2 & c_\emptyset^2 \end{pmatrix}.$$

Theorem 2. *There is a bijection between the mechanisms for D and the mechanisms of $\lambda D + c$. That is the mechanism with the same allocation and payments $p' = \lambda \cdot p + c$ is also truthful for $\lambda D + c$. This holds for any number of players n.*

Threshold Mechanisms and Their Payments. The characterization in [4] reveals the class of threshold mechanisms, which are truthful, very simple in their description, and not (necessarily) affine maximizers. The immediate question is whether there exist other domains for which threshold mechanisms are truthful. We describe here the truthful threshold mechanisms for the translated domain $\lambda D + c$. A threshold mechanism for the additive combinatorial auctions (/scheduling) domain is one for which there are threshold functions h_{ij} such that the mechanism allocates item j to player i if and only if $v_i(\{j\}) \geq h_{ij}(v_{-i})$.

Theorem 3. *If D is the domain of additive valuations and a_i is the set of items allocated to player i, then a mechanism for the domain $\lambda D + c$ is a threshold mechanism if and only if it satisfies $p_i(a_i, v_{-i}) - c_{a_i}^i = \sum_{j \in a_i} \left(p_i(\{j\}, v_{-i}) - c_{\{j\}}^i \right)$.*

How to Vanish Threshold Mechanisms. Here we show how starting from the additive domain and slightly enriching the domain of possible valuations we obtain a domain that does not admit any truthful threshold mechanisms. This shows that truthful threshold mechanisms are specific for the domain of additive valuations and its affine transformations and that they cannot be generalized for richer domains.

Let S_i be the set of all valuation functions v_i that are additive. We define the set of valuation functions $S_i + \delta$ as follows: $S_i + \delta$ contains all valuation functions v_i' with $v_i'(a_i) = \sum_{j \in a_i}^m v_i(\{j\}) + (|a_i| - 1) \cdot \delta$ where $\delta \neq 0$ is some tiny constant. That is $v_i \in S_i$ and $v_i' \in S_i + \delta$ agree only on the valuation for getting singletons and the emptyset and differ by $\delta \times$ (size of the bundle -1) for bigger bundles. There exist many choices of valuations for which our proofs hold. Of course if you would like to obtain the characterization, say, of sub-modular auctions, you should mind to make a choice of valuations that are submodular.

We start with two domains, that differ slightly in the valuations one of the players. Each one separately admits truthful threshold mechanisms, but their union does not:

Lemma 3. *Consider a truthful mechanism for the domain $\left(S_1 \cup (S_1 + \delta)\right) \times S_2 \times \ldots \times S_2$. If it is a threshold mechanism when restricted to $S_1 \times S_2 \times \ldots \times S_n$, then it is non-threshold when restricted to $(S_1 + \delta) \times S_2 \times \ldots \times S_2$. Consequently for the domain $\left(S_1 \cup (S_1 + \delta)\right) \times S_2 \times \ldots \times S_2$ threshold mechanisms are non-truthful.*

Theorem 4. *If the only truthful mechanisms for the domain $S_1 \times S_2 \times \ldots \times S_2$ are either affine maximizers or threshold mechanisms, then the only truthful stable mechanisms, for the domain $\left(S_1 \cup (S_1 + \delta)\right) \times S_2 \times \ldots \times S_2$, or any super-domain of it, are affine maximizers.*

Applying our Tools for the Known Characterization. The machinery we just developed opts for a characterization of stable truthful mechanisms for additive combinatorial auctions/scheduling for n players, but this is an important open problem. We only have one [4] for 2-player mechanisms, that are decisive and allocate all items.

The characterization in [4] is only for additive valuations, applying Theorem 2 it also applies to any affine transformation of the domain of additive valuations. We can now state our main Theorem:

Theorem 5. *The only possible truthful mechanisms, for $S_1 \cup (S_1 + \delta) \times S_2$ or any super-domain of it, that have at least 3 outcomes, are decisive and allocate all items are the affine maximizers. Consequently the only truthful 2-player mechanisms that are decisive and have at least 3 outcomes for: (a) Combinatorial Auctions with Submodular or Subadditive or Superadditive valuations that allocate all items, as well as for (b) the Unrestricted domain and Combinatorial Public Projects, are the affine maximizers.*

3   Conclusion and Future Directions

Submodular combinatorial auctions is an important domain [6,9,18], whose characterization (assuming decisiveness and that all items are allocated) we obtain in this work for the first time almost for free. Although we characterize at once the very rich class of superdomains of additive combinatorial auctions, the most important aspect of our work is not in characterizing new domains, but in classifying them in terms of which domain's characterization we can use as a black box in order to obtain the characterization of all of it's super-domains and obtaining unified proofs and a unified understanding. An important reason why we used this specific characterization [4] as a black box is that it is the only one that involves truthful mechanisms that are not affine maximizers. We enrich the domain very slightly and these mechanisms seize to be truthful, thus the domain of additive combinatorial auctions is the transition domain [18,14] where the affine maximizers are not any more the only truthful mechanisms.

Of course the big open question still remains to obtain characterizations of domains that admit non-stable mechanisms. However the approach of classifying domains in the way we propose provides a more thorough understanding of the existing techniques and results and adds rigor to an intuition that was on the same time helpful and misleading.

Acknowledgements. I would like to thank Giorgos Christodoulou, Elias Koutsoupias and Annamária Kovács for many helpful comments.

References

1. Bikhchandani, S., Chatterji, S., Lavi, R., Mu'alem, A., Nisan, N., Sen, A.: Weak monotonicity characterizes deterministic dominant-strategy implementation. Econometrica 74(4), 1109–1132 (2006)
2. Buchfuhrer, D., Dughmi, S., Fu, H., Kleinberg, R., Mossel, E., Papadimitriou, C., Schapira, M., Singer, Y., Umans, C.: Inapproximability for vcg-based combinatorial auctions. In: SODA (2010)
3. Christodoulou, G., Koutsoupias, E.: Mechanism design for scheduling. BEATCS 97, 39–59 (2009)
4. Christodoulou, G., Koutsoupias, E., Vidali, A.: A Characterization of 2-Player Mechanisms for Scheduling. In: Halperin, D., Mehlhorn, K. (eds.) Esa 2008. LNCS, vol. 5193, pp. 297–307. Springer, Heidelberg (2008)
5. Dobzinski, S.: A note on the power of truthful approximation mechanisms. CoRR, abs/0907.5219 (2009)
6. Dobzinski, S.: An impossibility result for truthful combinatorial auctions with submodular valuations. In: STOC, pp. 139–148 (2011)
7. Dobzinski, S., Nisan, N.: Limitations of vcg-based mechanisms. In: STOC (2007)
8. Dobzinski, S., Nisan, N.: A Modular Approach to Roberts' Theorem. In: Mavronicolas, M., Papadopoulou, V.G. (eds.) SAGT 2009. LNCS, vol. 5814, pp. 14–23. Springer, Heidelberg (2009)
9. Dobzinski, S., Nisan, N., Schapira, M.: Approximation algorithms for combinatorial auctions with complement-free bidders. In: STOC, pp. 610–618 (2005)
10. Dobzinski, S., Sundararajan, M.: On characterizations of truthful mechanisms for combinatorial auctions and scheduling. In: EC, pp. 38–47 (2008)
11. Kevin, R.: The characterization of implementable choice rules. Aggregation and Revelation of Preferences, 321–348 (1979)
12. Koutsoupias, E., Vidali, A.: A Lower Bound of $1+\phi$ for Truthful Scheduling Mechanisms. In: Kučera, L., Kučera, A. (eds.) MFCS 2007. LNCS, vol. 4708, pp. 454–464. Springer, Heidelberg (2007)
13. Krysta, P., Ventre, C.: Combinatorial Auctions with Verification are Tractable. In: de Berg, M., Meyer, U. (eds.) ESA 2010. LNCS, vol. 6347, pp. 39–50. Springer, Heidelberg (2010)
14. Lavi, R.: Searching for the possibility-impossibility border of truthful mechanism design. SIGecom Exch. (2007)
15. Lavi, R., Mu'alem, A., Nisan, N.: Towards a characterization of truthful combinatorial auctions. In: FOCS, pp. 574–583 (2003)
16. Lavi, R., Mualem, A., Nisan, N.: Two simplified proofs for roberts theorem. Social Choice and Welfare 32(3), 407–423 (2009)
17. Nisan, N., Ronen, A.: Algorithmic mechanism design. In: STOC (1999)
18. Nisan, N., Roughgarden, T., Tardos, E., Vazirani, V.: Algorithmic Game Theory. Cambridge University Press (2007)
19. Papadimitriou, C.H., Schapira, M., Singer, Y.: On the hardness of being truthful. In: FOCS, pp. 250–259 (2008)
20. Vidali, A.: The Geometry of Truthfullness. In: Leonardi, S. (ed.) WINE 2009. LNCS, vol. 5929, pp. 340–350. Springer, Heidelberg (2009)

Optimal Multi-period Pricing
with Service Guarantees
Working Paper

Christian Borgs[1], Ozan Candogan[2], Jennifer Chayes[1],
Ilan Lobel[3], and Hamid Nazerzadeh[4]

[1] Microsoft Research New England Lab
{borgs,jchayes}@microsoft.com
[2] Lab for Information and Decision Systems, Massachusetts Institute of Technology
candogan@mit.edu
[3] Stern School of Business, New York University
ilobel@stern.nyu.edu
[4] Marshall School of Business, University of Southern California
hamidnz@marshall.usc.edu

Abstract. We consider the multi-period pricing problem of a service firm facing time-varying capacity levels. Customers are assumed to be fully strategic with respect to their purchasing decisions, and heterogeneous with respect to their valuations, and arrival-departure periods. The firm's objective is to set a sequence of prices that maximizes its revenue while guaranteeing service to all paying customers. Although the corresponding optimization problem is non-convex, we provide a polynomial-time algorithm that computes the optimal sequence of prices. We show that due to the presence of strategic customers, available service capacity at a time period may bind the price offered at another time period. Consequently, when customers are more patient for service, the firm offers higher prices. This leads to the underutilization of capacity, lower revenues, and reduced customer welfare. Variants of the pricing algorithm we propose can be used in more general settings, such as a robust optimization formulation of the pricing problem.

N. Chen, E. Elkind, and E. Koutsoupias (Eds.): WINE 2011, LNCS 7090, p. 415, 2011.

Pricing and Efficiency in the Market
for IP Addresses
Working Paper

Benjamin Edelman[1] and Michael Schwarz[2]

[1] Harvard Business School
[2] Yahoo Research

Abstract. We consider market rules for the transfer of IP addresses, numeric identifiers required by all computers connected to the Internet. Excessive fragmentation of IP address blocks causes growth in the Internet's routing table, which is socially costly, so an IP address market should discourage subdividing IP address blocks more than necessary. Yet IP address transfer rules also need to facilitate purchase by the networks that need the addresses most, from the networks who value them least. We propose a market rule that avoids excessive fragmentation while almost achieving social efficiency, and we argue that implementation of this rule is feasible despite the limited powers of central authorities. We also offer a framework for the price trajectory of IPv4 addresses. In a world without uncertainty, the unit price of IPv4 is constant before the first time when all blocks of IPv4 addresses are in use and decreasing after that time. With uncertainty, the price before that time is a martingale, and the price trajectory afterwards is a supermartingale.

A full version of this paper is available as HBS Working Paper 12-020 at http://www.hbs.edu/research/pdf/12-020.pdf.

N. Chen, E. Elkind, and E. Koutsoupias (Eds.): WINE 2011, LNCS 7090, p. 416, 2011.
© Springer-Verlag Berlin Heidelberg 2011

A Note on the Incompatibility
of Strategy-Proofness and Pareto-Optimality
in Quasi-Linear Settings with Public Budgets
Working Paper

Ron Lavi* and Marina May**

Faculty of Industrial Engineering and Management
Technion, Israel Institute of Technology
ronlavi@ie.technion.ac.il, didra@techunix.technion.ac.il

Abstract. We study the problem of allocating multiple identical items that may be complements to budget-constrained bidders with private values. We show that there does not exist a deterministic mechanism that is individually rational, strategy-proof, Pareto-efficient, and that does not make positive transfers. This is true even if there are only two players, two items, and the budgets are common knowledge. The same impossibility naturally extends to more abstract social choice settings with an arbitrary outcome set, assuming players with quasi-linear utilities and public budget limits. Thus, the case of infinite budgets (in which the VCG mechanism satisfies all these properties) is really the exception.

* Supported in part by grants from the Israeli Science Foundation, the Bi-national Science Foundation, the Israeli ministry of science, and by the Google Inter-university center for Electronic Markets and Auctions.
** Supported in part by grants from the Israeli Science Foundation, the Bi-national Science Foundation, the Israeli ministry of science, and by the Google Inter-university center for Electronic Markets and Auctions.

N. Chen, E. Elkind, and E. Koutsoupias (Eds.): WINE 2011, LNCS 7090, p. 417, 2011.

Author Index